Deep Space

The NASA Mission Reports

compiled from the archives & edited
by Robert Godwin & Steve Whitfield

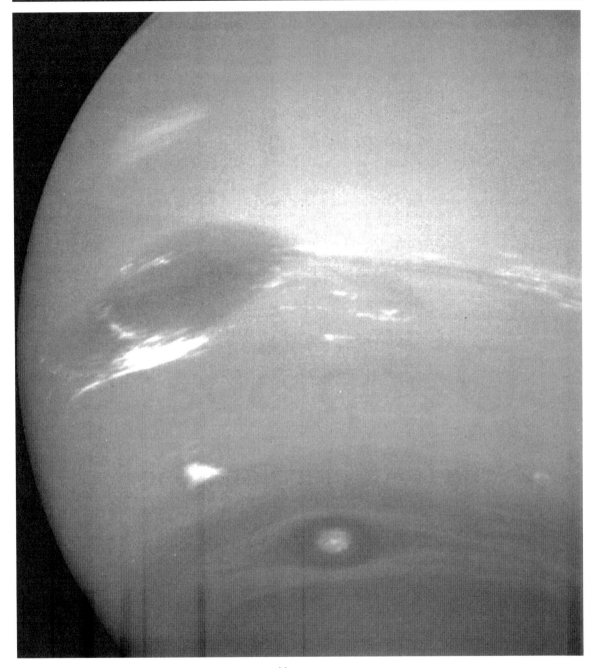

Neptune

All rights reserved under article two of the Berne Copyright Convention (1971).
We acknowledge the financial support of the Government of Canada through the
Book Publishing Industry Development Program for our publishing activities.
Published by Collector's Guide Publishing Inc., Box 62034,
Burlington, Ontario, Canada, L7R 4K2
Printed and bound in Canada
DEEP SPACE – THE NASA MISSION REPORTS
by Robert Godwin & Steve Whitfield
ISBN 1-894959-15-9
ISSN 1496-6921
Apogee Books Space Series No. 48
©2005 Apogee Books
All photos courtesy of NASA,
except as shown on pages 379 and 380.

Voyager 1

Deep Space Network dish antenna, Goldstone, California

Table of Contents

Introduction

A long-running debate within the space community concerns the value derived from manned versus unmanned space missions. Proponents of both sides have produced endless convincing arguments as to why the limited funds available would be better spent in their own arena.

With a little longer-term thinking, however, it's obvious that both mission types are necessary and are, in fact, completely interdependent, each mission type building upon and relying on the other. The *Explorer* probes in early 1958 discovered and measured the Van Allen radiation belts surrounding the Earth – a potential danger to manned flights. The Apollo astronauts determined that the lunar surface was solid with a surety that a lander and short-distance rover could never match. *Pioneer 10*, the first mission described in this book, tested the validity and conditions of a planetary gravity assist, whereby a spacecraft acquires a velocity change (speed and direction) by flying close by a planet. Gravity assist has since become an essential component of solar system exploration missions, and will certainly be used in future manned missions.

Pioneer 10 (closely followed by *Pioneer 11*) was the first man-made object to travel past the orbit of Mars, heading for Jupiter, the primary object of its investigations. After being sling-shotted away from Jupiter, *Pioneer 10* continued on to become the first human artifact to leave our solar system. Although *Pioneer 10*'s power source has been dead for some years, it continues to perform its secondary duty – humanity's ambassador to the universe, a precedent that it set for subsequent missions that left the solar system.

Following after the Pioneer 10 and 11 missions (launched in 1972 and 1973, respectively) were *Voyagers 1* and *2* (launched in 1977), *Galileo* (1989), and the currently active mission, Cassini-Huygens (1997). Each mission, building on the results of its predecessors, has extended our knowledge further. Additionally, each mission has placed importance on confirming and clarifying measurements made on earlier missions. *Pioneer 11* exceeded the achievements of Pioneer 10 by continuing on (with a gravity assist from Jupiter) to Saturn – the first manmade object to reach the ringed planet.

The Voyager 1 mission visited Jupiter and Saturn, expanding upon the *Pioneer 11* achievements. *Voyager 2* greatly extended man's conquest of the solar system by continuing on past Jupiter and Saturn to fly by both Uranus and Neptune, the first and only spacecraft to do so. This four-planet mission, referred to as the Grand Tour, took advantage of a rare alignment of the planets whereby the spacecraft's velocity allowed it to intersect each planet's orbit at a time when the planet was at the same location. This planetary alignment repeats every 176 years; the joke among mission scientists was that the last time this mission was possible, Jefferson was president, but he blew his chance.

Galileo arrived at Jupiter in December 1995, after a six year journey. *Galileo* was the first spacecraft to actually orbit Jupiter (or any outer solar system planet). The *Galileo* spacecraft released a probe into the Jovian atmosphere. For a brief period, until it succumbed to Jupiter's immense atmospheric pressure, this probe provided our first direct measurements of temperature, pressure and composition of the Jovian atmosphere.

After a nearly seven-year flight, including four gravity assist maneuvers, the *Cassini* spacecraft – the largest and most massive spacecraft ever launched by the U.S. – went into orbit around Saturn in 2004. *Cassini* carried with it the European Space Agency's *Huygens* probe, which it released on an intersect course with Saturn's largest moon, Titan. *Huygens* entered the atmosphere of Titan and parachuted to the surface on January 14, 2005. *Huygens* radioed scientific data to *Cassini* during its descent and for a short period from Titan's surface. By design, *Huygens*' batteries would be depleted after a maximum of half an hour on the surface. *Huygens* appears to have performed exactly as intended. As this book goes to print the *Huygens* data is being compiled and processed. One of the major questions to be answered by the *Huygens* probe was whether the surface of Titan is solid, liquid, or both – would *Huygens* land on a hard surface or a hydrocarbon ocean? Early indications are that it split the difference and splatted down in what is described as mud.

As well as being one of the largest moons in our solar system, Titan is of special interest to scientists because its atmosphere, and perhaps its surface, may be very much like that of the primordial Earth. If this indeed turns out to be the case, investigations of Titan may shed light on the evolution of life on Earth, and perhaps in the universe at large. Titan's discoverer, Christiaan Huygens a 17th century Dutch scientist, could not possibly have imagined that one day his name would be attached to a disc-shaped machine that would travel billions of miles to send back close-up snapshots of his discovery. Huygens not only discovered Titan but he was also the first scientist to recognize that life in the universe may be totally unfamiliar and different to our own terrestrial brand, and although this thought repelled him he was the first to make allowances for such bizarre life. Huygens also understood that life needed a solvent to flourish.

Just over three hundred years later his namesake appears to have discovered the first extra-terrestrial body to have clear evidence of such a solvent – albeit not the one we would like to have found – water. Evidence suggests that hydrocarbons may be sloshing about on Titan, although other environmental conditions on Titan are certainly too harsh to support life as we understand it. The Huygens probe certainly went where no person could easily have gone with the state of today's technology.

Not all of NASA's deep space missions have been planetary missions.

The primary purpose of the Deep Space 1 mission (launched October 1998) was to test advanced, high-risk technologies in space. This successful mission was extended to perform an encounter with Comet Borrelly, and returned the best images and other science data ever gathered from a comet.

The Stardust mission (launched February 1999) was the first U.S. mission dedicated solely to a comet, and will be the first to return extraterrestrial material from outside the orbit of the Moon. Stardust's main objective is to capture particle samples from a well-preserved comet called Wild-2 (pronounced Vilt-2), which it did in January 2004. Stardust is due to return these samples to Earth, along with samples of interplanetary particles, in January 2006.

The Deep Impact mission (launched January 2005) is due to encounter Comet Tempel 1, 83 million miles from Earth, on July 4, 2005. The flyby spacecraft will release an "impactor" which will create a large crater in the comet. Images from cameras and a spectrometer recording ice and dust debris ejected from the crater will be sent to Earth for analysis.

The Dawn mission (due to launch June 2006) is a nine-year mission to investigate the asteroids Ceres (2010-2011) and Vesta (2014-2015). Ceres and Vesta have been chosen because they are large and have remained intact since their formations, thereby characterizing the conditions and processes of the solar system's earliest epoch.

The proposed Jupiter Icy Moons Orbiter (JIMO) is a spacecraft designed to orbit Jupiter's moons Callisto, Ganymede and Europa, which may have oceans beneath their icy surfaces. The mission would launch in 2015 or later.

Something about these robotic emissaries still attracts our attention in record numbers. The evidence can be seen by the amount of traffic generated at the ESA and JPL web sites where billions of "hits" are recorded for robotic missions like Cassini-Huygens. They also continue to increase our knowledge of the solar system, including Earth. Some people refer to this as "abstract" knowledge, implying that it's of little practical value. In the long term, however, this same knowledge points towards the resources and technologies that can guarantee us a future.

In the final analysis, both manned and unmanned missions will continue to be necessary if either is to progress. The point is often made that a scientific instrument can't make a judgment call or adapt itself to changing conditions in the manner of a human being with hands and a brain. At the same time, with neither hands nor brain can a human being measure radiation levels, record a spectrogram, or determine a chemical composition. But if we employ both men and machines, each according to their strengths, we can achieve more than the sum of what each can accomplish separately.

For obvious reasons, unmanned spacecraft have always gone to any destination before manned missions. The *Pioneers*, the *Voyagers*, *Galileo* and *Cassini* have opened the road to the outer solar system, showing us what's there and what more we need to understand before considering more ambitious missions. Whereas manned missions tend to last days or months, the unmanned deep space missions, of necessity, are years long. Scientists, engineers, technicians and managers have often committed a significant portion of their entire careers to a single program. For those who were a part of both Pioneer and Voyager, that's 30 years or more dedicated to a single undertaking – a "commitment" of the same caliber as most marriages.

In view of our continually rising global population, and the increasing strain on and depletion of Earth's resources, this "abstract" knowledge, gained by the deep space missions of the 20th and 21st centuries, will probably turn out to have been our first steps toward preserving ourselves. Long before looking off Earth for real estate, we'll be going to the asteroids, and perhaps the planets and comets as well, for resources – resources upon which we will have become absolutely dependent. The unmanned deep space missions don't feature "the right stuff" of Apollo, and never have. Just the same, in the long run, they are perhaps "the important stuff."

Robert Godwin and Steve Whitfield
(Editors)

National Aeronautics and
Space Administration

Jet Propulsion Laboratory
California Institute of Technology
Pasadena, California 91109
Office of Public Information
Telephone (213) 354-5011

Prepared by Betty Shultz

Historical Summary of Unmanned Space Probes

SPACECRAFT	LAUNCH DATE	ENCOUNTER DATE	S/C WEIGHT	No. OF PICTURES	COST OF PROJECT
Explorer 1	Jan. 31, 1958	N/A	30.8 lbs	None	—
Pioneer 3	Dec. 6, 1958	N/A	12.95 lbs	None	—
Pioneer 4	Mar. 3, 1959	N/A	13.40 lbs	None	—
Ranger 1	Aug. 23, 1961	None	675 lbs	None	$12 million
Ranger 2	Nov. 18, 1961	None	675 lbs	None	—
Ranger 3	Jan. 26, 1962	Jan. 28, 1962	727 lbs	None	$13 million
Ranger 4	Apr. 23, 1962	Apr. 26, 1962	730 lbs	None	—
Mariner 1	Jul. 22, 1962	None	447 lbs	None	$44 million (1 & 2)
Mariner 2	Aug. 27, 1962	Dec. 14, 1962	447 lbs	None	$44 million (1 & 2)
Ranger 5	Oct. 18, 1962	Oct. 21, 1962	755 lbs	None	—
Ranger 6	Jan. 30, 1964	Feb. 2, 1964	800 lbs	None	$28 million
Ranger 7	Jul. 28, 1964	Jul. 31, 1964	805 lbs	4,316	—
Mariner 3	Nov. 5, 1964	None	575 lbs	None	$105 million (3 & 4)
Mariner 4	Nov. 28, 1964	Jul. 14, 1965	575 lbs	21	$105 million (3 & 4)
Ranger 8	Feb. 17, 1965	Feb. 20, 1965	805 lbs	7,137	$30 million
Ranger 9	Mar. 21, 1965	Mar. 24, 1965	805 lbs	5,814	—
Surveyor 1	May 31, 1966	June 1, 1966 (Landing)	600 lbs	11,240	$426 million (7 S/C)
Lunar Orbiter 1	Aug. 10, 1966	Oct. 29, 1966 (Impact)	853 lbs	207	$163 million (5 S/C)
Surveyor 2	Sept. 20, 1966	Sept. 22, 1966 ((Landing)	620 lbs	None	$426 million (7 S/C)
Lunar Orbiter 2	Nov. 6, 1966	Oct. 11, 1967 (Impact)	1861 lbs	211	$163 million (5 S/C)
Lunar Orbiter 3	Feb. 4, 1967	Oct. 9, 1967 (Impact)	850 lbs	182	$163 million (5 S/C)
Surveyor 3	Apr. 16, 1967	Apr. 19, 1967 (Landing)	1620 lbs	6,326	$426 million (7 S/C)
Lunar Orbiter 4	May 4, 1967	Oct. 6, 1967 (Impact)	860 lbs	163	$163 million (5 S/C)
Mariner 5	June 12, 1967	Oct. 19, 1967	1540 lbs	None	$38 million
Surveyor 4	Jul. 14, 1967	Jul. 18, 1967	625 lbs	None	$426 million (7 S/C)
Lunar Orbiter 5	Aug. 1, 1967	Jan. 31, 1968	860 lbs	212	$163 million (5 S/C)
Surveyor 5	Sept. 8, 1967	Sept. 10, 1967 (Landing)	616 lbs	19,118	$426 million (7 S/C)
Surveyor 6	Nov. 6, 1967	Nov. 9, 1967 (Landing)	600 lbs	29,952	$426 million (7 S/C)
Surveyor 7	Jan. 6, 1968	Jan. 9, 1968	637 lbs	21,038	$426 million (7 S/C)
Mariner 6	Feb. 24, 1969	Jul. 30, 1969	850 lbs	75	$152 million (6 & 7)
Mariner 7	Mar. 27, 1969	Aug. 4, 1969	850 lbs	126	$152 million (6 & 7)
Mariner 8	May 8, 1971	None	2,150 lbs	None	$129 million (8 & 9)
Mariner 9	May 30, 1971	Nov. 13, 1971	2,150 lbs	7,329	$129 million (8 & 9)
Pioneer 10	Mar. 2, 1972	Dec. 3, 1973	570 lbs		$165 million (10 & 11)
Pioneer 11	Apr. 5, 1973	Dec. 2, 1974 (Jupiter) Sept. 1, 1979 (Saturn)	570 lbs		$165 million (10 & 11)
Mariner 10	Nov. 3, 1973	Feb. 5, 1974 (Venus) Mar. 29, 1974, Sept. 21, 1974, Mar. 16, 1975 (Mercury)	1,108 lbs	8,200	$98 million
Viking 1	Aug. 20, 1975	June 19, 1976 (Orbit) Jul. 20, 1976 (Landing)	7,750 lbs	More than 50,000 from both orbiters & landers	$1 billion (1 & 2)
Viking 2	Sept. 9, 1975	Aug. 7, 1976 (Orbit) Sept. 3, 1976 (Landing)	7,750 lbs	More than 50,000 from both orbiters & landers	$1 billion (1 & 2)
Voyager 2	Aug. 20, 1977	Jul. 9, 1979 Aug. 25, 1981	1,797 lbs	More than 67,000 from both 1 & 2	$338 million (1 & 2)
Voyager 1	Sept. 5, 1977	Mar. 5, 1979 Nov. 12, 1980	1,797 lbs	More than 67,000 from both 1 & 2)	$338 million (1 & 2)
Pioneer Venus 1	May 20, 1978	Dec. 4, 1978 (Orbit)	1,250 lbs		$225 million (1 & 2)
Pioneer Venus 2	Aug. 8, 1978	Dec. 9, 1978 (Landing)	1,950 lbs		$225 million (1 & 2)

National Aeronautics and
Space Administration

Prepared by Betty Shultz

Jet Propulsion Laboratory
California Institute of Technology
Pasadena, California 91109
Office of Public Information
Telephone (213) 354-5011

Planetary Data

(AS OF 1980)

	Mercury	Venus	Earth	Mars	Jupiter	Saturn	Uranus	Neptune	Pluto
Mean Distance from Sun (millions of km)	57.9	108.2	149.6	227.9	778.3	1,425.0	2,867.0	4,496.6	5,890.0
Surface Gravity (Earth=1)	0.37	0.88	1	0.38	2.64	1.15	1.17	1.18	?
Equatorial Diameter (km)	4,878	12,104	12,756	6,786	142,796	120,000	50,800	48,600	3,000
Period of Revolution (d=days, y=years)	87.96 d	224.7 d	365.26 d	1.88 d	11.86 y	29.46 y	84.01 y	164.79 y	247.7 y
Rotation Period	58.56 d	-243.01 d Retrograde	23h 37m 4s	24h 37m 23s	9h 55m 30s	10h 39m 24s	-15.57 h Retrograde	18.43 h	6.39 d
Inclination of Axis	<28°	3°	23° 27'	23° 59'	3° 05'	26° 44'	82° 5'	28° 48'	?
Inclination of Orbit to Ecliptic	7°	3.4°	0°	1.9°	1.3°	2.5°	0.8°	1.8°	17.2°
Eccentricity of Orbit	0.206	0.007	0.017	0.93	0.048	0.056	0.047	0.009	0.25
Volume (Earth=1)	0.06	0.88	1	0.15	1,316	755	67	57	0.1 (?)
Density (Water=1)	5.4	5.2	5.5	3.9	1.3	0.7	1.2	1.7	4.9
Mean Temperature at visible Surface (Degrees Celsius) S=solid, C=clouds	350 (S) day -170 (S) night	-33 (C) 480(S)	22 (S)	-23 (S)	-150 (C)	-180 (C)	-210 (C)	-220 (C)	-230 (?)
Number of Satellites	0	0	1	2	15	*	5	2	1
Number of Rings	0	0	0	0	1	1,000+	5	0	0

MERCURY 4,878 KM (3,024 MI)

VENUS 12,100 KM (7,502 MI)

EARTH 12,756 KM (7,909 MI)

MARS 6,796 KM (4,214 MI)

MOON (EARTH) 3,476 KM (2,155 MI)

IO (JUPITER) 3,630 KM (2,251 MI)

EUROPA (JUPITER) 3,138 KM (1,946 MI)

GANYMEDE (JUPITER) 5,262 KM (3,262 MI)

CALLISTO (JUPITER) 4,800 KM (2,976 MI)

TITAN (SATURN) 5,150 KM (3,193 MI)

National Aeronautics and
Space Administration
Prepared by Betty Shultz

Jet Propulsion Laboratory
California Institute of Technology
Pasadena, California 91109
Office of Public Information
Telephone (213) 354-5011

Pioneer Missions Data Summary

	Pioneer 10	Pioneer 11
1. Objective	Study Jupiter, the asteroid belt, the heliosphere. Pioneer 10 and 11 were the first spacecraft beyond Mars, and the first to encounter Jupiter. Pioneer 10 will be the first spacecraft to leave the solar system; Pioneer 11 was the first spacecraft to reach Saturn.	
2. Project Manager Project Scientist	Charles F. Hall, Ames Research Center Dr. John Wolfe, Ames Research Center	Same Same
3. Launch Date	March 2, 1972	April 5, 1973
4. Encounter Dates	Jupiter: December 3, 1973	Jupiter: December 2, 1974 Saturn: September 1, 1979
5. Spacecraft-to-Earth Distance At Arrival	Jupiter: 514 million miles	Jupiter: 514 million miles Saturn: 963 million miles
6. Flyby Distance	Jupiter: 82,000 miles	Jupiter: 26,600 miles Saturn: 13,300 miles
7. Arc Distance	Jupiter: 620 million miles	Jupiter: 620 million miles Saturn: 2 billion miles
8. Built By	TRW	Same
9. Weight of Spacecraft	570 lbs; Instruments: 67 lbs	Same
10. Instruments on Spacecraft	a) Vector helium magnetometer (P.I.: Edward J. Smith, JPL) b) Fluxgate magnetometer (*Pioneer 11* only) (P.I.: Mario Acuna, Goddard Space Flight Center) c) Plasma analyzer (P.I.: John H. Wolfe, Ames Research Center) d) Charged particle (P.I.: John A. Simpson, University of Chicago) e) Cosmic ray telescope (P.I.: Frank B. McDonald, Goddard Space Flight Center) f) Geiger tube telescope (P.I.: James A. Van Allen, University of Iowa) g) Trapped radiation detector (P.I.: R. Walker Fillius, University of California, San Diego) h) Asteroid/meteoroid detector (P.I.: Robert K Soberman, General Electric Co. and Philadelphia Drexel University) i) Meteoroid detector (P.I.: William H. Kinard, Langley Research Center) j) Ultraviolet photometer (P.I.: Darrell L. Judge, University of Southern California, Los Angeles) k) Imaging photopolarimeter (P.I.: Tom Gehrels, University of Arizona, Tucson) l) Infrared radiometer (P.I.: Andrew Ingersoll, Caltech)	
11. Cost of Project	Cost of *Pioneer 10 & 11*: $165 million	
12. Pictures Returned By Spacecraft	*Pioneer 10* at Jupiter = 180	*Pioneer 11* at Jupiter = 111 *Pioneer 11* at Saturn = 113
13. Responsible Managing Center	Ames Research Center	
14. Written Material Available	Pioneer Saturn Encounter	
15. Refer To	Ames Research Center, TRW	

NATIONAL AERONAUTICS AND
SPACE ADMINISTRATION
Washington, D. C. 20546
202-755-8370

FOR RELEASE:

FEBRUARY 20, 1972

PROJECT: PIONEER 10

P
R
E
S
S

K
I
T

CONTENTS

**NATIONAL AERONAUTICS AND
SPACE ADMINISTRATION**
Washington, D. C. 20546
202-755-8370

Howard Allaway   Phone: (202) 755-3680
Peter W. Waller (Ames) Phone: (415) 961-2671

FOR RELEASE:
FEBRUARY 20, 1972

RELEASE No: 72-25

Pioneer F Mission To Jupiter

Man will reach out beyond Mars to take the first close look at the planet Jupiter on the mission of the unmanned *Pioneer F* spacecraft, to be launched by the National Aeronautics and Space Administration from Cape Kennedy, Florida, between February 27 and March 13, 1972.

The trip to Jupiter will last less than two years, for most launch dates, with most arrival times before December 31, 1973.

Jupiter is a spectacular planet. It appears to have its own internal energy source and is so massive that it is almost a small star. It may have the necessary ingredients to produce life. Its volume is 1,000 times that of Earth, and it has more than twice the mass of all the other planets combined. Striped in glowing yellow-orange and blue-gray, it floats in space like a bright-colored rubber ball. It has a huge red "eye" in its southern hemisphere and spins more than twice as fast as Earth.

The mission includes a number of other firsts. *Pioneer F* is expected to make the first reconnaissance of the Asteroid Belt between the orbits of Mars and Jupiter. It is planned to be the first man-made object to escape the solar system, and the first to use the orbital velocity and powerful gravity of Jupiter for this escape. It is also the first NASA spacecraft to draw its electrical power entirely from nuclear generators, four radioisotope thermoelectric generators (RTGs) developed by the Atomic Energy Commission.

The Atlas-Centaur-TE-M-364-4 launch vehicle will drive the spacecraft away from the Earth initially at 51,800 km/hr (32,000 mph) – faster than any man-made object has flown before. At this time the spacecraft will be named *Pioneer 10*. For the first week, it will travel an average of 800,000 kilometers (a half-million miles) a day. It will pass the Moon's orbit in about 11 hours.

Pioneer's 13 scientific experiments are expected to provide new knowledge about Jupiter and many aspects of the outer solar system and our galaxy. It will return the first close-up images of Jupiter, and will make the first measurements of Jupiter's twilight side, never seen from the Earth.

The mission opens the era of exploration of the outer planets, since it is intended in part to develop technology for other outer planet missions.

The spacecraft will test out the hazards of cosmic debris in the Asteroid Belt. It will probe Jupiter's radiation belts, which could cripple or destroy a spacecraft approaching too closely. The belts are estimated to be as much as one million times more intense than Earth's Van Allen radiation belts.

Jupiter is so far away that radio messages moving at the speed of light will take 45 minutes to reach the spacecraft there, with a round trip time of 90 minutes. This will demand precisely planned command operations. Although *Pioneer* can store five commands, it will be controlled mostly by frequent instructions from Earth.

To carry out the mission, the advanced communications technology of NASA's Deep Space Network (DSN) will be strained to the limit. The DSN's 64-meter (210-foot) "big dish" antennas, one of which now hears the *Mariner 9* spacecraft in Mars orbit, will have to hear seven times as far as *Pioneer* approaches Jupiter.

Pioneer's eight-watt signal, transmitted from Jupiter, will reach DSN antennas with a power of $1/100,000,000,000,000,000$ watts. Collected for 19 million years, this energy would light a 7.5-watt Christmas tree bulb for one-thousandth of a second.

Pioneer F is a new design for the outer solar system, but it retains many tested subsystems of its predecessors, the *Pioneer 6 to 9* spacecraft. All four are still operating in interplanetary space. *Pioneer 6* is in its seventh year.

The 260-kilogram (570-pound) *Pioneer F* is spin-stabilized, giving its instruments a full-circle scan. It uses nuclear sources for electric power because solar radiation is too weak at Jupiter for an efficient solar-powered system.

Its 2.75-meter (nine-foot) dish antenna will be locked on the Earth like a big eye throughout the mission – changing its view direction as the home planet moves to and fro in its orbit around the Sun. The entire flight path is in, or very close to, the plane of Earth's orbit, the ecliptic.

Jupiter itself is little understood.

It broadcasts predictably modulated radio signals of enormous power. Though it has only $1/1,000$th the mass of the Sun, it may have Sun-like internal processes, apparently radiating about four times as much energy as it receives from solar radiation.

In addition to helium, the planet's atmosphere contains ammonia, methane, hydrogen, and probably water, the same ingredients believed to have produced life on Earth about four billion years ago. Because of the planet's internal heat source, many scientists believe that large regions below the frigid cloud layer are around room temperature. These conditions could allow the planet to produce living organisms despite the fact that it receives only $1/27$th of the solar energy received by the Earth.

Jupiter is probably more than 75 percent hydrogen, the main constituent of the universe. The planet may have no solid surface. Due to its high gravity, it may go from a thick gaseous atmosphere down to oceans of liquid hydrogen, to a slushy layer, and then to a solid hydrogen core. Ideas of how deep beneath its striped cloud layers any solid hydrogen "icebergs" or "continents" might lie vary by thousands of kilometers.

Astronomers have long seen violent circulation of the planet's large-scale cloud features. A point on Jupiter's equator moves at 35,400 km/hr (22,000 mph), compared with 1,600 km/hr (1,000 mph) for a similar point on Earth's equator.

The most bizarre feature of the planet is the Great Red Spot, known as the "Eye of Jupiter." This huge oval is 48,000 kilometers (30,000 miles) long and 13,000 kilometers (8,000 miles) wide, large enough to swallow up several Earths with ease. The Red Spot may be an enormous standing column of gas, or, says one scientist, a "raft" of hydrogen ice floating on a bubble of warm hydrogen in the cooler hydrogen atmosphere, and bobbing up and down at 30-year intervals, so that the Spot disappears and reappears. The Spot appears to rotate at a different speed from the planet. Its red color may be due to the presence of organic compounds found in a gigantic lightning charge in the Jovian atmosphere, according to one theory.

Potential benefits of the Pioneer Jupiter mission and others like it include increased knowledge of "collisionless plasmas" of the solar wind. This bears directly on the "ultimate" clean system for electric power production, controlled hydrogen fusion. The findings may also lead to better understanding of Earth's weather cycles, and to insights into Earth's atmosphere circulation through study of Jupiter's rapidly rotating atmosphere. There may also be indications of Jovian resources, such as perhaps a quantity of petrochemicals equivalent to Earth's consumption for a million years.

Pioneer F spacecraft will carry a 30-kilogram (65-pound) experiment payload. It will make 20 types of measurements of Jupiter's atmosphere, radiation belts, heat balance, magnetic field, moons, and other phenomena. It also will characterize the heliosphere (solar atmosphere); perhaps the interstellar gas; cosmic rays; asteroids; and meteoroids between the Earth and 2.4 billion kilometers (1.5 billion miles) from the Sun.

A second, almost identical spacecraft, *Pioneer G*, will be launched to Jupiter in early April of 1973.

NASA's Office of Space Science assigned project management for the two *Pioneer Jupiter* spacecraft to NASA's Ames Research Center, Mountain View, California, near San Francisco. The spacecraft are built by TRW Systems, Redondo Beach, California. The scientific instruments are supplied by NASA Centers, universities and private industry.

Tracking is by NASA's Deep Space Network, operated by the Jet Propulsion Laboratory, Pasadena, California. NASA's Lewis Research Center, Cleveland, manages the launch vehicle, which is built by General Dynamics, San Diego, California.

Cost of two *Pioneer Jupiter* spacecraft, scientific instruments, and data processing and analysis is about $100 million. This does not include costs of launch vehicles and data acquisition.

The 30-minute evening launch window opens progressively earlier each day – approximately 9:00 p.m. EST, on February 27, and at 7:00 p.m. by March 13, 1972.

Depending on launch date, the trip to Jupiter will take from 630 to 795 days with arrival dates between November 21, 1973, and July 27, 1974.

Mission Profile

Pioneer F will be launched toward Jupiter and eventual escape from the solar system on a direct-ascent trajectory from Cape Kennedy in a direction 4.5 to 20 degrees south of straight east, passing over South Africa shortly after launch vehicle burnout.

The trip will follow a curving path about a billion kilometers long (620 million miles) between the orbits of Earth and Jupiter. The path will cover about 160 degrees going around the Sun between launch point and Jupiter. Because of the changing positions of the Earth and Jupiter, the shortest trip times are for launches during the early days of the two-week launch period. These early dates would put the Earth-Jupiter line far from the Sun during flyby, resulting in less interference by the Sun with data transmission from the spacecraft and with simultaneous telescope observations of Jupiter from Earth.

The high-energy launch marks the-first use of a third stage, the TE-M-364-4, with the Atlas-Centaur launch vehicle.

After liftoff, burnout of the 1,829,000-newton (411,000-pound)-thrust, stage-and-a-half Atlas booster will occur in about four minutes. Stage separation and ignition of the 130,000-newton (30,000-pound)-thrust Centaur second stage will then take place and the hydrogen-fueled Centaur engine will burn for about 7.5 minutes. The 10.7-meter (35-foot)-long aerodynamic shroud covering the third stage and the spacecraft will be jettisoned after leaving the atmosphere, about 12 seconds after Centaur engine ignition.

At about 13 minutes after liftoff, small solid-fuel rockets will spin up the 65,860-newton (15,000-pound)-thrust third stage and the attached spacecraft to 60 rpm for stability during stage firing. The third stage will then ignite and burn for about 44 seconds.

About two minutes after third stage burnout (16 minutes after liftoff), *Pioneer F* will separate from the third stage and be on Jupiter trajectory.

During powered flight, launch vehicle and spacecraft will be monitored from the Mission Control Center at Cape Kennedy via Eastern Test Range tracking stations and by a tracking ship in mid-Atlantic.

Seventeen minutes after launch, NASA's Ascension Island tracking station will be able to send commands, if necessary, to the spacecraft.

Mission control will shift from the Mission Director at Cape Kennedy to the Flight Director at the Pioneer Mission Support Area, (PMSA) at the Space Flight Operations Facility (SFOF) at the Jet Propulsion Laboratory, Pasadena, California.

At about 25 minutes after launch, the DSN's Johannesburg, South Africa station will lock on the spacecraft and can send commands. By then, *Pioneer F* will have emerged from Earth's shadow and begin to get timing data from its Sun-sensor. Johannesburg may turn on the Jovian Trapped Radiation instrument in order to calibrate it against measurements of the known radiation in Earth's Van Allen belts.

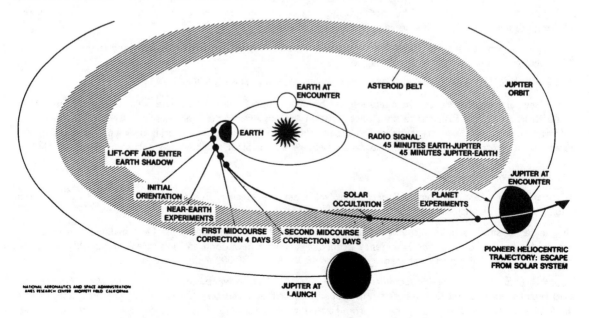

Instrument Turn-On

About 32 minutes after launch, *Pioneer F*'s onboard sequencer will initiate despin down to about 21 rpm from the 60-rpm spin imparted before third stage ignition. The sequencer then will start deployment of the four nuclear power sources, the Radioisotope Thermoelectric Generators (RTGs), using spacecraft spin to slide them out on their trusses three meters (10 feet) from the center of the spacecraft. The sequencer will next initiate deployment of the magnetometer to 6.6 meters (21.5 feet) from the spacecraft center by unfolding its lightweight boom. When boom deployments are complete, spacecraft spin rate will be down to five rpm, allowing the scientific instruments to scan a full circle five times a minute.

At launch plus 43 minutes, turn-on of the remaining 10 scientific instruments will begin. Because of the possibility of high-voltage arcing, some instruments will be allowed to outgas. Flight directors will turn them on the second and third day of the mission.

About three hours after liftoff, controllers will command the spacecraft to turn itself so that its high-gain dish antenna points at Earth. This will greatly increase signal strength. Before this, the spacecraft will have been communicating via its rear-facing, low-gain antenna. This maneuver will take about two hours to complete. It also puts the spacecraft in position for proper heat balance.

At eight hours after launch, *Pioneer F* will "rise" at the DSN's Goldstone, California station. Goldstone will trim spin rate to the normal 4.8 rpm. The automatic Conscan system (whereby the spacecraft uses the pattern of the incoming radio signal to refine its Earth-pointing) will be tested and used.

In the first few days after launch, DSN specialists at Pasadena will perform intensive computer calculations to establish the precise trajectory.

During the early days of the mission, the imaging photopolarimeter experiment will make maps of the brightness of the sky, as the spacecraft rotates, in order to correlate *Pioneer*'s zodiacal light measurements with ground-based observations. Simultaneous measurements of zodiacal light will be made by Dudley Observatory, Albany, New York, from Mount Haleakala, Hawaii.

Midcourse Corrections

Four days of accurate tracking will have found any errors in launch velocity and direction. A second Conscan maneuver will establish precise Earth-point for reference. The DSN will then command change of spacecraft attitude, firing of thrusters for the first midcourse velocity change to eliminate errors in aiming at Jupiter, and return to Earth-point.

At about a week after launch, the spacecraft and its operations team will have settled into the interplanetary cruise phase of the mission. Analysts will regularly assess performance of systems and instruments and develop strategy to get around any malfunctions.

The solar wind and magnetic field instruments will be mapping the interplanetary medium. The particle experimenters will map distribution of solar and galactic cosmic ray particles, and the ultraviolet photometer will measure neutral hydrogen. The meteoroid and dust detectors will gather data on sizes and distribution of interplanetary matter. Zodiacal light measurements will be made periodically by the photopolarimeter instrument and the meteoroid telescopes to help determine amounts of interplanetary material.

Conscan maneuvers to sharpen Earth-point will be commanded routinely every three days, and later in the mission every week or two.

At the end of the first month, a second midcourse velocity change will be made if needed.

Within several months, if spacecraft operation has become routine, control personnel will move from the Pioneer Mission Support Area at Pasadena to the Pioneer Mission Analysis Area at the Ames Research Center, Mountain View, California. Control of Pioneer is at the Jet Propulsion Laboratory (JPL) in Pasadena during critical phases of the mission because of the greater computer and display capability there and the SFOF's ability to support many systems analysts.

Asteroid Traverse

At about four months after launch, *Pioneer Jupiter* will enter the Asteroid Belt and begin attempts to observe light scattered by asteroid material with the four onboard telescopes.

The number of small particles in the belt is uncertain by several orders of magnitude. Determining this number is in fact one of the key objectives. Emergency procedures will be developed and practices for possible subsystem malfunctions due to impact of a high-velocity – around 48,300 km/hr (30,000 mph) – dust particle.

During early parts of the mission, command, tracking, and data return will be primarily by the DSN's 26-meter (85-foot) antennas located at 120 degree intervals around the Earth at Goldstone, California; Johannesburg, South Africa; and Canberra, Australia. At greater ranges from Earth, the 64-meter (210-foot) dish antenna, at Goldstone will track the spacecraft. In mid-1973, when the two large overseas dishes become operational, these will also be used. The highly sensitive 64-meter dishes will also be used for critical maneuvers, such as midcourse velocity corrections.

At Jupiter distance, the 26-meter dish antennas will be able to receive 64 data bits per second (bps), the 64-meter dishes 1,024 bps.

At about 300 days from launch, the spacecraft will pass almost directly behind the Sun, causing communication difficulties for a few days because of the Sun's radio noise.

A year of tracking may show need for a further small velocity change to alter flyby point slightly, or change arrival time by a few minutes.

Factors in Planet Encounter

Halfway to Jupiter (315 to 400 days), detailed planning for the week-long flyby will begin. Jupiter's radiation belts, perhaps a million times stronger than Earth's belts, present a serious hazard. Most future outer planet missions will be shaped by this problem of Jupiter's radiation belts — missions to Jupiter and also those using Jupiter's gravity and orbital speed to increase velocity for trips to planets farther out. Mission planners must know how close they can safely approach the giant planet and what radiation hazard they will have to protect against.

Pioneer F / G Encounter With Jupiter

SOLAR SYSTEM
ESCAPE
TRAJECTORY

MAGNETIC FIELD LINES

EXPERIMENT MEASUREMENTS

- MAGNETOSPHERE BOUNDARY
- TRAPPED RADIATION
- MAGNETIC FIELD

- IONOSPHERE
- ATMOSPHERE COMPOSITION AND MIXING
- CLOUD STRUCTURE

- THERMAL BALANCE
- HYDROGEN / HELIUM RATIO
- MASS: JUPITER / MOONS

JUPITER ORBIT

TO SUN

TO EARTH

ASSUMED LIMIT OF
JUPITER'S DISTURBANCE
IN THE
INTERPLANETARY MEDIUM

PIONEER
TRACK

NATIONAL AERONAUTICS AND SPACE ADMINISTRATION
AMES RESEARCH CENTER MOFFETT FIELD CALIFORNIA

It is planned that *Pioneer F* will fly through the belts at one Jupiter diameter, 140,000 kilometers (87,000 miles) from the planet's visible surface. High-energy protons (hydrogen nuclei) and electrons in the belts could penetrate the spacecraft and degrade the functioning of vital transistors or circuitry. Mission controllers may have to operate a crippled spacecraft after flight through Jupiter's radiation belts; or spacecraft data transmission could be cut off entirely and the mission terminated at Jupiter, but determination of the severity of the hazard is one of *Pioneer*'s objectives.

A minor hazard is Jupiter's massive radio emissions, which may interfere with *Pioneer*'s radio signals at times but do no permanent damage to the equipment.

Tests of the spacecraft in magnetic fields higher than Jupiter's have shown slight effects but no serious problems.

Jupiter's large electric fields are not expected to affect the spacecraft.

Because Jupiter's gravity will alter spacecraft course, the actual flyby will be a wide loop around the planet. The spacecraft will approach from the planet's sunlit side, then swing almost completely around its dark side. Planet rotation aside, *Pioneer F* will pass over about two thirds of Jupiter's visible surface at various altitudes.

It will approach Jupiter in a counterclockwise direction (looking down at the North Pole). Its path on the planet's surface will pass over part of Jupiter's southern hemisphere, cross the equator at an angle of 14 degrees, and exit over the northern hemisphere.

Targeting

The target point at Jupiter has been selected so that the spacecraft will achieve: a good survey of Jupiter's strong magnetic field and its intense radiation belt; good view angles for the instruments that look at the planet; and a short passage behind the planet.

Other targeting considerations are to have a post-encounter trajectory resulting in solar system escape, to pass close to one of Jupiter's moons, and to avoid biological contamination.

A basic flyby factor is to have simultaneous tracking from two Earth stations during the critical five hours before periapsis (closest approach).

Experimenters would, in addition, like to have one of the Jovian moons pass between the spacecraft and Earth to use effects on the spacecraft radio signal to look for an atmosphere on that moon. They also would like to view the Great Red Spot and the shadow of a moon on Jupiter's surface to measure temperatures.

Not all of these secondary options can be taken during one flyby. Which ones will be selected will be determined after launch by flyby timing, spacecraft health, and scientific interest.

Flyby Operations

At several weeks before periapsis, mission control will move back to the Pioneer Mission Support Area, Pasadena.

The greater DSN capabilities will be especially needed to handle the increased volume of spacecraft command and analysis activity. Because communication time to Jupiter will be around 45 minutes, commands must be precisely timed in advance for performance at a particular point over the planet. Five commands can be stored in advance on the spacecraft.

Ten days before periapsis, controllers will command a final Conscan maneuver to establish a precise spacecraft attitude for the flyby. Biases may be included for observed effects of Jovian gravity and the planet's magnetic field, which could cause drift in spacecraft attitude. After this there will be no further thrusting by the spacecraft engines because these might invalidate several of the scientific measurements made near Jupiter.

As the spacecraft is sucked in by the giant planet's gravity, its velocity relative to Jupiter will increase in 20 days from an approach speed of 33,000 km/hr (20,000 mph) to 126,000 km/hr (78,000 mph) at periapsis.

Tracking stations will prepare to handle effects of this four-fold increase in speed on the two-way doppler measurements used to track the spacecraft. Doppler shift will grow very large, and this will be complicated by loss of communication when the spacecraft passes behind the planet after periapsis.

Experimenters will calibrate their instruments before flyby, and will change sensitivity ranges of the high-energy particle, solar wind and magnetic field instruments for the far more intense phenomena at the planet. These experiments will then look for the first Jupiter effects, the bow shock wave in the solar wind, Jupiter's magnetic field, and Jovian effects on streams of high-energy solar particles.

For the flyby itself, *Pioneer F* will be close to Jupiter for about four days (100 hours) with distance from Jupiter's cloud tops at periapsis 140,000 kilometers (87,000 miles). At 50 hours before periapsis, *Pioneer* instruments will see the planet in almost full sunlight. At 40 minutes before periapsis, Jupiter will be half dark, and at periapsis itself about 60 percent of Jupiter will be dark. After periapsis, the spacecraft will continue to see less than a half-phase Jupiter.

The imaging photopolarimeter will have begun making polarization and intensity measurements of Jupiter's reflected light several weeks out from the planet. Starting at about 20 hours, 1.3 million kilometers (800,000 miles), from periapsis, the instrument will begin to alternate these measurements with the taking of images. The polarization and intensity measurements will provide clues on the physical properties of Jupiter's clouds and atmosphere.

The ultraviolet photometer will examine Jupiter's upper atmosphere for hydrogen / helium ratio, atmosphere mixing rate, temperature, and for evidence of an auroral oval near the poles on the planet's day side. It will have a two-hour viewing period 30 hours before periapsis and a second two hours of observation between 26 and eight hours from periapsis.

The infrared radiometer will look for Jovian radiation from an internal heat source, hot spots in the outer atmosphere, and cold spots at the poles indicating ice caps of frozen methane. It will measure the atmosphere's hydrogen / helium ratio. Its observation period will be the last 80 to 90 minutes leading up to periapsis. Throughout its observations, it will have a good view of the boundary between light and dark (the terminator).

Beyond Jupiter

As the spacecraft passes behind the planet for an hour and re-emerges, ground stations will record diffusion effects on its radio signals to calculate the density and composition of Jupiter's atmosphere.

After periapsis, solar wind, particle, and magnetometer experiments will look for planetary effects on the interplanetary medium behind the planet. Going away from the planet, the polarimeter will continue to measure characteristics of sunlight scattered by Jupiter's atmosphere.

After the planet encounter, *Pioneer* will return to its interplanetary cruise mode of operation.

The basic Pioneer F mission is defined to last through encounter and about three months thereafter. (The costs of about $100 million cover only this basic mission.) Although the spacecraft has been designed for the basic mission, it is not possible to predict exactly how long it will continue to function after encounter. Besides the uncertainty of the particle and radiation damage, the basic life expectancy of the parts and the exact rate of power system degradation are also not known for sure. Another variable is the amount of reorientation and velocity adjustment maneuvers that will be needed.

Due to the swingby effect, Jupiter's gravity and orbital velocity will have speeded up the spacecraft to solar system escape speed, and will have bent the spacecraft trajectory inward toward Jupiter's orbital path. Cruise velocity relative to the Sun (not the planet) of 38,500 km/hr (24,000 mph) will have increased to 79,200 km/hr (49,320 mph).

Although *Pioneer*'s objective is to make measurements of Jupiter, it is possible that some "bonus" data will be obtained after leaving the planet.

The most interesting experimental questions beyond Jupiter will be: What is the flux of galactic cosmic rays and the distribution of neutral hydrogen, non-solar-wind plasmas and interstellar hydrogen and helium? What do these tell about the interstellar space beyond the boundary of the heliosphere (the Sun's atmosphere)?

Of equal interest will be the search for the heliosphere boundary itself, "where the solar wind stops blowing" and interstellar space begins. The plasma, magnetic field, high-energy particle, and ultraviolet photometer experiments will share these two searches.

During this post-encounter period, communications distance will grow steadily longer. Incoming signals will become almost infinitesimal and the time required to command the spacecraft and get a response will lengthen to hours.

Only the most intensive and sophisticated efforts by the Deep Space Network will allow communication with the spacecraft at all as it penetrates far out beyond any areas previously explored directly by man and his machines.

The spacecraft will cross the orbit of Saturn at about 1.5 billion kilometers (950 million miles) from the Sun, about five years after launch, and will cross Uranus' orbit at about 2.9 billion kilometers (1.8 billion miles) from the Sun eight years after launch – but communication will not be possible at this great distance.

Pioneer will then continue into interstellar space. Six years after Jupiter encounter, at around 3.2 billion kilometers (two billion miles), the spacecraft flight path will have curved 87 degrees further around the Sun than the Jupiter flyby point. Its velocity relative to the Sun at this point will be 53,300 km/hr (33,100 mph). From then on, its flight path will curve still farther around the Sun, another 25 degrees. At this point, far out in interstellar space it will proceed away from the Sun in essentially a straight line, and its velocity will have dropped to a permanent 41,400 km/hr (25,700 mph) away from the Sun.

The Asteroids

The asteroids travel around the Sun in elliptical orbits like small planets. The Asteroid Belt lies between the orbits of Mars and Jupiter, between distances from the Sun of 270 million to 555 million kilometers (170 million to 345 million miles).

The Belt is a region roughly 280 million kilometers (175 million miles) wide circling the Sun and extending about 40 million kilometers (25 million miles) above and below the plane of the Earth's orbit.

Scientists believe the asteroids condensed individually from the primordial gas cloud which formed the Sun and planets, or that they are debris from the break up of a very small planet. Clearly, they contain important information on the origin of the solar system.

Passage of *Pioneer Jupiter* through the Asteroid Belt will allow the first survey of the density of asteroids too small to be seen by telescope and of fragments and dust in the Belt. This will be of scientific interest but is even more important to exploration missions. Since the Belt is too thick to fly over or under, all outer planet missions must fly through it. And assessment, by an actual flight, of the Belt's hazard to future spacecraft will be the first consideration.

There is estimated to be enough material in the Belt to make a planet with a volume about $1/1,000^{th}$ that of the Earth.

Astronomers have identified and calculated orbits for 1,776 asteroids, and there may be 50,000 in the size range from the largest 770-kilometer (480-mile)-diameter Ceres, down to bodies one mile in diameter.

In addition to asteroids, the Belt is presumed to contain hundreds of thousands of asteroid fragments, and uncountable billions of dust particles ranging down to millionths and billionths of a gram. Two zones of heavier concentrations of fragments and dust are believed to exist at distances from the Sun of 400 and 480 million kilometers (250 and 300 million miles).

While concentrations and sizes of these smaller particles and dust are unknown, one estimate is that an area as large as the United States placed in the Belt would receive impacts by eight particles with a mass of one gram or greater every second; or one pinhead-sized particle with a mass of one millionth of a gram would pass through one square meter every month.

In the center of the Belt, asteroids and particles orbit the Sun at about 17 kilometers (10 miles) per second. These particles would impact the spacecraft (which has its own velocity in somewhat the same direction) at about 48,000 km/hr (about 30,000 mph). In short, asteroidal material is thinly spread, but penetrating.

A few asteroids stray far beyond the Belt. Hermes can come within about 350,000 kilometers (220,000 miles) of the Earth, or closer than the Moon. Icarus, another asteroid comes to within nine million miles of the Sun.

Many meteorites which survive atmosphere entry and land on Earth are believe to be asteroidal material. These meteorites are mostly stony, but some are iron and some contain large amounts of carbon.

How severe a hazard to *Pioneer F* does the Asteroid Belt pose?

One study examined the orbits of 1,735 known asteroids relative to typical flight paths of *Pioneer F* through the Belt. In general, about ten asteroids came within about 14,959,900 kilometers (9,275,138 miles) of the spacecraft.

The closest approach was by an asteroid seven kilometers (4.4 miles) in diameter which came within 4,370,000 kilometers (2,715,000 miles) of the spacecraft. The largest asteroid encountered 23 kilometers (14.3 miles) in diameter, came no closer than 40 million kilometers (25 million miles). Even if the estimated number of "large" asteroids in the Belt is increased 50 times, from 1,700 to 100,000, likelihood of a close approach is insignificant.

In short, the threat of baseball, green pea, or even BB-sized asteroids is probably negligible. The most serious hazard comes from particles of $1/10$ to $1/1,000^{th}$ of a gram mass. Smaller particles are too small to do damage.

Meteoroids with a mass of $1/100^{th}$ gram, which travel at about 54,000 km/hr (33,600 mph) relative to the spacecraft near Earth, can penetrate a single sheet of aluminum one centimeter thick. At Jupiter, these particles travel only about 25,200 km/hr (15,660 mph).

Total numbers of particles in the $1/10$ to $1/1,000^{th}$ gram size range are unknown and can only be estimated. Spacecraft designers believe they have given *Pioneer F* a good safety margin.

Comets

Throughout the Pioneer F mission, cometary particles present some hazard.

At Earth, about half of cometary particles travel in streams which follow the orbital paths of existing or dissipated comets. When these streams (the Leonids, the Perseids, etc.) intersect Earth's orbit, meteor showers result as the particles burn up in the atmosphere. Cometary particle streams orbit the Sun in long ovals. Usually these orbits are tilted out of the ecliptic by 30 degrees or more.

The other half of the cometary particles near the Earth appear sporadically.

Cometary particles have average speeds relative to the spacecraft of 72,000 km/hr (45,000 mph), and a $1/100^{th}$ gram cometary particle can penetrate a one centimeter-thick aluminum sheet. Their speed at Jupiter will be down to 32,400 km/hr (20,000 mph).

Jupiter

What We Know

Though many mysteries concerning this big, brilliant planet remain to be solved, hundreds of years of astronomical observations and analysis have provided a stock of information. Galileo made the first telescopic observations and discovered Jupiter's four larger moons in 1610.

Seen from Earth, Jupiter is the second brightest planet, and fourth brightest object, in the sky. It is 773 million kilometers (480 million miles) from the Sun, and circles the Sun once in just under 12 years. The planet has 12 moons, the four outer ones in backward orbit from the direction most moons go. Two of the moons, Ganymede and Callisto, are about the size of the planet Mercury. Two others, Io and Europa, are similar in size to the Earth's moon.

Jupiter completes a rotation once every 10 hours, the shortest day of any of the nine planets. Because of Jupiter's size, this means that a point at the equator on its visible surface (cloud tops) races along at 35,400 km/hr (22,000 mph), compared to a speed of 1600 km/hr (1,000 mph) for a similar point on Earth.

This tremendous rotational speed (and fluid character of the planet) makes Jupiter bulge at its equator. Jupiter's polar diameter of about 124,000 kilometers (77,000 miles) is 19,000 kilometers (11,800 miles) smaller than its equatorial diameter of about 143,000 kilometers (88,900 miles).

Jupiter's visible surface (cloud top area) is 62 billion square kilometers (23.9 billion square miles). The planet's gravity at cloud top is 2.36 times that of Earth.

The mass of the planet is 318 times the mass of Earth. Its volume is 1,000 times Earth's. Because of the resulting low density (one-fourth of the Earth's or 1.3 times the density of water), most scientists are sure that the planet is made up of a mixture of elements similar to that in the Sun or the primordial gas cloud which formed the Sun and planets. This means there are very large proportions (at least three quarters) of the light gases hydrogen and helium. Scientists have identified hydrogen, deuterium (the heavy isotope of hydrogen), methane (carbon and hydrogen), and ammonia (nitrogen and hydrogen) by spectroscopic studies of Jupiter's clouds.

Jupiter's Visible Surface

NORTH POLAR REGION
NORTH NORTH NORTH TEMPERATE BELT
NORTH NORTH TEMPERATE ZONE
NORTH NORTH TEMPERATE BELT
NORTH TEMPERATE ZONE
NORTH TEMPERATE BELT
NORTH TROPICAL ZONE
NORTH EQUATORIAL BELT
EQUATORIAL ZONE
EQUATORIAL BAND
EQUATORIAL ZONE
NORTH COMPONENT OF SOUTH EQUATORIAL BELT
SOUTH EQUATORIAL BELT
SOUTH COMPONENT OF SOUTH EQUATORIAL BELT
SOUTH TROPICAL ZONE
GREAT RED SPOT
SOUTH TEMPERATE BELT
SOUTH TEMPERATE ZONE
SOUTH SOUTH TEMPERATE BELT
SOUTH SOUTH TEMPERATE ZONE
SOUTH POLAR REGION

NATIONAL AERONAUTICS AND SPACE ADMINISTRATION
AMES RESEARCH CENTER, MOFFETT FIELD, CALIFORNIA

Clouds, Currents, and Visual Appearance. Seen through a telescope, the lighted hemisphere of Jupiter (the only one seen from the Earth) is almost certainly a view of the tops of gigantic regions of towering multi-colored clouds.

Over all, due to its rotation, the planet is striped or banded, parallel with its equator, with large, dusky gray regions at both poles. Usually between the two polar regions are five permanent, bright salmon-colored stripes, known as Zones, and four darker, slate-gray stripes, known as Belts – the South Equatorial Belt, for example. The planet as a whole changes hue periodically, possibly as a result of the Sun's 11-year activity cycles.

The Great Red Spot in the southern hemisphere is frequently bright red, and since 1665 has disappeared completely several times. It seems to brighten and darken at 30-year intervals.

Scientists agree that the cold tops in the Zones are probably largely ammonia vapor and crystals, and the gray polar regions may be condensed methane. The bright cloud Zones have a complete range of colors from yellow and delicate gold to red and bronze. Clouds in the Belts range from gray to blue-gray.

In addition to the Belts and Zones, many smaller features – streaks, wisps, arches, loops, patches, lumps, and spots – can be seen. Most are hundreds of thousands of kilometers in size.

Circulation of these cloud features has been identified in a number of observations. The Great Equatorial Current (The Equatorial Zone), 20 degrees wide, sweeps around the planet 410 km/hr (255 mph) faster than the cloud regions on either side of it, and is like similar atmospheric jet streams on Earth. The South Tropical circulating current is a well-known feature, as is a cloud current which sweeps completely around the Great Red Spot.

Many other features of the circulation of the atmosphere have been observed.

When Jupiter passed in front of stars in 1953 and again in 1971, astronomers were able to calculate roughly the molecular weight of its upper atmosphere by the way it refracted the stars' light. They found a molecular weight of around 3.3 which means a large proportion of hydrogen (molecular weight 2) because all other elements are far heavier. (Helium is 4, carbon 12, and nitrogen 14.)

Under Jovian gravity, atmospheric pressure at the cloud tops is calculated to be up to ten times one atmosphere on Earth.

The transparent atmosphere above the clouds can be observed spectroscopically and in polarized light. It is believed to be at least 60 kilometers (35 miles) thick.

Scientists have suggested from cloud-top observations by telescope that the general circulation pattern of Jupiter's atmosphere is like that of Earth, with circulation zones corresponding to Earth's equatorial, tropical, sub-tropical, temperate, subpolar, and polar regions. However, Jupiter's polar regions (from an atmospheric circulation standpoint) appear to begin at about 26 degree latitude from the equator, instead of at 60 degrees as on Earth.

Magnetic Fields and Radiation Belts. Among the nine planets only Jupiter and Earth are known to have magnetic fields. Evidence for Jupiter's magnetic field and radiation belts comes from its radio emissions. The only phenomenon known that could produce the planet's decimetric (very high frequency) radio waves is trapped electrons gyrating around the lines of such a magnetic field. When such electrons approach the speed of light, they emit radio waves.

Radio emissions indicate that Jupiter's magnetic field is toroidal (doughnut-shaped) with north and south poles like the Earth's. It appears around 20 times as strong as Earth's field and presumably contains high-energy protons (hydrogen nuclei) and electrons trapped from the solar wind. The field's center appears to be near the planet's axis of rotation and south of the equatorial plane. Jupiter's powerful magnetic field can hold more particles trapped from the solar wind than can Earth's field, and it can increase particle energies. As a result, particle concentrations and energies could be up to a million times higher than for Earth's radiation belts, a flux of a billion particles or more per square centimeter per second.

Because Jupiter has such high gravity and is so cold at the top of its atmosphere, the transition region between dense atmosphere and vacuum is very narrow. As a result, the radiation belts may come much closer to the planet than for Earth.

Jovian Radio Signals. Earth receives more radio noise from Jupiter than from any source except the Sun. Jupiter broadcasts three kinds of radio noise: (1) thermal – from the temperature-induced motions of the molecules in its atmosphere (typical wavelength, 3 centimeters); (2) decimetric (centimeter range) – from the gyrations of electrons around the lines of force of the planet's magnetic field (typical wavelength 3-70 centimeters); and (3) decametric (up to 10s of meters) – believed from huge discharges of electricity (like lightning flashes) in Jupiter's ionosphere (wavelength 70 centimeters to 60 meters).

The powerful decametric radio waves originate at known longitudes and have been shown to be related to passages of Jupiter's close moon, Io, whose orbit is 2.5 planet diameters, 350,000 kilometers (about 218,000 miles), above the tops of Jupiter's cloud layer. Some scientists believe that conductivity of Io is sufficient to link up magnetic lines of force to the planet's ionosphere, allowing huge electrical discharges of built up electrical potential.

The power of these regular decametric radio bursts is equal to the power of several hydrogen bombs. Their average peak value is 10,000 times greater than the power of Jupiter's decimetric signals.

Temperature. The average temperature at the tops of Jupiter's clouds appears to be about -145 degrees C (-229 degrees F), based on many observations of Jupiter's infrared radiation. But recent stellar occultation studies indicate that much of the diffuse outer atmosphere is close to room temperature, and that the top layer is at about 20 degrees C (68 degrees F). Only about $1/27$th as much heat from the Sun arrives at Jupiter as arrives at Earth. Recent infrared measurements made from high altitude aircraft suggest that the giant planet radiates about 2.5 to three times more energy than it absorbs from the sun. The question is, what is the source of this energy? The shadows of Jupiter's moons on the planet appear to measure hotter than surrounding sunlit regions.

Jupiter Unknowns

Scientists are reasonably sure of most of the preceding phenomena and observations, though it will be important to check them at close range with *Pioneer F*. They have few explanations of these observed phenomena, and they know very little about other aspects of the planet. What is hidden under the heavy Jovian clouds? How intense are the radiation belts? Most of the statements which follow are scientific guesses or very general approximations.

Life. Perhaps the most intriguing unknown is the possible presence of life in Jupiter's atmosphere.

Estimates of the depth of the Jovian atmosphere beneath the cloud layer vary from 100 to 6,000 kilometers (60 to 3,600 miles). The compositions and interactions of the gases making up the atmosphere are unknown. If the atmosphere is deep, it must also be dense. By one estimate, with an atmospheric depth of 4,200 kilometers (2,600 miles), pressure at the Jovian "surface" would be 200,000 times Earth's atmospheric pressure due to the total weight of gas in the high Jovian gravity. One source cites eight different proposed models of Jupiter's atmosphere.

However, scientists do appear to agree on the presence of liquid water droplets in the atmosphere. Since the planet is believed to have a mixture of elements similar to that found in the Sun, it is almost sure to have abundant oxygen. And most of this oxygen has probably combined with the abundant Jovian hydrogen as water.

If large regions of Jupiter's atmosphere come close to room temperature, both liquid water and water ice should be present.

Jupiter's atmosphere contains ammonia, methane, and hydrogen. These constituents, along with water, are the chemical ingredients of the primordial "soup" believed to have produced the first life on Earth by chemical evolution. On this evidence, Jupiter could contain the building blocks of life.

Some scientists suggest that the planet may be like a huge factory turning out vast amounts of life-supporting chemicals (complex carbon-based compounds) from these raw materials, using its own internal energy. If so, life could exist without photosynthesis. Any solar photosynthesis would have to be at a very low level since Jupiter receives only $1/27th$ of Earth's solar energy. It would probably be low-energy life forms at most (plants and microorganisms) because there is believed to be no free oxygen. Life forms would float or swim because a solid surface, if any, would be deep within Jupiter at very high pressures.

Planet Structure. While there are wide differences among scientists on planet structure, most proposed models contain elements like the following:

Going down, it is believed that temperature rises steadily. The cloud tops may consist of super-cold ammonia crystals, underlaid by a layer of ammonia droplets, under which may be a region of ammonia vapor. Below this may be layers of ice crystals, water droplets, and water vapor. Below this is either the planet's solid surface, or liquid hydrogen oceans. Still lower is a region of metallic hydrogen created by high Jovian gravity with perhaps a core of rocky silicates and metallic elements. The core might be ten times the mass of the Earth by one estimate.

Some theorists doubt that the planet has any solid material at all, but is entirely liquid. Others propose a gradual thickening from slush to more rigid material.

Most planetologists think the Great Red Spot may be a column of gas, the center of an enormous vortex, rising from Jupiter's surface to the top of the atmosphere. Known in physics as a Taylor column, such a vortex would have to be anchored by a prominent surface feature, either a high spot or huge depression. The Red Spot has circled the planet more than once relative to visible cloud features in the past hundred years.

Model of Jupiter Interior

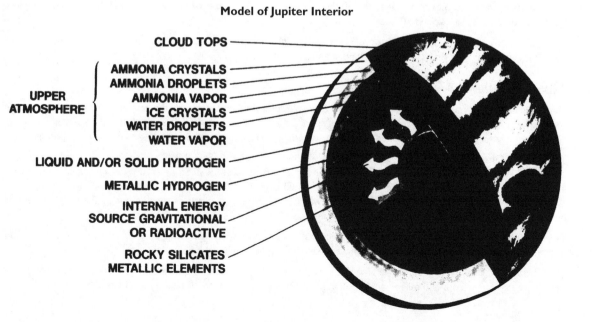

CLOUD TOPS

UPPER ATMOSPHERE
- AMMONIA CRYSTALS
- AMMONIA DROPLETS
- AMMONIA VAPOR
- ICE CRYSTALS
- WATER DROPLETS
- WATER VAPOR

LIQUID AND/OR SOLID HYDROGEN

METALLIC HYDROGEN

INTERNAL ENERGY SOURCE GRAVITATIONAL OR RADIOACTIVE

ROCKY SILICATES METALLIC ELEMENTS

Some scientists suggest that Jupiter's surface itself may be rotating at varying rates of speed relative to the atmosphere, thus moving the Spot. The rotation of the planet's magnetic field, tracked by the timing of its radio emissions, is believed the best measure of rotation rate of the planet.

A Hot Planet. Theorists suggest that Jupiter is almost large enough to be a small star. Because of measurements of excess heat radiated by the planet, current Jupiter theories call for a relatively hot core, compared with earlier ideas of a super-cold interior.

One hypothesis holds that despite the five billion years since formation of the planets, Jupiter has not completed its gravitational condensation. This continued settling toward the center (as little as one millimeter per year) could produce the required heat energy. If the planet has a rocky core, some of the heating could be from decay of radioactive material in the core.

Magnetic Field. Internal heat also could explain the magnetic field. A hot core might be a fluid core. Convective and rotational motion of electrically conducting fluids at temperatures of 10,000 degrees C (18,000 degrees F) may generate the field. Or it could be generated by conductive atmosphere layers below the clouds which would store and later release energy.

With knowledge of Jupiter's magnetic field, scientists should be able to make inferences about the planet's internal structure, particularly its fluid component.

Moons. Jupiter's 12 natural satellites have some odd characteristics. The second moon, Io, appears to be brighter for 10 minutes after emerging from Jupiter's shadow. If so, the simplest explanation, supported by recent stellar occultation observations, is that Io has an atmosphere (probably nitrogen or methane) which "snows out" on the surface when Io is on the cold, dark side and revaporizes when back in sunlight. Io also is distinctly orange in color and has odd reflecting properties. Most of the sunlight reaching the moon's surface is strongly scattered back, making Io extremely bright. This reflection property is believed to be more pronounced for Io than for any other known object in the solar system.

The inner moons in order of distance from the planet are: tiny Amalthea, diameter 160 kilometers (100 miles), which orbits Jupiter twice a day at only 1.5 planet diameters, 106,000 kilometers (66,000 miles), above the cloud tops; the four large moons Io, Europa, Ganymede, and Callisto, whose orbits lie between 422,000 kilometers (262,000 miles) and 1,882,000 kilometers (1,117,000 miles) from Jupiter.

Beyond those are the seven tiny outer moons. The inner three of these, Hestia, Hera, and Demeter, have orbits which lie between 11.5 and 11.7 million kilometers (7.2 and 7.3 million miles) from Jupiter.

Orbits of the four outermost, Andrastea, Pan, Poseidon, and Hades, lie between 20.7 million and 23.7 million kilometers (12.9 and 14.7 million miles) of Jupiter. All are in retrograde orbits, moving counter to the usual direction of planet rotation. This suggests that they may be asteroids captured by Jupiter's powerful gravity. Diameters of six of the outer moons range from 15 to 40 kilometers (9 to 24 miles), with Hestia, the seventh, having a diameter of around 130 kilometers (80 miles).

Orbital periods of the four large inner moons range from 1.7 days (for Io) to 16.7 days. Orbital periods of the inner three of the outer seven moons are around 250 days. While the four farthest-out, backward-orbiting moons complete their circuits of Jupiter in around 700 days.

The backward orbit of the far outer moon, Poseidon, is highly inclined to the equator, and wanders so much due to various gravitational pulls that astronomers have a difficult time finding it.

The Heliosphere

Pioneer F may explore much of the heliosphere, the atmosphere of the Sun created by the gases flowing outward from the solar surface. The spacecraft may survey a region of space 2.4 billion kilometers (1.5 billion miles) wide between the orbits of Earth and Uranus. It has a good chance of learning something about the interstellar space outside the heliosphere.

The thinly diffused solar atmosphere is hundreds of times less dense than the best vacuums on Earth. Yet it is important. The ionized gas known as the solar wind, a 50-50 mixture of protons (hydrogen nuclei) and electrons, flows out from the 3,600,000 degree F (2,000,000 degree C) corona of the Sun in all directions at average speeds of 1.6 million km/hr (one million mph). Solar cosmic rays which are high-energy particles thrown out by the huge explosions on the Sun's rotation. It contains complex electric fields.

The solar wind interacts with itself through collisions of individual solar wind streams. It interacts with solar and galactic cosmic ray particles, and with the planets, their atmospheres, magnetic fields, and satellites.

Where does the solar wind stop blowing? Where is the boundary between the solar atmosphere and the interstellar gas? Estimates range from 300 million to 16 billion kilometers (200 million to 10 billion miles) from the Sun.

In addition to the planets, moons, asteroids, comets and dust, the heliosphere is traversed by electromagnetic radiation from the Sun, radio waves, infrared and ultraviolet waves, and visible light. Earth gets most of its energy from this radiation.

Interplanetary space also contains cosmic ray particles coming from elsewhere in the galaxy at near light speeds. This means enormous particle energies, up to 10^{20} (1 followed by 20 zeros) electron-volts. There are also neutral hydrogen atoms, believed part of the interstellar gas from which the Sun and planets formed.

The study of these interplanetary phenomena has a variety of practical effects and research applications. For example, storms of solar particles striking Earth interrupt radio communications and sometimes electric power transmission.

The heliosphere can be thought of as a huge particle physics laboratory where many phenomena occur that are impossible to produce and study on Earth. For example, man is unable to accelerate nuclear particles in Earth laboratories to the near light speeds reached by galactic cosmic ray particles. These particles will be studied by *Pioneer* instruments.

Solar wind particles are so thinly spread that they rarely collide and only the magnetic field and electrostatic fields link them together as a collisionless ionized gas (collisionless plasma). Even though collisionless, these plasmas transmit waves of many kinds because of this magnetic field linkage.

Knowledge of these plasmas sheds light on an area of current high interest – how to contain plasmas in magnetic fields. Doing this would allow control of the hydrogen thermonuclear reaction continuously on a small scale to generate virtually unlimited electric power by a clean process.

All this involves specific relationships such as the fact that temperatures of solar wind ions and electrons near Earth are almost the same, while theory predicts that the electrons should be several hundred times hotter.

How is energy transferred between them to equalize the temperatures? How far out from the Sun do these temperatures remain equal? Why? Knowledge of plasmas also helps us understand such cosmic phenomena as star formation.

Solar Wind, Magnetic Field, and Solar Cosmic Rays

Near the Earth, the speed of the solar wind varies from one to three million km/hr (600,000 to 2,000,000 mph), depending on activity of the Sun. Its temperature varies from 10,000 degrees to 1,000,000 degrees C (18,000 to 1,800,000 degrees F). Near the Earth, collisions between streams of the solar wind use up about 25 percent of its energy. The wind also fluctuates due to features of the rotating solar corona, where it originates, and because of various wave phenomena.

The fastest solar cosmic ray particles jet out from the Sun in streams which travel in nearly straight lines, covering the 150 million kilometers (93 million miles) to the Earth in as little as 20 minutes. Slower particle streams take one or more hours to reach Earth and tend to follow the curving interplanetary magnetic field.

The positive ions are 90 percent protons and ten percent helium nuclei, with occasional nuclei of heavier elements.

There are from none to 20 flare events on the Sun each year which produce high-energy solar particles, with the largest number of flares at the peak of the 11-year cycle.

Depending on the effect of the heliosphere, *Pioneer F* may find whether the boundary of the solar atmosphere (the end of the wind) is a shock wave or a turbulent region. It will seek to learn if the structure of the solar magnetic field breaks down beyond Mars' orbit.

Planetary Interactions and Neutral Hydrogen

Since Jupiter is expected to have a magnetic field like that of Earth, a bow shock wave should form in the solar wind in front of the planet. There should be a magnetic envelope around the planet, shutting out the solar wind, and a trailing magnetic tail.

Pioneer F instruments should easily characterize ionized particles swept from the atmospheres of Jupiter, its Moons, or the asteroids because their properties will greatly differ from those of particles of solar origin.

Pioneer will look for helium, as well as neutral hydrogen. Helium would be part of the interstellar gas which has forced its way into the heliosphere as the solar system moves through it at 72,000 km/hr (45,000 mph). Hydrogen is believed to be an indirect result of the collision of interplanetary hydrogen with the heliosphere. *Pioneer F* also will measure the ultraviolet glow of the interstellar gas beyond the heliosphere boundary which penetrates the inner solar system.

Galactic Cosmic Rays

Galactic cosmic ray particles usually have far higher energies (velocities) than solar cosmic rays. These particles may get their tremendous energies from the explosion of stars (supernovas), the collapse of stars (pulsars), or acceleration in the colliding magnetic fields of two stars. Pioneer studies of these particles may settle questions of their origin in our galaxy and important features of the origin and evolution of the galaxy itself. These studies should answer such questions as the chemical composition of stellar sources of cosmic ray particles in the galaxy.

Near Earth, numbers of these galactic cosmic ray particles can vary up to 50 percent as solar activity pushes them out of the inner solar system at the peak of the 11-year solar cycle. Particle intensity varies up to 30 percent due to individual solar flares. With increases in solar activity, the solar wind and magnetic field push these particles out of the inner solar system.

Galactic cosmic rays consist of: protons (hydrogen nuclei), 85 percent; helium nuclei, 13 percent; nuclei of other elements, two percent; and high-energy electrons, one percent.

Particles range in energy from 100 million electron-volts (MeV) to 10^{14} MeV.

Near Earth the average flux of these particles is four per square centimeter per second, with most particles in the 1,000 MeV range. Presumably as the spacecraft moves toward the edge of the blocking solar atmosphere, more of these medium-energy particles will be observed.

The Spacecraft

Pioneer Jupiter is the first spacecraft designed to travel into the outer solar system and operate there, possibly for as long as seven years and as far from the Sun as 2.4 billion kilometers (1.5 billion miles).

For this mission, the spacecraft must have extreme reliability, be of very light weight, have a communications system for extreme distances, and employ a non-solar power source.

The spacecraft is stabilized in space like a gyroscope by its five-rpm rotation, so that its scientific instruments scan a full circle five times a minute. Designers chose spin stabilization for its simplicity and effectiveness.

Since the orbit planes of Earth and Jupiter coincide to about one degree, the spacecraft will be in or near Earth's orbit plane (the ecliptic) throughout its flight. To maintain a known orientation in this plane, the spacecraft spin axis points constantly at Earth. The spin axis always coincides with the center line of the radio beam, which is locked constantly on Earth.

Spacecraft navigation is handled on Earth by two-way doppler tracking and by angle-tracking.

For mid-course corrections, the *Pioneer* propulsion system can make changes in velocity totaling 720 km/hr (420 mph).

Pioneer F / G Spacecraft

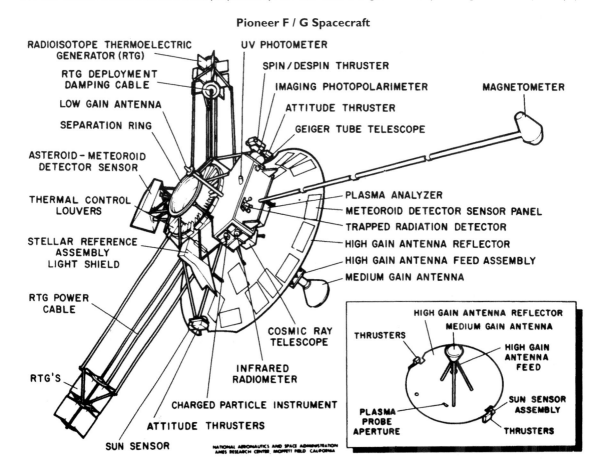

The spacecraft can return a maximum of 2,048 data bits per second (bps) to the Earth, 1,024 bps from Jupiter distance, to the 64-meter (210-foot) antennas of the Deep Space Network. It can store up to 49,152 data bits while other data is being transmitted.

Pioneer F is controlled largely from the Earth rather than by sequences of commands stored in onboard computers.

Launch energy requirements to reach Jupiter are far higher than for a Mars mission, so the spacecraft had to be very light. *Pioneer F* weighs only 270 kilograms (570 pounds). This includes 30 kilograms (65 pounds) of scientific instruments, and 27 kilograms (60 pounds) of propellant for attitude changes and mid-course corrections.

Because solar energy at Jupiter is only four percent of energy received at the Earth and grows steadily weaker beyond the planet, designers selected a nuclear power source over solar cells.

The mission's 13 experiments are carried out by 11 onboard scientific instruments. Two experiments use the spacecraft and its radio signal as instruments.

Pioneer Jupiter employs the reliable internal systems proved out in the *Pioneer 6 to 9* spacecraft. All four are still operating in interplanetary space. *Pioneer 6* is in its seventh year. *Pioneers 6 to 9* were built and are operated by the same team which is carrying out the Jupiter missions.

For reliability, spacecraft builders have employed an intensive screening and testing program for parts and materials. They have selected components designed to withstand radiation from the spacecraft's nuclear power source, and from Jupiter's radiation belts. In addition, key systems are redundant. (That is, two of the same component or subsystem are provided in case one fails.) Communications, command and data return systems, propulsion electronics, thrusters, and attitude sensors are redundant. Virtually all spacecraft systems reflect the need to survive and return data for many years a long way from the Sun and the Earth.

Pioneer F Description

The spacecraft fits within the 3-meter (10-foot) diameter shroud of the Atlas-Centaur launch vehicle with booms retracted, and with its dish antenna facing forward (upward). The Earth- facing dish antenna is designated the forward end of the spacecraft. *Pioneer F* is 2.9 meters (9.5 feet) long, measuring from its farthest forward component, the medium-gain antenna horn, to its farthest rearward point, the tip of the aft-facing omnidirectional antenna. Exclusive of booms, its widest crosswise dimension is the 2.7-meter (9-foot) diameter of the dish antenna.

The axis of spacecraft rotation and the center-line of the dish antenna are parallel, and *Pioneer* spins constantly for stability.

The spacecraft equipment compartment consists of a flat box, top and bottom of which are regular hexagons. This hexagonal box is roughly 35.5 cm (14 inches) deep and each of its six sides is 71 cm (28 inches) long. One side joins to a smaller box also 35.5 cm (14 inches) deep, whose top and bottom are irregular hexagons.

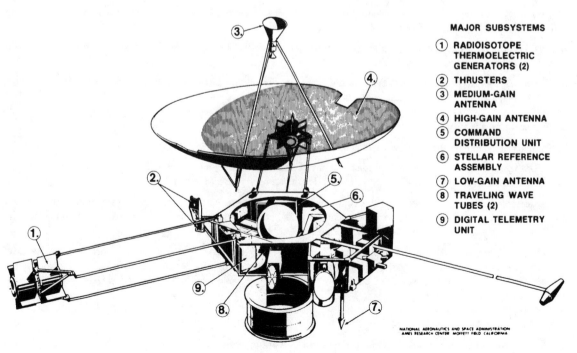

MAJOR SUBSYSTEMS

1. RADIOISOTOPE THERMOELECTRIC GENERATORS (2)
2. THRUSTERS
3. MEDIUM-GAIN ANTENNA
4. HIGH-GAIN ANTENNA
5. COMMAND DISTRIBUTION UNIT
6. STELLAR REFERENCE ASSEMBLY
7. LOW-GAIN ANTENNA
8. TRAVELING WAVE TUBES (2)
9. DIGITAL TELEMETRY UNIT

NATIONAL AERONAUTICS AND SPACE ADMINISTRATION
AMES RESEARCH CENTER MOFFETT FIELD CALIFORNIA

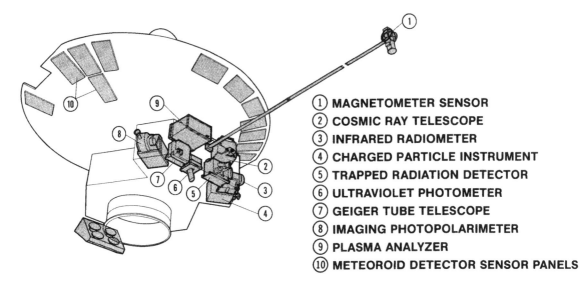

1. MAGNETOMETER SENSOR
2. COSMIC RAY TELESCOPE
3. INFRARED RADIOMETER
4. CHARGED PARTICLE INSTRUMENT
5. TRAPPED RADIATION DETECTOR
6. ULTRAVIOLET PHOTOMETER
7. GEIGER TUBE TELESCOPE
8. IMAGING PHOTOPOLARIMETER
9. PLASMA ANALYZER
10. METEOROID DETECTOR SENSOR PANELS

This smaller compartment contains most of the 11 onboard scientific experiments. However, 12.7 kilograms (28 pounds) of the 30 kilograms (65 pounds) of scientific instruments (the plasma analyzer, cosmic ray telescope, asteroid-meteoroid telescopes, meteoroid sensors, and the magnetometer sensor) are mounted outside the instrument compartment. The other experiments have openings cut for their sensors to look out. Together both compartments provide 1.4 square meters (16 square feet) of platform area.

Attached to the hexagonal front face of the equipment compartment is the 2.7-meter (9-foot)-diameter 46-centimeter (18-inch)-deep dish antenna.

The high-gain antenna feed and the medium-gain antenna horn are mounted at the focal point of the antenna dish on three struts projecting about 1.2 meters (4 feet) forward of the rim of the dish. The low-gain, omnidirectional spiral antenna extends about 0.76 meters (2.5 feet) behind the equipment compartment.

Two three-rod trusses, 120° apart, project from two sides of the equipment compartment, deploying the spacecraft's nuclear electric power generator about 3 meters (10 feet) from the center of the spacecraft. A third boom, 120° from each of the other two, projects from the experiment compartment and positions the magnetometer sensor 6.6 meters (21.5 feet) from the spacecraft center. The booms are extended after launch.

At the rim of the antenna dish, a Sun sensor is mounted. A star sensor looks through an opening in the equipment compartment and is protected from sunlight by a hood.

Both compartments have aluminum frames with bottoms and side walls of aluminum honeycomb. The dish antenna is made of aluminum honeycomb sandwich.

Rigid external tubular trusswork supports the dish antenna, three pairs of thrusters located near the rim of the dish, boom mounts, and launch vehicle attachment ring.

Orientation and Navigation

The spacecraft communications system also is used to orient the *Pioneer* in space.

Heart of the communications system is the spacecraft's fixed dish antenna. This antenna is as large as diameter of the launch vehicle permits. It focuses the radio signal on Earth in a narrow beam.

The spacecraft spin axis is aligned with the center lines of its dish antenna and of the spacecraft radio beam. Except during course changes, all three are locked on Earth throughout the mission. *Pioneer F* maintains a known attitude in the Earth's orbit plane by this continuous Earth point.

For navigation, analysts will use the doppler shift in frequency of the *Pioneer* radio signal and angle-tracking by DSN antennas to calculate continuously the speed, distance, and direction of the spacecraft from Earth. Motion of the spacecraft away from Earth causes the frequency of the spacecraft radio signals to drop and wavelength to increase. This is known as the doppler shift.

Propulsion and Attitude Control

The propulsion and attitude control system provides three types of maneuvers.

Navigation

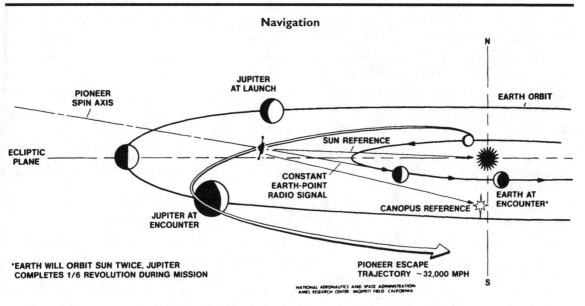

*EARTH WILL ORBIT SUN TWICE, JUPITER
COMPLETES 1/6 REVOLUTION DURING MISSION

PIONEER ESCAPE
TRAJECTORY ~32,000 MPH

NATIONAL AERONAUTICS AND SPACE ADMINISTRATION
AMES RESEARCH CENTER MOFFETT FIELD CALIFORNIA

Pioneer / Jupiter Attitude Control and Propulsion

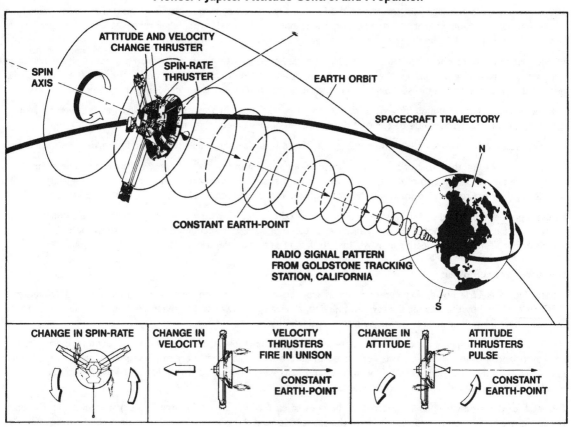

It can change velocity, thus altering course to adjust the place and time of arrival at Jupiter.

It can change the attitude of the spacecraft in space, either to point thrusters in the right direction for velocity-change thrusts or to keep the spacecraft narrow-beam antenna pointed precisely at the Earth. Controllers will command about 150 of these Earth-point adjustments as *Pioneer F* and Earth constantly change positions in space.

And the system maintains spacecraft spin at 4.8 rpm.

Over the entire mission, the propulsion and attitude control system can make changes in spacecraft velocity totaling 720 km/hr (420 mph), attitude changes totaling 1200° (almost four full rotations of the spin axis), and total spin-rate adjustments of 50 rpm. These capacities should be substantially more than needed.

The system employs six thruster nozzles which can be fired steadily or pulsed and have 1.4 to 0.4 pounds of thrust each. Electronics, thrusters, and attitude sensors are fully redundant.

For both attitude and velocity changes, two thruster pairs have been placed on opposite sides of the dish antenna rim. One thruster of each pair points forward, the other aft.

To change attitude, the spacecraft spin axis can be rotated in any desired direction. This is done using two nozzles, one on each side the dish antenna.

One is fired forward, one aft, in momentary thrust pulses at just one point on the circle of spacecraft rotation. Each thrust pulse is timed by a signal which comes once each spacecraft rotation. These signals originate either from a sensor which sees the star Canopus, once per rotation, or from one of the two Sun-sensors which see the Sun once per rotation. Each thrust pulse turns the spacecraft and its spin axis a few tenths of a degree, until the desired attitude is reached.

For velocity changes, the spin axis is rotated until it points in the direction of the desired velocity change. Then two thruster nozzles, one on each side of the dish, but both facing forward or both facing aft, are fired continuously.

These thrusts can thus be in any direction relative to *Pioneer*'s direction of travel, depending on the velocity change needed.

Flight directors can command these velocity change maneuvers directly in real time. Or they can put commands for the maneuvers into the system's 5-command storage register. Rotation of the spin axis, velocity change thrust, and rotation back to Earth-point all take place automatically. This automatic sequence may be important at very long distances where, if Earth-point were lost, recovery of communication with the spacecraft might be difficult.

Lock-on of the spacecraft high-gain antenna to the Earth is maintained by making use of the fact that the movable feed for the antenna dish can be offset one degree from the spin axis. This means that when the spin axis is slightly off exact Earth-point, signals received by the spacecraft from Earth vary up and down in strength.

An automatic system on the spacecraft, known as Conscan, changes attitude in a direction to reduce this variation in signal strength, returning the spin axis again to precise Earth-point. An onboard signal processor uses the varying radio signal to time attitude change thrusts. Spatial reference for these maneuvers is provided by either Sun or star sights.

Earth-point also can be trimmed by ground command, using variations in strength of the signal as received by ground stations. This variation also is due to offset of the spacecraft antenna feed.

The spacecraft medium-gain horn is permanently offset 9° to the spin axis and can also be used for Earth-point maneuvers.

To change spin rate, a third pair of thrusters, also set along the rim of the dish antenna, will be used. These thrusters are on a line tangent to the rim of the antenna (the circle of spacecraft rotation). One thruster points against the spin direction and is fired to increase spin rate; the other points with the direction of spin and is fired to reduce spin.

Thrust is provided by liquid hydrazine, which is decomposed into gas by a catalyst in the chamber of each thruster and then ejected from thruster nozzles. The hydrazine is stored in a single 42-centimeter (16.5-inch)-diameter pressurized spherical tank. Tank and connecting lines will be kept from freezing by electric heaters.

Nuclear-Electric Power

Nuclear-fueled electric power for *Pioneer F* comes from four SNAP-19 type radioisotope thermoelectric generators (RTGs), developed by the Atomic Energy Commission, similar to those used to power the Nimbus-3 satellite. These units turn heat from their nuclear power source into electricity.

SNAP-19 has been designed for maximum safety in case of a launch pad explosion, abort, or reentry. Before its use was authorized, a thorough review was conducted to assure the health and safety of personnel involved in the launch and of the general public. Extensive safety analyses and tests were conducted which demonstrated that the fuel would be safely contained under almost all credible accident conditions.

Two each of these power units are located at the ends of each of the two 2.7-meter (9-foot) spacecraft RTG booms. The RTGs are located on the opposite side of the spacecraft from the experiment compartment to minimize effects of their neutron radiation on the scientific instruments. Design of experiments and spacecraft equipment has been such as to counter effects of RTG radiation.

The four RTGs are expected to provide 155 watts of power at launch, about 140 watts at Jupiter, more than 100 watts five years after launch. Spacecraft power requirements at Jupiter will be slightly over 100 watts, of which 26 watts is for the scientific instruments.

The RTGs are fueled with plutonium-238 dioxide. Thermoelectric converters consist of banks of thermoelectric couples surrounding the cylindrical fuel compartment, 90 for each of the four RTGs. The couples are made of lead telluride for negative elements and an alloy of tellurium, silver, germanium, and antimony for the positive elements.

SNAP-19 / Pioneer Radioisotope Thermoelectric Generator

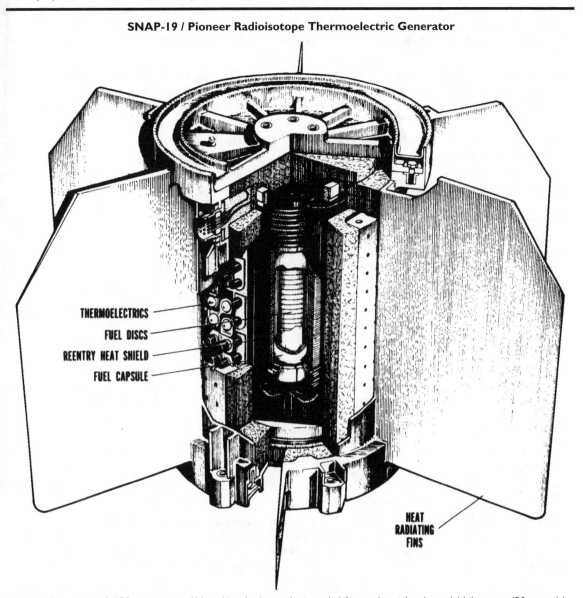

THERMOELECTRICS

FUEL DISCS

REENTRY HEAT SHIELD

FUEL CAPSULE

HEAT
RADIATING
FINS

Each RTG consists of a 28-centimeter (11-inch) cylinder with six radial fins and weighs about 14 kilograms (30 pounds).

Electric power is distributed throughout the spacecraft. It goes from each RTG through one of four inverters to form a main alternating current (AC) bus. Most of this AC power is rectified to supply the main 28-volt DC bus, with excess dumped through an external heat radiator. A battery automatically carries overloads, and is recharged when power is available. The scientific instruments and the radio transmitter's traveling wave tube amplifiers receive power from the main DC bus. Most other spacecraft systems are supplied from the central transformer, which receives power from the AC bus and provides various DC output voltages.

Communications

The communications system provides for two-way communication between Earth and the spacecraft. For reliability, this system is fully redundant.

The system depends on the sensitivity of the DSN's 64-meter (210-foot) antennas and their receivers which can hear the very low amounts of energy emanating from *Pioneer* at Jupiter. When used in reverse as transmitters, the 64-meter antennas have such precision and radiated power (up to 400 million watts) that outgoing commands are still of sufficient strength to be received when they reach the spacecraft.

The spacecraft system consists of high-gain, medium-gain, and low-gain antennas, used for both sending and receiving. The high-gain antenna is the spacecraft's parabolic reflector dish; the medium-gain antenna is an Earth-facing horn mounted along with the high-gain antenna feed on struts at the focus of the dish. The low-gain antenna is a spiral, pointed to the rear, designed to provide communication at the few times when the aft end of the spacecraft is pointing toward Earth.

Each antenna is always connected to one of the two spacecraft radio receivers, and the two receivers are interchangeable by command, or automatically after a certain period of inactivity, so that if one receiver fails the other can take over.

For transmitting, the high-gain antenna produces a maximum gain of 38 decibels and has a 3.3° beam width; the medium-gain antenna, a gain of 12 decibels and beam width of 32°. Other components for transmitting are two radio transmitters and two traveling-wave-tube power amplifiers (TWTs). The TWTs weigh only 1.75 kilograms (3.85 pounds) but produce 8 watts of power in S-band.

In addition to commands and data return, the communications system provides data for doppler tracking and maintaining spacecraft Earth-point.

Communication is on S-band frequencies, uplink to the spacecraft at 2110 and downlink at 2292 MHz. In the Asteroid Belt, the system can return 2,048 data bits per second to the DSN's 64-meter dish antennas, and at Jupiter range 1,024 bps.

Command System

Controllers use 222 different commands to operate the spacecraft. The command system consists of two Command Decoders and a Command Distribution Unit. For reliability, two redundant Command Decoders are provided, and redundant circuits are provided throughout the logic of the Command Distribution Unit.

Commands are transmitted to the spacecraft at a rate of one data bit per second. A single command message has 22 bits.

The Command Distribution Unit routes commands to any one of 222 destinations on the spacecraft. Seventy-three of these operate the experiments, the remaining 149, spacecraft systems. The science commands include moving the photometer telescope for pictures of the planet, calibrating instruments, and changing data types. Up to five commands can be stored for later execution.

If a command is not properly verified by the Decoder, the Command Distribution Unit does not act on the command, thus reducing the possibility of executing wrong commands. Commands also are verified by computer on the ground before transmission.

Data Handling

The spacecraft data system turns science and engineering information into an organized stream of data bits for radio transmission back to Earth.

The system also can store up to 49,152 data bits for later transmission to Earth. Storage can take place while other data is being transmitted. This allows for taking data faster than the spacecraft can transmit it.

System components include the Digital Telemetry Unit, which prepares the data for transmission in one of 11 data formats at one of eight bit rates of from 16 to 2,048 bits per second; the Data Storage Unit; and the Convolutional Coder, which rearranges the data in sequential form that allows identification and correction of errors by computer on the ground. This coding allows higher bit rates to be retained for greater spacecraft distances without loss of accuracy.

The 11 data formats are divided into science and engineering groups. The science group includes two basic science formats and three special-purpose science formats. The basic science formats each contain 192 bits, 144 of which are assigned to the scientific instruments; the remainder are for operating instructions of various kinds. One of the basic science formats is primarily for interplanetary flight, the other for Jupiter encounter.

The three special-purpose science formats transmit all 192 bits of data from only one or two instruments. This allows high data rates from one instrument such as the photometer during Jupiter picture taking, or the infrared radiometer whose principal purpose is to take a careful look at Jupiter's heat radiation just before closest approach.

There are four engineering formats, each of 192 bits. Each of the engineering formats specializes in one of the following areas: data handling, electrical, communications, and orientation and propulsion. These specialized data formats provide high bit rates during critical spacecraft events.

Short samples of science data can be inserted into engineering formats and vice versa.

The Digital Telemetry Unit, which puts all data together, has redundant circuits for critical functions. It contains a stable crystal-controlled clock for all spacecraft timing signals, including those for various orientation maneuvers and experiments.

Temperature Control

Temperature on the spacecraft is controlled at between -25° and 40° C (-10° and 100° F) inside the scientific instrument compartment, and at various other levels elsewhere for satisfactory operation of the spacecraft equipment.

The temperature control system will have to cope with gradually decreasing heating as the spacecraft moves away from the Sun and with two frigid periods, one when *Pioneer F* passes through Earth's shadow at launch; the other during passage through Jupiter's shadow during fly-by. The system also will have to handle heat from the third stage engine, atmosphere friction, spacecraft nuclear electric power generators, and other equipment.

The equipment compartments are insulated by multi-layered blankets of aluminized Mylar or Kapton. Temperature-responsive louvers at the bottom of the equipment compartment, opened by bimetallic springs, allow controlled escape of excess heat. Other equipment has individual thermal insulation and is warmed by electric heaters and 12 radioisotope heaters, 3 fueled with plutonium-238, developed by the Atomic Energy Commission.

Magnetic Cleanliness

Pioneer F is among the most magnetically "clean" spacecraft ever built. It has been made so to provide the least interference with the magnetometer as it measures the weak interplanetary magnetic field. Since virtually any electric current flow within the spacecraft can produce a magnetic field, demagnetizing is difficult. A wide range of special components and techniques have been used in the electrical systems to achieve this low spacecraft magnetism of 0.03 gamma at the magnetometer sensor.

Reliability

In addition to reliability measures already described, the spacecraft has these added design characteristics:

It can do most mission tasks in a variety of ways. For example, several combinations of thrusters can be used for attitude and velocity changes.

Besides redundant components and systems, many other components have redundant internal circuitry. Virtually all parts and components have been selected from those with a known and documented history of successful operation in space.

Virtually all electronic components have been "burned in" by having current applied to them before installation. For example, all transistors have been run for 200 hours. Engineers have tracked all test anomalies and failures to their source via a rigid, formal control system. NASA project engineers have intensively monitored all contractor test and system integration activities.

The Experiments

The basic measurements made by the 13 *Pioneer F* experiments will produce important findings about Jupiter, the Asteroid Belt, and the heliosphere.

Specific mission objectives for the 13 experiments are:

For interplanetary space:

- To map the interplanetary magnetic field
- To study the radial gradient of the solar wind and its fluctuations and structure
- To study the radial and transverse gradients and arrival directions of high-energy charged particles (solar and galactic cosmic rays)
- To investigate the relationships between the solar wind, magnetic field, and cosmic rays
- To search for the boundary and shape of the heliosphere (solar atmosphere)
- To determine the density of neutral hydrogen
- To determine the properties of interplanetary dust

For the Asteroid Belt:

- To determine the size, mass, flux, velocity, and orbital characteristics of the smaller particles in the Belt
- To determine Asteroid Belt hazard to spacecraft

For Jupiter:

- To map the magnetic field
- To measure distributions of high-energy electrons and protons in the radiation belts, and look for auroras
- To find a basis for interpreting the decimetric and decametric radio emission from Jupiter
- To detect and measure the bow shock and magnetospheric boundary and their interactions with the solar wind
- To verify the thermal balance and to determine temperature distribution of the outer atmosphere
- To measure the hydrogen / helium ratio in the atmosphere
- To measure the structure of the ionosphere and atmosphere
- To measure the brightness, color and polarization of Jupiter's reflected light
- To perform two-color visible light imaging
- To increase the accuracy of orbit predictions and masses of Jupiter and its moons.

Some of the instruments will return data on all three areas of *Pioneer* exploration, but they fall roughly into the categories that follow:

Magnetic Fields

Magnetometer (Jet Propulsion Laboratory). This sensitive instrument will measure the interplanetary magnetic field in three axes from the orbit of Earth out to the limits of spacecraft communication. It will study solar wind interaction with Jupiter, map Jupiter's strong magnetic fields at all longitudes and many latitudes. It may be able to use these data to infer the fluid composition and other characteristics of Jupiter's interior. It will study the relationship of the fields to Jupiter's Moons, and will look for the heliosphere boundary. The magnetic field is carried by the solar wind and measurements of both field and wind are needed to understand either.

The instrument is a helium vector magnetometer similar to that flown successfully on *Mariners* 4 and 5. Its sensor is mounted on a light-weight mast extending 6.5 meters (21.5 feet) from the center of the spacecraft to minimize interference of spacecraft fields.

The instrument operates in any one of eight different ranges, the lowest covering magnetic fields up to 2.5 gamma; the highest, fields up to 1.4 gauss (140,000 gamma). Earth's surface field is 50,000 gamma. These ranges are selected by ground command or automatically by the instrument itself. The magnetometer can measure fields as weak as 0.01 gamma.

Its sensor is a cell filled with helium, excited by radio frequencies and infrared optical pumping. Changes in the helium caused by fields the spacecraft passes through are measured. It weighs 2.6 kilograms (5.7 pounds) and uses up to five watts of power.

Interstellar Solar Wind and Heliosphere

Plasma Analyzer (Ames Research Center). The plasma or solar wind instrument will map the density and energy of the solar wind (ions and electrons flowing out from the Sun). It will determine solar wind interactions with Jupiter, including the planet's bow shock wave, and will look for the boundary of the heliosphere.

The instrument consists of a high-resolution and medium-resolution analyzer. It will look toward the Sun through an opening in the spacecraft dish antenna, and the solar wind will enter like the electron beam in a TV tube. The instrument measures direction of travel, energy (speed), and numbers of ions and electrons.

The particles enter between curved metal plates and strike detectors, which count their numbers. Their energy is found by the fact that when a voltage is applied across the plates in one of 64 steps, only particles in that energy range can enter. Direction of particle travel is found from orientation of the instrument and which target detected it.

In the high-resolution analyzer, the targets are 26 continuous-channel multipliers, which measure the ion flux in energy ranges from 100 to 18,000 electron-volts. Detectors in the medium-resolution detector are five electrometers, which measure ions only in ranges from 100 to 8,000 electron-volts and electrons from one to 500 electron-volts.

The plasma analyzer weighs 5.5 kilograms (12.1 pounds) and uses four watts of power.

Cosmic Rays, Jupiter's Radiation Belts, and Radio Emissions

Charged Particle Composition Instrument (University of Chicago). This instrument has a family of four measuring systems. Two systems are particle telescopes operating primarily in interplanetary space between Earth and Jupiter and beyond Jupiter to the limit of spacecraft communications. The other two systems measure trapped electrons and protons inside the Jovian magnetic field.

During the interplanetary phase of the mission (before and after Jupiter encounter) two telescopes will identify the nuclei of all eight chemical elements from hydrogen to oxygen and separate the isotopes hydrogen, deuterium, helium-3 and helium-4. Because of their differences in isotopic and chemical composition and spectra it will be possible to separate galactic from solar radiation as the spacecraft moves outward from under the influence of the Sun towards the nearby interstellar space. The instrument also measures how streams of high-energy particles escape from the Sun and travel through interplanetary space. There is a main telescope of seven solid state detectors which measures the composition from 1 to 500 million electron-volt (MeV) particles and a 3-element telescope which measures 0.4 to 10 MeV protons, and helium nuclei. If a key detector element should be destroyed in space, there exist diagnostic procedures and commands from Earth which will remove the defective element from the telescope operation.

For the magnetosphere of Jupiter, two new types of sensors were developed to cope with the extremely high intensities of trapped radiations. A solid state ion chamber operating below -40° C (-40° F) measures only those electrons that generate the radio waves which reach Earth. The trapped proton detector contains a foil of thorium which undergoes nuclear fission from protons above 30 million electron-volts, but is not sensitive to the presence of the intense electron radiation. The instrument weighs 3 kilograms (7.3 pounds) and uses 2.4 watts of power.

Cosmic Ray Telescope (Goddard Space Flight Center). The Cosmic Ray Telescope also will monitor solar and galactic cosmic ray particles. It will track the twisting paths of high-energy particles from the Sun and measure bending effects of the solar magnetic field on particles from the galaxy, some traveling at near light speeds. The instrument can distinguish which of the ten lightest elements make up these particles. It also will measure high-energy particles in Jupiter's radiation belts.

The instrument consists of three three-element, solid state telescopes. A high-energy telescope measures the flux of protons between 56 and 800 MeV. A medium-energy telescope measures protons with energies between three and 22 MeV, and identifies the ten elements from hydrogen to oxygen. The low-energy telescope will study the flux of electrons between 50,000 electron-volts and one MeV and protons between 50,000 electron-volts and 20 MeV.

The instrument weighs 3.2 kilograms (7 pounds) and uses 2.2 watts of power.

Jupiter's Charged Particles

Geiger Tube Telescope (University of Iowa). This experiment will attempt to characterize Jupiter's radiation belts. The instrument employs seven Geiger-Mueller tubes to survey the intensities, energy spectra, and angular distributions of electrons and protons along *Pioneer*'s path through the magnetosphere of Jupiter.

The tubes are small cylinders containing gas that generates electrical signals from charged particles.

Three tubes (considered a telescope) are parallel. Three others will be in a triangular array to measure the number of multiparticle events (showers) occurring. The combination of a telescope and shower detector will enable experimenters to compare primary with secondary events in the Jovian radiation belts.

Another telescope will detect low-energy electrons (those above 40,000 electron-volts), and together with the others should shed light on Jupiter's shock front in the solar wind and the tail of its magnetosphere.

The instrument can count protons with energies above 5 million electron-volts (MeV) and electrons with energies between two and 50 MeV.

The instrument weighs 1.6 kilograms (3.6 pounds) and uses 0.7 watts of power.

Jovian Trapped Radiation Detector (University of California at San Diego). The nature of particles trapped by Jupiter, particle species, angular distributions, and intensities will be determined by this instrument. It will have a very broad range of energies from 0.01 to 100 MeV for electrons and from 0.15 to 350 MeV for protons (hydrogen nuclei).

Experimenters will attempt to correlate particle data with Jupiter's mysterious radio signals.

They will use five detectors to cover the planned energy range. An unfocused Cerenkov counter that measures direction of particle travel by light emitted in a particular direction will detect electrons of energy above 1 MeV and protons above 450 MeV. A second detector measures electrons at 100, 200 and 400 thousand electron-volts.

An omnidirectional counter is a solid state diode, which discriminates minimum ionizing particles at 400,000 electron-volts (eV), and high-energy protons at 1.8 MeV.

Twin DC scintillation detectors for low-energy particles distinguish roughly between protons and electrons because of different scintillation material in each. Their energy thresholds are about 10,000 eV for electrons and 150,000 eV for protons.

The instrument weighs 1.7 kilograms (3.9 pounds) and uses 2.9 watts of power.

Asteroids, Meteoroids, Interplanetary Dust

Asteroid-Meteoroid Detector (General Electric Company). A survey of the solid material between the Earth's orbit and 2.4 billion kilometers (1.5 billion miles) will be made by this experiment, labeled Sisyphus because of its near-impossible task. (Sisyphus was a mythical Corinthian King, whose job in Hades was to roll a stone up a hill, only to have it roll back forever.)

By measuring the orbits of the material in the vicinity of the spacecraft, experimenters hope to learn about the origins of meteoroids, asteroids, and comets and the distribution of these bodies in the solar system.

Four nonimaging telescopes will characterize objects, ranging from several hundred miles in diameter (asteroids) down to particles with a mass of one millionth of a gram, by measuring sunlight reflected from them. The telescopes will measure numbers of solid particles, and individual particle size, velocity, and direction of travel. Billions of such particles strike the Earth's atmosphere each day.

This experiment (as does the Photopolarimeter) will periodically measure zodiacal light (sunlight scattered by the total mass of meteoroids and dust in deep space). Comparisons of zodiacal light measurements with particle measurements will allow conclusions about the relationship of the zodiacal cloud brightness with size distribution of particles.

Experimenters also will study effects of secondary forces such as solar radiation and planetary gravity on orbits of small particles in the solar system.

These data will be fundamental to design of all future outer planet spacecraft. They will allow estimates to be made of chances of penetration and spacecraft failure by impact of larger particles, and of surface erosion from bombardment by smaller particles.

Each of its four telescopes consists of a 20-centimeter (8-inch) mirror, an 8.4-centimeter (3.3-inch) secondary mirror, coupling optics, and a photomultiplier tube. Each telescope has an 8° view cone. The four cones overlap in part, and particle distance and speed is measured by timing entry and exit of view cones. Photomultiplied pulses are counted to find particle numbers.

The instrument weighs 3.3 kilograms (7.2 pounds) and uses 2.2 watts of power.

Meteoroid Detector (Langley Research Center). To detect the distribution in space of tiny particles (masses of greater than one millionth of a gram), scientists will use a system of 234 pressure cells mounted on the back of the spacecraft dish antenna. The pressure cells come in 20 by 30 centimeter (8 by 12 inch) panels, 18 to a panel, and there will be 13 panels. In construction, the panels are something like a stainless steel air mattress.

Each of the pressure cells will be filled with a gas mixture of 75 percent argon, and 25 percent nitrogen, and each time a particle penetrates a cell, it will be counted by a transducer as the gas escapes the cell.

The average mass and energy of particles penetrating the cells will be calculated using laboratory impact test data. These data show the combined mass and velocity required for a particle to penetrate a cell. By combining these data with trajectory information, the experimenters can calculate spatial distribution of the tiny meteoroids.

The exterior surface of the cells will be one-thousandth-of-an-inch-thick stainless steel sheet, and this is thin enough to allow penetrations of particles with a mass of one billionth of a gram or more.

The total weight of the instrument (panels and electronics) is 1.7 kilograms (3.7 pounds), and it uses 0.7 watts of power.

Celestial Mechanics

Celestial Mechanics (Jet Propulsion Laboratory). This experiment uses the spacecraft itself as a sensitive instrument to better determine the mass of Jupiter and its satellites and to determine the harmonics and possible anomalies in Jupiter's gravity field.

Experimenters will measure the gravity effects of Jupiter and its satellites on the spacecraft flight trajectory.

Deep Space Network doppler tracking of the spacecraft determines its velocity along the Earth-spacecraft line down to a fraction of a millimeter per second, once per minute. These data are further augmented by optical and radar position measurements of the planets. Computer calculations using the spacecraft trajectory and known planet and satellite orbital characteristics should allow a five-fold improvement in the accuracy of current calculations of Jupiter's mass. Masses of Jupiter's four large (Galilean) moons will be determined to an accuracy of about one percent. Experimenters expect to find the planet's dynamical polar flattening to within one half mile, and to make estimates of the mass of the planet's surface layers.

Interplanetary Hydrogen, Helium, and Dust; Jupiter's Atmosphere, Temperatures, Auroras, Moons

Ultraviolet Photometer (University of Southern California). This instrument will address some basic questions about interplanetary and interstellar space as well as about Jupiter.

During the planet flyby the ultraviolet photometer instrument will measure the scattering by Jupiter's atmosphere of ultraviolet (UV) light from the Sun in two wavelengths, one for hydrogen and one for helium. The experimenters will use these data to find the amount of atomic hydrogen in Jupiter's upper atmosphere, the mixing rate of Jupiter's atmosphere, the amount of helium in its atmosphere, and the ratio of helium to molecular hydrogen. So far helium has not been identified on Jupiter.

Virtually all theories of Jupiter's origin make assumptions about amounts of helium in the planet's makeup.

By measuring changes in intensity of ultraviolet light glow as it scans across the planet, the instrument should also be able to identify Jovian auroras – the bright areas caused by an auroral oval similar to the auroral zone on Earth – if they exist.

Radio telescope measurements tell us that the solar system is immersed in an interstellar gas of cold neutral hydrogen. By measuring the scattering of the Sun's ultraviolet light the instrument can measure amounts of this neutral hydrogen within the heliosphere. Presence of neutral hydrogen (already measured near Earth) would be a result of neutralization of fast solar wind hydrogen ions at the heliosphere boundary, their conversion into fast hydrogen atoms, and their diffusion back into the heliosphere.

Neutral helium may penetrate the heliosphere directly as the solar system moves through the interstellar gas at 72,000 km/hr (45,000 mph). From such helium measurements, the experimenters may be able to calculate the percent of helium in the interstellar gas. The "big bang" theory of the origin of the universe suggests that it should be about seven percent.

The instrument's measurements also should allow calculation of concentrations of hydrogen and dust in the interstellar space outside the heliosphere from measurements of the interstellar ultraviolet glow from starlight, at the hydrogen wavelength. The UV glow from scattered sunlight will diminish greatly as the spacecraft proceeds out from the Sun.

The instrument has two photo cathodes, one of which measures ultraviolet radiation at 1216 angstroms, the other at 584 angstroms the wavelengths at which hydrogen and helium scatter solar ultraviolet.

The instrument has a fixed viewing angle and will use the spacecraft spin to scan the planet. It weighs 0.7 kilograms (1.5 pounds) and uses 0.7 watts of power.

Imaging Photopolarimeter (University of Arizona). This instrument will provide data in a number of areas, using photometry (measurement of light intensity) and polarimetry (photometry measurements of the linear polarization of light), and imaging.

En route to Jupiter, the instrument will measure brightness and polarization of zodiacal light (sunlight scattered by interplanetary dust and solid matter) several times a month to determine the amount and character of interplanetary solid material.

The instrument will take about ten images of Jupiter in the last 20 hours before closest approach to the planet (periapsis).

Photo Imaging System

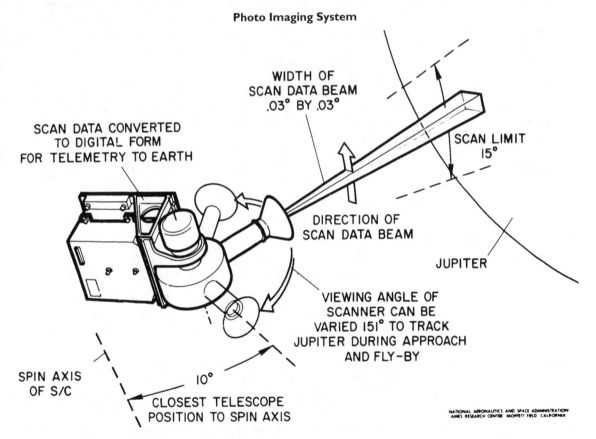

Although resolution of the images should be better than that from the best Earth telescopes, the most important fact is that the images will be taken from viewing angles impossible to get from Earth. At least one will show Jupiter's terminator (the line between sunlit and dark hemispheres), which is never seen from Earth. This should show texture and shapes of cloud regions.

Imaging will be done in both red and blue light, and it may be possible to superimpose these, providing "color pictures" of the planet in red and blue.

The instrument uses a photoelectric sensor which measures changes in light intensity, much like the light sensor for a television camera. But unlike a TV camera it will employ the five rpm spin of the spacecraft to electronically scan the planet, in narrow strips 0.03° wide.

This electronic imaging system will complete the scans for a picture in from 25 to 110 minutes, depending on distance from the planet. Scan data will be converted to digital form on the spacecraft and radioed to Earth by telemetry. Engineers then will build up the images, using various computer techniques.

One of the planet pictures may be omitted if the spacecraft trajectory allows a good view of one of Jupiter's Moons.

Controllers can vary the viewing angle of the instrument's 8.64-centimeter (3.4-inch) focal length telescope by about 150° relative to the spacecraft's spin axis, which is fixed on the Earth-spacecraft line. The telescope can see to within 10° of the spin axis, looking away from Earth.

For the first pictures, the telescope will be looking almost down the spacecraft spin axis toward Jupiter. For the last pictures, close to the planet, it will be looking almost at right angles to the spin axis.

Picture reconstruction by computer on Earth will be complicated by the fact that picture scans will be curved as the telescope rotates with the spacecraft, while looking out at various angles to the spin axis. The only straight-line scans will be those looking straight out from the spin axis. A second factor is that Jupiter's equatorial zone rotates at 35,400 km/hr (22,000 mph). At periapsis the spacecraft will be traveling in roughly the same direction as the planet's rotation at 126,000 km/hr (78,000 mph) relative to Jupiter.

These two factors mean that computer techniques will have to sort out scan paths whose curvature changes steadily with spacecraft movement. They will also have to compensate for smear caused by high-speed motion of the planet and the spacecraft.

During Jupiter fly-by, experimenters will use its measurements to seek an improved understanding of the structure and composition of Jupiter's clouds and determine the amount and nature of atmospheric gas above the clouds and aerosols in this gas. These data also will shed light on Jupiter's heat balance. The instrument will, if possible, make light scattering measurements of the Jovian satellites and any large asteroids. These data may provide understanding of the surface properties, possible atmospheres, and origins of these bodies.

The instrument includes a 2.5-centimeter (1-inch)-aperture, 8.6-centimeter (3.4-inch)-focal length telescope which can be moved 150° in the plane of the spacecraft spin axis by ground command or automatically. Incoming light is split by a prism according to polarization into two separate beams. Each beam is further split by going through a red filter (5800-7000 angstroms), and through a blue filter (3900-4900 angstroms). Channeltron detectors turn the light into electrical impulses, which are telemetered to Earth.

The instrument uses three viewing apertures: one 40 by 40 milliradians (mR) for zodiacal light measurements, a second 8 by 8 mR for nonimaging light measurements of Jupiter, and a third 0.5 by 0.5 mR for scans of the planet from which pictures will be reconstructed.

The instrument weighs 4.3 kilograms (9.5 pounds) and uses 2.2 watts of power.

Jupiter's Atmosphere, Ionosphere, Temperature

Infrared Radiometer (California Institute of Technology). Perhaps the most important mystery of Jupiter involves its heat radiation. Is the planet in fact radiating about three times more energy than it absorbs from the Sun?

Experimenters will use the Infrared Radiometer to seek the answer by measuring Jupiter's net heat energy output.

The two-channel radiometer, used successfully on two *Mariner Mars* spacecraft, will make measurements in the 14-25 and 29-56 micron wavelengths to study the net energy flux, its distribution over the Jovian disc, and the thermal structure and chemical composition of Jupiter's atmosphere.

Experimenters will use these data to try to find the temperature distribution in Jupiter's outer atmosphere, the existence of a polar ice cap of frozen methane, the brightness temperature of Jupiter's dark hemisphere, hot or cold spots in the atmosphere, the reason for "hot shadows" of Jupiter's Moons, temperature of the Great Red Spot, and the overall hydrogen / helium ratio. (Transparency of Jupiter's atmosphere to heat depends on the amount of helium there.)

Like the Ultraviolet Photometer, it uses a fixed telescope, so that its scans of the planet are made by rotation of the spacecraft. Because of this fixed viewing angle, the instrument will view the planet only during the several hours before periapsis.

The instrument has a 7.2-centimeter (3-inch) Cassegrain telescope, and the detectors in its two channels are 88-element, thin film, bimetallic thermopiles. Its field of view is about 2,400 by 700 kilometers on Jupiter's cloud surface at closest approach of one Jupiter diameter. At this distance, it has a resolution of about 2,400 kilometers (1,500 miles).

The instrument weighs 2 kilograms (4.4 pounds) and uses 1.3 watts of power.

Occultation Experiment (Jet Propulsion Laboratory). Passage of the spacecraft radio signal through Jupiter's atmosphere as *Pioneer F* swings behind the planet for an hour will be used to measure Jupiter's ionosphere and density of the planet's atmosphere down to a pressure level of about one Earth atmosphere.

Experimenters will use computer analysis of the incoming radio signals recorded on tape to determine the refractive index profile of Jupiter's atmosphere.

This sort of analysis has been done with stars passing behind the planet, but *Pioneer* S-band telemetry is a precisely known signal source. These refraction data should allow measurements of the electron density of Jupiter's ionosphere and, used with temperature measurements, will allow inferences about the hydrogen / helium ratio in the atmosphere. Experimenters also will measure the absorption profile of the atmosphere, which should allow calculation of ammonia abundance.

If the spacecraft trajectory allows, the experimenters also will look for an atmosphere on the Jovian satellite, Io.

Experiments and Investigators

Magnetic Fields Experiment

Instrument: Magnetometer
Principal Investigator: Edward J. Smith, Jet Propulsion Laboratory, Pasadena, California
Co-Investigators: Palmer Dyal, NASA-Ames Research Center, Mountain View, California
                         David S. Colburn, NASA-Ames Research Center, Mountain View, California
                         Charles P. Sonett, NASA-Ames Research Center, Mountain View, California
                         Douglas E. Jones, Brigham Young University, Provo, Utah
                         Paul J. Coleman, Jr., University of California at Los Angeles
                         Leverett Davis, Jr., California Institute of Technology, Pasadena

Plasma Analyzer Experiment

Instrument: Plasma Analyzer
Principal Investigator: John H. Wolfe, NASA-Ames Research Center
Co-Investigators: Louis A. Frank, University of Iowa, Iowa City
                         Reimar Lust, Max-Planck-Institute fur Physik und Astrophysik
                         Institute fur Extraterrestrische Physik, Munchen, Germany
                         Devrie Intriligator, University of California at Los Angeles
                         William C. Feldman, Los Alamos Scientific Laboratory, New Mexico

Charged Particle Composition Experiment

Instrument: Charged Particle Instrument
Principal Investigator: John A. Simpson, University of Chicago
Co-Investigators: Joseph J. O'Gallagher, University of Maryland, College Park
                         Anthony J. Tuzzolino, University of Chicago

Cosmic Ray Energy Spectra Experiment

Instrument: Cosmic Ray Telescope
Principal Investigator: Frank B. McDonald, NASA-Goddard Space Flight Center, Greenbelt, Maryland
Co-Investigators: Kenneth G. McCracken, Minerals Research Laboratory, North Ryde, Australia
                         William R. Webber, University of New Hampshire, Durham
                         Edmond C. Roelof, University of New Hampshire, Durham
                         Bonnard J. Teegarden, NASA-Goddard Space Flight Center
                         James H. Trainor, NASA-Goddard Space Flight Center

Jovian Charged Particles Experiment

Instrument: Geiger Tube Telescope
Principal Investigator: James A. Van Allen, University of Iowa, Iowa City

Jovian Trapped Radiation Experiment

Instrument: Trapped Radiation Detector
Principal Investigator: R. Walker Fillius, University of California at San Diego
Co-Investigator: Carl E. McIlwain, University of California at San Diego

Asteroid-Meteoroid Astronomy Experiment

Instrument: Asteroid-Meteoroid Detector
Principal Investigator: Robert K. Soberman, General Electric Company, Philadelphia, Drexel University, Philadelphia
Co-Investigator: Herbert A. Zook, NASA-Manned Spacecraft Center, Houston, Texas

Meteoroid Detection Experiment

Instrument: Meteoroid Detector
Principal Investigator: William H. Kinard, NASA-Langley Research Center, Hampton, Virginia
Co-Investigators: Robert L. O'Neal, NASA-Langley Research Center
                                    Jose M. Alvarez, NASA-Langley Research Center
                                    Donald H. Humes, NASA-Langley Research Center
                                    Richard E. Turner, NASA-Langley Research Center

Celestial Mechanics Experiment

Instrument: *Pioneer F* and the DSN
Principal Investigator: John D. Anderson, Jet Propulsion Laboratory
Co-Investigator: George W. Null, Jet Propulsion Laboratory

Ultraviolet Photometry Experiment

Instrument: Ultraviolet Photometer
Principal Investigator: Darrell L. Judge, University of Southern California, Los Angeles
Co-Investigator: Robert W. Carlson, University of Southern California

Imaging Photopolarimetry Experiment

Instrument: Imaging Photopolarimeter
Principal Investigator: Tom Gehrels, University of Arizona, Tucson
Co-Investigators: David L. Coffeen, University of Arizona
                                    William Swindell, University of Arizona
                                    Jyrki Hameen-Anttila, University of Arizona
                                    Charles E. KenKnight, University of Arizona
                                    Robert F. Hummer, Santa Barbara Research Center
                                    Jerry Weinberg, Dudley Observatory, Albany, New York

Jovian Infrared Thermal Structure Experiment

Instrument: Infrared Radiometer
Principal Investigator: Guido Munch, California Institute of Technology
Co-Investigators: Gerry Neugebauer, California Institute of Technology
                                    Stillman C. Chase, Santa Barbara Research Center
                                    Laurence M. Trafton, University of Texas, Austin, Texas

S-Band Occultation Experiment

Instrument: The Spacecraft Radio Transmitter and the DSN
Principal Investigator: Arvydas J. Kliore, Jet Propulsion Laboratory
Co-Investigators: Gunnar Fjeldbo, Jet Propulsion Laboratory
                                    Dan L. Cain, Jet Propulsion Laboratory
                                    Boris L. Seidel, Jet Propulsion Laboratory
                                    S. Ichtiaque Rasool, NASA-Headquarters, Washington, D.C.

The Project Team

Office of Space Science

Dr. John E. Naugle   Associate Administrator for OSS
Vincent L. Johnson   Deputy Associate Administrator for OSS
Robert S. Kraemer   Director, Planetary Programs
S. Ichtiaque Rasool   Deputy Director, Planetary Programs
Fred D. Kockendorfer   Pioneer Program Manager
Thomas P. Dallow   Pioneer Program Engineer
Dr. Albert G. Opp   Pioneer Program Scientist
Joseph B. Mahon   Director, Launch Vehicle Programs
T. Bland Norris   Manager, Medium Launch Vehicles
F. Robert Schmidt   Manager, Atlas-Centaur

Office of Tracking and Data Acquisition

Gerald M. Truszynski   Associate Administrator for OTDA
Arnold C. Belcher   Network Operations
Maurice E. Binkley   Network Support

Ames Research Center, Mountain View, California

Dr. Hans Mark   Center Director
C.A. Syvertson   Deputy Director
John V. Foster   Director of Development
Charles F. Hall   Pioneer Project Manager
Dr. John H. Wolfe   Pioneer Project Scientist
Ralph W. Holtzclaw   Pioneer Spacecraft System Manager
Joseph E. Lepetich   Pioneer Experiment System Manager
Robert R. Nunamaker   Pioneer Mission Operations System Manager
J. Richard Spahr   Management Control
Robert U. Hofstetter   Mission Analysis and Launch Coordination
Arthur C. Wilbur   Nuclear Power
Eldon W. Kaser   Contracts
J.R. Mulkern   Reliability and Quality Assurance
R.O. Convertino   Reliability and Quality Assurance
Ernest J. Iufer   Magnetics

Deep Space Network, Jet Propulsion Laboratory, Pasadena, California

Dr. William H. Pickering   Director, Jet Propulsion Laboratory
Dr. Nicholas A. Renzetti   Tracking and Data System Manager
Alfred J. Siegmeth   Deep Space Network Manager for Pioneer

Lewis Research Center, Cleveland, Ohio

Bruce T. Lundin   Center Director
Edmund R. Jonash   Chief, Launch Vehicles Division
Daniel J. Shramo   Atlas-Centaur Project Manager
Edwin Muckley   Centaur Project Engineer

Kennedy Space Center, Florida

Dr. Kurt H. Debus   Center Director
John J. Neilon   Director, Unmanned Launch Operations (ULO)
John D. Gossett   Chief, Centaur Operations Branch, ULO
Donald C. Sheppard   Chief, Spacecraft Operations Branch, ULO
James Johnson   (Manager, Pioneer Operations, ULO)

AEC Space Nuclear Systems Division

David S. Gabriel   Division Director
Glenn A. Newby   Assistant Director for Space Electric Power
Harold Jaffe   Chief, Isotope Power Systems Projects Branch
Bernard Rock   SNAP 19 / Pioneer RTG Program Manager

TRW Systems Group

Bernard J. O'Brien   Pioneer Project Manager

Pioneer F Contractors

Contractor	Location	Item
EMR Telemetry Division, Weston Instruments Inc.	Sarasota, Florida	Telemetry Decommutation, Display Equipment
Edcliff Instrument Division, Systron Donner	Monrovia, California	Despin Sensor Assembly
Electronic Memories, Division of Electronic Memories and Magnetics Corp.	Hawthorne, California	Memory Storage Units
Jet Propulsion Laboratory	Pasadena, California	Helium Vector Magnetometer
Time Zero Corporation	Torrance, California	Plasma Analyzer and Magnetometer Electronics
University of Chicago	Chicago, Illinois	Charged Particle Instrument
University of Iowa	Iowa City, Iowa	Geiger Tube Telescope
University of California at San Diego	San Diego, California	Trapped Radiation Detector
Analog Technology Corporation	Pasadena, California	Ultraviolet Photometry
Santa Barbara Research Center	Santa Barbara, California	Imaging Photopolarimeter and Infrared Radiometer
General Electric Company	Philadelphia, Pennsylvania	Asteroid / Meteoroid Detector
Teledyne Isotopes	Germantown, Maryland	Radioisotope Thermoelectric Generators (RTGs)
Mound Laboratories	Miamisburg, Ohio	Radioisotope Heater Unit Capsules, RTG Fuel and Capsules
Los Alamos Scientific Laboratory	Los Alamos, New Mexico	RTG Fuel Discs
Bendix Corporation	Columbia, Maryland	Software
General Dynamics, Convair Division	San Diego, California	Launch Vehicle – First and Second Stages
Thiokol Chemical Company	Elkton, Maryland	Launch Vehicle – Third Stage Motor
McDonnell-Douglas Corp., Astronautics Company	Huntington Beach, California	Launch Vehicle – Third Stage Motor
North American Rockwell Corp., Rocketdyne Division	Canoga Park, California	Launch Vehicle – First Stage Motor
Pratt and Whitney Aircraft Co.	East Hartford, Connecticut	Launch Vehicle – Second Stage Motor
TRW Systems Group, TRW Inc.	Redondo Beach, California	Spacecraft
Frequency Electronics Inc.	New Hyde Park, New York	Oscillator (TCXO)
United Detector Technology Inc.	Santa Monica, California	Silicon Photo Detectors
Holex Inc.	Hollister, California	Explosive Cartridge
Allen Design	Burbank, California	Propellant Valves
Electra Midland Corp., Cermatrik Division	San Diego, California	Current Limiters
Bendix Mosaic Fabrication Division	Sturbridge, Massachusetts	Fiber Optics
Pressure Systems Inc.	Los Angeles, California	Propellant Tanks
Xerox Data Systems	El Segundo, California	Computer Systems
Wavecom Inc.	Chatsworth, California	Diplexer Assemblies
Teledyne Microwave	Sunnyvale, California	RF Transfer Switch
Yardney Electric Corp.	Pawcatuck, Connecticut	Silver-Cadmium Battery Cells
Siliconix Inc.	Santa Clara, California	Integrated Circuits
Amelco Semiconductor	Mountain View, California	Integrated Circuits
Watkins-Johnson Co.	Palo Alto, California	Traveling Wave Tube Amplifier
Texas Instruments	Dallas, Texas	Integrated Circuits
Data Products Corp.	Woodland Hills, California	ADP Line Printer
Computer Communications Inc.	Inglewood, California	Communication Stations
Honeywell, Inc., Radiation Center	Lexington, Massachusetts	Sun Sensor Assemblies

Launch Vehicle

The launch of *Pioneer F* marks the first use of the Atlas-Centaur vehicle in the Pioneer program. It is also the first time Atlas-Centaur has been used with a third stage, the 66,750-newton (15,000-pound)-thrust, solid fuel TE-M-364-4 engine.

The AC-27 vehicle has been assigned to the Pioneer F mission.

Centaur was developed under the direction of the National Aeronautics and Space Administration's Lewis Research Center and is in the process of being integrated with the Titan III to launch the *Viking* spacecraft to Mars in 1975.

AC-27 consists of an Atlas SLV-3C booster combined with a Centaur second stage, and the TE-M-364-4 third stage. The first two stages are 3.05 meters (10 feet) in diameter and are connected by an interstage adapter. Both Atlas and Centaur stages rely on internal pressurization for structural integrity.

The Atlas booster develops 1,830,520 newtons (411,353 pounds) of thrust at liftoff, using two 778,042-newton (174,841-pound)-thrust booster engines, one 268,410-newton (60,317-pound)-thrust sustainer engine and two vernier engines developing 3,008 newtons (676 pounds) thrust each.

Centaur carries insulation panels which are jettisoned just before the vehicle leaves the Earth's atmosphere. The insulation panels, weighing about 523 kilograms (1,154 pounds), surround the second stage propellant tanks to prevent heat or air friction from causing boil-off of liquid hydrogen during flight through the atmosphere.

To date, Centaur has successfully launched the seven unmanned *Surveyor* spacecraft to the Moon, *Mariners* 6, 7, and 9, and Orbiting Astronomical Observatory 2, Applications Technology Satellite 5, and three Intelsat 4s.

The solid-fueled TE-M-364-4 third stage develops almost 66,750 newtons (15,000 pounds) of thrust. It is an uprated version of the retromotor used for the *Surveyor* moon-landing vehicle.

Both third stage and spacecraft will be enclosed in an 8.7-meter (29-foot)-long, 3-meter (10-foot)-diameter fiberglass shroud, which is jettisoned after leaving the atmosphere.

Launch Vehicle Characteristics

*Liftoff weight including spacecraft: 146,673 kilograms (323,415 pounds)
  Liftoff height: 40.3 meters (132 feet)
  Launch complex: 36A
  Launch azimuth sector: 94.5-110 degrees

	SLV-3C Booster	Centaur Stage	TE-M-364-4
Weight**	128,920 kilograms (284,269 pounds)	17,734 kilograms (39,104 pounds)	1,142 kilograms (2,510 pounds)
Height	22.9 meters (75 feet) (including interstage adapter)	14.6 meters (48 feet) (with payload fairing)	2.1 meters (6.75 feet) (to top interstage ring)
Thrust	1,830,520 newtons (411,353 pounds) sea level	130,000 newtons (29,200 pounds) vacuum	65,860 newtons (14,800 pounds)
Propellants	Liquid oxygen and RP-1	Liquid hydrogen and liquid oxygen	Solid chemical fuel TP-H 3062
Propulsion	MA-5 system (two 778,042-newton (174,841-pound)-thrust engines, one 268,410-newton (60,317-pound)-thrust sustainer engine, and two 3,008-newton (676-pound)-thrust vernier engines)	Two 65,000-newton (14,600-pound)-thrust RL-10 engines. Ten small hydrogen peroxide thrusters	One solid-fuel engine
Velocity	2,534 m/sec (5,665 mph) at BECO. 3,543 m/sec (7,926 mph) at SECO.	10,262 m/s (22,948 mph) at MECO.	13,913 m/s (31,122 mph) at spacecraft separation.
Guidance	Pre-programmed pitch rates through BECO. Switch to Centaur inertial guidance for sustainer phase.	Inertial guidance	Spin (60 rpm) imparted by spin table on top of Centaur.

*   Measured at 5.08 centimeters (two inches) of rise.
** Weights are based on AC-27 configuration.

Flight Sequence

Atlas / Centaur / TE-M-364-4 Flight Sequence (AC-27)

	Nominal Time (seconds)	Altitude (kilometers)	Altitude (statute miles)	Surface Range (kilometers)	Surface Range (statute miles)	Relative Velocity (km/sec)	Relative Velocity (miles/hour)
Liftoff	0	0	0	0	0	0	0
Booster Engine Cutoff	148.3	58.3	36.2	86.4	53.7	2,534	5,665.0
Jettison Booster	151.4	61.3	38.1	94.1	58.5	2,563	5,733.0
Admit Guidance Steering	156.3	65.9	40.9	105.9	65.8	2,616	5,851.4
Jettison Insulation Panels	193.3	98.2	61.0	202.2	125.7	2,927	6,547.0
Sustainer / Vernier Cutoff	243.3	136.7	85.0	355.6	220.9	3,544	7,926.8
Atlas / Centaur Separation	245.2	138.2	85.9	362.3	225.1	3,534	7,920.0
Centaur Ignition	254.8	144.9	90.0	394.5	245.1	3,541	7,905.7
Enable Guidance Steering	258.8	147.6	91.7	408.2	253.6	3,546	7,931.6
Jettison Nose Fairing	266.8	152.7	94.9	435.6	270.7	3;587	8,024.9
Centaur Main Engine Cutoff	705.8	161.5	100.3	2,993.2	1,859.9	10,262	22,948.0
Ignite Spin Rockets	775.8	176.6	109.7	3,686.0	2,290.4	10,251	22,933.0
Centaur / TE-M-364-4 Separation	777.8	177.7	110.4	3,706.6	2,303.1	10,250	22,931.0
TE-M-364 4 Ignition	790.8	184.9	114.9	3,834.8	2,382.8	10,243	22,915.0
TE-M-364-4 Burnout	834.7	218.2	135.6	4,327.8	2,689.1	14,042	31,413.0
TE-M-364-4 / Spacecraft Separation	934.7	449.6	279.3	5,637.5	3,502.9	13,896	31,088.0

Launch Vehicle

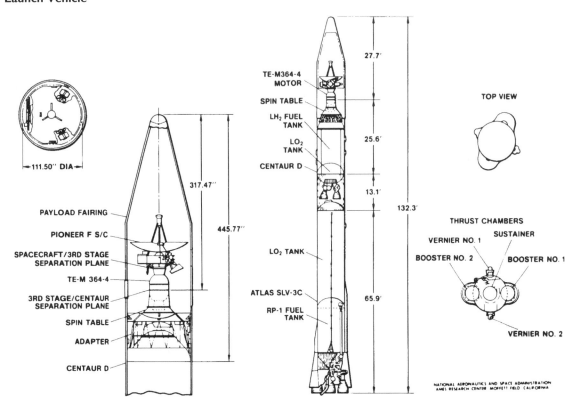

Atlas Phase

After liftoff, AC-27 will rise vertically for about 15 seconds before beginning its pitch program. Starting at two seconds after liftoff and continuing to T+15 seconds, the vehicle will roll to the desired flight azimuth.

After 148 seconds of flight, the booster engines are shut down (BECO) and jettisoned. BECO occurs when an acceleration of 5.7 g's is sensed by accelerometers on the Centaur and the signal is issued by the Centaur guidance system. The booster package is jettisoned 3.1 seconds after BECO. The Atlas sustainer engine continues to burn for approximately another minute and 32 seconds propelling the vehicle to an altitude of about 137 kilometers (85 miles), attaining a speed of 3,543 m/s (7,926 mph).

Sustainer engine cutoff (SECO) occurs at propellant depletion. Centaur insulation panels are jettisoned prior to SECO.

The Atlas and Centaur stages are then separated. An explosive shaped charge slices through the interstage adapter. Retrorockets mounted on the Atlas slow the spent stage.

Centaur Phase

At four minutes, 15 seconds into the flight, the Centaur's two RL-10 engines ignite for a planned 7.5 minute burn. After MECO, the Centaur stage and spacecraft are reoriented with the Centaur attitude control thrusters to place the third stage and spacecraft on the proper trajectory after separation.

Third Stage Phase

A spin table mounted on the top of the Centaur carries the third stage and the spacecraft. This is spun up to 60 rpm by small thruster rockets. Explosive bolts separate the spinning third stage and spacecraft. The TE-M-364-4 engine then ignites and burns for 44 seconds.

Separation of the *Pioneer* spacecraft is achieved by firing explosive bolts on the metal ring holding the spacecraft to the third stage. Compressed springs then push the spacecraft away from the third stage at a rate of 0.6 meters-per-second (2.1 feet-per-second).

Retromaneuver

Requirements for the Centaur retrosystem call for a 7.6-meter (25-foot) separation distance to be achieved within 13 seconds after the third stage has separated from Centaur. This separation distance is required to prevent spacecraft damage from reflected motor exhaust products and to minimize third stage ignition transients. At one second after third stage separation Centaur retrothrust is initiated to provide the required clearance.

Launch Operations

The Unmanned Launch Operations (ULO) Directorate at the John F. Kennedy Space Center (KSC) is responsible for the preparation and launch of unmanned spacecraft from the Air Force Cape Kennedy Station, Florida.

The *Pioneer F* spacecraft was delivered to the Cape from TRW Systems about 1½ months prior to launch and placed in the spacecraft checkout area for final verification tests.

In providing launch operations, KSC handles scheduling of test milestones and review of data to assure that the launch vehicle has met all of its test requirements and is ready for launch.

Atlas-Centaur No. 27 was erected on Pad 36A in December 1971.

The Terminal Countdown Demonstration (TCD) is conducted about four weeks prior to launch. The TCD primarily demonstrates that all of the functions leading to the actual countdown can be performed. It is an end-to-end check of all systems and includes propellant loading of launch vehicle stages to verify that tanks and facilities were ready for countdown.

Following this, the Flight Acceptance Composite Test (FACT) is conducted about three weeks before launch to assure that the vehicle is electrically ready for final launch preparations. The FACT includes running the computer and programmer through post-flight events and monitoring the data to assure correct response to all signals with the umbilical ejected.

About two weeks before planned launch the spacecraft is mated on Atlas-Centaur and an electrical test conducted.

The Countdown Readiness Test is scheduled for the space vehicle about four days before launch. It verifies the ability of the launch vehicle to go through post-liftoff events. The range support elements participate along with the spacecraft and launch vehicle just as during a launch.

The T-1 Day Functional Test involves final preparations in getting the entire space vehicle ready for launch, preparing ground support equipment, completing readiness procedures and installing ordnance on the launch vehicle.

The countdown is picked up at T-340 minutes. All systems are checked against readiness procedures, establishing the integrity of the vehicle and ground support equipment interface prior to tower removal at T-120 minutes. Loading of cryogenic propellants (liquid oxygen and liquid hydrogen) begins at T-80 minutes, culminating in complete vehicle readiness at T-5 minutes. The terminal count begins monitoring all systems and topping off and venting propellant and purge systems. At T-10 seconds, the automatic release sequence is initiated and the space vehicle is clear for liftoff.

Mission Operations

Pioneer F is controlled primarily by ground commands rather than by command sequences stored in onboard computers. This means 24-hour-a-day spacecraft operations activity, continuous real-time analysis of incoming science and spacecraft telemetry, and careful planning of command sequences.

The fully-ground-controlled spacecraft concept allows for great flexibility in changing plans and objectives during flight. It requires a continuously effective communications system.

Ground-commanded spacecraft probably are less expensive to build and can easily be adapted to a variety of missions.

Pioneer Control and spacecraft operations will be at the Pioneer Mission Support Area (PMSA), Deep Space Network Headquarters, JPL, Pasadena, during the critical launch and planet encounter phases of the mission (two to three months). Control will be at the Pioneer Mission Analysis Area (PMAA) at the Ames Research Center, Mountain View, California, during cruise phases of the mission. The Pioneer Project is managed by Ames.

Design development, operation, and control of the Pioneer flight mission is the responsibility of Pioneer Project Mission Operations System personnel. They have worked with NASA's tracking and data acquisition organizations to develop facilities, computers, displays, and computer programs to control the flight of the Pioneer mission, and to analyze science data.

The ground data reception and transmission system for the mission is provided primarily by the DSN.

Several organizations will direct and support the launch and mission operations.

The Pioneer Mission Operations Team (PMOT) consists of personnel from many government and contractor organizations, and will operate under control of the Flight Director.

In addition to several assistant flight directors, the PMOT will include the following groups:

Members of the Spacecraft Performance Analysis Team will analyze and evaluate spacecraft performance and predict spacecraft responses to commands. They will do this through analysis of incoming telemetry data. They will develop alternative methods of using the spacecraft to satisfy mission requirements if this should be required due to a spacecraft malfunction.

The Navigation and Maneuvers Team will handle spacecraft navigation, and analysis of the spacecraft attitude in space.

The Science Performance Analysis Team will determine the status of the onboard scientific instruments. Team engineers will verify that these instruments are returning correct values for their various measurements. They will formulate the command sequences to achieve the mission's scientific objectives.

Many other support groups both at Ames and the DSN will assist the mission operations team in order to insure the effective services of computers, computer displays, computer programs, communications, physical facilities, and the operation of the DSN with its worldwide net of tracking stations.

Tracking and Data Retrieval

With facilities located at 120° intervals around the Earth, NASA's Deep Space Network will be able to support the *Pioneer Jupiter* spacecraft continuously on its two-year flight to Jupiter. As the spacecraft "sets" at one station due to the Earth's rotation, it will rise at the next one.

DSN stations also can provide overlapping coverage during critical events, and they will be able to track the spacecraft for the one billion miles of communications distance, and the five years of flight time beyond Jupiter, if the mission continues that long.

The DSN maintains seven operational 26-meter-diameter (85-foot) dish antenna stations. These are at Madrid (Robledo de Chavela and Cebreros) Spain; Canberra and Woomera, Australia; and Johannesburg, South Africa, with two stations at Goldstone, California. In addition, the DSN has one 64-meter (210-foot) antenna at Goldstone. Two more 64-meter dishes at Madrid and Canberra will be complete by mid-1973, in time to support the Jupiter encounter. This will provide continuous coverage by the big, sensitive 64-meter antennas. A 1.2-meter (four-foot) antenna at Cape Kennedy covers prelaunch and launch phases of the flight.

All DSN stations have been redesigned and are now equipped with general purpose telemetry equipment capable of receiving, data-synchronizing, decoding and processing data at high transmission rates.

The tracking and data acquisition network is tied to the Space Flight Operations Facility (SFOF), the DSN central control facility at Pasadena, by NASA's Communications Network (NASCOM).

Engineers and scientists at consoles in the Pioneer Mission Support Area (PMSA) at JPL will have push-button control of the large data processing computers and can display information either on television screens in the consoles or on high-speed printers.

The critical phases of the flight will be operated from the PMSA, where a capability exists to house spacecraft subsystem experts and provide at their fingertips processed data for display. Data analyses of spacecraft status or possible malfunctions also can be performed in real time using this "instant" data.

After several months as more is known about the spacecraft, and its day-to-day status becomes more predictable, the mission operation teams will be reduced in size. Mission control will shift to the Pioneer Analysis Area (PMAA) at Ames. There the same subsystem experts can be quickly assembled if surprises in the spacecraft performance occur. However, as the mission continues, they also will work on preparation for the next launch to Jupiter in April 1973.

In the launch phase of the mission, tracking will be carried out by the DSN with aid of other facilities. These are tracking antennas of Air Force Eastern Test Range and downrange elements of NASA's Manned Space Flight Network (MSFN) together with the tracking ship *Vanguard* and an instrumented jet aircraft. Both ship and jet are operated by MSFN.

The mission is complicated by the fact that Jupiter will be 500 million miles from the Earth at the time of first encounter, and by the end of communication capability, five years after encounter, *Pioneer* will be 2.4 billion kilometers (1.5 billion miles) away. At these distances, the spacecraft radio signal becomes extremely faint, and round-trip communications times are two or three hours.

For all of NASA's unmanned missions in deep space, such as planetary and Sun-orbiting spacecraft, the network provides the tracking information on course and direction of the flight, velocity and range from Earth. It also receives engineering and science telemetry, including planetary television coverage, and sends commands for spacecraft operations. All communication links are in the S-band frequency.

The 64-meter antenna at Goldstone, and those under construction at Madrid and Canberra, can receive eight times the volume of data of the other antennas.

The DSN will support the operation by acquiring data telemetered from the spacecraft at from 2,048 down to 16 bits per second through the 26-meter antennas in early parts of the mission, and the 64-meter antennas later on. Data will be routed immediately to the SFOF, distributed to computers and other specialized processing equipment and then displayed in the PMSA at JPL or the PMAA at Ames for analysis and spacecraft control.

At the same time DSN stations will relay spacecraft doppler tracking information to the SFOF where other large computer systems will calculate orbits in support of navigation and planetary target planning. High-speed data links from all stations are capable of 4,800 bps. These new data links allow real-time transmission of all data from spacecraft to PMSA and on to the PMAA at Ames. Throughout the mission, recorded scientific data will be sent from DSN stations to Ames where it will be reduced and converted into magnetic tapes for transmission to the individual experimenters.

The planning and analysis of tracking and data acquisition functions of the Pioneer Mission are carried out by special teams of engineers, including a DSN Mission Operations Team and a DSN Project Engineering Team. The Operations team is responsible not only for the operation of the Network but coordination of near-Earth phase assistance and NASCOM support. The other team has charge of resources, operations planning and configuration control.

All of NASA's networks are under the direction of the Office of Tracking and Data Acquisition, NASA Headquarters, Washington, D.C. The Jet Propulsion Laboratory manages the DSN, while the MSFN facilities and NASCOM are managed by NASA's Goddard Space Flight Center, Greenbelt, Maryland. The Goldstone DSN stations are operated and maintained by JPL with the assistance of the Philco-Ford Corporation.

The Woomera and Canberra stations are operated by the Australian Department of Supply. The Johannesburg station is operated by the South African government through the National Institute for Telecommunications Research. The two facilities near Madrid are operated by the Spanish government's Instituto Nacional de Tecnica Aeroespacial (INTA).

Pioneer Plaque Location

NATIONAL AERONAUTICS AND SPACE ADMINISTRATION
Washington, D. C. 20546
202-755-8370

FOR RELEASE:
November 19, 1973
12:00 Noon

RELEASE No: 73-243K

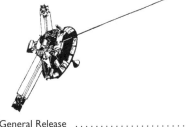

PROJECT: Pioneer 10
Jupiter Encounter

P
R
E
S
S

K
I
T

CONTENTS

**NATIONAL AERONAUTICS AND
SPACE ADMINISTRATION**
Washington, D. C. 20546
202-755-8370

November 7, 1973

Nicholas Panagakos   (202) 755-3680
Headquarters, Washington, D.C.

Peter Waller   (415) 965-5091
Ames Research Center, Mountain View, California

RELEASE No: 73-243

FOR RELEASE:
November 19, 1973
12:00 Noon

Pioneer Will Reach Jupiter December 3

Man's first spacecraft to Jupiter will reach the giant planet on December 3, after a billion-kilometer (620-million)-mile journey that began nearly two years ago.

Pioneer 10 will glide by at a distance of 131,400 kilometers (81,000 miles), taking pictures of the brightly-colored planet and making measurements that should provide new clues to old Jovian mysteries.

Pioneer will return color pictures of Jupiter and its inner moons, and provide information on its turbulent atmosphere and cloud currents, its bizarre Red Spot, its murky interior, and the surrounding magnetic, electrical, and radiation environment.

After Jupiter, the spacecraft will continue into deep space, eventually becoming the first manmade object to leave the solar system.

A few hours before closest approach, *Pioneer 10* will begin to test the second major hazard of its flight, Jupiter's intense radiation belts.

These are believed by some to be a million times stronger than Earth's belts and to grow 100 times more intense for each Jupiter radius a spacecraft moves closer to the planet. The radiation belts may cripple or destroy *Pioneer*, and their intensity and extent must be known for future missions to Jupiter.

Most immediately, information on the belts will be used to shape the voyage of *Pioneer 10*'s sister spacecraft, *Pioneer 11*. It will determine whether *Pioneer 11* can safely approach Jupiter even more closely than *Pioneer 10* – to within 32,000 kilometers (20,000 miles) – close enough to use Jupiter's gravity to make the first trip to the next planet, Saturn.

Pioneer 10 overcame the first major hurdle of its 21-month mission last summer when it successfully navigated the asteroid belt, the mass of rock and space debris that lies between the orbits of Mars and Jupiter. The spacecraft established that the high-velocity asteroidal particles in the Belt are infrequent and pose little hazard to outer planet missions.

Pioneer 11, scheduled to encounter Jupiter a year from now, entered the asteroid belt last August. It will exit in March.

The Pioneer 10 mission marks the beginning of the exploration of the outer planets – Jupiter, Saturn, Uranus, Neptune and Pluto.

The mission has already set an array of records.

- *Pioneer* has traveled farther and faster than any other man-made object. Its launch speed of 51,704 km/hr (32,114 mph), a record in itself, will be surpassed by its speed at closest approach to Jupiter of 132,000 km/hr (82,000 mph).
- It has communicated many times farther than ever before, and may communicate from as far out as Uranus' orbit, almost two billion miles away.
- It was first to fly beyond Mars' orbit, first to cross the asteroid belt, and will become the first man-made object to escape the solar system.
- It will be first to use Jupiter swingby for this escape.
- It is the first NASA spacecraft to use all-nuclear power.

Pioneer is remote in time as well as distance. Its radio signals, traveling at the speed of light (186,000 miles per second), take 46 minutes to reach Earth from Jupiter.

Pioneer will arrive at Jupiter within a minute of the planned time – one minute in almost two years of flight.

Pioneer 10 will return information on the mass and orbits of many of Jupiter's moons. It will take low-resolution pictures of the four, large inner moons, measure their temperatures, indicate some of their surface characteristics, and look for atmospheres. It will look especially for an ionosphere and atmosphere on the next-to-closest moon, Io. Orange-colored Io is the most reflective object known in the solar system.

Pioneer will locate the beginning of Jupiter's strong magnetic field by studying the bow shock wave the field makes in the "solar wind" which streams out from the Sun. It will determine the structure of the field, and whether it is offset as far to the south as some believe. It will try to find whether Jupiter's moons and magnetic field are related to the planet's massive bursts of radio noise. It will study the electromagnetic field relationships, which produce one kind of radio signal from Jupiter.

The spacecraft also will map Jupiter's radiation belts to determine kinds and energies of high-energy particles they contain. It will look for a possible invisible dust ring in Jupiter's equatorial plane. It also will look for auroras similar to the Earth's, and will measure the structure and density of Jupiter's ionosphere.

Looking into the layer of transparent atmosphere, above the planet's cloud tops, *Pioneer* will estimate its depth and amount, its characteristics as a gas, its mixing rate and density. It will look for liquid droplets, aerosols, and proportions of major constituents in this upper atmosphere – hydrogen, helium, and ammonia. It also will look for hot and cold spots (temperature distribution) in the outer atmosphere, and measure the brightness temperature of Jupiter's dark hemisphere.

A spacecraft instrument will take pictures of the planet in two colors. They may be substantially better than the best photographs taken with Earth-based telescopes. The pictures and other measurements should determine much about structure, temperature, and dynamics of prominent features, such as the Great Red Spot and the cloud regions and currents making up Jupiter's major belts and zones.

Scientists will use spacecraft data to determine some things about the regions below Jupiter's dense cloud layers and in the planet's interior. The *Pioneer* instruments will look for ice caps of frozen methane or ammonia. They will measure Jupiter's net heat output from its internal heat source. (Jupiter appears to radiate three times as much energy as it receives from the Sun.)

Pioneer will measure the total mass of the planet, determine polar flattening, due to Jupiter's high rotational speed, to within one-half mile, and the mass distribution in Jupiter's surface layers. Magnetic field data will be used to calculate the fluid composition and structure of Jupiter's interior.

Spacecraft operations during the encounter will be complicated by the 92 minutes of round-trip communications time, and the need to send 10,000 commands to the spacecraft in the two weeks centered on the closest approach – the most commands ever handled by the Deep Space Network. Operations strategy is to set most systems in one standard mode throughout encounter with most of the commands going to the spacecraft imaging system.

If operational difficulties occur, they will probably come from effects on the spacecraft of unknowns in the Jovian environment. These may include radiation damage, torques induced by Jupiter's magnetic field, arc-discharge of electrostatic charges, attitude changes due to high gravity, and possible loss of command after passage behind the Moon, Io.

Pioneer's target planet, Jupiter, is unique in the solar system. With more than twice the mass of all the other planets combined, it has more than 1,000 times the volume of the Earth. If Mars were placed on the face of Jupiter, it would look like a dime on a dinner plate.

Jupiter's atmosphere contains ammonia, methane, hydrogen, and probably water, the same ingredients believed to have produced life on Earth by chemical evolution about four billion years ago. Many scientists believe that large regions below the frigid cloud layer may be at room temperature. These conditions could allow the planet to produce living organisms despite the fact that it receives only $1/27^{th}$ of the solar energy received by Earth.

Jupiter probably is composed mostly of the two lightest elements, hydrogen and helium, the main constituents of the universe. The planet may have no solid surface. Due to its high gravity, it may go from a thick gaseous atmosphere down to oceans of liquid hydrogen, to a slushy layer, and then perhaps to a solid hydrogen core. Ideas of how deep beneath its striped cloud layers any solid hydrogen "icebergs" or "continents" might lie vary by thousands of kilometers.

Astronomers have long seen violent circulation of the planet's large-scale cloud currents and other features. A point on Jupiter's equator rotates at 35,400 km/hr (22,000 mph) during its ten-hour day, compared with 1,600 km/hr (1,000 mph) for a similar point on Earth's equator.

The Great Red Spot, the "Eye of Jupiter," is a huge oval 48,000 kilometers (30,000 miles) long and 13,000 kilometers (8,000 miles) wide, large enough to swallow up several Earths. The Red Spot may be an enormous stagnant column of gas, or even a raft of hydrogen ice floating on a bubble of warm hydrogen in the cooler hydrogen atmosphere and bobbing up and down at 30-year intervals, disappearing and reappearing.

Pioneer 10 weighs 260 kilograms (570 pounds) and is spin-stabilized, giving its instruments a full circle scan five times a minute. It uses nuclear sources for electric power because sunlight is too weak at Jupiter for an efficient solar-powered system.

Its 2.75-meter (9-foot) dish antenna looks back at Earth throughout the mission – adjusting its view by changes in spacecraft attitude as the home planet moves in its orbit around the Sun.

Pioneer carries a 30-kilogram (65-pound) scientific payload of 11 instruments.

Partners with the spacecraft are the three incredibly sensitive "big dish" antennas of NASA's Deep Space Network, which retrieve *Pioneer* data. *Pioneer*'s 8-watt radio signal reaches these antennas from Jupiter with a power of $1/100,000,000,000,000,000$ watts. Collected for 19 million years, this energy would light a 7.5-watt Christmas tree bulb for one-thousandth of a second.

In addition to its findings en route to Jupiter about the Asteroid Belt, *Pioneer 10* has found a variety of things about the solar atmosphere. One example is that, contrary to predictions, the solar atmosphere shields the solar system from cosmic rays at least as far out as Jupiter's orbit.

After Jupiter, *Pioneer* will head out of the solar system, crossing Saturn's orbit in 1976, Uranus' orbit in 1979, Neptune's orbit in 1983, and in 1987, 15 years after launch, the orbit of Pluto, the boundary of the solar system. *Pioneer*'s destination among the stars of the galaxy will then be somewhere in the constellation Taurus. The spacecraft carries a plaque telling any intelligent species who may find it several million years from now, who sent it and where it came from.

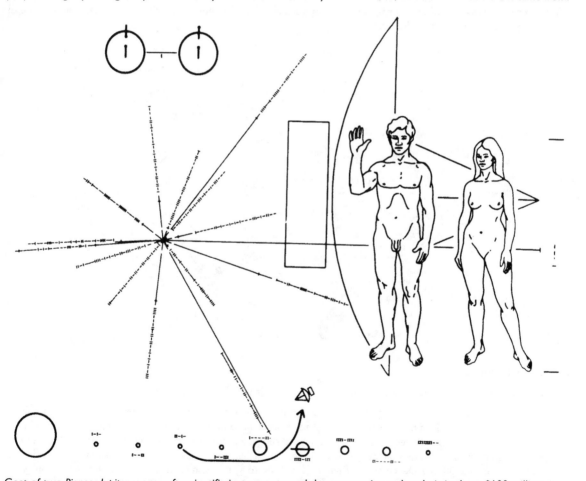

Cost of two *Pioneer Jupiter* spacecraft, scientific instruments, and data processing and analysis is about $100 million.

NASA's Office of Space Science has assigned management of the Pioneer Jupiter project to NASA's Ames Research Center, Mountain View, California, near San Francisco. The spacecraft were built by TRW Systems, Redondo Beach, California. The scientific instruments were provided by NASA centers, universities, and private industry.

Tracking and data acquisition is the responsibility of NASA's Deep Space Network, operated by the Jet Propulsion Laboratory, Pasadena, California.

(End of General Release. Background information follows.)

Encounter Profile

Pioneer 10 began its journey to Jupiter and eventual escape from the solar system on March 2, 1972.

On its 641-day trip to Jupiter, it has followed a curving path about one billion kilometers (620 million miles) long, covering about a million miles a day.

Pioneer 10 will assess the mission's principal remaining hazard when it determines the numbers and energy flux of particles in Jupiter's radiation belts by flying through them.

High energy protons (hydrogen nuclei) and electrons in the belts could penetrate deeply into the spacecraft and degrade the functioning of vital solid state circuitry, crippling or destroying it.

But even if this happens, *Pioneer 10* will have accomplished a major objective because questions on the severity of the radiation belts will have been answered. Exploration of the asteroid belt and return of much other interplanetary and planetary science, and incoming pictures of the planet also will have been achieved.

Because Jupiter's gravity will bend the spacecraft's course, the actual flyby will be a wide loop around the planet. It will approach from Jupiter's sunlit side then swing almost completely around its dark side. *Pioneer 10* will fly about two-thirds of the way around Jupiter at various distances. During the 48 hours of closest approach, Jupiter will have completed five rotations.

Pioneer 10 will approach in a counterclockwise direction (looking down from the north pole) – the same direction as Jupiter's rotation. The spacecraft will pass over part of Jupiter's southern hemisphere, cross the equator at an angle of 14°, and depart over the northern hemisphere.

Pioneer 10 Mission

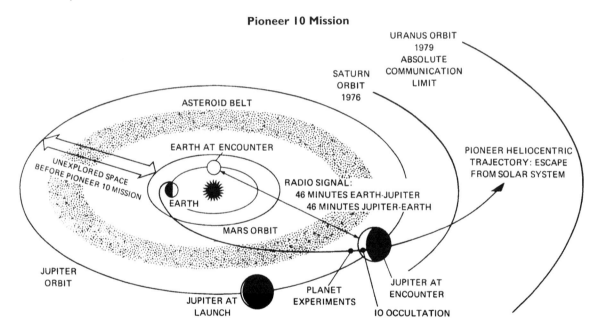

Encounter Sequence

The two-month *Pioneer* encounter period, November 3, 1973 to January 3, 1974, breaks down as follows:

- During the three weeks, November 3 to 24, the spacecraft (still in interplanetary space) will move in from about 24 million kilometers (15 million miles) from the planet to about 9 million kilometers (5.7 million miles).
- From November 25 to 29, *Pioneer* is expected to pass through the bow shock wave in the solar wind created by Jupiter's magnetic field and enter Jupiter's magnetosphere, approaching to about 4.7 million kilometers (3 million miles).
- Between November 30 and December 2, *Pioneer 10* will fly to within 2 million kilometers (1.2 million miles) of Jupiter and will pass through the outer magnetosphere, while viewing the planet from a distance.
- During the last day and a half (38 hours), before closest approach on December 3, *Pioneer* will reach its closest distance of 130,000 kilometers (81,000 miles) above the planet's cloud tops (periapsis). During this last 38 hours, it will make its most important planet measurements and pictures, and will pass through the most intense parts of the radiation belts. It also will traverse Jupiter's dust belt (if any) in the planet's equatorial plane.

The spacecraft will repeat this sequence as it exits the planet environment.

Pioneer 10 And 11 Events

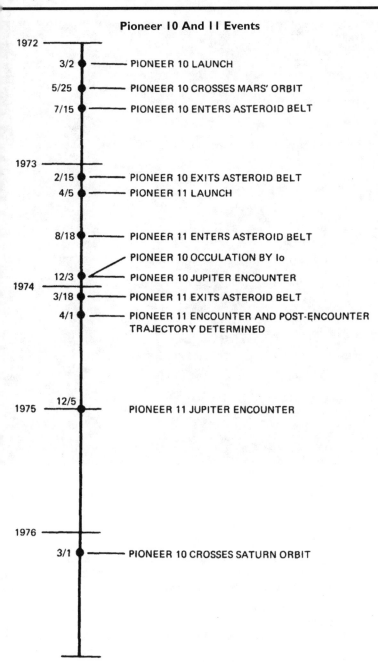

1972	
3/2	PIONEER 10 LAUNCH
5/25	PIONEER 10 CROSSES MARS' ORBIT
7/15	PIONEER 10 ENTERS ASTEROID BELT
1973	
2/15	PIONEER 10 EXITS ASTEROID BELT
4/5	PIONEER 11 LAUNCH
8/18	PIONEER 11 ENTERS ASTEROID BELT
	PIONEER 10 OCCULATION BY Io
12/3	PIONEER 10 JUPITER ENCOUNTER
1974	
3/18	PIONEER 11 EXITS ASTEROID BELT
4/1	PIONEER 11 ENCOUNTER AND POST-ENCOUNTER TRAJECTORY DETERMINED
12/5	PIONEER 11 JUPITER ENCOUNTER
1975	
1976	
3/1	PIONEER 10 CROSSES SATURN ORBIT

Passage of Jupiter's Moons

Early in the encounter, on November 8, *Pioneer* crossed the orbit of Hades, first of Jupiter's four other moons; the following day, it crossed the orbit of Pan.

On November 11, it crossed Amdrastea's orbit. On November 22, eleven days later, *Pioneer* will reach and cross the orbits of the three middle moons, Demeter, Hera, and Hestia, located 11.3 million kilometers (7 million miles) from Jupiter.

On December 3, the day of encounter, the spacecraft will make its closest approach to all five of Jupiter's inner moons. It will come within 1,392,300 kilometers (860,000 miles) of Callisto, which is as large as the planet Mercury, 14 hours before periapsis. It will sweep by Ganymede at a distance of 446,250 kilometers (280,000 miles), 12.5 hours from periapsis. Ganymede, Jupiter's largest moon, is larger than Mercury, although not nearly as dense. At 7 hours out, *Pioneer* will come within 320,000 kilometers of Europa, which is about the size of the Earth's moon. It will most closely approach Io, which is larger than our moon, at about 6 hours from periapsis.

The spacecraft will come nearest to the orbit of Amalthea, Jupiter's tiny inner moon, at periapsis itself. Amalthea's orbit will still be 18,500 kilometers (11,420 miles) inside *Pioneer*'s closest approach distance.

Pictures

Imaging of Jupiter and its four large moons began on November 5, 29 days from periapsis. Early images were largely for calibration. On November 26, imaging activity went on a 23-hour-a-day basis.

At periapsis minus 48 hours, on December 1, imaging results should become equal to, or better than, those of Earth telescopes.

At periapsis minus about 8 hours, on December 3, the planet will just fill the spacecraft camera's view field. This fourth-from-last picture before periapsis may actually be the last one because, after this, the possibility that Jupiter's hard radiation may cripple the spacecraft camera is greatly increased.

Numbers of pictures:

- December 2. *Pioneer 10* is expected to take 23 images of Jupiter, and two images each of Io, Ganymede, and Callisto.
- December 3. Plans call for 18 pictures of Jupiter, and one each of Io and Europa.
- December 4. After closest approach, 23 images of the planet are scheduled, looking back toward the dark side at a crescent Jupiter.
- December 5. There will be 22 pictures.
- December 10. Imaging will return to a schedule of 3 to 8 hours a day.

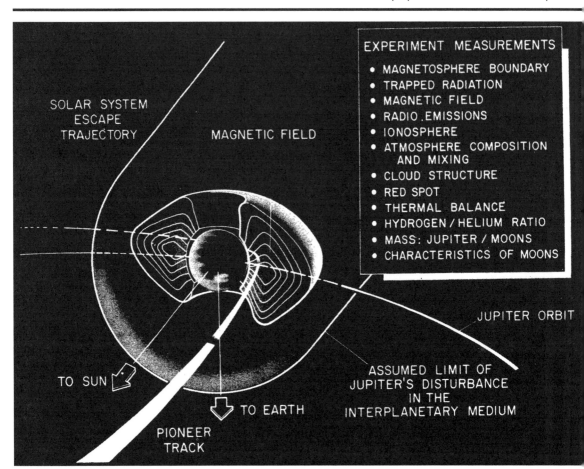

The Imaging Photopolarimeter (IPP), which takes the pictures, will use half the spacecraft's data return capacity of 1024 bits per second during encounter, largely for picture transmission. About 60 percent of the IPP observation time will be used for pictures.

Encounter Science Sequence

The first evidence at the spacecraft of Jupiter's presence may come on November 25. This is the earliest date for passage through the bow shock wave created as the solar wind strikes Jupiter's magnetic field, and for measuring the effects of streams of high-energy solar particles on Jupiter's field. *Pioneer* will cross the bow shock wave between November 25 and December 1.

On November 26, *Pioneer 10* will make its final attitude change to assure precise pointing of the spacecraft at Earth for optimum communications. To calculate Jupiter's mass distribution, and masses and orbits of its moons during encounter, it is necessary to entirely eliminate the spacecraft thrusts required for attitude changes because these would alter *Pioneer*'s course slightly. The investigator will use even tiny course changes caused by varying gravity fields to make mass and orbit calculations.

The earliest date for entering Jupiter's magnetic field is November 29. Latest date is December 2.

Studies of the Jovian magnetic field should provide information on Jupiter's interior and on its massive radio signals.

The first ultraviolet viewing of Jupiter will begin on November 30, continuing for 38 hours. These studies will examine Jupiter's upper atmosphere for hydrogen / helium ratio, mixing rate, and auroras. A second viewing period, closer to the planet, will last for 5 hours, beginning at 9 hours before periapsis. The earlier view period may be the most important because the ultraviolet instrument may be saturated with radiation by the second view time.

Infrared viewing to determine temperature occurs from 18 hours to 11 minutes before periapsis. It begins on December 2 with measurements of the moon Callisto. *Pioneer* will also "take the temperature" of Ganymede, Europa, Io, and tiny Amalthea.

The most important infrared measurements will be of the planet itself, during an 82-minute viewing period starting at 2 hours and 11 minutes before periapsis. These observations should produce a thermal map, temperature of the dark side and prominent features, and data on thermal balance from Jupiter's internal heat source.

The infrared instrument will have a good view of the boundary between Jupiter's light and dark hemispheres.

Since November 5, the Imaging Photopolarimeter has been studying the nature of atmospheric gas above the planet's atmosphere, and the amount of aerosols it contains, as well as structure and composition of Jupiter's upper cloud layers. The IPP will provide similar information on Jupiter's moons.

About six hours from periapsis, *Pioneer* may begin encountering hard radiation, and four high-energy particle counters will chart the radiation belts, determining types and energies of particles.

This should determine the radiation belt hazard at Jupiter. The cumulative effect of 12 hours of intense radiation on *Pioneer* will further help define this hazard.

Three hours before periapsis, imaging and polarimetry measurements of Jupiter's Great Red Spot will begin. Combined with temperature measurements and other findings, these may help man understand this enormous region.

Periapsis will occur at 6:26 p.m. PST, and 10 minutes later *Pioneer* will pass from the southern hemisphere through Jupiter's equatorial plane. Most of Jupiter's moons are in this plane, and if the planet has an invisible dust belt, it is expected there. Experimenters expect a possible increase in dust concentrations of 10 to 1,000 times from the almost dustless interplanetary environment, probably not enough dust to pose a hazard. The micrometeroid detector will measure impacts of dust particles.

Fifteen minutes after periapsis, the spacecraft will pass behind Io, and experimenters will look for refraction effects on its radio signal as the signal grazes the moon. Experimenters hope to find an ionosphere and traces of an atmosphere on Io. Io's ionosphere is believed to link up lines of force in Jupiter's magnetosphere, allowing enormous discharges of electricity to produce Jupiter's most intense radio signals.

An hour after passage behind Io, the spacecraft will pass behind Jupiter itself for 65 minutes. Experimenters on Earth will make refraction studies similar to those at Io. These should tell the electron density of Jupiter's ionosphere and give information on the density and composition of its atmosphere.

Measurements will be made both going in and coming out from behind Jupiter.

Two hours and 51 minutes after periapsis, the spacecraft will enter the Jovian night for 51 minutes. Returning into sunlight, *Pioneer* will begin its exit from Jupiter's environment, continuing to make measurements as it goes out.

Post-Encounter

After encounter, solar wind, particle, and magnetometer experiments will look for planetary effects on the interplanetary medium behind Jupiter.

The basic Pioneer 10 mission is defined to last through encounter plus 3 months. (The costs of about $100 million for the Pioneer 10 and 11 flights cover only this basic mission.) It is not possible to predict exactly how long *Pioneer* will function after encounter.

The most interesting experimental questions beyond Jupiter will be: What is the solar wind and solar magnetic field expansion process? What is the flux of galactic cosmic rays and the distribution of neutral hydrogen, non-solar-wind plasmas and interstellar hydrogen and helium out beyond Jupiter's orbit? What do these tell about the interstellar space beyond the boundary of the heliosphere (the Sun's atmosphere)?

Of equal interest will be the search for the heliosphere boundary itself, "where the solar wind stops blowing" and interstellar space begins. The plasma, magnetic field, high-energy particle, and ultraviolet photometer experimenters will share in this.

During the post-encounter period, communications distance will grow steadily longer. Incoming signals will become almost infinitesimal and the time required for Earth-based scientists to command the spacecraft and get a response will lengthen to hours.

Jupiter's gravity and orbital velocity will have speeded up the spacecraft to solar system escape speed, and will have bent the spacecraft trajectory inward toward Jupiter's orbital path. Cruise velocity relative to the Sun of 38,500 km/hr (24,000 mph) before encounter will have increased to 73,100 km/hr (45,400 mph) a year afterwards.

In 1976, the spacecraft will cross the orbit of Saturn at about 1.5 billion kilometers (950 million miles) from the Sun, and will cross Uranus' orbit at about 2.9 billion kilometers (1.8 billion miles) from the Sun in 1979 – but communication will only be marginally possible at Uranus' great distance.

Pioneer will then continue into interstellar space. Six years after Jupiter encounter, about 3.2 billion kilometers (2 billion miles) away, the spacecraft's flight path will have curved 87° farther around the Sun than the Jupiter flyby point. Its velocity relative to the Sun at this point will be 53,300 km/hr (33,100 mph). From then on, its flight path will curve still more around the Sun, another 25°. At this point, far out in interstellar space, it will proceed away from the Sun in essentially a straight line, and its velocity will have dropped to a permanent 41,400 km/hr (25,000 mph).

Jupiter

What We Know

See *What We Know* in the **Pioneer 10 Press Kit**, page 22.

Clouds, Currents, and Visual Appearance. Seen through a telescope, the lighted hemisphere of Jupiter is almost certainly a view of the tops of gigantic regions of towering multi-colored clouds.

Overall, due to its rotation, the planet is striped or banded, parallel with its equator, with large, dusky gray regions at both poles. Between the two polar regions are five bright, salmon-colored stripes, known as zones, and four darker, slate-gray stripes, known as belts – the South Equatorial Belt, for example. The planet as a whole changes hue periodically, possibly as a result of the Sun's 11-year activity cycles.

The Great Red Spot in the southern hemisphere is frequently bright red, and since 1665 has disappeared completely several times. It seems to brighten and darken at 30-year intervals.

Scientists agree that the cold cloud tops in the zones are probably largely ammonia vapor and crystals, and the gray polar regions many be condensed methane. The bright cloud zones have a complete range of colors from yellow and delicate gold to red and bronze. Clouds in the belts range from gray to blue-gray.

Cloud Markings

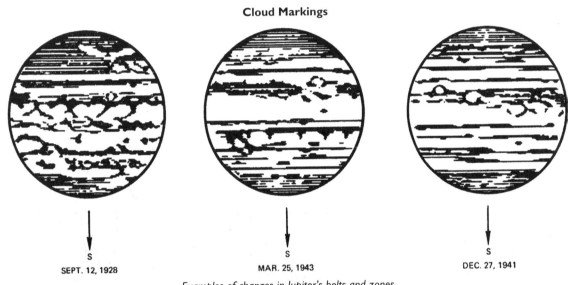

| S | S | S |
| SEPT. 12, 1928 | MAR. 25, 1943 | DEC. 27, 1941 |

Examples of changes in Jupiter's belts and zones

In addition to the belts and zones, many smaller features – streaks, wisps, arches, loops, patches, lumps, and spots - can be seen. Most are hundreds of thousands of kilometers in size.

Circulation of these cloud features has been identified in a number of observations. The Great Equatorial Current (The Equatorial Zone), 20 degrees wide, sweeps around the planet 410 km/hr (225 mph) faster than the cloud region on either side of it, and is like similar atmospheric jet streams on Earth. The South Tropical circulating current is a well-known feature, as is a cloud current which sweeps completely around the Great Red Spot.

When Jupiter passed in front of stars in 1953 and again in 1971, astronomers were able to calculate roughly the molecular weight of its upper atmosphere by the way it refracted the stars' light. They found a molecular weight of around 3.3 which means a large proportion of hydrogen (molecular weight 2) because all other elements are far heavier. (Helium is 4, carbon 12, and nitrogen 14.)

Under Jovian gravity, atmospheric pressure at the cloud tops is calculated to be up to ten times one atmosphere on Earth.

The transparent atmosphere above the clouds can be observed spectroscopically and in polarized light. It is believed to be at least 60 kilometers (35 miles) thick.

Scientists have suggested from cloud-top observations by telescope that the general circulation pattern of Jupiter's atmosphere is like that of Earth, with circulation zones corresponding to Earth's equatorial, tropical, subtropical, temperate, subpolar, and polar regions. However, Jupiter's polar regions (from an atmospheric circulation standpoint) appear to begin at about 26 degree latitude from the equator, instead of at 60 degrees as on Earth.

Magnetic Fields and Radiation Belts. See *Magnetic Fields and Radiation Belts* in the **Pioneer 10 Press Kit**, page 24.

Jovian Radio Signals. See *Jovian Radio Signals* in the **Pioneer 10 Press Kit**, page 24.

Temperature. See *Temperature* in the **Pioneer 10 Press Kit**, page 24.

Jupiter Unknowns

See *Jupiter Unknowns* in the **Pioneer 10 Press Kit**, page 24.

Life. Perhaps the most intriguing unknown is the possible presence of life in Jupiter's atmosphere.

Estimates of the depth of the Jovian atmosphere beneath the cloud layer vary from 100 to 6,000 kilometers (60 to 3,600 miles). The compositions and interactions of the gases making up the atmosphere are unknown. If the atmosphere is deep, it must also be dense. By one estimate, with an atmospheric depth of 4,200 kilometers (2,600 miles), pressure at the Jovian "surface" would be 200,000 times Earth's atmospheric pressure due to the total weight of gas in the high Jovian gravity. One source cites eight different proposed models of Jupiter's atmosphere.

However, scientists do appear to agree on the presence of liquid water droplets in the atmosphere. Since the planet is believed to have a mixture of elements similar to that found in the Sun, it is almost sure to have abundant oxygen. And most of this oxygen has probably combined with the abundant Jovian hydrogen as water.

If large regions of Jupiter's atmosphere come close to room temperature, both liquid water and water ice should be present.

Temperature

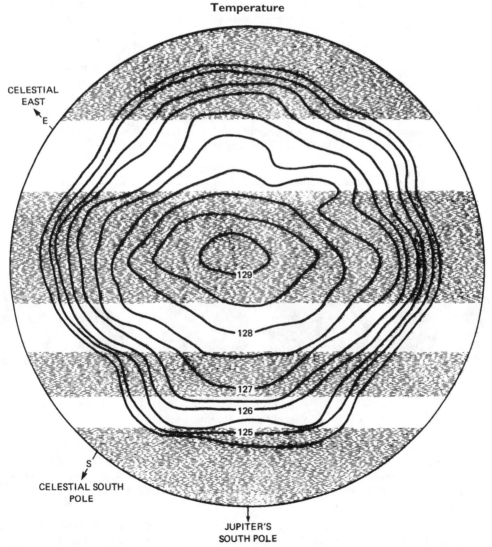

Composite map of 8 to 14 micron brightness temperature distribution (°K) over the Jovian disk.

Jupiter's atmosphere contains ammonia, methane, and hydrogen. These constituents, along with water, are the ingredients of the primordial "soup" believed to have produced the first life on Earth by chemical evolution. On this evidence, Jupiter could contain the building blocks of life.

Some scientists suggest that the planet may be like a huge factory, turning out vast amounts of life-supporting chemicals (complex carbon-based compounds) from these raw materials, using its own internal energy. If so, life could exist without photosynthesis. Any solar photosynthesis would have to be at a very low level since Jupiter receives only 1/27th of Earth's solar energy. It would probably be low-energy life forms at most (plants and microorganisms) because there is believed to be no free oxygen. Life forms would float or swim because a solid surface, if any, would be deep within Jupiter at very high pressures.

Planet Structure. See *Planet Structure* in the **Pioneer 10 Press Kit**, page 25.

A Hot Planet. See *A Hot Planet* in the **Pioneer 10 Press Kit**, page 26.

Magnetic Field. See *Magnetic Field* in the **Pioneer 10 Press Kit**, page 26.

Jupiter's Moons

Jupiter's 12 natural satellites have some odd characteristics. The second moon, Io, appears to be brighter for 10 minutes after emerging from Jupiter's shadow. If so, the simplest explanation, supported by recent stellar occultation observations, is that Io has an atmosphere (probably nitrogen or methane) which "snows out" on the surface when Io is on the cold, dark side and revaporizes when back in sunlight. Io also is distinctly orange in color and has odd reflecting properties. Most of the sunlight reaching the moon's surface is strongly scattered back, making Io extremely bright. This reflection is believed to be more pronounced for Io than for any other known object in the solar system. Infrared measurements show Europa and Ganymede to have water ice on their surfaces.

The inner moons in order of distance from the planet are: tiny Amalthea, diameter 160 kilometers (100 miles), which orbits Jupiter twice a day at only 1.5 planet diameters, 106,000 kilometers (66,000 miles), above the cloud tops; and the four large moons Io, Europa, Ganymede, and Callisto, whose orbits lie between 422,000 kilometers (262,000 miles) and 1,882,000 kilometers (1,169,000 miles) from Jupiter.

Beyond these are the seven tiny outer moons. The inner three of these, Hestia, Hera, and Demeter, have orbits which lie between 11.5 and 11.7 million kilometers (7.2 and 7.3 million miles) from Jupiter.

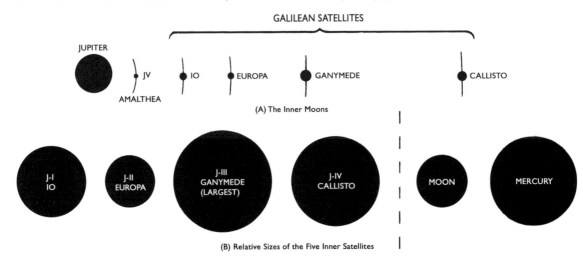

The satellites of Jupiter – distances and sizes

Orbits of the four outermost, Andrastea, Pan, Poseidon, and Hades, lie between 20.7 million and 23.7 million kilometers (12.9 and 14.7 million miles) of Jupiter. All are in retrograde orbits, moving counter to the usual direction of planet rotation. This suggest that they may be asteroids captured by Jupiter's powerful gravity. Diameters of six of the outer moons range from 15 by 40 kilometers (9 to 24 miles), with Hestia, the seventh, having a diameter of about 130 kilometers (81 miles).

Orbital periods of the four large inner moons range from 1.7 days (for Io) to 16.7 days. Orbital periods of the inner three of the outer seven moons are around 250 days. While the four farthest-out, backward-orbiting moons complete their circuits of Jupiter in around 700 days.

The backward orbit of the far outer moon, Poseidon, is highly inclined to the equator, and wanders so much due to various gravitational pulls that astronomers have a difficult time finding it.

The Experiments and Projected Findings

The 13 *Pioneer 10* experiments are designed to produce findings about Jupiter and the heliosphere. Their objectives are:

For Jupiter:

- To map the magnetic field.
- To measure distributions of high-energy electrons and protons in the radiation belts, and look for auroras.
- To find a basis for interpreting the decimetric and dacametric radio emission from Jupiter.
- To detect and measure the bow shock and magnetospheric boundary and their interactions with the solar wind.
- To verify the thermal balance and to determine temperature distribution of the outer atmosphere.
- To measure the hydrogen / helium ratio in the atmosphere.
- To measure the structure of the ionosphere and atmosphere.
- To measure the brightness, color, and polarization of Jupiter's reflected light.
- To perform two-color visible light imaging.
- To increase the accuracy of orbit predictions and masses of Jupiter and its moons.

For Interplanetary Space:

- To map the interplanetary magnetic field.
- To study the radial gradient of the solar wind and its fluctuations and structure.
- To study the radial and transverse gradients and arrival directions of high-energy charged particles (solar and galactic cosmic rays).
- To investigate the relationships between the solar wind, magnetic field, and cosmic rays.
- To search for the boundary and extent of the heliosphere (solar atmosphere).
- To determine the density of neutral hydrogen.
- To determine the properties of interplanetary dust.

The instruments aboard *Pioneer 10* return data in roughly the following categories:

Magnetic Fields

Magnetometer (Jet Propulsion Laboratory). See *Magnetometer* in the **Pioneer 10 Press Kit**, page 36.

Interplanetary Solar Wind and Heliosphere

Plasma Analyzer (Ames Research Center). See *Plasma Analyzer* in the **Pioneer 10 Press Kit**, page 36.

Jupiter's Radiation Belts, Radio Emissions, and Cosmic Rays

Charged Particle Composition Instrument (University of Chicago). See *Charged Particle Composition Instrument* in the **Pioneer 10 Press Kit**, page 36.

Cosmic Ray Telescope (Goddard Space Flight Center). See *Cosmic Ray Telescope* in the **Pioneer 10 Press Kit**, page 37.

Jupiter's Charged Particles

Geiger Tube Telescopes (University of Iowa). See *Geiger Tube Telescopes* in the **Pioneer 10 Press Kit**, page 37.

Jovian Trapped Radiation Detector (University of California at San Diego). See *Jovian Trapped Radiation Detector* in the **Pioneer 10 Press Kit**, page 37.

Jupiter's Dust Belts, Meteoroids, and Interplanetary Dust

Asteroid-Meteoroid Detector (Sisyphus) (General Electric Company). See *Asteroid-Meteoroid Detector* in the **Pioneer 10 Press Kit**, page 37.

Meteoroid Detector (Langley Research Center). See *Meteoroid Detector* in the **Pioneer 10 Press Kit**, page 38.

Celestial Mechanics

Celestial Mechanics (Jet Propulsion Laboratory). See *Celestial Mechanics* in the **Pioneer 10 Press Kit**, page 38.

Jupiter's Atmosphere, Temperature, Auroras, Moons, Interplanetary Hydrogen, Helium, and Dust

Ultraviolet Photometer (University of Southern California). See *Ultraviolet Photometer* in the **Pioneer 10 Press Kit,** page 38.

Pictures of Jupiter, the Planet's Atmosphere

Imaging Photopolarimeter (University of Arizona). This instrument provides data in a number of areas using photometry (measurement of light intensity) and polarimetry (photometry measurements of the linear polarization of light), and imaging.

The instrument will take pictures of Jupiter as the spacecraft approaches the planet.

Resolution of many of the pictures should be better than that from the best Earth-based telescopes. The pictures also will be taken from viewing angles impossible to obtain from Earth. At least one will show Jupiter's terminator (the line between sunlit and dark hemispheres), never seen from Earth. This should show texture and shapes of cloud regions.

Images will be made in both red and blue light, and these will be superimposed, providing "color pictures" in red and blue.

Photo Imaging System

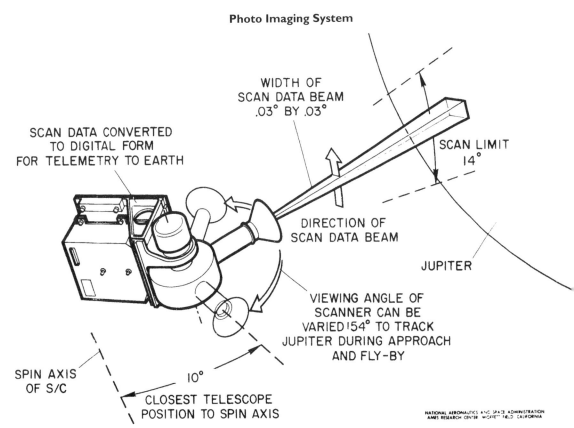

WIDTH OF
SCAN DATA BEAM
.03° BY .03°

SCAN DATA CONVERTED
TO DIGITAL FORM
FOR TELEMETRY TO EARTH

SCAN LIMIT
14°

DIRECTION OF
SCAN DATA BEAM

JUPITER

VIEWING ANGLE OF
SCANNER CAN BE
VARIED 154° TO TRACK
JUPITER DURING APPROACH
AND FLY-BY

SPIN AXIS
OF S/C

10°

CLOSEST TELESCOPE
POSITION TO SPIN AXIS

NATIONAL AERONAUTICS AND SPACE ADMINISTRATION
AMES RESEARCH CENTER MOFFETT FIELD CALIFORNIA

The instrument uses a photoelectric sensor which measures changes in light intensity, something like the light sensor for a television camera. But unlike a TV camera it will employ the 5 rpm spin of the spacecraft to scan the planet, in narrow strips 0.03° wide.

This electronic imaging system will complete the scans for a picture in from 25 to 110 minutes, depending on distance from the planet. Scan data are converted to digital form on the spacecraft and radioed to Earth by telemetry. Engineers then will build up the images, using various computer techniques.

Pictures also will be taken of Jupiter's four large moons.

Controllers can vary the viewing angle of the instrument's 8.64-cm (3.4-inch)-focal length telescope by about 160° relative to the spacecraft's spin axis, which is fixed on the Earth-spacecraft line. The telescope can see to within 10° of the spin axis, looking away from the Earth.

For the first pictures, the telescope will be looking almost down the spacecraft spin axis toward Jupiter. For the last pictures, close to the planet, it will be looking almost at right angles to the spin axis.

Picture reconstruction will be complicated by the fact that picture scans will be curved as the telescope rotates with the spacecraft, while looking out at various angles to the spin axis. The only straight-line scans will be those looking straight out from the spin axis. A second factor is that Jupiter's equatorial zone rotates at 35,400 km/hr (22,000 mph). Depending on flyby trajectory, at periapsis the spacecraft may be traveling in roughly the same direction as the planet's rotation at 132,000 km/hr (82,000 mph) relative to Jupiter, or much faster.

These two factors mean that computers will have to sort out scan paths whose curvature changes steadily with spacecraft movement. They will also have to compensate for smear caused by high-speed motion of the planet and the spacecraft.

During Jupiter flyby, experimenters also will use the instrument to obtain an improved understanding of the structure and composition of Jupiter's clouds, and determine the amount and nature of atmospheric gas above the clouds and aerosols in this gas. These data also will shed light on Jupiter's heat balance. The instrument will make light scattering measurements of the Jovian satellites. These data may lead to an understanding of the surface properties, possible atmospheres, and origins of the moons.

The instrument includes a 2.5-cm (1-inch)-aperture, 8.6-cm (3.4-inch)-focal length telescope which can be moved 160° in the plane of the spacecraft spin axis by ground command or automatically. Incoming light is split by a prism according to polarization into two separate beams. Each beam is further split by going through a red filter (5,940-7,200 angstroms), and through a blue filter (3,900-5,000 angstroms). Channeltron detectors turn the light into electrical impulses, which are telemetered to Earth. The instrument uses three viewing apertures: one 40x40 milliradians (mR) for zodiacal light measurements, a second 8x8 mR for nonimaging light measurements of Jupiter, and a third 0.5x0.5 mR for scans of the planet from which pictures will be reconstructed.

The instrument weighs 4.3 kilograms (9.5 pounds) and uses 2.2 watts of power.

Jupiter's Atmosphere, Ionosphere, Temperature

Infrared Radiometer (California Institute of Technology). See *Infrared Radiometer* in the **Pioneer 10 Press Kit**, page 40.

Occultation Experiment (Jet Propulsion Laboratory). See *Occultation Experiment* in the **Pioneer 10 Press Kit**, page 40.

Encounter Time Line

Notes:
1. All distances measured from Jupiter's surface (cloud tops).
2. P* Periapsis (closest approach)
3. R_J stands for Jupiter radius.
4. R_J = 71,372 kilometers (44,350 miles).
5. All times PST.

11-03-73   P-30 days
            Start *Pioneer* encounter
11-04-73   P-29 days
            First imaging and polarimetry of Jupiter by University of Arizona Imaging Photopolarimeter (IPP) instrument. After this date, from 3 to 8 hours of imaging and polarimetry every day through November 25, 1973, as well as from December 11, 1973 to January 2, 1974.
11-08-73   P-25 days, 16 hours
 2:26 a.m. Cross orbit of Hades, Jupiter's outermost moon, at 329 RJ, 23,630,000 kilometers (14,680,000 miles) from Jupiter.
19:26 a.m. P-25 days, 8 hours
            Cross orbit of Poseidon at 325 RJ, 23,202,000 kilometers (14,420,000 miles) from Jupiter.
11-09-73   P-24 days, 6 hours
12:26 a.m. Cross orbit of Pan, 312 RJ, 22,273,600 kilometers (13,840,000 miles) from Jupiter
11-11-73   P-22 days, 8 hours
10:26 a.m. Cross orbit of Andrastea at 289 RJ, 20,632,000 kilometers (12,821,000 miles) from Jupiter.
11-22-73   P-11 days, 4 hours
 2:26 p.m. Cross orbits of Demeter and Hera at 163 RJ, 11,636,600 kilometers (7,231,000 miles) and 11,655,600 kilometers (7,243,600 miles) from Jupiter.
 6:26 a.m. P-11 days, 12 hours
            Cross orbit of Hestia at 159 RJ, 11,400,600 kilometers (7,084,000 miles) from Jupiter.
11-24-73   P-9 days
 6:26 p.m. Earliest date for entering Jupiter's bow shock wave in the solar wind. Should occur sometime in six-day period 11-24-73 to 11-30-73, P-9 days to P-3 days. This event shows presence of Jupiter's magnetic field. The bow shock will be measured by the Jet Propulsion Laboratory (JPL) magnetometer and Ames Research Center plasma (solar wind) instrument.

11-25-73   P-8 days
  6:26 p.m. *Pioneer 10* is 8,333,580 kilometers (5,180,000 miles) from Jupiter.

11-26-73
    All day 12 images of Jupiter. Polarimetry of Jupiter and Io.
  6:26 a.m. P-7 days, 12 hours to P-7 days, 4 hours, 6:26 a.m. to 2:26 p.m.
        Last spacecraft attitude change before periapsis, and until 8:26 p.m. December 5 (P+2 days, 2 hours). This
        is the last adjustment of spacecraft pointing at Earth (for communications) prior to encounter. For the JPL
        celestial mechanics experiment, it is essential to eliminate spacecraft thrusts during encounter. The
        experimenter uses gravity effects on the spacecraft trajectory to calculate Jupiter's mass distribution and
        mass and orbits of moons.
  6:00 p.m. Begin 23-hour-a-day imaging and polarimetry for 15 days through December 10, P-7.5 days. Both imaging
        and polarimetry of Jupiter will occur every day through this period, imaging a little more than half the time.
  6:26 p.m. P-7 days
        Pioneer 10 is 7,455,000 kilometers (4,633,000 miles) from Jupiter.

11-27-73
    All day 26 images of Jupiter. Polarimetry of Io, Ganymede, Jupiter.
    All day Jupiter marble-sized in 8x8 inch (14°x14°) view-frame of imaging instrument.
  4:00 a.m. Trim maneuver if greater pointing accuracy required than achieved in yesterday's maneuver. 6 hour period,
        4 a.m. to 10 a.m.
  6:26 p.m. P-6 days
        Pioneer 10 is 6,560,550 kilometers (4,077,000 miles) from Jupiter.

11-28-73   P-5 days
  6:26 p.m. *Pioneer* is 5,645,300 kilometers (3,508,000 miles) from Jupiter.
  6 :26 p.m. Earliest time for passage of magnetopause, actual entrance to Jupiter's magnetic field. This will occur in a
        3.5 day period between 11-28-73 and 12-1-73, from P-5 days to P-1 day, 12 hours.

11-29-73
    All day 25 images of Jupiter. Polarimetry of Ganymede, Europa, Jupiter.
  6:26 p.m. P-4 days
        Pioneer is 4,703,400 kilometers (2,923,000 miles) from Jupiter.

11-30-73
    All day 20 images of Jupiter. Polarimetry of Callisto, Io, Europa, Jupiter.
  2:26 p.m. Begin first view of Jupiter in ultraviolet (UV) light with University of Southern California ultraviolet
        photometer. Will last 38 hours, 3 minutes (from 2:26 p.m. November 30 to 4:29 a.m. December 2). P-76
        hours to P-37 hours, 57 minutes.
        First UV period considered prime because of potential radiation background interference during second
        closer view on December 3. Will investigate atmosphere composition and search for possible auroral
        activity. Observations cover 1,673,919 kilometers (1,040,000 miles) of flight, from 3,892,700 kilometers
        (2,419,000 miles) from Jupiter to 2,218,500 kilometers (1,378,800 miles) from Jupiter.
  6:26 p.m. P-3 days
        Latest time to enter Jupiter's bow shock wave.
  6:26 p.m. *Pioneer* is 3,722,400 kilometers (2,313,000 miles) from Jupiter.

12-01-73
    All day 26 images of Jupiter, 2 images of Io. Polarimetry of Ganymede, Callisto, Jupiter.
    All day Jupiter golf-ball-sized in 8x8 inch (14°x14°) view-frame of imaging system.
  6:26 p.m. P-48 hours
        Point on incoming trajectory where spacecraft imaging resolution becomes better than resolution of largest
        Earth telescopes. Based on 100 or more picture elements across the planetary disc. Best Earth telescopes
        get from 80 to 150 elements, but over 100 is rare.
  6:26 p.m. P-48 hours
        Pioneer is 2,681,000 kilometers (1,666,000 miles) from Jupiter.

12-02-73
    All day 23 images of Jupiter, 2 images of Io, 2 images of Callisto, 2 images of Ganymede. Polarimetry of Ganymede,
        Callisto, Jupiter.
  1:00 a.m. P-17.2 hours
        Jupiter is 3 inches in diameter in 8x8 inch (14°x14°) view-frame of imaging system.
  4:28 a.m. End first UV view of Jupiter (see 11-30-73).
  6:26 a.m. P-1 day, 12 hours
        Latest time for passage of Jupiter's magnetopause, actual entry into the planet's magnetic field.

7:05 a.m. P-35 hours, 21 minutes
      Infrared (IR) view of Callisto with California Institute of Technology Infrared Radiometer. Will determine Callisto's heat characteristics.

7:26 a.m. P-35 hours
      Pioneer is 29 RJ, 2,070,000 kilometers (1,286,000 miles) from Jupiter.

12:26 p.m. P-30 hours
      Cross orbit of Callisto at 25.4 RJ, 1,810,400 kilometers (1,125,000 miles) from Jupiter.

6:26 p.m. P-1 day
      Pioneer is 1,526,200 kilometers (948,380 miles) from Jupiter.

12-03-73

All day 18 images of Jupiter, 1 image of Io, 1 image of Europa. Polarimetry of Jupiter.

3:17 a.m. P-15 hours, 19 minutes
      Infrared view of Ganymede.

4:26 a.m. P-14 hours
      Closest approach to Callisto 1,392,300 kilometers (865,200 miles).

4:26 a.m. P-14 hours
      Cross orbit of Ganymede at 14RJ, 798,800 kilometers (496,400 miles) from Jupiter. Ganymede is Jupiter's largest moon.

5:56 a.m. P-12 hours, 30 minutes
      Closest approach to Ganymede, 446,250 kilometers (277,300 miles).

8:40 a.m. Start second ultraviolet view of Jupiter. View period 4 hours, 48 minutes from 8:40 a.m. to 1:28 p.m., P-9 hours, 46 minutes to P-4 hours, 58 minutes. Radiation may saturate instrument during this second view period.

10:26 a.m. P-8 hours
      Cross Europa's orbit at 8.4 RJ, 599,400 kilometers (372,500 miles) from Jupiter.

10:26 a.m. P-8 hours
      Jupiter just fills 8x8 inch (14°x14°) view-frame of imaging system.

11:08 a.m. P-7 hours, 18 minutes
      Infrared view of Europa

11:26 a.m. P-7 hours
      Closest approach to Europa at 321,000 kilometers (193,470 miles)

12:26 p.m. P-6 hours
      Enter region of intense radiation. Hard radiation continues for 12 hours until 12:26 a.m. 12-4-73 from P-6 hours to P+6 hours.
      Critical region for the spacecraft and future missions with regard to radiation hazards. Hard radiation belts begin at 5 RJ from planet. They will be measured by high-energy particle detectors from Universities of California-San Diego, Iowa, and Chicago; and NASA-Goddard Space Flight Center.

12:38 p.m. P-5 hours, 48 minutes
      Infrared view of Io, Jupiter's closest large moon, and most reflective known object in the solar system.

2:26 p.m. Crossing of Io's orbit at P-4 hours – 4.9 RJ, 356,200 kilometers (221,300 miles) from Jupiter.

2:26 p.m. P-4 hours
      Jupiter overlaps the 8x8-inch (14°x14°) view-frame of the imaging system. Has about a 12-inch diameter relative to frame.

2:56 p.m. P-3 hours, 30 minutes
      Closest approach to Io at 357,000 kilometers (221,840 miles).

3:45 p.m. Begin imaging of Red Spot for about 1 hour, 2:45 a.m. to 4:45 a.m. This takes place between 142,830 kilometers (88,740 miles) and 196,380 kilometers (122,000 miles) from Jupiter, 2 RJ to 2.75 RJ.

4:02 p.m. P-2 hours, 24 minutes
      The 8x8-inch (14°x14°) view-field of the imaging system covers about a quarter of the full disc of Jupiter.

4:15 p.m. Begin 82-minute viewing period of Jupiter in infrared light, continues until 5:37 a.m. (P-2 hours, 11 minutes to P-49 minutes).
      Investigate Jupiter's atmospheric composition, temperature structure, and thermal balance. This is the only opportunity to determine these through IR throughout the whole mission. View period is 3 RJ to 2 RJ from Jupiter.

6:15 p.m. P-11 minutes
      IR view of Amalthea.

6:26 p.m. P-0 days
      Periapsis – 130,000 kilometers (81,000 miles) from Jupiter's cloud tops.

6:26 p.m. Closest approach to Malthea's orbit, 18,500 kilometers (11,470 miles).
      Amalthea is Jupiter's smallest and closest Moon.

6:36 p.m. P+10 minutes
> Crossing Jupiter's equatorial plane – Location of a potentially hazardous dust ring. NASA-Langley Research Center meteoroid detector will count dust particles. General Electric Co. Asteroid-Meteoroid Telescope will make dust observations during encounter, if not saturated by radiation.

6:41 p.m. Begin 1 minute, 31 second occultation by Io, 6:41:45 to 6:43:16 p.m., P+15 minutes, 45 seconds to P+17 minutes, 16 seconds. Jet Propulsion Laboratory experimenter will determine whether Io has an atmosphere and an ionosphere by studying effects on spacecraft radio signal of its passage by the moon. Io's atmosphere is not measurable from the Earth.

7:42 p.m. Begin 65 minute occultation by Jupiter, from 7:42 to 8:47 p.m. (P+1 hour, 16 minutes to 2 hours, 21 minutes). Determine planet's ionospheric structure and investigate nature of neutral atmosphere via studies on Earth of spacecraft radio signal effects. (See Io occultation just above.) During this time, *Pioneer* is out of contact with Earth.

8:16 p.m. Begin solar eclipse of spacecraft – Flight through Jovian night (shadow).
> Duration is 51 minutes, 8:16 p.m. to 9:07 p.m. (P+1 hour, 50 minutes to P+2 hours, 41 minutes.)

12-04-73
> All day 23 images of planet looking back at thin crescent Jupiter. Polarimetry of Jupiter, Io.

12:26 a.m. Exit hard radiation region.

6:26 p.m. P+1 day
> *Pioneer* is 1,526,200 kilometers (948,256 miles) from Jupiter.

12-05-73
> All day 22 images of Jupiter. Polarimetry of Europa, Ganymede, Callisto, Jupiter.

6:26 a.m. P+2 days
> *Pioneer* is 2,681,000 kilometers (1,660,000 miles) from Jupiter.

6:26 a.m. Earliest time to exit magnetopause outbound.

8:26 p.m. P+2 days, 2 hours

12-06-73 First precession (adjustment of spacecraft attitude) after periapsis. 8-hour period, 4:00 a.m. to 12 noon.
> All day 16 images of Jupiter. Polarimetry of Io, Jupiter.

6:26 p.m. P+3 days
> *Pioneer* is 3,722,000 kilometers (2,313,000 miles) from Jupiter.

12-07-73
> All day 31 images of Jupiter. Polarimetry of Callisto, Ganymede, Europa, Jupiter.

6:26 p.m. P+4 days
> *Pioneer* is 4,703,000 kilometers (2,922,000 miles) from Jupiter.

12-08-73
> All day 25 images of Jupiter. Polarimetry of Io, Ganymede, Europa, Jupiter.

6:26 p.m. P+5 days
> *Pioneer* is 5,645,000 kilometers (3,508,000 miles) from Jupiter.

12-09-73
> All day 38 images of Jupiter. Polarimetry of Callisto, Io, Jupiter.

6:26 a.m. Earliest time to exit through Jupiter's bow shock in the solar wind. Begins 10.5 day period when shock front exit could take place. Latest time of exit is 12-20-73. (Period is P+5.5 to P+20 days.)

6:26 p.m. P+6 days
> *Pioneer* is 6,560,300 kilometers (4,076,000 miles) from Jupiter.

12-10-73
> All day 20 images of Jupiter. Polarimetry of Callisto, Jupiter.

6:26 p.m. P+7 days
> *Pioneer* is 7,455,100 kilometers (4,633,000 miles) from Jupiter.

12-11-73 Return to imaging and polarimetry only 3 to 8 hours per day each day through January 2, 1974.

6:26 p.m. P+8 days
> *Pioneer* is 8,334,000 kilometers (5,180,030 miles) from Jupiter.

12-19-73 P+16 days
> Latest time to exit bow shock.

01-02-74 End Jupiter encounter.

Jupiter Pictures – Quantity and Quality

Pioneer 10's imaging system (see Experiments – Imaging Photopolarimeter) will take pictures in red and blue light – 336 views of Jupiter and 10 of its moons in the 2-week period centered on periapsis. If the system works well, about 80 of these pictures, taken from 48 hours before to 48 hours after periapsis, could have resolution equal to or better than that of Earth telescopes.

In addition, *Pioneer* will fly from Jupiter's day side to its night side — providing view-angles impossible from Earth, and shadow-definition of features. Fifty hours before periapsis, the Imaging Photopolarimeter will see Jupiter in almost full sunlight. Forty minutes before periapsis, Jupiter will be half-dark, and at periapsis itself, about 60 percent of Jupiter will be dark.

In the last few pictures before periapsis, there should be shadow definition of clouds, cloud currents and eddies, the Great Red Spot, and other features, providing information that cannot be obtained from Earth. Greatest shadow definition should be within a few degrees of the boundary between Jupiter's light and dark hemispheres.

On the pessimistic side, some of the best views of Jupiter may never reach the Earth because of temporary saturation or damage to the channeltrons in the imaging instrument due to Jupiter's intense radiation belts.

By one estimate, the last good image, free of radiation effects, may well be the fourth before closest approach, taken at ten Jupiter radii, 700,000 kilometers (440,000 miles) from the planet. This would occur about 8 hours before periapsis and show the planet almost filling the 14° instrument field of view. During this view the instrument might be able to define objects from 700 to 1,000 kilometers (435 to 620 miles) in length.

The third closest picture will be taken from about 350,000 kilometers (five Jupiter radii) about four hours before periapsis, and will show an area on Jupiter's surface of 89,000 kilometers (55,300 miles) on a side. This picture would show objects from 350 to 525 kilometers (220 to 325 miles) in length.

The second picture before periapsis will be taken at 215,000 kilometers (three Jupiter radii) about an hour before periapsis, and will cover an area on Jupiter's surface roughly 53,340 kilometers (33,145 miles) on a side. Here, objects from 200 to 300 kilometers (124 to 186 miles) long might be seen.

The last picture, a long narrow one showing the terminator, is planned at periapsis, at 130,000 kilometers (81,000 miles) out. This picture may resolve objects 130 to 195 kilometers (80 to 120 miles) long.

In addition to radiation damage to the camera, other factors may reduce quality of images produced by *Pioneer*. First of these is possible radiation damage to various parts of the spacecraft communications system.

Beyond this, since the instrument "sees" the planet in narrow scans or strips, overlap and underlap of these strips may reduce quality. Because everything is moving quickly — the spacecraft at 82,000 mph in three dimensions; and the planet's surface, at 22,900 mph — there may be computer difficulties in unscrambling the camera scans.

Picture Time Schedule

Schedule of Times For Build-Up of Jupiter Pictures at Ames Research Center

* Times are PST. Each time period represents one picture.

Jupiter Imaging			
December 2:		6:21 p.m. to	6:57 p.m.
Callisto Imaging			
December 2:	1.	8:32 p.m. to	8:47 p.m.
	2.	8:55 p.m. to	9:16 p.m.
Ganymede Imaging			
December 2:		9:38 p.m. to	10:26 p.m.
Jupiter Imaging			
December 3:	1.	12:17 a.m. to	12:42 a.m.
	2.	12:42 a.m. to	1:07 a.m.
	3.	1:07 a.m. to	1:32 a.m.
	4.	1:32 a.m. to	1:58 a.m.
	5.	1:53 a.m. to	2:23 a.m.
	6.	2:23 a.m. to	2:48 a.m.
	7.	2:48 a.m. to	3:14 a.m.
	8.	3:14 a.m. to	3:39 a.m.
	9.	3:39 a.m. to	4:05 a.m.
	10.	4:05 a.m. to	4:30 a.m.
	11.	4:30 a.m. to	4:55 a.m.
	12.	4:55 a.m. to	5:21 a.m.
	13.	5:21 a.m. to	5:46 a.m.
	14.	5:46 a.m. to	6:23 a.m.
	15.	6:23 a.m. to	7:01 a.m.
	16.	7:01 a.m. to	7:49 a.m.
	17.	7:49 a.m. to	8:37 a.m.
	18.	8:37 a.m. to	9:47 a.m.
	19.	9:47 a.m. to	11:01 a.m.
Europa Imaging			
December 3:		11:26 a.m. to	11:52 a.m.
Io Imaging			
December 3:		12:02 p.m. to	12:22 p.m.
Jupiter Imaging			
December 3:	1.	2:28 p.m. to	3:30 p.m.
	2.	5:15 p.m. to	6:14 p.m.
	3.	6:59 p.m. to	7:11 p.m.

(periapsis 6:26 p.m., plus 46 minutes or 7:12 p.m. on Earth PST)

December 4:	4.	12:06 a.m. to	1:13 a.m.
	5.	1:20 a.m. to	3:00 a.m.
	6.	3:40 a.m. to	5:16 a.m.
	7.	5:22 a.m. to	6:26 a.m.
	8.	6:30 a.m. to	7:47 a.m.
	9.	8:19 a.m. to	8:52 a.m.
	10.	8:57 a.m. to	9:32 a.m.
	11.	2:05 p.m. to	2:37 p.m.
	12.	2:44 p.m. to	3:16 p.m.
	13.	3:26 p.m. to	4:10 p.m.
	14.	4:17 p.m. to	4:58 p.m.
	15.	5:05 p.m. to	6:14 p.m.

"Noise" in all the electronic steps between the planet and the final television tube can reduce quality. Since the system is digital, it reports light intensities at 64 levels, not in continuous tones as does the eye. The system may have some difficulty defining low-contrast objects.

Nonetheless, it should be able to resolve objects of the sizes discussed above.

Assuming that the photopolarimeter survives Jupiter's radiation and can resolve objects 200 to 300 kilometers long in its final full picture, a resolution five to seven times that obtainable through Earth telescopes can be expected.

Radiation Hazard

The scientific instruments are susceptible to damage from Jupiter's radiation belts in the following order: The Asteroid-Meteoroid Detector is especially sensitive to radiation. Three of the four high-energy radiation counters are considered to be next most vulnerable. The solar wind, ultraviolet, and photopolarimeter are third most likely to be damaged. The magnetometer, the fourth high-energy radiation counter, the infrared and meteoroid instruments are least likely to suffer radiation damage.

Among spacecraft subsystems, all are equally susceptible to radiation damage.

Encounter Operations

During the two weeks of *Pioneer 10*'s encounter with Jupiter, the team of spacecraft controllers, analysts, and scientists at Ames Research Center, and personnel of NASA's Deep Space Network (DSN) will run a three-shift, 24-hour-a-day operation.

They will be sending from 10,000 to 15,000 commands to the spacecraft – by far the most intensive command activity the DSN has ever handled. They will have to be constantly alert to analyze and recover from expected but unpredictable problems with spacecraft systems or with scientific instruments caused by the unknown environment of Jupiter.

Their job will be complicated by the 92 minutes of required time for round-trip communications to Jupiter. They will have to wait an hour and a half between the sending of a command and verification of its execution.

The *Pioneers* are controlled primarily by instructions from Earth – not by operations sequences stored in onboard computers, though five commands can be stored on board. The spacecraft also can store up to 49,152 bits of science data.

The fully ground-controlled spacecraft allows great flexibility for changing plans during flight. It requires continuously effective communications. Ground-commanded spacecraft probably are less expensive to build, and easily adapted to a variety of missions. *Pioneer*'s Jupiter encounter will be a rigorous test of the commandable spacecraft concept.

Mission planners have simplified the encounter operation.

Operations strategy is to turn on, well in advance, most spacecraft subsystems and scientific instruments in a standard operating mode, and leave them this way throughout the encounter.

If all goes well, only a few routine spacecraft operations will need to be commanded. These are: daily checks of pointing direction of the spin axis (spacecraft attitude), changes in telemetry data formats either twice or three times a day, and readjustments of spacecraft attitude on the third day after periapsis.

Only 5 of the 11 scientific instruments will require commands. Three of the five will need only a few commands per day. Of the other two, the infrared radiometer will require several hundred commands on the two days before periapsis to scan Jupiter's five close moons. The bulk of the 10,000-plus commands will direct the imaging photopolarimeter in scanning the planet or its moons, for pictures of polarimetry.

If *Pioneer* were not a half billion miles away, remote in both time and distance, linked to Earth only by a thread of radio signal, and traveling through an unknown but highly active environment, the planned strategy would be nearly foolproof.

As it is, many types of problems have had to be anticipated. Some eight major rehearsals and training tests have been held since July, involving the entire world-wide operations team.

Controllers have tested all redundant, backup systems on the spacecraft. The known problems are on Earth. With the large amount of DSN equipment, some failures are almost certain. Rehearsals have involved: swaps of computers in case of failure, loss of high-speed data links, and coping with the 10-minute loss of command during switch-over between DSN stations.

At Jupiter, there are a number of potential problems:

- Damage to subsystems and scientific instruments by the radiation belts.
- "Sticking" of charged particles to the spacecraft in the radiation belts, especially in the Jovian night zone where there is no photon scouring from sunlight. This could cause charge build-ups and arcing between parts of the spacecraft insulated from each other, could do damage, and also generate radio signals which could produce unplanned commands.
- During the Io occultation, communications will be cut for 90 seconds. This could cause loss of ability to command since the frequency of the spacecraft receiver is constantly tuned to the Earth signal. A minute and a half loss of this signal causes the spacecraft receiver frequency to drift, and might mean failure to receive commands during the hour between Io and Jupiter occultations. Therefore, no important commands are planned in this period.
- High Jovian gravity pulling unevenly on spacecraft parts of different mass (such as the magnetometer at the end of its long, lightweight boom) may slightly alter spacecraft position, changing scan paths for the pictures.

- Passage of *Pioneer* through Jupiter's strong magnetic field will set up eddy currents in the spinning spacecraft, creating magnetic torques and slowing spacecraft spin somewhat.
- If Jupiter has dust belts, these might mean high-velocity particle penetrations of *Pioneer* and damage to subsystems.

The six noncommanded scientific instruments include: the ultraviolet photometer, which is "on" throughout the encounter. (Its two viewing periods result from the combination of the instrument's look angle and changes in spacecraft attitude as it moves around Jupiter.)

Other noncommand instruments are: the magnetometer, two of the four high-energy particle instruments, the asteroid-meteoroid telescope, and the meteoroid detector. The three instruments needing several commands a day for calibration or to react to unexpected levels of planetary phenomena are the solar wind instrument, and the other two high-energy particle counters.

During encounter operations, the many other uncommanded functions on the spacecraft will be monitored in several ways. *Pioneer* returns complete data on its subsystems every 48 seconds. This is scanned by computer. With uncommanded changes, or functions beyond normal limits, the computer presents alarm signals on controllers' data displays.

In addition, analysts will review spacecraft data every 10 minutes.

Pioneer control and spacecraft operations are at the Pioneer Mission Operations Center (PMOC), Ames Research Center, Mountain View, California.

The PMOC has computing capability both for commanding the spacecraft and to interpret the data stream as it comes in from the DSN stations.

Several Ames organizations direct and support the encounter.

The Pioneer Mission Operations Team (PMOT) consists of personnel from many government and contractor organizations, and operates under control of the Flight Director.

In addition to several assistant flight directors, the PMOT includes the following groups:

- The Spacecraft Performance Analysis Team analyzes and evaluates spacecraft performance and predicts spacecraft responses to commands.
- The Navigation and Maneuvers Team handles spacecraft navigation, and attitude control in space.
- The Science Analysis Team determines the status of the onboard scientific instruments and formulates command sequences for the instruments.

Analysis teams from 10 of the 11 experimenter organizations will be at Ames to monitor performance of their instruments. The University of Arizona team monitoring responses of its Imaging Photopolarimeter to 10,000-plus commands, to catch improper picture scans, will be especially active.

Tracking and Data Retrieval

With facilities located at 120° intervals around the Earth, NASA's Deep Space Network (DSN) will support *Pioneer* continuously during encounter. As the spacecraft "sets" at one station, due to the Earth's rotation, it will "rise" at the next one.

The DSN maintains three highly sensitive 64-meter (210-foot) dish antennas at Goldstone, California; Madrid, Spain; and Canberra, Australia. As a backup, DSN has six 26-meter (85-foot) dish antenna stations at Madrid (Robledo de Chavela and Cebreros); Canberra; and Johannesburg, South Africa; with two stations at Goldstone.

The 26-meter stations have enough power to command *Pioneer* at Jupiter. Where necessary they can receive data, but at the extremely low rate of 128 bits per second (bps), compared with the 1024 bps reception rate of the 64-meter antennas.

At periapsis, there will be a 4-hour overlap between the Goldstone and Canberra 64-meter stations, with both receiving this vital data. However, Goldstone loses the spacecraft just after closest approach. Canberra will handle the last part of periapsis operations. This will mean long-distance communications from Australia to Pioneer Control at Ames.

As *Pioneer* is drawn in by Jupiter's gravity, speed increases from 20,000 mph to 81,000 mph will cause the doppler shift to grow very large. This will require "ramping" operations at tracking stations, complicated by communications cut-off when *Pioneer* passes behind the planet.

DSN stations can track the spacecraft for the 1.5 billion miles of communications distance, and the five years of flight time beyond Jupiter, if the mission continues that long. At these distances, the radio signal becomes incredibly faint, and round-trip communications times are 2 to 4 hours.

All DSN stations have general-purpose telemetry equipment capable of receiving, data-synchronizing, decoding, and processing data at high transmission rates.

The tracking and data acquisition network is tied to the Mission Control and Computing Center (MCCC), JPL's central data processing facility at Pasadena, by NASA's Communications Network (NASCOM).

During encounter, some Ames controllers will be at the MCCC to provide a backup command capability in case of failure of high-speed data links between DSN stations and Ames.

For all of NASA's unmanned missions in deep space, *Pioneers* and *Mariners*, the DSN provides tracking information on course and direction of the flight, velocity, and range from Earth. It also receives engineering and science telemetry, including planetary television coverage, and sends commands for spacecraft operations. All communications links are in the S-band frequency.

DSN stations relay spacecraft doppler tracking information to the MCCC, where computers calculate orbits in support of navigation and planetary target planning. High-speed data links from all stations are capable of 4,800 bps, permitting real time transmission of all data from spacecraft to Pioneer control centers at Ames or JPL. Throughout the mission, scientific data recorded on magnetic tape will be sent from DSN stations to Ames where it will be divided into individual magnetic tapes for each experimenter.

All of NASA's networks are under the direction of the Office of Tracking and Data Acquisition, NASA Headquarters, Washington, D.C. The Jet Propulsion Laboratory manages the DSN, while the STDN facilities and NASCOM are managed by NASA's Goddard Space Flight Center, Greenbelt, Maryland.

The Goldstone DSN stations are operated and maintained by JPL with the assistance of the Philco-Ford Corporation.

The Canberra station is operated by the Australian Department of Supply. The Johannesburg station is operated by the South African government through the National Institute for Telecommunications Research. The two facilities near Madrid are operated by the Spanish government's Instituto Macional de Tecnica Aerospacial (INTA).

The Spacecraft

Pioneer 10 is the first spacecraft designed to travel into the outer solar system and operate there, possibly for as long as 7 years and as far from the Sun as 2.4 billion kilometers (1.5 billion miles) or more.

For this mission, the spacecraft must have extreme reliability, be of very light weight, have a communications system for extreme distances, and employ a nonsolar power source.

The spacecraft is stabilized in space like a gyroscope by its 5-rpm rotation, so that its scientific instruments scan a full circle five times a minute. Designers chose spin stabilization for its simplicity and effectiveness.

Since the orbit planes of Earth and Jupiter coincide to about 1° the spacecraft will be in or near the plane of Earth's orbit (the ecliptic) throughout its flight. To maintain a known orientation in this plane, controllers regularly adjust spacecraft position so that the spin axis points constantly at Earth. The spin axis coincides with the center line of the radio beam, which also points constantly at Earth.

Spacecraft navigation is handled on Earth using two-way doppler tracking and angle-tracking.

For course corrections, the *Pioneer* propulsion system makes changes in velocity.

The spacecraft can return 1,024 data bits per second (bps) from Jupiter to the 64-meter (210-foot) antennas of the Deep Space Network. It can store up to 49,152 data bits while other data is being transmitted.

Pioneer 10 is controlled largely from Earth, rather than by sequences of commands stored in onboard computers.

Because launch energy requirements to reach Jupiter are far higher than for shorter missions, the spacecraft is very light. *Pioneer 10* weighed only 270 kilograms (570 pounds) at launch. This included 30 kilograms (65 pounds) of scientific instruments.

Because solar energy at Jupiter is only 4 percent of energy received at the Earth and grows steadily weaker beyond the planet, designers selected a nuclear power source over solar cells.

The mission's 13 experiments are carried out by 11 onboard scientific instruments. Two experiments use the spacecraft and its radio signal as instruments.

For reliability, spacecraft builders have employed an intensive screening and testing program for parts and materials. They have selected components designed to withstand radiation from the spacecraft's nuclear power source, and from Jupiter's radiation belts. In addition, key systems are redundant. (That is, two of the same component or subsystem are provided in case one fails.) Communications, command and data return systems, propulsion electronics, thrusters, and attitude sensors are largely redundant.

Pioneer / Jupiter Spacecraft

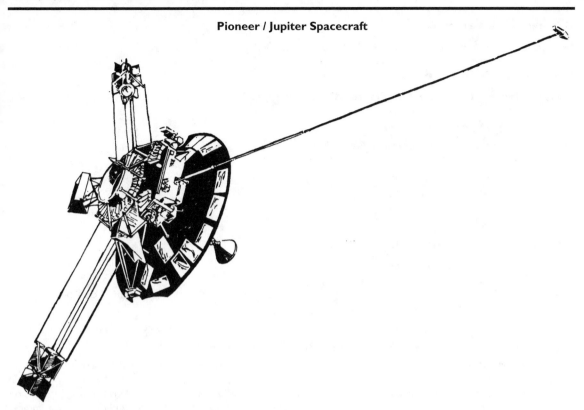

Pioneer 10 Description

Pioneer's Earth-facing dish antenna is in effect the forward end of the spacecraft. *Pioneer 10* is 2.9 meters (9.5 feet) long, measuring from its farthest forward component, the medium-gain antenna horn, to its farthest rearward point, the tip of the aft-facing omnidirectional antenna. Exclusive of booms, its widest crosswise direction is the 2.7-meter (9 foot) diameter of the dish antenna.

The axis of spacecraft rotation and the center-line of the dish antenna are parallel, and *Pioneer* spins constantly for stability.

The spacecraft equipment compartment is a hexagon box, roughly 35.5 cm (14 inch) deep. Each of its six sides is 71 cm (28 inch) long. One side joins to a smaller box also 35.5 cm (14 inch) deep, whose top and bottom are irregular hexagons.

The smaller compartment contains most of the 12 onboard scientific experiments. However, 12.7 kilograms (28 pounds) of the 30 kilograms (65 pounds) of scientific instruments (the plasma analyzer, cosmic ray telescope, four asteroid-meteoroid telescopes, meteoroid sensors, and the magnetometer sensors) are mounted outside the instrument compartment. The other experiments have openings for their sensors. Together, both compartments provide 1.4 square meters (16 square feet) of platform area.

Attached to the hexagonal front face of the equipment compartment is the 2.7-meter (9-foot)-diameter. 46-cm (18-inch)-deep dish antenna.

The high-gain antenna feed and the medium-gain antenna horn are mounted at the focal point of the antenna dish on three struts projecting about 1.2 meters (4 feet) forward of the rim of the dish. The low-gain, omnidirectional spiral antenna extends about 0.76 meters (2.5 feet) behind the equipment compartment.

Two three-rod trusses, 120° apart, project from two sides of the equipment compartment, deploying the spacecraft's nuclear electric power generator about 3 meters (10 feet) from the center of the spacecraft. A boom, 120° from each of the two trusses, projects from the experiment compartment and positions the helium vector magnetometer sensor 6.6 meters (21.5 feet) from the spacecraft center. The booms are extended after launch.

At the rim of the antenna dish, two Sun sensors are mounted. A star sensor looks through an opening in the equipment compartment and is protected from sunlight by a hood.

Both compartments have aluminum frames with bottoms and side walls of aluminum honeycomb. The dish antenna is made of aluminum honeycomb.

Rigid external tubular trusswork supports the dish antenna, three pairs of thrusters located near the rim of the dish, radioisotope thermoelectric generator trusses, and launch vehicle attachment ring. The plaque with its message for other intelligent species also is attached to this trusswork.

Orientation and Navigation

See *Orientation and Navigation* in the **Pioneer 10 Press Kit**, page 30.

Propulsion and Attitude Control

See *Propulsion and Attitude Control* in the **Pioneer 10 Press Kit**, page 30.

Nuclear-Electric Power

See *Nuclear-Electric Power* in the **Pioneer 10 Press Kit**, page 32.

Communications

See *Communications* in the **Pioneer 10 Press Kit**, page 33.

Command System

See *Command System* in the **Pioneer 10 Press Kit**, page 34.

Data Handling

The spacecraft data system turns science and engineering information into an organized stream of data bits for radio transmission back to Earth.

The system also can store up to 49,152 data bits for later transmission to Earth. Storage can take place while other data are being transmitted. This allows for taking data faster than the spacecraft can transmit it.

System components include the digital telemetry unit, which prepares the data for transmission in one of nine data formats at one of eight bit rates of from 16 to 2,048 bits per second; the data storage unit; and the convolutional coder, which rearranges the data in sequential form that allows identification and correction of errors by computers on the ground. This coding allows higher bit rates to be retained for greater spacecraft distances without loss of accuracy.

The nine data formats are divided into science and engineering groups. The science group includes two basic science formats and three special-purpose science formats. The basic science formats each contain 192 bits, 144 of which are assigned to the scientific instruments; the remainder are for operating instructions of various kinds. One of the basic science formats is primarily for interplanetary flight, the other for Jupiter encounter

The three special-purpose science formats transmit all 192 bits of data from only one or two instruments. This allows high data rates from one instrument such as the photopolarimeter during Jupiter picture taking, or the infrared radiometer whose principal purpose is to take a careful look at Jupiter's heat radiation just before closest approach.

There are four engineering formats, each of 192 bits. Each of the engineering formats specializes in one of the following areas: data handling, electrical, communications and orientation, and propulsion. These specialized data formats provide high bit rates during critical spacecraft events.

Short samples of science data can be inserted into engineering formats and vice versa.

The digital telemetry unit, which puts all data together, has redundant circuits for critical functions.

Timing

The digital telemetry unit contains a stable crystal-controlled clock for all spacecraft timing signals, including those for various orientation maneuvers and experiments.

Temperature Control

See *Temperature Control* in the **Pioneer 10 Press Kit**, page 34.

Reliability

See *Reliability* in the **Pioneer 10 Press Kit**, page 35.

The Encounter Team

See *The Project Team* in the **Pioneer 10 Press Kit**, page 42.

The Experimenters

See *Experiments and Investigators* in the **Pioneer 10 Press Kit**, page 41.

Pioneer 10 Contractors

See *Pioneer F Contractors* in the **Pioneer 10 Press Kit**, page 44.

The Heliosphere

Pioneer 10 may explore much of the heliosphere, the atmosphere of the Sun. The spacecraft may survey a region of space 2.4 billion kilometers (1.5 billion miles) wide between the orbits of Earth and Uranus. It may also learn something about the interstellar space outside the heliosphere.

The thinly diffused solar atmosphere is hundreds of times less dense than the best vacuums on Earth. Yet it is important because it contains:

- The ionized gas known as the solar wind, roughly a 50-50 mixture of protons (hydrogen nuclei) and electrons. It flows out from the 3,600,000 degrees F (2,000,000 degrees C) corona of the Sun in all directions at average speeds of 1.6 million km/hr (one million mph). Where does the solar wind stop blowing? Where is the boundary between the solar atmosphere and the interstellar gas? Estimates range from 560 million to 16 billion kilometers (350 million to 10 billion miles) from the Sun.
- Complex magnetic and electric fields, carried out from the Sun by the solar wind.
- Solar cosmic rays, high-energy particles thrown out by the huge explosions on the Sun's surface at up to 480 million km/hr (300 million mph).

The heliosphere also encompasses planets, moons, asteroids, comets and dust. It is traversed by electromagnetic radiation from the Sun: radio waves, infrared, ultraviolet, and visible light. Earth gets most of its energy from this radiation.

The heliosphere further contains cosmic ray particles from within and beyond our galaxy, traveling at nearly the speed of light. This means enormous particle energies, up to 10^{14} million electron-volts (MeV). The expression 10^{14} is 1 followed by 14 zeros. There are also neutral hydrogen atoms, believed to be part of the interstellar gas from which the Sun and planets were formed.

Study of these phenomena has a variety of applications. For example, storms of solar particles striking Earth interrupt radio communications and sometimes electrical power transmission.

There is evidence that solar storm particles travel through the heliosphere and trigger the Earth's long-term weather cycles.

The heliosphere can be thought of as a huge laboratory where phenomena occur that cannot be simulated on Earth. For example, man cannot accelerate particles in Earth laboratories to the near-light speeds reached by galactic cosmic ray particles. These particles are observed by *Pioneer* instruments.

Solar wind particles are so thinly spread that they rarely collide, and only the magnetic field and electrostatic fields link them together as a collisionless ionized gas (collisionless plasma). Even though collisionless, these plasmas transmit waves of many kinds because of this magnetic field linkage.

Knowledge of these plasmas can shed light on an area of current high interest – how to contain plasmas in magnetic fields. Accomplishing this would allow control of the hydrogen thermonuclear reaction continuously on a small scale to generate virtually unlimited electric power by a clean process.

All this involves such relationships as that of the temperatures of solar wind protons and electrons near Earth: they are almost the same, even though theory predicts that the electrons should be several hundred times hotter.

How is energy transferred between them to equalize the temperature? How far out from the Sun do these temperatures remain equal? Why?

Solar Wind, Magnetic Field, and Solar Cosmic Rays

Near the Earth, the speed of the solar wind varies from one to three million km/hr (600,000 to 2,000,000 mph), depending on activity of the Sun. Its temperature varies from 10,000 to 1,000,000 degrees C (18,000 to 1,800,000 degrees F). Near the Earth, collisions between streams of the solar wind use up about 25 percent of its energy. The wind also fluctuates due to features of the rotating solar corona, where it originates, and because of various wave phenomena.

The fastest solar cosmic ray particles jet out from the Sun streams which can cover the 150 million kilometers (93 million miles) to Earth in as little as 20 minutes. Slower particle streams take one or more hours to reach Earth. Both types tend to follow the curving interplanetary magnetic field.

The positive ions are 90 percent protons and ten percent helium nuclei, with occasional nuclei of heavier elements.

There are from none to hundreds of flare events on the Sun each year which produce high-energy solar particles, with the largest number of flares at the peak of the 11-year cycle.

Planetary Interactions and the Interstellar Gas

See *Planetary Interactions and the Interstellar Gas* in the **Pioneer 10 Press Kit**, page 27.

Galactic Cosmic Rays

See *Galactic Cosmic Rays* in the **Pioneer 10 Press Kit**, page 27.

Asteroids and Comets

Asteroids

Asteroids travel around the Sun in elliptical orbits like small planets. The Asteroid Belt forms a doughnut shaped region, lying between the orbits of Mars and Jupiter, 300 million to 545 million kilometers (186 million to 338 million miles) from the Sun.

The Belt is roughly 245 million kilometers (152 million miles) wide and extends about 40 million kilometers (25 million miles) above and below the plane of the Earth's orbit.

Scientists believe the asteroids either condensed individually from the primordial gas cloud which formed the Sun and planets, or somewhat less likely, that they are debris from the break-up of a very small planet. Clearly, they contain important information on the origin of the solar system.

Passage of *Pioneer 10* through the Asteroid Belt appears to have shown that high-velocity, smaller asteroid particles pose only a minor hazard to spacecraft. This is an important discovery because the belt is too thick to fly over or under, so all outer planet missions must fly through it.

Some estimates suggest that there is enough material in the Belt to make a planet with a volume about $1/1,000$th that of the Earth.

Astronomers have identified and calculated orbits for 1,776 asteroids. There may be 50,000 in the size range from the largest, Ceres, which has a diameter of 770 kilometers (480 miles), to bodies 1 mile in diameter.

In addition, the Belt is presumed to contain hundreds of thousands of asteroid fragments.

In the center of the Belt, asteroids and particles orbit the Sun at about 61,200 kilometers (38,000 miles) per hour.

A few asteroids stray far beyond the Belt. Hermes can come within about 354,000 kilometers (220,000 miles) of the Earth, or closer than the Moon. Icarus comes within 9 million miles of the Sun.

Many meteorites which survive atmosphere entry and land on Earth are believed to be asteroidal material. These meteorites are mostly stony, but some are iron. Some contain large amounts of carbon, including organic chemicals like amino acids.

Comets

See *Comets* in the **Pioneer 10 Press Kit**, page 22.

Pioneer Findings to Date

Pioneer 10 findings so far include the fact that the distribution of dust particles between the Earth's orbit and the far side of the Asteroid Belt seems to depend on particle size. There are no more of the very small particles in the belt than outside it, and there may be fewer of the smallest particles in the belt than near the Earth.

Because turbulence of the solar wind and magnetic field do not decline with solar distance, large amounts of the low-energy cosmic ray particles from the galaxy appear to be shut out of the solar system perhaps as far as beyond Jupiter. The solar wind appears to slow down while its gases heat up.

Among solar high-energy particles, experimenters have found the elements sodium and aluminum for the first time.

They have found that the neutral hydrogen of the interstellar wind appears to enter the heliosphere in the plane of the Earth's orbit, about 60° away from the expected direction of travel of the solar system through space. They have measured interstellar helium for the first time.

Asteroid Belt and Interplanetary Dust

Distributions with particle size were as follows:

Going out from the Earth's orbit, the smallest particles (around $1/1,000$ mm diameter) actually appear to decline in numbers. Somewhat larger particles ($1/100$ to $1/10$ mm diameter) seemed to be evenly distributed all the way from the Earth's orbit through the far side of the Belt with no increase in the Belt. (There are 25 mm to an inch.)

Still larger particles ($1/10$ to 1 mm diameter) were found all the way out but where almost three times as frequent in the center of the Belt as outside it.

Particles larger than 1 mm diameter appear to be very thinly spread (as many scientists expected).

For the smallest particles (around $1/1,000$ mm diameter), measurements of the zodiacal light reflected from these particles as one looks away from the Sun have shown that going away from the Sun zodiacal light declines steadily. If these very small particles were mostly in the Asteroid Belt, the reflected sunlight corrected for distance would have remained relatively constant until the belt was reached, and then would have dropped rapidly as the spacecraft passed through the belt.

The rate of decline of this reflected light suggests that concentrations of these tiny particles decline as the square of the distance from the Sun.

Since the counter glow (gegenschein), which is seen from the Earth as a brightening in the night sky at the point opposite the Sun, was also observed from *Pioneer 10* when it was far out in space, the gegenschein is now proved to be an interplanetary phenomenon rather than Earth associated.

In the very small asteroid particles $1/100$ to $1/10$ mm diameter (one billionth to one millionth of a gram), there was no increase in the Asteroid Belt and the experimenter believes the particles measured are cometary. For these particles, the one-quarter square meter of gas-cell detector surface received 25 penetrations between the Earth's orbit and the Belt's near side, and 17 penetrations in the Belt.

For the larger asteroid particles $1/10$ mm to 1 mm diameter (one ten millionth to one ten thousandth gram), analyses suggest that there are almost three times more particles in the Belt than outside it.

The Solar Wind and Magnetic Field

As viewed near the Earth's orbit, the solar wind is most dramatically seen as containing high-speed, up to 2,520,000 km/hr (1,566,000 mph) streams embedded in slower-speed, 1,000,000 km/hr (624,000 mph) areas of solar wind. These high-speed streams are presumably caused by high temperature regions in the solar corona. Between the Earth and Jupiter these high-speed streams speed up the gas ahead of them and they in turn are slowed down. This accounts for the large amount of turbulence in the solar wind and magnetic field still observed as far out as 510 million kilometers (420 million miles). In addition, this stream interaction also produces thermal energy so that the solar wind is not cooling to the extent expected for a simple expansion model. Present estimates indicate that the solar wind will ultimately become steady (constant speed) somewhere beyond the orbit of Jupiter, perhaps as far out as the orbit of Saturn.

Galactic Cosmic Rays

Galactic cosmic rays are high-energy particles from the stars of the galaxy. Because of the expected decrease in shielding by the solar wind and magnetic field, numbers of low-energy (0 to 100 million electron-volts) cosmic ray particles were expected to increase out to Jupiter. This did not occur apparently because, despite the decline of solar wind density and magnetic field strength, turbulence remained as high as near Earth. This turbulence apparently can shut out a large part of the low-energy galactic particles from the inner solar system, perhaps as far out as beyond Jupiter.

High-Energy Solar Particles

The best resolution yet seen of elements making up the solar high-energy particles has been obtained by *Pioneer 10*. It identified sodium and aluminum for the first time, determined relative abundances of helium, carbon, nitrogen, oxygen, fluorine, neon, sodium, magnesium, aluminum, and silicon nuclei coming from the Sun.

Interstellar Gas

The ultraviolet instrument has been able to locate concentrations of neutral hydrogen and helium and to make separations between gas originating in the Sun and that penetrating the solar system from interstellar space.

The experimenters have measured the ultraviolet light glow emitted by interplanetary neutral hydrogen and helium atoms. Preliminary findings are that the interstellar neutral hydrogen appears to be entering the heliosphere in the plane of the Earth's orbit at around 100,000 km/hr (62,000 mph).

The direction of movement of the solar system through space is thought to be about 60° above the ecliptic, and the interstellar hydrogen might well have been expected to enter from this direction.

**NATIONAL AERONAUTICS AND
SPACE ADMINISTRATION**
Washington, D. C. 20546
202-755-8370

FOR RELEASE:
September 10, 1974
12 Noon

74-238
August 28, 1974

A New Look at Jupiter

Results from the Pioneer 10 Mission to Jupiter

S
P
E
C
I
A
L

R
E
P
O
R
T

NATIONAL AERONAUTICS AND SPACE ADMINISTRATION
Washington, D. C. 20546
202-755-8370

Nicholas Panagakos   Phone: (202) 755-3680
Headquarters, Washington, D.C.

Peter Waller   Phone: (415) 965-5091
Ames Research Center, Moffett Field, California

RELEASE No: 74-238

FOR RELEASE:
12 Noon Tuesday, September 10

Pioneer Findings Paint New Picture of Jupiter

Information returned by the *Pioneer Jupiter* spacecraft seems to support the theory that the largest planet in the solar system is a spinning ball of liquid hydrogen, without any detectable solid surface.

At best, Jupiter has only a small rocky core, thousands of miles below the heavily clouded atmosphere.

And its mysterious Great Red Spot is probably the vortex of a gigantic storm that has been raging along a 40,000-kilometer (25,000-mile) front for at least 700 years.

These are among the findings of *Pioneer 10* experimenters and other scientists who have been analyzing data from the spacecraft since its visit to the colossal planet last December.

Much of the new information contradicts many previous theories on the nature of Jupiter.

The brightly-banded planet, big and bizarre, by itself containing more than two-thirds of all the planetary material in the solar system, now seems to have these characteristics:

- Jupiter appears to be almost entirely a liquid planet, without any solid surface.
- Jupiter seems to have a turbulent interior, much hotter than previously thought.
- Jupiter's magnetic field is much larger than some models predicted, and its radiation belts are far more intense than many expected.
- Jupiter turns out to be a source of high-energy particle radiation, the only one in the solar system besides the Sun. Some of these particles may be measurable at Earth.
- At birth, four and a half billion years ago, the solar system giant was far hotter than scientists had believed.
- Jupiter's atmospheric circulation pattern is far different from Earth's. For example, the familiar circular cyclones and anti-cyclones (highs and lows) of the Earth are stretched completely around Jupiter. This is probably due to the effects of heat radiation from the planet's interior, and Jupiter's 35,200-kilometer (22,000-mile) per hour rotational speed. This "weather-stretching" accounts for Jupiter's semi-permanent, planet-girdling, alternating brown-red and gray-white cloud bands.
- *Pioneer* pictures show huge and striking cloud features which indicate strong atmospheric upwelling.
- The density of Jupiter's four planet-sized moons is directly proportional to their distance from the planet. It decreases from the density of rock for the closest large moons, Io and Europa, to the density of a water-ice and rock mixture for the outer Ganymede and Callisto. This outward decrease is probably due to the great amounts of heat that were radiated by the planet at its formation.
- Io has a tenuous atmosphere, perhaps containing hydrogen, nitrogen and sodium. Ganymede is known to have an atmosphere, and the other two big moons probably also have atmospheres. *Pioneer* pictures of the largest moon, Ganymede, are still being processed, but seem to show maria and highlands, somewhat like those on the Earth's Moon and on Mars.
- While *Pioneer 10* findings do not explain for certain Jupiter's Great Red Spot, they indicate that the Spot is the vortex of an intense storm – hundreds of years old, something like a 25,000-mile-long mass of whirling clouds, towering some eight kilometers (five miles) above the surrounding cloud deck.
- *Pioneer 10* has shown that Jupiter's gray-white zones are cloud ridges of rising atmosphere circling the planet, and looming 20 kilometers (12 miles) above the cloud belts. Conversely, its orange-brown belts are cloud troughs of descending atmosphere probably 20 kilometers (12 miles) deep. A number of individual cloud features seen for the first time by *Pioneer* also show this rising-descending circulation pattern.

In some cases, this first consensus among Pioneer experimenters and other scientists verifies pre-existing theories about Jupiter. In other cases, the findings are completely new. In both instances, however, *Pioneer 10*'s visit to the planet has provided the first firm ideas of what Jupiter is like, says Dr. John Wolfe, Pioneer Project scientist. All previous Jupiter measurements had been made from Earth, a half billion miles away.

Dr. Wolfe is at NASA's Ames Research Center, Mountain View, California, which manages the Pioneer project for the space agency. The *Pioneer* spacecraft was built by TRW Systems, Redondo Beach, California.

Pioneer 10 provided man with his first close-up look at the giant planet on December 5, 1973 when the spacecraft flew within 81,000 miles of the "surface," making measurements and taking photographs.

A sister spacecraft, *Pioneer 11*, launched April 6, 1973, is scheduled to fly by Jupiter on December 3, 1974 passing within 46,400 kilometers (29,000 miles). It will then continue on to Saturn.

(End of General Release. Background information follows.)

A New Look at Jupiter

Based on *Pioneer 10* findings, Jupiter now looks like this:

Jupiter's Interior

Jupiter is almost certainly a liquid planet for it is too hot to solidify, even with its enormous internal pressures of millions of atmospheres.

Both temperature and pressure rise as one goes deeper. At the transition zone to liquid, 1,000 kilometers (600 miles) below the top of the atmosphere, the temperature is calculated to be 2,000° C (3,600° F). At 3,000 kilometers (1,800 miles) down, temperature is 5,500° C (10,000 °F) and pressure is 90,000 Earth atmospheres. At this point, the weight of the Jovian atmosphere has compressed the hydrogen gradually into the form of a liquid about one quarter as dense as water. At 25,000 kilometers (15,000 miles) down, the temperature reached 11,000 °C (20,000° F) and the pressure is three million atmospheres. At this level, liquid hydrogen turns to liquid metallic hydrogen.

The Jovian temperature profile was calculated by Dr. William B. Hubbard, University of Arizona, using measurements obtained by the gravity-sensing experiment of Dr. John Anderson, NASA's Jet Propulsion Laboratory, Pasadena, California. The experiment, used the *Pioneer 10* spacecraft itself as a sensor.

The gravity-sensing measurements so far show that, unlike the Earth, Jupiter has no concentrations of mass, such as a rigid crust or other solid areas. The measurements show that the planet is in "hydrostatic equilibrium," or largely liquid. At Jupiter's center, theoreticians believe, there may be a small rocky core, perhaps containing some iron. The gravity analysis, not yet complete, may well show this small core.

Pioneer measurements indicate that Jupiter radiates two to three times more heat than it receives from the Sun. This means that Jupiter is losing heat at a tremendously rapid rate. Calculations by Dr. Hubbard, based on this heat balance, indicate that the temperature at Jupiter's center may be about 30,000° C (54,000° F) (six times the temperature on the surface of the Sun). From the center outward to the surface, temperatures decrease steadily to perhaps -17°C (10° F) above zero, at a point somewhat below the cloud tops.

While Jupiter has no solid region, except for its probable small, rocky core, the weight of its atmosphere compresses its hydrogen gradually into the form of a liquid about one quarter as dense as water at around 3,000 kilometers (1,800 miles) below the cloud tops.

The best explanation for Jupiter's tremendous internal heat is that it is primordial heat left over from the heat of formation of the planet at the time the solar system was formed.

This high heat of formation is confirmed by *Pioneer* measurements of the closest large moons, Io and Europa, which, unlike the planet-sized "ice" moons lying farther out, are rocky. Dr. James Pollack of NASA's Ames Research Center believes that a hot primordial Jupiter radiated enough heat to prevent water vapor from condensing into ice during formation of Io and Europa.

(An alternative explanation by some scientists for part of Jupiter's internal heat is that it is energy released by the fractionation of hydrogen and helium, a process believed to be currently underway somewhere near the center of the planet.)

The displacement of the center of Jupiter's magnetic field from the center of the planet, and tilt of the field further confirm that Jupiter is a huge, flattened, fast-spinning ball of liquid hydrogen, says Dr. Edward Smith, JPL, magnetic field experimenter. Jupiter's field, like the Earth's, is believed to result from a "dynamo effect" within the planet, in which eddies (circular currents) within the liquid interior generate electric currents, and hence magnetic fields. Only a planet with a tremendously active interior, scientists believe, could produce a magnetic field as far offset from its center as Jupiter's.

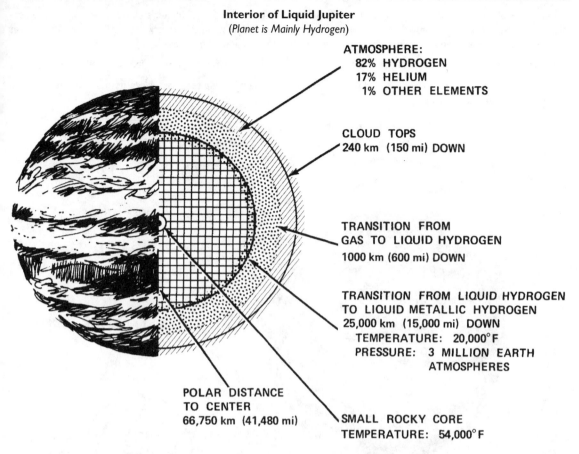

Interior of Liquid Jupiter
(Planet is Mainly Hydrogen)

ATMOSPHERE:
82% HYDROGEN
17% HELIUM
1% OTHER ELEMENTS

CLOUD TOPS
240 km (150 mi) DOWN

TRANSITION FROM
GAS TO LIQUID HYDROGEN
1000 km (600 mi) DOWN

TRANSITION FROM LIQUID HYDROGEN
TO LIQUID METALLIC HYDROGEN
25,000 km (15,000 mi) DOWN
TEMPERATURE: 20,000°F
PRESSURE: 3 MILLION EARTH
ATMOSPHERES

POLAR DISTANCE
TO CENTER
66,750 km (41,480 mi)

SMALL ROCKY CORE
TEMPERATURE: 54,000°F

Jupiter's hot interior is believed to literally seethe as convective currents circle and eddy, carrying the heat to the surface. Dr. Hubbard estimates that hydrogen moving up from Jupiter's center covers the 70,400 kilometers (44,000 miles) from the center of the planet to the top of the atmosphere in 10 to 100 years. Interior currents move as fast as 2,400 kilometers (1,500 miles) per year, he estimates.

The rapid convection of heat is reflected in the constant rise and fall of the atmosphere. This is shown by a number of prominent, semi-permanent features seen in the planet's clouds, such as striking white ovals of rising atmosphere surrounded by darker borders of descending atmosphere. Such features have been recorded in the *Pioneer 10* pictures of Dr. Tom Gehrels, University of Arizona, Pioneer imaging experimenter, and by long-time telescope observers such as Dr. Bradford Smith, New Mexico State University.

Dr. Anderson's gravity-sensing experiment also shows that the planet is more oblate or "squashed down" than visual measurements make it appear. The planet is 9,280 kilometers (5,800 miles) thinner measured through the poles than through the equator. This gives it an equatorial diameter of 142,796 kilometers (88,298 miles), and a diameter through the poles of 133,516 kilometers (82,976 miles), the most exact measurements yet made. The *Pioneer* occultation measurements are almost exactly the same. Jupiter is 10 times as flattened as the Earth, due to its high-speed rotation. Jupiter's exact mass is 317.8 times Earth's mass, one lunar mass heavier than had been thought.

Jupiter's Atmosphere

Jupiter's atmosphere, the planet's 1,000-kilometer (600-mile)-deep outer layer, consists primarily of hydrogen and helium gas with very small amounts of the other elements.

Pioneer found that the atmosphere accounts for about 1 percent of Jupiter's total mass.

Calculations based on *Pioneer 10* findings put the ratio of hydrogen to helium in Jupiter's upper atmosphere close to 80 percent to 20 percent, with less than 1 percent for all the other elements. This is similar to the ratio of elements found in the Sun.

Jupiter's Clouds and Heat Balance

Jupiter's cloud layers form in the planet's predominantly hydrogen-helium atmosphere. They are believed by theoretical astronomers to make up a kind of four-decker sandwich.

Ammonia "ice" crystals, measured from Earth, are believed to form the top of Jupiter's clouds. Below this are red-brown clouds, probably of ammonia hydrosulfide crystals, and below that, water-ice crystals. Still lower, liquid water droplets containing ammonia in solution may be present. (Water has never been observed.)

Just what the "weather" is in these lower cloud layers is unknown, aside from the up-down circulation of the atmosphere.

There may be lightning, storm activity, and other phenomena, because Jupiter has a large amount of energy and a violently circulating atmosphere.

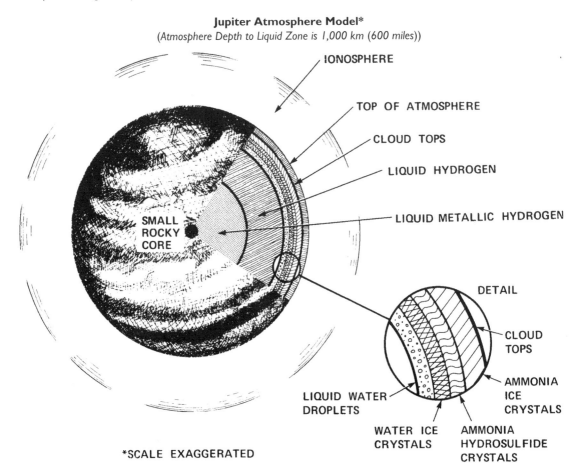

Jupiter Atmosphere Model*
(Atmosphere Depth to Liquid Zone is 1,000 km (600 miles))

IONOSPHERE

TOP OF ATMOSPHERE

CLOUD TOPS

LIQUID HYDROGEN

LIQUID METALLIC HYDROGEN

SMALL ROCKY CORE

DETAIL

CLOUD TOPS

AMMONIA ICE CRYSTALS

LIQUID WATER DROPLETS

WATER ICE CRYSTALS

AMMONIA HYDROSULFIDE CRYSTALS

*SCALE EXAGGERATED

Dr. Guido Munch, California Institute of Technology, Pioneer infrared (heat) experimenter, puts the topmost cloud layer of ammonia ice crystals at a level where pressure would be 700 millibars and temperature -120° C (-184° F). This is the temperature-pressure combination at which ammonia would condense out.

The transparent outer atmosphere above the clouds, Dr. Munch believes, contains some ammonia, plus some well-mixed methane.

Fourteen kilometers (8 miles) above the cloud tops, at the 300 millibar pressure level, the temperature falls to -145° C (-229° F). Farther out, a haze of ammonia crystals may exist, extending up to a layer where solar heat is absorbed by the atmospheric methane. This would produce an inversion layer in which the atmosphere is warmer. The inversion layer appears to be 35 kilometers (21 miles) above the visible clouds at a pressure level of 100 millibars (1/10th of an Earth atmosphere), and the temperature is -155° C (-247° F).

Still higher, there appears to be a haze layer of aerosols (droplets) and hydrocarbons (such as the ethane and acetylene recently identified from Earth), which also may absorb sunlight and heat up the atmosphere.

By some interpretations, *Pioneer 10*'s infrared and radio occultation experiments do not agree on temperature levels at the top of the atmosphere. If the occultation findings are correct, the cloud regions and zones of Earth-like temperature may be far higher than believed near the very top of the atmosphere – or else estimates of the amount of ammonia in Jupiter's atmosphere may have to be revised drastically downward.

However, several other calculations place the clouds at about 240 kilometers (150 miles) below the atmosphere top. Experimenters expect to resolve this difference through further analysis, or eventually by sending a probe into Jupiter's atmosphere.

Pioneer showed that Jupiter's atmosphere is somewhat warmer near the equator than the poles, though there also are warm belts near the poles. Jupiter has the same temperature in northern and southern hemispheres and on the day and night sides.

Pioneer also found that cloud particles in the cloud tops tend to be quite small (a few tenths of a micron) in both the dark belts and light zones. Particles are smaller than those in Earth's cumulostratus clouds – the most common Earth clouds.

The cloud particles tended to be very reflective in both belts and zones, which could mean either "ice" particles or shiny droplets.

Atmosphere gas, which would move toward the equator by convection, instead, due to Coriolis force, moves around the planet against the direction of rotation. Gas which would move toward the poles instead moves around Jupiter in the rotation direction.

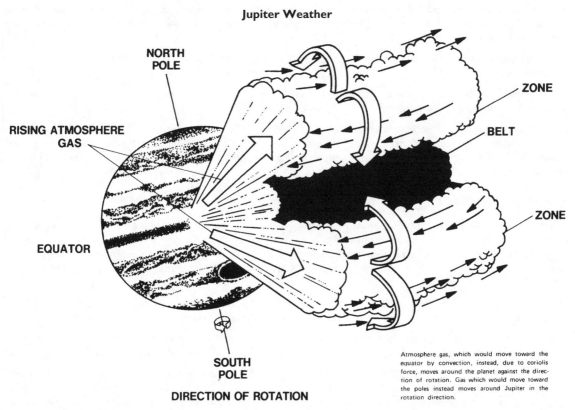

Jupiter Weather

Atmosphere gas, which would move toward the equator by convection, instead, due to coriolis force, moves around the planet against the direction of rotation. Gas which would move toward the poles instead moves around Jupiter in the rotation direction.

Jupiter Weather

Jupiter's 17 relatively permanent belts and zones appear to be comparable to the continent-spanning cyclones and anticyclones which produce most of the weather in the Earth's temperate zones. On both planets these phenomena are huge regions of rising or falling atmosphere gas, powered by the Sun (and in Jupiter's case also by its internal heat source).

On Earth, huge masses of warm, light gas rise to high altitudes, cool off, get heavier, and then roll down the sides of new rising columns of gas. General direction of this atmospheric heat flow on Earth is from the tropics toward the poles.

Coriolis forces, (produced by planetary rotation) cause the descending gas, which would normally move north or south, to flow around the planet west to east. However, on Earth, unstable flow converts this west-east motion into the enormous spirals known as cyclones and anticyclones.

With Jupiter's high-speed, 22,000 mph rotation, these round-the-planet coriolis forces are very strong.

Because of instabilities like those on Earth, flow should be in even more violent spirals than Earth's. But several factors appear to have a "calming effect," so that the motion is mostly linear. These factors include Jupiter's internal heat source and heat circulation, lack of a solid surface and liquid character.

As a result, the combination of convection due to the Sun's heat plus internal heat, and coriolis due to Jupiter's rapid rotation, stretches the planet's large permanent weather features completely around the planet, forming the belts and zones.

One aspect of the "calming effect" in Jupiter's weather is that heat radiated from Jupiter's center reaches all parts of the planet's surface in equal amounts. By contrast, the Earth's heat from the Sun is received mostly in the tropics and circulates toward the icy poles.

Pioneer 10 found the tops of Jupiter's bright zones to be 9° C (15° F) cooler, and an estimated 20 kilometers (12 miles) higher in the atmosphere than the tops of the dark belts. This, plus observations from Earth and *Pioneer 10* pictures, confirms Jupiter's overall circulation pattern. The gray-white zones are warm, upward-rising "stretched" weather cells, whose tops are probably high clouds of ammonia crystals. The lower red-brown areas are cooler, descending weather cells, believed to be clouds of ammonia hydrosulfide crystals.

The tops of the bands of warm, lightweight, upward rising white clouds flow from the center to the edges. Cloud material in the cooler, heavier dark belts flows downward from the edges to the center of the belts. Because of the forces just described, this produces atmospheric streams along the bands – going in opposite directions on either side of each zone, with the greatest velocities at zone edges.

These same mechanisms mean that Jupiter's bands are streams of atmosphere whose speeds of flow can differ by as much as 600 km/hr (360 mph). This amounts to 360 mph winds. Large numbers of huge eddies, thousands of miles across, created when one stream has a high velocity relative to another, are clearly visible in the detailed *Pioneer* pictures of the planet.

In general, these atmosphere streams flow most rapidly near the equator. The greatest contrasts in flow speeds occur at mid-latitudes.

Strong thermal vortices, such as the Great Red Spot, and many other similar features, are stretched into long ovals by the same forces that make the streams.

High activity of Jupiter's atmospheric circulation is further shown by ultraviolet measurements of atomic versus molecular hydrogen, according to Dr. Darrell Judge, University of Southern California, Pioneer ultraviolet experimenter. This ratio shows that the atmospheric mixing rate in the top of the atmosphere is 10 to 100 times as high as in the top of the Earth's atmosphere.

The Great Red Spot

While *Pioneer 10* has not completely explained the Great Red Spot, it appears to be a centuries-old, 40,000-kilometer (25,000-mile)-wide vortex of a violent storm, as first proposed by the late Dr. Gerard P. Kuiper. The Spot is calculated to rise some 8 kilometers (5 miles) above the surrounding cloud deck. This is shown by the fact that the clouds at the top of the Spot have less atmosphere above them and are cooler (and hence higher) than surrounding clouds, according to Dr. Gehrels.

Jupiter's liquid character and the fact that the Spot was not "seen" by the gravity-sensing experiment (which should have seen even small variations in density) seem to eliminate most previous theories on the nature of the centuries-old mystery.

Pioneer 10 pictures show that the Spot's internal structure appears to be a pinwheel-like vortex and that its rapid circulation pattern makes it rigid enough to displace the clouds of the South Tropical Zone strongly northward.

A second "red spot," about one third the size of the Great Red Spot, can be seen in the *Pioneer* pictures. This is located in the northern counterpart of the southern hemisphere zone containing the Great Red Spot. The little red spot also is cooler than its surrounding clouds, and is believed to rise as high as the Great Red Spot. This second spot lends support to the idea that "red spots" are occasional meteorological phenomena on Jupiter, found in the middle of the planet's bright zones where the atmosphere is rising, says Dr. Andrew Ingersoll, California Institute of Technology.

Jupiter's Ionosphere

Jupiter's ionosphere rises 3,000 kilometers (1,800 miles) above the $^1/_{10}$ millibar level. It is ten times thicker and five times hotter than had been predicted. Average temperature of the ionosphere is about 1,000° C (1,800° F). It has at least three sharply-defined layers of differing density, and its unexpected depth is due to the diffusion of its ionized gas by high temperatures. High temperature of the ionosphere is believed due to impacts of high-energy particles, and to hydromagnetic waves from Jupiter's magnetosphere.

Jupiter's Moons

Going out from the planet, Jupiter's four planet-sized moons show a progression of densities. The closest moon, Io, is 3.5 times the density of water; Europa, 3.14 times; Ganymede, 1.94 times; and Callisto, 1.62 times.

Thus, the two inner moons, having a density a little more and a little less respectively than that of the Earth's Moon, must be primarily rocky. The outer two probably consist largely of water-ice, as indicated both by their density and by Jupiter's heat characteristics.

Pioneer measurements have shown that the masses of the moons are as follows: Io, 1.22 Earth-Moon masses; Europa, 0.67 lunar masses; Ganymede, 2.02 lunar masses; and Callisto, 1.44 lunar masses.

Ganymede is definitely larger than the planet Mercury; Callisto, about the same size; and Io and Europa, somewhat smaller.

The four large moons have an average surface temperature on the sunlit side of -145 degrees C (-230 degrees F).

A *Pioneer* picture with 400-kilometer (240-mile) resolution of Ganymede needs further processing but it appears to show a south polar mare and another central mare, 800 kilometers (480 miles) in diameter, plus various large meteorite craters and a bright north polar region, according to Dr. Gehrels.

Io, the closest large moon, which turns out to be 23 percent heavier than previously thought, has a tenuous atmosphere. This makes it the smallest known celestial body with an atmosphere. Io also has an extended ionosphere almost as dense as that of Venus, reaching to about 700 kilometers (420 miles) on the day side. This dense ionosphere may mean an unusual gas mixture, and Dr. Arvydas Kliore, JPL, the Pioneer radio occultation experimenter, says that an ionosphere with these characteristics may indicate the presence of elements such as sodium, hydrogen, and nitrogen on Io. Earth measurements show the presence of sodium on Io, and *Pioneer* measurements indicate the presence of hydrogen.

When Io emerges from Jupiter's shadow, it is for about 10 minutes the most reflective object in the solar system. Then it begins to turn from white to its pronounced orange color again. Apparently during its 21 hours in the freezing Jovian night, methane snowflakes form from Io's atmosphere, which then evaporate with return to sunlight. There appears to be enough atmosphere to make this happen, according to Dr. Kliore.

The radio occultation experiment found a density of 60,000 electrons per cubic centimeter in the ionosphere on the day side of Io, compared with only 9,000 electrons per cubic centimeter on the night side. *Pioneer* also found Io is embedded in a hydrogen cloud extending a third of the way around its orbit – an entirely new and unexpected phenomenon.

Pioneer pinpointed Io's position more precisely by 50 kilometers (30 miles).

Jupiter's Magnetic Field

Pioneer 10 found that the strength of Jupiter's magnetic field at the planet's cloud tops is more than 10 times the strength of the Earth's field at the Earth's surface. Total energy in the Jovian field is 400 million times that of Earth's.

Jupiter's inner magnetic field extends into space about 1,280,000 kilometers (800,000 miles) from the planet's cloud tops. The outer field extends a minimum distance of 3.4 million kilometers (2.1 million miles) from Jupiter's cloud tops, and sometimes reaches out three times as far, 10.4 million kilometers (6.5 million miles). The poles of Jupiter's field are reversed from Earth's; a Jovian compass would point south.

Jupiter's inner field is tilted about 10 degrees to the planet's axis of rotation, and the center of the field and center of the planet do not coincide. The field's center lies about 2,100 kilometers (1,320 miles) north of the center of the planet and 7,740 kilometers (4,840 miles) outward from the rotational axis in a direction parallel with the equator.

Because the field's center is not in the center of the planet, the strength of the field as it emerges from the cloud tops is calculated to vary over the clouds' surface from 3 to 10 Gauss. At its closest approach, 129,600 kilometers (81,000 miles) above the cloud tops, *Pioneer* measured a field strength of 0.2 Gauss. (Earth's field at the surface is 0.35 Gauss.)

Because Jupiter's magnetic field is tilted 10 degrees to the planet's axis of rotation, the inner field, as seen from space, wobbles up and down through an arc of 20 degrees once every 10-hour rotation of Jupiter.

Jupiter's Radiation Belts

As on Earth, the high-energy particles which form Jupiter's inner radiation belt are trapped within the planet's inner magnetic field. In the weak outer magnetic field, particles bounce around, but eventually make their way to the field's outer edge. There they are spun off into space by the high-speed rotation of the planet and by the radiation belts themselves. Particles also escape in other ways.

Jupiter's radiation belts have the highest intensity ever observed by space measurements, comments Dr. James Van Allen, University of Iowa, one of the four energetic particle experimenters. Total energy of particles in the belts is many millions of times the total energy of those in the Earth's belts.

The belts' intense particle radiation pose the greatest threat to spacecraft flying close to Jupiter, or orbiting the planet. Three of the four planet-sized moons lie within the intense inner radiation belt. Hence a manned landing on one of them would be impossible without very advanced technology to protect the men.

The fourth big moon, Callisto, lies outside the region of intense radiation, and would be somewhat more feasible for a manned landing.

High-energy electrons in the inner radiation belt were found to be 100 times more intense than the more damaging high-energy protons (hydrogen nuclei). This low level of protons was a surprise to many scientists, and is being scrutinized further.

The inner radiation belt has by far the highest radiation intensities, and forms a doughnut-shaped ring around the planet. Like the inner magnetic field, which contains it, the inner belt extends out about 1,280,000 kilometers (800,000 miles) from the top of Jupiter's ionosphere.

The outer radiation belt extends out beyond the inner belt in a relatively flat ring a minimum distance of another 2.1 million kilometers (1.3 million miles), or a maximum of four times that far to 9.6 million kilometers (almost 6 million miles).

The outer radiation belt has an average thickness of about 711,000 kilometers (445,000 miles), explains Pioneer energetic particle experimenter Dr. John Simpson, University of Chicago. However, the zone of high radiation intensity in the outer belt forms a thin sheet, which bisects the outer belt horizontally, and lies parallel with the planet's equator. High-energy particles in the outer belt are mostly electrons. Their maximum intensity is several hundred times less than the maximum intensity levels in the inner belt.

The inner belt is contained in and shaped by Jupiter's internal magnetic field, and since the field wobbles up and down 20 degrees every 10 hours with each Jupiter rotation, so does the belt. The greatest radiation intensities in the inner radiation belt coincide with the equatorial plane of Jupiter's magnetic field. Going either north or south from the magnetic equator, radiation levels quickly decline. Spacecraft trajectories could be planned to pass rapidly through this region of intense radiation.

From the magnetic equator, radiation intensity fell three times in the first 20 degrees of Jovian latitude and ten times in 40 degrees of latitude north or south. The closer *Pioneer 10* came to Jupiter, the more tightly particles were confined to the magnetic equator.

Particles making up the radiation in both the inner and outer belts are believed to come from the solar wind, the planet's ionosphere, or both. They consist of light electrons and heavier protons.

Peak intensity of electrons in the inner belt was measured by *Pioneer* at 10,000 times the peak intensity of the Earth's Van Allen belts. Peak intensity for protons in the inner belt was several thousand times the peak intensity in the Earth's belts.

The intensity of electrons increases tenfold for every 211,000 kilometers (132,000 miles) one moves closer to the planet. For protons, there is a similar intensity increase.

Electron intensities increased up to *Pioneer*'s closest approach to Jupiter 130,000 kilometers (81,000 miles) above the cloud tops – and probably continue to increase to the top of Jupiter's atmosphere.

Proton intensities, on the other hand, appear to peak at 182,000 kilometers (114,000 miles) above Jupiter's cloud tops, and measurements indicate that for unknown reasons they may decrease with proximity to Jupiter. *Pioneer 11* which will fly 41,600 kilometers (26,000 miles) above the cloud tops, may confirm this.

The total radiation measured by *Pioneer 10* was about 100 times the lethal dose for humans, and was near the limit of radiation tolerance for spacecraft systems.

The big moons, Io, Europa and Ganymede, appear to carve out troughs in the radiation belts by soaking up particles. This "taking-out-of-circulation" of high-energy particles by these moons appears to reduce total radiation close to Jupiter by from 10 to 100 times.

Pioneer measurements of peak radiation intensity were an enormous one billion electrons per square centimeter per second striking the skin of the spacecraft. Of the total electrons, 90 percent were in the energy range from 3 million to 30 million electron-volts (MeV).

For all the protons with energies above 35 MeV, intensity was 70 million protons per square centimeter per second. Numbers of low-energy protons (0.3 to 35 MeV) were not as great as would have been expected from totals of higher energy protons.

In the outer radiation belt, intensities of all electrons over 60,000 electron-volts reached three million per square centimeter per second at times. For all the protons with energies higher than 0.5 MeV intensity at times reached several hundred thousand protons per square centimeter per second.

Jupiter's Outer Environment

Bow Shock Wave

Like the Earth, Jupiter has a bow shock wave produced when the million-mile-an-hour solar wind rushing out from the Sun strikes the planet's magnetosphere and flows around it. There is also a turbulent transition region like Earth's inside the bow shock front, and within that an enclosed magnetosphere or magnetic envelope somewhat like Earth's created by Jupiter's magnetic field. The solar wind cannot directly penetrate this magnetosphere.

These phenomena at Jupiter are on a scale never before seen.

Pioneer 10 crossed Jupiter's bow shock wave outbound at 13,200,000 (8,240,000 miles). This means that a line from one side of the shock to the other passing through the planet's day-night boundary would be 26.4 million kilometers (16.5 million miles) long, about 80 percent of the distance between the orbits of the Earth and Venus.

Jupiter's magnetosphere is equally huge. It has an average diameter of nine million miles. If it could be seen from Earth, half a billion miles away, it would occupy 2 degrees of the sky, compared with only half a degree of the sky occupied by the Sun.

When *Pioneer 10* crossed the bow shock, the solar wind changed direction 40 degrees as it flowed around the planet. The "wind" increased density three times and temperatures 100 times as it crossed the shock front.

Magnetosphere

Jupiter's magnetosphere is the region of space occupied by the planet's magnetic field. It is quite different from Earth's magnetosphere, and is 100 times bigger in diameter, and a million times larger in volume. It rotates at several hundred thousand mph along with the planet (like a big wheel with Jupiter as the hub).

Like the radiation belts which it contains, the magnetosphere consists of two regions. The inner magnetosphere is created by the planet's internal magnetic field, and is shaped like a doughnut with the planet in the hole.

The highly unstable outer magnetosphere is, in one sense, an extension of the inner magnetosphere, and is shaped as though the outer part of the doughnut had been squashed or flattened.

The outer magnetosphere is created by Jupiter's internal field, plus its "current-sheet" magnetic field.

The outer magnetosphere is "spongy." It pulsates in the solar wind like a huge jellyfish, and often shrinks to one third of its largest size.

Like Jupiter's inner and outer radiation belts, the inner magnetosphere is about 2.9 million kilometers (1.8 million miles) across. The outer magnetosphere has a diameter of 7 million kilometers (4.4 million miles), when squashed in, and 21.1 million kilometers (13.2 million miles) when extended.

Pioneer 10 first crossed the magnetosphere's sharply-defined outer boundary in a few minutes at 6.7 million kilometers (4.2 million miles) from Jupiter. Then as the magnetosphere abruptly shrank in size, the spacecraft emerged for 11 hours, and then reentered at 3.5 million kilometers (2.2 million miles). This constant change in magnetosphere size is further shown by the fact that *Pioneer* crossed the constantly moving bow shock 17 times as it left Jupiter.

Magnetosphere Dynamics

Jupiter's magnetosphere is unlike any ever seen before, and may produce some new physics.

The magnetosphere's rapid rotation constantly forces ionized particles outward, moving the outer magnetosphere's boundary farther and farther from the planet. Strong gusts of solar wind then squash the magnetosphere inward. As solar wind pressure declines, the outward-moving ionized particles push it out again; the cycle repeats continuously.

The charged particles spun out from the planet have the greatest outward velocities at Jupiter's equator, and hence the particles flow out as a flat equatorial ring or sheet.

Since moving ionized (electrified) particles produce an electric current, the particle sheet is also a sheet of intense electric current. And, as an electric current also produces a magnetic field, the current sheet produces a flat, stretched magnetic field – a flat ring-field. This field, plus Jupiter's internal field, combine to produce the planet's outer magnetosphere. It is a relatively feeble, constantly varying field, averaging 5 gamma ($1/20,000$th gauss) in strength. It is easily pushed around by the solar wind.

The thin disc of intense radiation in the outer magnetosphere coincides with the current sheet because the particles follow the current sheet magnetic field. Both above and below the sheet is the outer radiation belt, 1,424,000 kilometers (890,000 miles) thick. Above and below the belt is the rest of the outer magnetosphere. This region appears to be devoid of energetic particles. In total, the outer magnetosphere is believed to be about 7 million kilometers (4.5 million miles) thick, says Dr. John Wolfe, solar wind experimenter.

The current sheet would be parallel with Jupiter's equator (where the rotation motion is greatest) except for Jupiter's magnetic field. In the inner magnetosphere, the strong inner field forces the electrified particles outward along Jupiter's tilted magnetic equator.

However, in the outer magnetosphere, Jupiter's field becomes so weak that the particles take control. They continue to move out in a flat sheet, but the sheet lies parallel to the planet's geographical equator. Because the inner magnetosphere is tilted, and the outer one parallel, Dr. Edward Smith, JPL, the magnetometer experimenter, calls the entire current sheet the "fedora hat" model, with the front brim lower than the back one.

Jupiter as a Radiation Source

High-energy particles spiral along the stretched-out magnetic field lines created by the current sheet. At the magnetosphere boundary these particles are spun off from Jupiter's magnetosphere by various mechanisms related to kinking-up of stretched field lines.

These masses of high-energy particles are a completely new discovery. They make Jupiter a second source of high-energy particle radiation in the solar system, the only one except the Sun. Dr. James Trainor of NASA's Goddard Space Flight Center, Greenbelt, Maryland, points out that *Pioneer 10* saw these spun-off Jovian high-energy particles 224 million kilometers (140 million miles) out from Jupiter. Further studies, he says, have shown that these particles probably have been seen for several years from Earth orbit, half a billion miles away, though scientists did not know their source. Behavior of these Jupiter particles, some of which travel into the Sun's magnetic field, instead of away from it, is of great interest to scientists.

Despite Jupiter's high-speed spin, its magnetic field unexpectedly spirals back very little. As at Earth, Jupiter's magnetosphere is believed to have a magnetic tail streaming away from the Sun, and *Pioneer 10* will have an opportunity to look for this Jupiter tail in March 1976, when it should cross it at around 704 million kilometers (440 million miles) outside Jupiter's orbit.

Jupiter's fast-spinning outer radiation belts, says Dr. Van Allen, have some analogy to pulsars (rapidly rotating, super-dense neutron stars). Both throw out particles, and a study of Jupiter's belts could provide some clues to the nature of pulsars.

NATIONAL AERONAUTICS AND SPACE ADMINISTRATION
Washington, D. C. 20546
202-755-8370

FOR RELEASE:

April 1, 1973

PROJECT: PIONEER G

P
R
E
S
S

K
I
T

CONTENTS

NATIONAL AERONAUTICS AND SPACE ADMINISTRATION
Washington, D. C. 20546
202-755-8370

March 14, 1973   **FOR RELEASE:**
                                                                                  Sunday April 1, 1973

Nick Panagakos   Phone: (202) 755-3680

Peter Waller (Ames) Phone: (415) 965-5091

RELEASE No: 73-41

Pioneer G Readied for Launch to Jupiter

Man will send his second spacecraft to Jupiter with the launch of *Pioneer G* by the National Aeronautics and Space Administration from Kennedy Space Center, Florida, between April 5 and 26, 1973.

Pioneer G's flight to the largest and most active of the planets will take less than two years for most launch dates.

The unmanned spacecraft, to be named *Pioneer 11* after successful launch, will have a number of mission options – both at Jupiter and far beyond. These options will depend on results returned by *Pioneer G*'s predecessor, *Pioneer 10*, now three-quarters of the way to Jupiter. *Pioneer 10* will make the first reconnaissance of the giant planet during the period between December 1 and 6, 1973.

If *Pioneer 10* is unsuccessful, *Pioneer G* will repeat its mission. Otherwise, *Pioneer G* may fly a different course over the planet's surface, making measurements and taking pictures that will complement those of *Pioneer 10*. The spacecraft may also investigate another of Jupiter's planet-sized moons. After flyby, it may follow *Pioneer 10* and become the second man-made object to escape the solar system, or mission directors may choose a solar orbit near Jupiter's orbit. One possible trajectory could take it to Saturn in 1980.

One of *Pioneer G*'s objectives, Jupiter, is unique among planets. It is enormous, colorful, and the most dynamic planet in the solar system. It appears to have its own internal energy source and is so massive that it is almost a small star. Its dense, cloudy, and turbulent atmosphere may have the necessary ingredients to produce life. Jupiter's volume is more than 1,000 times that of Earth, and it has more than twice the mass of all the other planets combined. Striped in glowing yellow-orange and blue-gray, it floats in space like a bright colored rubber ball. It has a huge red "spot" in its southern hemisphere and spins more than twice as fast as Earth. Of its 12 moons, two, Ganymede and Callisto, are the size of the planet Mercury, and two others, Io and Europa, are as large as Earth's Moon.

On its way to Jupiter, *Pioneer G* will make the second spacecraft passage through the Asteroid Belt between August 1973 and February 1974.

It will carry a plaque identical to one on *Pioneer 10* for possible identification of its origin by another intelligent species, if any exist, should *Pioneer G* follow *Pioneer 10* out of the solar system into the galaxy.

Spacecraft controllers can accomplish the various options open to *Pioneer G* by varying the spacecraft's swingby trajectory around Jupiter – using the planet's gravity and orbital motion to change spacecraft speed and direction.

The trajectory will be chosen on the basis of scientific questions raised by *Pioneer 10*'s flight past the planet. Some proposed flyby trajectories call for an approach as close as one-quarter of *Pioneer 10*'s closest distance. This would be only 35,000 kilometers (27,000 miles) above Jupiter's striped cloud tops.

Nearness of approach to Jupiter may depend on *Pioneer 10* findings about the planet's radiation belts. The belts are believed by many scientists to be as much as a million times stronger than the Earth's, with intensity increasing rapidly as one approaches the planet. If this is correct, the belts could disable a spacecraft approaching too closely. Other unknown hazards near Jupiter also might affect approach distance and flyby trajectory.

Pioneer 10 so far has found that the Asteroid Belt between the orbits of Mars and Jupiter apparently offers no serious hazard to spacecraft. During its seven-month, 330-million-kilometer (205-million-mile) journey through the belt, *Pioneer 10* received no damaging hits by high-velocity asteroid particles.

Pioneer 10 also has found some surprising features of the solar atmosphere (the heliosphere), found various elements and isotopes among solar particles, and has made some findings about the "interstellar wind" outside the heliosphere.

Most conditions of the Pioneer G mission are at the extreme edge of space flight technology. The Atlas-Centaur-TE-M-364-4 launch vehicle will drive Pioneer G away from the Earth initially at 51,800 km/hr (32,000 mph) – equaling the speed of Pioneer 10, which flew faster than any previous man-made object. Pioneer G will pass the Moon in about 11 hours.

At Jupiter, Pioneer G will be so far from Earth that radio messages at the speed of light will require a round-trip time of 90 minutes. This will demand precisely planned command operations. Pioneer G will be controlled by frequent instructions from the Earth. The 64-meter (210-foot) "big dish" antennas of NASA's Deep Space Network (DSN) will have to listen over a distance of 800 million kilometers (a half billion miles) as Pioneer G approaches Jupiter.

The Pioneer G spacecraft is identical to Pioneer 10 except that a second magnetometer has been added to measure high magnetic fields close to Jupiter. Pioneer's scientific experiments should provide new knowledge about Jupiter, the Asteroid Belt, the outer solar system, and our galaxy. They will return images of Jupiter and will observe Jupiter's dusk side, a view never seen from the Earth.

The 14 experiments will make 20 types of measurements of Jupiter's atmosphere, radiation belts, heat balance, magnetic field, internal structure, moons, and other phenomena. They will characterize the heliosphere (the solar atmosphere), the interstellar gas, cosmic rays, asteroids, and meteoroids between Earth and beyond Jupiter.

Pioneers 10 and G are a new design for the outer solar system but are based on the subsystems of their predecessors, Pioneers 6 through 9. All four of these are still operating in interplanetary space, with Pioneer 6 in its eight year.

The 260-kilogram (570-pound) Pioneer G is spin-stabilized, giving its instruments a full-circle scan five times a minute. It uses nuclear sources for electric power because solar radiation is too weak at Jupiter for an efficient solar-powered system.

To maintain communications, Pioneer G's 2.75-meter (9-foot) dish antenna must point at Earth throughout the mission, changing its view by changes in spacecraft attitude, as the home planet moves about in its orbit around the Sun.

Potential benefits of the Pioneer Jupiter missions include increased knowledge of collisionless plasmas of the solar wind. The findings may also lead to better understanding of Earth's radiation belts, ionosphere and possibly weather cycles, and to insights into Earth's atmospheric circulation through study of Jupiter's rapidly rotating atmosphere.

NASA's Office of Space Science has assigned project management for the two Pioneer Jupiter spacecraft to NASA's Ames Research Center, Mountain View, California, near San Francisco. The spacecraft were built by TRW Systems Inc., Redondo Beach, California. The scientific instruments have been supplied by NASA Centers, universities and private industry.

Tracking is conducted by NASA's Deep Space Network, operated by the Jet Propulsion Laboratory, Pasadena, California. NASA's Lewis Research Center, Cleveland, manages the launch vehicle, which is built by General Dynamics, San Diego, California. Kennedy Space Center's Unmanned Launch Operations Directorate is responsible for the launch.

Cost of two Pioneer Jupiter spacecraft, scientific instruments, and data processing and analysis is about $100 million. This does not include costs of launch vehicles and data acquisition.

The 30- to 45-minute evening launch window opens progressively earlier each day – approximately 9:00 p.m. EST on April 5, and 6:00 p.m. by April 26.

Depending on launch date, the trip to Jupiter will take from 609 to 825 days, with arrival dates between December 5, 1974 and July 30, 1975.

Studies of Pioneer G's target, Jupiter, have produced many mysteries. The planet broadcasts predictably modulated radio signals of enormous power. It also apparently radiates about three times as much thermal energy as it receives from the Sun.

Jupiter's atmosphere contains ammonia, methane, hydrogen, and probably water, the same ingredients believed to have produced life on Earth about four billion years ago. Many scientists believe that large regions below the frigid cloud layer may be at room temperature. These conditions could allow the planet to produce living organisms despite the fact that it receives only $1/27^{th}$ of the solar energy received by Earth.

Jupiter is composed mostly of hydrogen and helium, the main constituents of the universe. The planet may have no solid surface. Due to its high gravity, it may go from a thick gaseous atmosphere down to oceans of liquid hydrogen, to a slushy layer, and then to a solid hydrogen core. Ideas of how deep beneath its striped cloud layers any solid hydrogen "icebergs" or "continents" might lie vary by thousands of kilometers.

Astronomers have long seen violent circulation of the planet's large-scale cloud currents and other features. A point on Jupiter's equator rotates at 45,600 km/hr (28,200 mph) during its ten-hour day, compared with 1,600 km/hr (1,000 mph) for a similar point on Earth's equator.

An odd feature of the planet is the Great Red Spot, the "Eye of Jupiter." This huge bright red oval is 48,000 kilometers (30,000 miles) long and 13,000 kilometers (8,000 miles) wide, large enough to swallow up several Earths. The Red Spot may be an enormous standing column of gas or even a raft of hydrogen ice floating on a bubble of warm hydrogen in the cooler hydrogen atmosphere and bobbing up and down at 30-year intervals, disappearing and reappearing. The spot's red color may be due to the presence of organic compounds, the building blocks of life, manufactured in Jupiter's atmosphere.

(End of General Release. Background information follows)

Mission Profile

Pioneer G will be launched toward Jupiter on a direct-ascent trajectory from Cape Kennedy in a direction 18 degrees south of straight east, passing over South Africa shortly after launch vehicle burnout.

The trip will follow a curving path about a billion kilometers (620 million miles) long between the orbits of Earth and Jupiter. The path will cover about 160 degrees going around the Sun between launch point and Jupiter. Because of the changing positions of the Earth and Jupiter, the shortest trip times are for launches during the early days of the 22-day launch period. These early dates would put the Earth-Jupiter line far from the Sun during flyby, resulting in less interference by the Sun with data transmission from the spacecraft and with simultaneous telescope observations of Jupiter from Earth.

The high-energy launch marks the second use of a third stage, the TE-M-364-4, with the Atlas-Centaur launch vehicle.

After liftoff, burnout of the 1,919,300-newton (431,500-pound)-thrust, stage-and-a-half Atlas booster will occur in about four minutes. Stage separation and ignition of the 131,200-newton (29,500-pound)-thrust Centaur second stage will then take place and the hydrogen-fueled Centaur engine will burn for about 7.5 minutes. The 10.7-meter (35 foot)-long aerodynamic shroud covering the third stage and the spacecraft will be jettisoned after leaving the atmosphere, about 12 seconds after Centaur engine ignition.

At about 13 minutes after liftoff, small solid-fuel rockets will spin up the 66,270-newton (14,900-pound)-thrust third stage and the attached spacecraft to 60 rpm for stability during stage firing. The third stage will then ignite and burn for about 44 seconds.

About two minutes after third stage burnout (16 minutes after liftoff), *Pioneer G* will separate from the third stage and be on Jupiter trajectory.

During powered flight, launch vehicle and spacecraft will be monitored from the Mission Control Center at Cape Kennedy via Eastern Test Range tracking stations and by a tracking ship in mid-Atlantic.

Twenty minutes after launch, mission control will shift from the Ames Research Center's Mission Director at Cape Kennedy to the Ames Flight Director at the Pioneer Mission Area, at the Jet Propulsion Laboratory (JPL), Pasadena, California.

Twenty-two minutes after launch, NASA's Ascension Island tracking station will acquire the spacecraft.

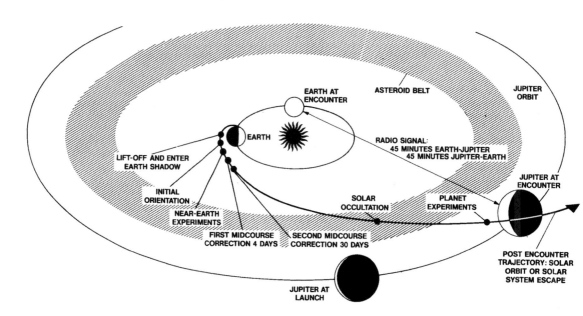

About 27 minutes after launch, the Johannesburg, South Africa station of the Deep Space Network (DSN) will lock on the spacecraft and will be able to send commands. By then *Pioneer G* will have emerged from the Earth's shadow and begin to get timing data from its Sun-sensor.

About 32 minutes after launch, *Pioneer G*'s onboard sequencer will initiate thruster firing for despin down to about 21 rpm from the 60-rpm spin imparted before third stage ignition. The sequencer then will start deployment of the four nuclear power sources, the radioisotope thermoelectric generators (RTG's), using spacecraft spin to slide them out on their trusses 3 meters (10 feet) from the center of the spacecraft. The sequencer will next initiate deployment of the magnetometer to 6.6 meters (21.5 feet) from the spacecraft center by unfolding its lightweight boom. When boom and truss deployments are complete, spacecraft spin rate will be down to five rpm.

Instrument Turn-On

At launch plus 40 minutes, the Johannesburg station will turn on the Geiger tube telescope and charged particle instruments to calibrate them against measurements of the known radiation in Earth's Van Allen Belts. At 50 minutes, the helium vector magnetometer will be turned on.

Because of the possibility of high-voltage arcing, some instruments will be allowed to outgas before turn-on. Flight directors will turn them on between six hours and two weeks after launch.

About two hours after liftoff, controllers will command the spacecraft to reposition itself so that its high-gain dish antenna points to within 20 degrees of Earth. This will greatly increase signal strength. Before this, the spacecraft will have been communicating via its rear-facing, low-gain antenna. This spacecraft turning maneuver will take about three hours to complete. The high-gain antenna will not be pointed directly at Earth, in order to avoid a spacecraft-Sun angle at which the intense solar radiation of space strikes the payload directly. Later in the mission, when the antenna must be pointed precisely at Earth for good communications, the Sun-angle will have changed greatly so that three meter dish antenna will shade most of the payload.

At eight hours after launch, *Pioneer G* will "rise" at the DSN's Goldstone, California, station. Goldstone will trim spin rate to the normal 4.8 rpm.

In the first few days after launch, JPL specialists at Pasadena will perform intensive computer calculations to establish the precise trajectory.

Also during the first days, the imaging photopolarimeter experiment will make maps of the brightness of the sky in order to correlate *Pioneer*'s measurements of zodiacal light with simultaneous measurements from Earth by Dudley Observatory, Albany, New York, and from Mount Haleakala, Hawaii.

Midcourse Correction

Four days of accurate tracking will have determined the precise launch velocity and direction. The automatic Conscan system (whereby the spacecraft uses the pattern of the incoming radio signal to refine its Earth-pointing) will be tested and then used to establish precise Earth-point for reference. Ames controllers will then command change of spacecraft attitude to point thrusters in the proper direction for the first midcourse velocity change to eliminate errors in aiming at Jupiter. Thrusters will then be fired and the spacecraft returned to its early-mission pointing direction of 20 degrees away from Earth.

At about a week after launch, the spacecraft and its operations team will have settled into the interplanetary operations and scientific data gathering phase of the mission. Analysts will regularly assess performance of systems and instruments.

The solar wind and magnetic field instruments will be mapping the interplanetary medium. The particle experiments will map distribution of solar and galactic cosmic ray particles, and the ultraviolet photometer will measure neutral hydrogen. The meteoroid and dust detectors will gather data on sizes and distribution of interplanetary matter. Zodiacal light measurements will be made periodically by the photopolarimeter instrument and the meteoroid telescopes to help determine amounts of interplanetary material.

Spacecraft attitude change maneuvers to obtain desired Earth-point will be commanded routinely every three to seven days, and later in the mission every week or two.

Near the end of the first month, a second midcourse velocity change will be made if needed.

When spacecraft operation has become routine, Ames personnel controlling the spacecraft will move from the Pioneer Mission Area at Pasadena to the Pioneer Mission Operations Center at the Ames Research Center, Mountain View, California. Ames personnel are at JPL in Pasadena during the first weeks after launch because of the greater computer and display capability there.

Asteroid Traverse

About four months after launch, *Pioneer G* will enter the Asteroid Belt and begin observations of light scattered by asteroid material with its four onboard telescopes, and the counting of particle penetrations with the gas-cell meteoroid detector.

Experimenters will compare numbers of asteroid particles with the relatively few particles measured by *Pioneer 10*. Controllers will ready emergency procedures for possible subsystem malfunctions due to impact of high velocity – around 48,000 km/hr (30,000 mph) – dust particles.

Early in the mission, command, tracking, and data return will be primarily by the DSN's 26-meter (85-foot) antennas located at 120-degree intervals around the Earth at Goldstone, California; Madrid, Spain, or Johannesburg, South Africa; and Canberra, Australia. At greater ranges from Earth, the DSN's 64-meter (210-foot) dish antennas at Goldstone, Madrid, and Canberra will track the spacecraft. The highly sensitive 64-meter dishes also will be used for critical maneuvers, such as midcourse velocity corrections.

At Jupiter distance, the 26-meter dish antennas will be able to receive 64 data bits per second (bps); the 64-meter dishes, 1,024 bps.

About 315 days from launch, the spacecraft will pass almost directly behind the Sun, causing communication difficulties for about a week because of the Sun's radio noise.

Factors in the Planet Encounter

About a year after launch (April 1974), scientists will have analyzed findings from *Pioneer 10*'s swing around Jupiter four months earlier (December 1973). Project officials will then decide on *Pioneer G*'s exact path past the planet and make the required course changes.

The question of Jupiter's radiation belts will be critical. The belts may be a million times stronger than Earth's belts. Number and energy flux of particles in the belts are believed to increase 100 times for each Jupiter radius closer to the planet. High-energy protons (hydrogen nuclei) and electrons could penetrate deeply into the spacecraft, damaging vital solid state electronic circuitry. The spacecraft could be crippled or destroyed.

For most future missions to Jupiter, the design of the spacecraft will be affected by the belts because spacecraft must approach the planet very closely through the most intense part. For a more distant approach, the spacecraft would have to carry exorbitant amounts of fuel for retrofire. Planners for these missions must know what particle radiation hazard to protect against. It could prove impossible to come closer to Jupiter than *Pioneer 10*'s 140,000 kilometers (87,000 miles).

Extremely high-velocity dust particles near Jupiter, pulled in by the planet's high gravity, may pose another danger. Though such particles are presumed to be few, their incoming velocities of up to 220,000 km/hr (137,000 mph) could be opposed to the 125,600 km/hr (78,000 mph) speed of the spacecraft at periapsis, possibly producing very penetrating impacts.

Because of Jupiter's gravity, any flyby course will be a turn around the planet. On a flyby trajectory like *Pioneer 10*'s, *Pioneer G* would approach from the sunlit side, then swing almost completely around its dark side traveling in a counterclockwise direction (looking down at the north pole). Planet rotation aside, it would fly about two thirds of the way around Jupiter at various altitudes in a week's time. It would pass over part of the southern hemisphere, cross the equator, and exit over the northern hemisphere.

With a different flyby point from *Pioneer 10*'s, flight paths could be entirely different and even include flight around Jupiter in the reverse direction (clockwise).

Targeting

As mentioned previously, prime consideration in selection of the target region is to get the best scientific data on Jupiter and its environment, especially the radiation belts. Hopefully, *Pioneer 10* results will permit narrowing the range of possible targets from points more distant than two planet radii, 140,000 kilometers (87,000 miles), down to about one-half radius above the cloud tops, 35,400 kilometers (22,000 miles). If a close approach is desirable scientifically, it could be possible for *Pioneer G* to fly to Saturn in 1980. For such a Saturn flyby, launch from Earth would have to have been accurate enough to leave thruster gas for necessary later course corrections.

Another targeting consideration will be close passage to one of Jupiter's moons.

A basic flyby factor will be the simultaneous tracking from two Earth stations during the critical five hours before periapsis (closest approach).

If possible, experimenters would also like to have one of the Jovian moons pass between the spacecraft and Earth, using effects on the spacecraft radio signal to detect an atmosphere on that moon. They also would like to view the Great Red Spot and the "hot shadow" of a moon on Jupiter's surface to measure temperatures.

Typical Pioneer 10 / G Encounter With Jupiter

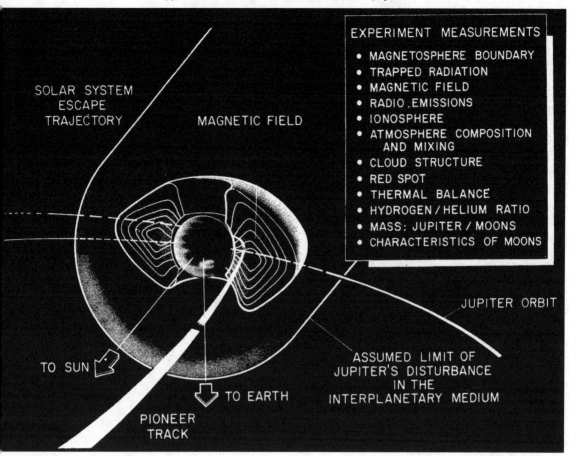

EXPERIMENT MEASUREMENTS

- MAGNETOSPHERE BOUNDARY
- TRAPPED RADIATION
- MAGNETIC FIELD
- RADIO EMISSIONS
- IONOSPHERE
- ATMOSPHERE COMPOSITION AND MIXING
- CLOUD STRUCTURE
- RED SPOT
- THERMAL BALANCE
- HYDROGEN / HELIUM RATIO
- MASS: JUPITER / MOONS
- CHARACTERISTICS OF MOONS

SOLAR SYSTEM ESCAPE TRAJECTORY

MAGNETIC FIELD

JUPITER ORBIT

TO SUN

TO EARTH

PIONEER TRACK

ASSUMED LIMIT OF JUPITER'S DISTURBANCE IN THE INTERPLANETARY MEDIUM

Flyby Operations

During flyby, spacecraft command and analysis activity will increase greatly. Because communication time from Earth to Jupiter will be around 45 minutes, commands must be precisely timed in advance for performance at a particular point over the planet. Five commands can be stored in advance on the spacecraft.

Ten days before periapsis (closest approach) controllers will command a final Earth point maneuver to establish a precise spacecraft attitude for the flyby. Biases may be included for effects of the Jovian magnetic field measured by *Pioneer 10*, which could cause drift in spacecraft attitude. After that, no further thruster firings will be made because these might distort several of the scientific measurements made near Jupiter.

As *Pioneer G* is drawn in by the giant planet's gravity, its velocity relative to Jupiter will soar. For a *Pioneer 10*-type trajectory, velocity would increase in 20 days from an approach speed of 33,000 km/hr (20,000 mph) to 125,600 km/hr (78,000 mph) at periapsis. For an approach to Jupiter's cloud tops as close as one-half a Jupiter radius, periapsis speed would be as high as 173,000 km/hr (107,000 mph).

Tracking stations will be prepared to handle effects of this four-to-five-fold increase in speed on the two-way doppler measurements used to track the spacecraft. Doppler shift will grow very large, and this will be complicated by loss of communication when the spacecraft is behind the planet.

Experimenters will calibrate their instruments before flyby, and will change sensitivity ranges of the high-energy particle, solar wind and magnetic field instruments for the far more intense phenomena at the planet.

These experiments will look for the first Jupiter effects – the bow shock wave in the solar wind, Jupiter's magnetic field, and Jovian effects on streams of high-energy solar particles.

For the flyby itself, on a *Pioneer 10*-type trajectory, *Pioneer G* would be close to Jupiter for about four days (100 hours). At 50 hours before periapsis, 5 million kilometers (3 million miles) away, *Pioneer* instruments would see the planet in almost full sunlight. At 40 minutes before periapsis, Jupiter would be half dark, and at periapsis itself about 60 percent of Jupiter would be dark. After periapsis, the spacecraft would continue to see less than a half-phase Jupiter.

For various other flyby trajectories, planet views and times will vary, but total length of time near Jupiter would be about the same, and the same types of observations will be made.

The imaging photopolarimeter will have begun making polarization and intensity measurements of Jupiter's reflected light several weeks out from the planet. Starting at about 1.3 million kilometers (about 800,000 miles) from periapsis, the instrument will begin to alternate these measurements with the imaging. The polarization and intensity measurements will provide information on the physical properties of Jupiter's clouds and atmosphere.

The ultraviolet photometer will examine Jupiter's upper atmosphere for hydrogen / helium ratio, atmosphere mixing rate, temperature, and evidence of an auroral oval near the poles on the planet's day side. On most trajectories, it will have several viewing periods during planet approach.

The infrared radiometer will look for Jovian radiation from an internal heat source, hot spots in the outer atmosphere, and cold spots at the poles indicating ice caps of frozen methane. It will measure the atmosphere's hydrogen / helium ratio. Its observation period will be close to periapsis. It will have good views of the boundary between Jupiter's light and dark hemispheres (the terminator).

As the spacecraft passes behind the planet and reemerges, ground stations will record diffusion effects on its radio signals to calculate the density and composition of Jupiter's atmosphere.

After periapsis, solar wind, particle, and magnetometer experiments will look for planetary effects on the interplanetary medium behind the planet. Going away from the planet, the polarimeter will again measure Jupiter's atmosphere.

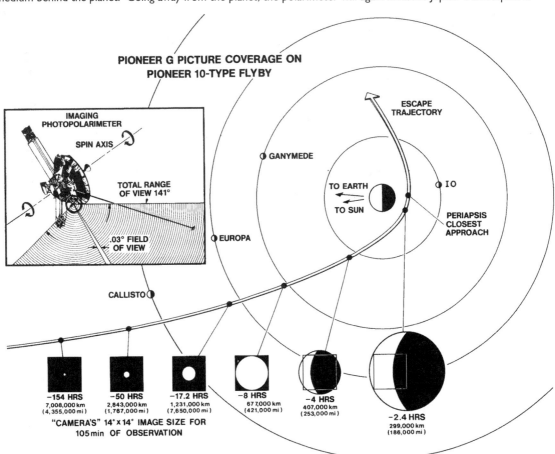

Beyond Jupiter

After planet encounter, *Pioneer* will return to its interplanetary data-gathering mode of operation.

As in the case of Pioneer 10, the basic Pioneer G mission is planned to last through encounter and about three months thereafter. (The costs of about $100 million for the Pioneer 10 and G flights cover only this basic mission.) After that it is not possible to predict exactly how long it will continue to function. Beside possible radiation damage, the life of parts and exact rate of power system degradation are not certain. Another variable is supply of thruster gas for Earth point and course-change maneuvers.

On a *Pioneer 10*-type flyby, Jupiter's gravity and orbital velocity would speed up *Pioneer G* to solar system escape speed, and bend the spacecraft trajectory inward toward Jupiter's orbital path. Cruise velocity relative to the Sun (not the planet) of 38,500 km/hr (24,000 mph) would have increased to 79,200 km/hr (49,320 mph).

The most interesting experimental questions for *Pioneers 10* and *G* beyond Jupiter will be: what is the variation in the flux of galactic cosmic rays; how do the solar wind and magnetic field change with increasing solar distance; and what is the distribution of interstellar neutral hydrogen and helium? What do these tell about the interstellar space beyond the boundary of the heliosphere (the Sun's atmosphere)? The plasma, magnetic field, high-energy particle, and ultraviolet photometer experiments will share these searches.

During the post-encounter period, for most flyby trajectories, strength of incoming signals will decline steadily, eventually becoming almost infinitesimal. Communications distance will grow steadily longer. Time required to command the spacecraft and get a response will lengthen to hours.

Only the most intensive and sophisticated efforts by the Deep Space Network will allow communications with the spacecraft at all.

On a *Pioneer 10*-type, solar system escape trajectory, *Pioneer G* would cross the orbit of Saturn, at about 1.5 billion kilometers (930 million miles) from the Sun, about four years after launch. It would reach Uranus' orbit at about 2.9 billion kilometers (1.8 billion miles) from the Sun 7.5 years after launch, but communication would be only marginally possible at Uranus' great distance.

Pioneer G would then continue into interstellar space. Six years after Jupiter encounter, at around 3.2 billion kilometers (2 billion miles), the spacecraft flight path would have curved 87 degrees further around the Sun than the Jupiter flyby point. Its velocity relative to the Sun at this point would be 53,300 km/hr (33,100 mph) From then on its flight path would curve still another 25 degrees around the Sun. At this point, far out in interstellar space, it would proceed away from the Sun in essentially a straight line, and its velocity would have dropped to a permanent 41,000 km/hr (25,500 mph) away from the Sun.

The Message Plaque

Like *Pioneer 10*, *Pioneer G* will carry a pictorial message intended for other intelligent species, if any exist, who might find the spacecraft thousands of years from now in some other star system.

The plaque tells when the *Pioneers* were launched, from where, and by whom. Location of the Sun is shown by the intersection point of signals from 14 pulsars (cosmic radio sources). Binary symbols show the frequencies of the pulsars today, and these could be used even a million years later to calculate the time of launch. The electron reversal in the hydrogen atom is shown to provide a measurement standard (its 8-inch radio wavelength) for both pulsar frequencies and the size of figures on the plaque. These are a man and woman shown standing in front of the *Pioneer* spacecraft. The Sun and nine planets also are shown, as is the spacecraft's trajectory, leaving the third planet, Earth, passing Mars and swinging by the fifth planet, Jupiter.

The plaque was designed by Dr. Carl Sagan, Director of the Laboratory for Planetary Studies, Cornell University; his wife, Linda Salsman Sagan, a painter and film maker: and Dr. Frank Drake, Director of the National Astronomy and Ionosphere Center, Cornell. A detailed description of it appears in the February 25, 1972, issue of the journal *Science*.

The Asteroids

The asteroids travel around the Sun in elliptical orbits like small planets. The Asteroid Belt forms a doughnut-shaped region, lying between the orbits of Mars and Jupiter, 300 million to 545 million kilometers (186 million to 338 million miles) from the Sun.

The Belt is roughly 245 million kilometers (152 million miles) wide and extends about 40 million kilometers (25 million miles) above and below the plane of the Earth's orbit.

Scientists believe the asteroids either condensed individually from the primordial gas cloud which formed the Sun and planets, or somewhat less likely, that they are debris from the break-up of a very small planet. Clearly, they contain important information on the origin of the solar system.

Passage of *Pioneer 10* through the Asteroid Belt appears to have shown that high-velocity, smaller asteroid particles pose only a minor hazard to spacecraft. This is an important discovery because the belt is too thick to fly over or under, so all outer planet missions must fly through it.

Some estimates suggest that there is enough material in the Belt to make a planet with a volume about $1/1,000^{th}$ of the Earth.

Astronomers have identified and calculated orbits for 1,776 asteroids. There may be 50,000 in the size range from the largest, Ceres, which has a diameter of 770 kilometers (480 miles), to bodies 1 mile in diameter.

In addition, the Belt is presumed to contain hundreds of thousands of asteroid fragments, and uncountable billions of dust particles.

In the center of the Belt, asteroids and particles orbit the Sun at about 61,200 km/hr (38,000 mph). These particles would impact the spacecraft (which has its own velocity in somewhat the same direction) at about 49,000 km/hr (about 30,000 mph). In short, asteroidal material appears to be thinly spread, but penetrating.

A few asteroids stray far beyond the Belt. Hermes can come within about 354,000 kilometers (220,000 miles) of the Earth, or closer than the Moon. Icarus comes within 9 million miles of the Sun.

Many meteorites which survive atmosphere entry and land on Earth are believed to be asteroidal material. These meteorites are mostly stony, but some are iron. Some contain large amounts of carbon, including organic chemicals like amino acids.

What effects could the Asteroid Belt have on *Pioneer G?*

None of the known large asteroids will come close to the spacecraft. The threat of baseball, green pea, or even BB sized asteroids is negligible. The most serious hazard comes from particles of $1/10^{th}$ to $1/1,000^{th}$ of a gram mass. Smaller particles are too tiny to do damage. (There are 28 grams to one ounce).

Meteoroids with a mass of $1/100^{th}$ gram, which travel at about 54,000 km/hr (33,500 mph) relative to the spacecraft near Earth, can penetrate a single sheet of aluminum one centimeter thick. At Jupiter's orbit these particles travel only about 25,200 km/hr (15,660 mph).

Results from *Pioneer 10* indicate that *Pioneer G* appears to have an excellent chance of passing through the Belt without a damaging hit by a particle in the dangerous size range of $1/10^{th}$ to $1/1,000^{th}$ gram.

Comets

Throughout the Pioneer G mission, cometary particles present some hazard.

In the vicinity of Earth, about half the cometary particles travel in streams which follow the orbital paths of existing or dissipated comets. When these streams (the Leonids, the Perseids, etc.) intersect Earth's orbit, meteor showers result as the particles burn up in the atmosphere. Cometary particle streams orbit the Sun in long ovals. Most of these orbits lie within 30 degrees of the ecliptic.

The other half of the cometary particles near the Earth appear sporadically.

Cometary particles near the Earth have average speeds relative to the spacecraft of 72,000 km/hr (45,000 mph), and a $1/100^{th}$ gram cometary particle can penetrate a one centimeter-thick aluminum sheet. Their speed at Jupiter's orbit will be down to 32,200 km/hr (20,000 mph).

Jupiter

What We Know

Though many mysteries concerning this big, brilliant planet remain to be solved, hundreds of years of astronomical observations and analysis have provided a stock of information. Galileo made the first telescopic observations and discovered Jupiter's four larger moons in 1610.

Seen from Earth, Jupiter is the second brightest planet, and fourth brightest object, in the sky. It is 773 million kilometers (480 million miles) from the Sun, and circles it once in just under 12 years. The planet has 12 moons; the four outer ones orbit the planet in a direction opposite to that of other known moons. Two of the moons, Ganymede and Callisto, are about the size of the planet Mercury. Two others, Io and Europa, are similar in size to the Earth's moon.

Jupiter spins once every 10 hours, the shortest day of any of the nine planets. Because of Jupiter's size, this means that a point at the equator on its visible surface (cloud tops) races along at 35,400 km/hr (22,000 mph), compared to a speed of 1,600 km/hr (1,000 mph) for a similar point on Earth.

This tremendous rotational speed (and fluid character of the planet) makes Jupiter bulge at its equator. Jupiter's polar diameter of about 124,000 kilometers (77,000 miles) is 19,000 kilometers (11,800 miles) smaller than its equatorial diameter of about 143,000 kilometers (88,900 miles).

Jupiter's visible surface (cloud top area) is 62 billion square kilometers (about 24 billion square miles). The planet's gravity at cloud top is 2.36 times that of Earth.

The mass of the planet is 318 times the mass of Earth. Its volume is 1,000 times Earth's. Because of the resulting low density (one-fourth of the Earth's or 1.3 times the density of water), most scientists are sure that the planet is made up of a mixture of elements similar to that in the Sun or the primordial gas cloud which formed the Sun and planets. This means there are very large proportions (at least three-quarters) of the light gases hydrogen and helium. Scientists have identified hydrogen, deuterium (the heavy isotope of hydrogen), methane (carbon and hydrogen), and ammonia (nitrogen and hydrogen) by spectroscopic studies of Jupiter's clouds.

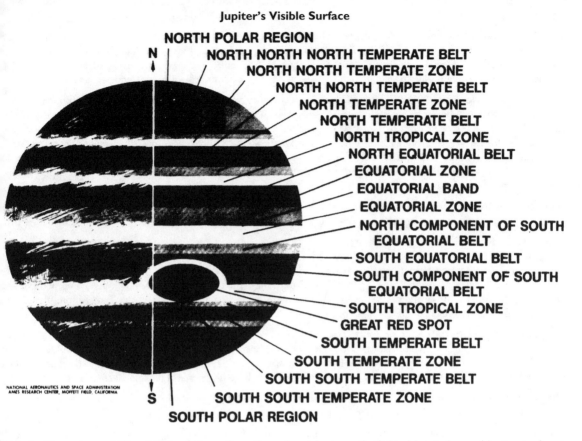

Jupiter's Visible Surface

NORTH POLAR REGION
NORTH NORTH NORTH TEMPERATE BELT
NORTH NORTH TEMPERATE ZONE
NORTH NORTH TEMPERATE BELT
NORTH TEMPERATE ZONE
NORTH TEMPERATE BELT
NORTH TROPICAL ZONE
NORTH EQUATORIAL BELT
EQUATORIAL ZONE
EQUATORIAL BAND
EQUATORIAL ZONE
NORTH COMPONENT OF SOUTH EQUATORIAL BELT
SOUTH EQUATORIAL BELT
SOUTH COMPONENT OF SOUTH EQUATORIAL BELT
SOUTH TROPICAL ZONE
GREAT RED SPOT
SOUTH TEMPERATE BELT
SOUTH TEMPERATE ZONE
SOUTH SOUTH TEMPERATE BELT
SOUTH SOUTH TEMPERATE ZONE
SOUTH POLAR REGION

NATIONAL AERONAUTICS AND SPACE ADMINISTRATION
AMES RESEARCH CENTER, MOFFETT FIELD, CALIFORNIA

Clouds, Currents, and Visual Appearance. Seen through a telescope, the lighted hemisphere of Jupiter is almost certainly a view of the tops of gigantic regions of towering multi-colored clouds.

Overall, due to its rotation, the planet is striped or banded, parallel with its equator, with large dusky, gray regions at both poles. Between the two polar regions are five, bright salmon-colored stripes, known as zones, and four darker, slate-gray stripes, known as belts – the South Equatorial Belt, for example. The planet as a whole changes hue periodically, possibly as a result of the Sun's 11-year activity cycles.

The Great Red Spot in the southern hemisphere is frequently bright red, and since 1665 has disappeared completely several times. It seems to brighten and darken at 30-year intervals.

Scientists agree that the cold cloud tops in the zones are probably largely ammonia vapor and crystals, and the gray polar regions many be condensed methane. The bright cloud zones have a complete range of colors from yellow and delicate gold to red and bronze. Clouds in the belts range from gray to blue-gray.

In addition to the belts and zones, many smaller features – streaks, wisps, arches, loops, patches, lumps, and spots – can be seen. Most are hundreds of thousands of kilometers in size.

Circulation of these cloud features has been identified in a number of observations. The Great Equatorial Current (The Equatorial Zone), 20 degrees wide, sweeps around the planet 410 km/hr (225 mph) faster than the cloud regions on either side of it, and is like similar atmospheric jet streams on Earth. The South Tropical circulating current is a well-known feature, as is a cloud current which sweeps completely around the Great Red Spot.

When Jupiter passed in front of stars in 1953 and again in 1971, astronomers were able to calculate roughly the molecular weight of its upper atmosphere by the way it refracted the stars' light. They found a molecular weight of around 3.3 which means a large proportion of hydrogen (molecular weight 2) because all other elements are far heavier. (Helium is 4, carbon 12, and nitrogen 14.)

Under Jovian gravity, atmospheric pressure at the cloud tops is calculated to be up to ten times one atmosphere on Earth.

The transparent atmosphere above the clouds can be observed spectroscopically and in polarized light. It is believed to be at least 60 kilometers (35 miles) thick.

Scientists have suggested from cloud-top observations by telescope that the general circulation pattern of Jupiter's atmosphere is like that of Earth, with circulation zones corresponding to Earth's equatorial, tropical, subtropical, temperate, subpolar, and polar regions. However, Jupiter's polar regions (from an atmospheric circulation standpoint) appear to begin at about 26 degree latitude from the equator, instead of at 60 degrees as on Earth.

Magnetic Fields and Radiation Belts. Among the nine planets, only Jupiter and Earth are known to have magnetic fields. Evidence for Jupiter's magnetic field and radiation belts comes from its radio emissions. The only phenomenon known that could produce the planet's decimetric (very high frequency) radio waves is trapped electrons gyrating around the lines of such a magnetic field. When such electrons approach the speed of light, they emit radio waves.

Radio emissions indicate that Jupiter's magnetic field is toroidal (doughnut-shaped) with north and south poles like the Earth's. It appears around 20 times as strong as Earth's field and presumably contains high-energy protons (hydrogen nuclei) and electrons trapped from the solar wind. The field's center appears to be near the planet's axis of rotation and south of the equatorial plane. Jupiter's powerful magnetic field can hold more particles trapped from the solar wind than can Earth's field, and with increased particle energies. As a result, particle concentrations and energies could be up to a million times higher than for Earth's radiation belts, a flux of a billion particles or more per square centimeter per second.

Because Jupiter has such high gravity and is so cold at the top of its atmosphere, the transition region between dense atmosphere and vacuum is very narrow. As a result, the radiation belts may come much closer to the planet than for Earth.

Jovian Radio Signals. Earth receives more radio noise from Jupiter than from any other extraterrestrial source except the Sun. Jupiter broadcasts three kinds of radio noise: (1) thermal – from the temperature-induced motions of the molecules in its atmosphere (typical wavelength, 3 centimeters); (2) decimetric (centimeter range) – from the gyrations of electrons around the lines of force of the planet's magnetic field (typical wavelength 3-70 centimeters); and (3) decametric (up to 10s of meters) – believed to be from huge discharges of electricity (like lighting flashes) in Jupiter's ionosphere (wavelength 70 centimeters to 60 meters).

The powerful decametric radio waves originate at known longitudes of Io with respect to Jupiter and have been shown to be modulated by passages of Jupiter's close moon, Io, whose orbit is 2.5 planet diameters, 350,000 kilometers (about 217,000 miles), above the tops of Jupiter's cloud layer. Some scientists believe that the conductivity of Io must be sufficient to link up magnetic lines of force to the planet's ionosphere, allowing huge discharges of built-up electrical potential.

The power of these decametric radio bursts is equal to the power of several hydrogen bombs. Their average peak value is 10,000 times greater than the power of Jupiter's decimetric signals.

Temperature. The average temperature at the tops of Jupiter's clouds appears to be about -145 degrees C (-229 degrees F), based on many observations of Jupiter's infrared radiation. But recent stellar occultation studies indicate that much of the diffuse outer atmosphere is close to room temperature, and that the top layer is at about 20 degrees C (68 degrees F). Only about $1/27^{th}$ as much heat from the Sun arrives at Jupiter as arrives at Earth. Recent infrared measurements made from high altitude aircraft suggest that the giant planet radiates about 2.5 to three times more energy than it absorbs from the Sun. The question is, what is the source of this energy? The shadows of Jupiter's moons on the planet appear to measure hotter than surrounding sunlit regions.

Jupiter Unknowns

Scientists are reasonably certain of most of the preceding phenomena and observations, though it will be important to check them at close range with *Pioneers 10* and *G*. They have few explanations for these observed phenomena, and they know very little about other aspects of the planet. What is hidden under the heavy Jovian clouds? How intense are the radiation belts?

Life. Perhaps the most intriguing unknown is the possible presence of life in Jupiter's atmosphere.

Estimates of the depth of the Jovian atmosphere beneath the cloud layer vary from 100 to 6,000 kilometers (60 to 3,600 miles). The compositions and interactions of the gases making up the atmosphere are unknown. If the atmosphere is deep, it must also be dense. By one estimate, with an atmospheric depth of 4,200 kilometers (2,600 miles), pressure at the Jovian "surface" would be 200,000 times Earth's atmospheric pressure due to the total weight of gas in the high Jovian gravity. One source cites eight different proposed models of Jupiter's atmosphere.

However, scientists do appear to agree on the presence of liquid water droplets in the atmosphere. Since the planet is believed to have a mixture of elements similar to that found in the Sun, it is almost sure to have abundant oxygen. And most of this oxygen has probably combined with the abundant Jovian hydrogen as water.

If large regions of Jupiter's atmosphere come close to room temperature, both liquid water and water ice should be present.

Jupiter's atmosphere contains ammonia, methane, and hydrogen. These constituents, along with water, are the ingredients of the primordial "soup" believed to have produced the first life on Earth by chemical evolution. On this evidence, Jupiter could contain the building blocks of life.

Some scientists suggest that the planet may be like a huge factory, turning out vast amounts of life-supporting chemicals (complex carbon-based compounds) from these raw materials, using its own internal energy. If so, life could exist without photosynthesis. Any solar photosynthesis would have to be at a very low level since Jupiter receives only $1/27$th of Earth's solar energy. It would probably be low-energy life forms at most (plants and microorganisms) because there is believed to be no free oxygen. Life forms would float or swim because a solid surface, if any, would be deep within Jupiter at very high pressures.

Planet Structure. While there are wide differences among scientists on planet structure, most proposed models contain elements like the following:

Going down, it is believed that temperature rises steadily. The cloud tops may consist of super-cold ammonia crystals, underlain by a layer of ammonia droplets, under which may be a region of ammonia vapor. Below this may be layers of ice crystals, water droplets, and water vapor. Below this is either the planet's solid surface, or liquid hydrogen oceans. Still lower is a region of metallic hydrogen created by high Jovian gravity with perhaps a core of rocky silicates and metallic elements. The core might be ten times the mass of the Earth by one estimate.

Some theorists doubt that the planet has any solid material at all, but is entirely liquid. Others propose a gradual thickening from slush to more rigid material.

Most planetologists think the Great Red Spot may be a column of gas, the center of an enormous vortex, rising from Jupiter's interior to the top of the atmosphere. Known in physics as a Taylor column, such a vortex would have to be anchored by a prominent surface feature, either a high spot or huge depression. The Red Spot has circled the planet more than once relative to visible cloud features in the past hundred years.

Some scientists suggest that Jupiter's surface itself may be rotating at varying rates of speed relative to the atmosphere, thus moving the Spot. The rotation of the planet's magnetic field, tracked by the timing of its radio emissions, is believed the best measure of rotation rate of the planet.

Model of Jupiter Interior

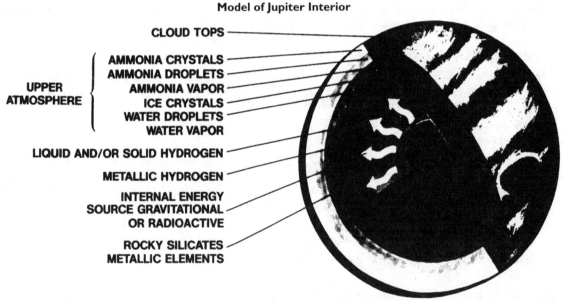

A Hot Planet. Theorists suggest that Jupiter is almost large enough to be a small star. Because of measurements of excess heat radiated by the planet, current Jupiter theories call for a relatively hot core, compared with earlier ideas of a super-cold interior.

One hypothesis holds that despite the five billion years since formation of the planets, Jupiter has not completed its gravitational condensation. This continued settling toward the center (as little as one millimeter per year) could produce the required heat energy. If the planet has a rocky core, some of the heating could be from decay of radioactive material in the core.

Magnetic Field. Internal heat also could explain the magnetic field. A hot core might be a fluid core. Convective and rotational motion of electrically conducting fluids at temperatures of 10,000 degrees C (18,000 degrees F) may generate the field. Or it could be generated by conductive atmosphere layers below the clouds which would store and later release energy.

With knowledge of Jupiter's magnetic field, scientists should be able to make inferences about the planet's internal structure, particularly its fluid component.

Jupiter's Moons

Jupiter's 12 natural satellites have some odd characteristics. The second moon, Io, appears to be brighter for 10 minutes after emerging from Jupiter's shadow. If so, the simplest explanation, supported by recent stellar occultation observations, is that Io has an atmosphere (probably nitrogen or methane) which "snows out" on the surface when Io is on the cold, dark side and revaporizes when back in sunlight. Io also is distinctly orange in color and has odd reflecting properties. Most of the sunlight reaching the moon's surface is strongly scattered back, making Io extremely bright. This reflection is believed to be more pronounced for Io than for any other known object in the solar system. Infrared measurements show Europa and Ganymede to have water ice on their surfaces.

The inner moons in order of distance from the planet are: tiny Amalthea, diameter 160 kilometers (100 miles), which orbits Jupiter twice a day at only 1.5 planet diameters, 106,000 kilometers (66,000 miles), above the cloud tops; and the four large moons Io, Europa, Ganymede, and Callisto, whose orbits lie between 422,000 kilometers (262,000 miles) and 1,882,000 kilometers (1,169,000 miles) from Jupiter.

Beyond these are the seven tiny outer moons. The inner three of these, Hestia, Hera, and Demeter, have orbits which lie between 11.5 and 11.7 million kilometers (7.2 and 7.3 million miles) from Jupiter.

Orbits of the four outermost, Andrastea, Pan, Poseidon, and Hades, lie between 20.7 million and 23.7 million kilometers (12.9 and 14.7 million miles) of Jupiter. All are in retrograde orbits, moving counter to the usual direction of planet rotation. This suggest that they may be asteroids captured by Jupiter's powerful gravity. Diameters of six of the outer moons range from 15 to 40 kilometers (9 to 24 miles), with Hestia, the seventh, having a diameter of about 130 kilometers (81 miles).

Orbital periods of the four large inner moons range from 1.7 days (for Io) to 16.7 days. Orbital periods of the inner three of the outer seven moons are around 250 days. While the four farthest-out, backward-orbiting moons complete their circuits of Jupiter in around 700 days.

The backward orbit of the far outer moon, Poseidon, is highly inclined to the equator, and wanders so much due to various gravitational pulls that astronomers have a difficult time finding it.

The Heliosphere

Pioneer G may follow *Pioneer 10* in exploring much of the heliosphere, the atmosphere of the Sun. The spacecraft may survey a region of space 2.4 billion kilometers (1.5 billion miles) wide between the orbits of Earth and Uranus. It may also learn something about the interstellar space outside the heliosphere.

The thinly diffused solar atmosphere is hundreds of times less dense than the best vacuums on Earth. Yet it is important because it contains:

- The ionized gas known as the solar wind, roughly a 50-50 mixture of protons (hydrogen nuclei) and electrons. It flows out from the 3,600,000 degrees F (2,000,000 degrees C) corona of the Sun in all directions at average speeds of 1.6 million km/hr (one million mph). Where does the solar wind stop blowing? Where is the boundary between the solar atmosphere and the interstellar gas? Estimates range from 560 million to 16 billion kilometers (350 million to 10 billion miles) from the Sun.

- Complex magnetic and electric fields, carried out from the Sun by the solar wind.

- Solar cosmic rays, high-energy particles thrown out by the huge explosions on the Sun's surface at up to 480 million km/hr (300 million mph).

The heliosphere also encompasses planets, moons, asteroids, comets and dust. It is traversed by electromagnetic radiation from the Sun – radio waves, infrared, ultraviolet, and visible light. Earth gets most of its energy from this radiation.

The heliosphere further contains cosmic ray particles from within and beyond our galaxy, traveling at nearly the speed of light. This means enormous particle energies, to 10^{14} million electron-volts (MeV). The expression 10^{14} is 1 followed by 14 zeros. There are also neutral hydrogen atoms, believed to be part of the interstellar gas from which the Sun and planets were formed.

Study of these phenomena has a variety of applications. For example, storms of solar particles striking Earth interrupt radio communications and sometimes electric power transmission.

The heliosphere can be thought of as a huge laboratory where phenomena occur that cannot be simulated on Earth. For example, man cannot accelerate particles in Earth laboratories to the near-light speeds reached by galactic cosmic ray particles. These particles are observed by *Pioneer* instruments.

Solar wind particles are so thinly spread that they rarely collide, and only the magnetic field and electrostatic fields link them together as a collisionless ionized gas (collisionless plasma). Even though collisionless, these plasmas transmit waves of many kinds because of this magnetic field linkage.

Knowledge of these plasmas can shed light on an area of current high interest – how to contain plasmas in magnetic fields. Accomplishing this would allow control of the hydrogen thermonuclear reaction continuously on a small scale to generate virtually unlimited electric power by a clean process.

All this involves such relationships as that of the temperatures of solar wind protons and electrons near Earth – they are almost the same, even though theory predicts that the electrons should be several hundred times hotter.

How is energy transferred between them to equalize the temperature? How far out from the Sun do these temperatures remain equal? Why?

Solar Wind, Magnetic Field, and Solar Cosmic Rays

Near the Earth, the speed of the solar wind varies from one to three million km/hr (600,000 to 2,000,000 mph), depending on activity of the Sun. Its temperature varies from 10,000 to 1,000,000 degrees C (18,000 to 1,800,000 degrees F). Near the Earth, collisions between streams of the solar wind use up about 25 percent of its energy. The wind also fluctuates due to features of the rotating solar corona, where it originates, and because of various wave phenomena.

The fastest solar cosmic ray particles jet out from the Sun in streams which can cover the 150 million kilometers (93 million miles) to Earth in as little as 20 minutes. Slower particle streams take one or more hours to reach Earth and tend to follow the curving interplanetary magnetic field.

The positive ions are 90 percent protons and ten percent helium nuclei, with occasional nuclei of heavier elements.

There are from none to 20 flare events on the Sun each year which produce high-energy solar particles, with the largest number of flares at the peak of the 11-year cycle.

Planetary Interactions and the Interstellar Gas

Since Jupiter is expected to have a magnetic field like that of Earth, a bow shock wave should form in the solar wind in front of the planet. There should be a magnetic envelope around the planet, shutting out the solar wind, and a trailing magnetic tail.

Pioneer G instruments should easily characterize ionized particles swept from the atmospheres of Jupiter, its moons, or the asteroids because their properties will greatly differ from those of particles of solar origin.

Pioneer G will measure helium, as well as interstellar neutral hydrogen. Both hydrogen and helium atoms are believed to be part of the interstellar gas which has forced its way into the heliosphere as the solar system moves through interstellar space at 72,000 km/hr (45,000 mph). *Pioneer G* also will measure the ultraviolet glow of the interstellar gas which can be seen out beyond the heliosphere boundary.

Galactic Cosmic Rays

Galactic cosmic ray particles usually have far higher energies (velocities) than solar cosmic rays. These particles may get their tremendous energies from the explosion of stars (supernovas), the collapse of stars (pulsars), or acceleration in the colliding magnetic fields of two stars. Pioneer studies of these particles may settle questions of their origin in our galaxy and important features of the origin and evolution of the galaxy itself. These studies should answer such questions as the chemical composition of stellar sources of cosmic ray particles in the galaxy.

Near Earth, numbers of these galactic cosmic ray particles can vary up to 50 percent as increased solar activity (a faster solar wind and stronger magnetic field) pushes them out of the inner solar system at the peak of the 11-year solar cycle. Cosmic ray particle intensity varies up to 30 percent during individual solar flares.

Galactic cosmic rays consist of protons (hydrogen nuclei), 85 percent; helium nuclei, 13 percent; nuclei of other elements, 2 percent; and high-energy electrons, 1 percent.

Particles usually range in energy from 100 million electron-volts (MeV) to 10^{14} MeV.

Near Earth the average flux of these particles is four per square centimeter per second, with most particles in the 1,000 MeV range. Presumably as the spacecraft moves toward the edge of the blocking solar atmosphere, more of these medium-energy particles will be observed.

Pioneer 10 Experiment Results

Pioneer 10 so far has crossed the region lying between 160 million and 544 million kilometers (100 million and 340 million miles) from the Sun over a flight path roughly 696 million kilometers (435 million miles) long. In crossing this 384-million-kilometer (240-million-mile)-wide region of space, it has surveyed the 243-million-kilometer (152-million-mile)-wide Asteroid Belt, and has measured the outward gradient of several phenomena.

The principal result so far is that the Asteroid Belt appears to contain somewhat less material than many scientists had believed, and does not offer a serious hazard to spacecraft. *Pioneer 10*'s instruments reported this, and the spacecraft itself served as an instrument, receiving no damaging hits by an asteroid particle on its seven-month, 338-million-kilometer (205-million-mile)-long flight through the belt.

Other major findings are that the distribution of dust particles between the Earth's orbit and the far side of the Asteroid Belt seems to depend on particle size. There appear to be no more of the very small particles in the belt than outside it, and there may be fewer of the smallest particles in the belt than near the Earth.

Researchers found that out to 560 million kilometers (350 million miles) from the Sun, solar magnetic field strength, solar wind density, and numbers of solar high-energy particles all decline roughly as the square of the distance from the Sun.

Because turbulence of the solar wind and magnetic field do not decline with solar distance, large parts of the low-energy cosmic ray particles from the galaxy appear to be shut out of the solar system perhaps as far as beyond Jupiter.

The gigantic solar storm of August 2, 1972 showed clearly that as it moves out the solar wind tends to slow down while its gases heat up.

Among solar high-energy particles, experimenters have found the elements sodium and aluminum for the first time.

Researchers found that the neutral hydrogen of the interstellar wind appears to enter the heliosphere in the plane of the Earth's orbit, about 60 degrees away from the direction of travel of the solar system through space, and measured interstellar helium for the first time.

Asteroid Belt and Interplanetary Dust

Distributions with particle size were as follows:

Going out from the Earth's orbit, the smallest particles (around $1/1,000$ mm diameter) actually appear to decline in numbers. Somewhat larger particles ($1/100$ to $1/10$ mm diameter) seemed to be evenly distributed all the way from the Earth's orbit through the far side of the Belt with no increase in the Belt. (There are 25 mm to an inch.)

Still larger particles ($1/10$ to 1 mm diameter) were found all the way out but were almost three times as frequent in the Belt as outside it.

Particles larger than 1 mm diameter appear to be very thinly spread (as many scientists expected). Preliminary analyses of observations by *Pioneer 10* asteroid telescopes have not produced certain identification of any particles larger than 1 mm diameter, though further analysis may well show some.

One explanation for the absence of very small particles in the Asteroid Belt would be that solar radiation may reduce the orbital speed of these particles. The larger particles with more mass would be less affected by solar radiation and would maintain their orbits.

Other explanations would be that original conditions of solar system formation led to this dust distribution, or that comets tend to break up relatively close to the Sun, leaving more dust there. Other findings:

For the smallest particles (around $1/1,000$ mm diameter) – measurements of the zodiacal light reflected from these particles as one looks away from the Sun, have shown that between the orbits of the Earth and Mars going away from the Sun zodiacal light declines steadily. If these very small reflective dust particles were mostly in the Asteroid Belt, the reflected sunlight would have remained relatively constant until the belt was reached, and then dropped rapidly as the spacecraft passed through the belt.

The rate of decline of this reflected light suggests that concentrations of these tiny particles may decline as the square of the distance from the Sun.

The gegenschein or counter glow seen from the Earth in the night sky at the point opposite the Sun was believed by some to be sunlight reflected from a "dust tail" of the Earth pointed away from the Sun. Since the glow remained the same when *Pioneer 10* was 9 million kilometers away, the gegenschein is now known to be an interplanetary phenomenon.

In the very small asteroid particles $1/100$ to $1/10$ mm diameter (one billionth to one millionth of a gram) – there was no increase in the Asteroid Belt and the experimenter believes the particles measured are cometary. For these particles, the one-quarter square meter of gas-cell detector surface received 25 penetrations between the Earth's orbit and the Belt's near side, and 17 penetrations in the Belt.

For the larger asteroid particles, $1/10$ mm to 1 mm diameter (one ten millionth to one ten thousandth gram) – incomplete data analyses suggest that there are almost three times more particles in the Belt than outside it.

Reflectivity of these particles is believed to be three times higher than expected, perhaps because of ice coating or solar wind polishing. The four Asteroid telescopes counted almost 200 particles in this size range. No "large" bodies (one meter diameter, for example) appear to have been seen.

The Solar Wind

Between 100 million and 350 million miles from the Sun:

- Density dropped by approximately the square of the distance from the Sun.
- Variations (turbulence) were about the same at 350 million miles as at 100 million miles.

Apparently the wind slows down while increasing in temperature as some of its outward motion energy is converted to thermal motion energy.

During the huge storm on the Sun of August 2, 1972, largest of the solar cycle, Pioneers 9 and 10 happened to be on the same line going out from the Sun. Pioneer 9 was near Earth's orbit, Pioneer 10, 213 million kilometers (132 million miles) farther out.

Over this distance, solar wind speed dropped from over 3,600,000 km/hr (2,240,000 mph) at Pioneer 9, the highest speed ever measured, to less than 2,520,000 km/hr (1,565,000 mph). But temperature went up from less than 1 million degrees K (about 1.8 million degrees F) to over 2 million degrees K (about 3.6 million degrees F) between Pioneers 9 and 10.

This temperature increase is believed to have occurred because the outward motion of the ionized solar wind particles was changed into thermal motion as the fast solar wind plasma bumped into slower plasma already there.

Magnetic Field

The field (carried by the solar wind) decreased in strength by roughly the square of the distance going out from the Sun. The field was measured over 9 solar rotations.

Changes in large-scale properties were consistent with extrapolation from near-Earth conditions outward.

Discontinuities (turbulence) were the same at 560 million kilometers (350 million miles) as at 160 million kilometers (100 million miles).

During the big storm, August 2, 1972, the field increased its strength three times in one minute (100 times over normal) at Pioneer 10. The shock wave in the solar wind shown by the field slowed down between Pioneers 9 and 10.

Galactic Cosmic Rays (high-energy particles from the stars of the galaxy)

Because of the expected decrease in blockage by the solar wind and magnetic field from the Sun, numbers of low-energy (0 to 100 million electron-volts) cosmic ray particles were expected to increase out to 350 million miles from the Sun. This did not occur, apparently because, despite the decline of solar wind density and magnetic field strength, solar wind and magnetic field turbulence remained as high as near Earth. This turbulence apparently can shut out a large part of the low-energy galactic particles from the inner solar system, perhaps as far out as beyond Jupiter.

The increases in the high-energy galactic particles for all energy ranges were less than 6 percent per 160 million kilometers (100 million miles) from the Sun, far less than the expected two to three times increase.

High-Energy Solar Particles

The best resolution yet seen of elements making up the solar high-energy particles has been obtained by the NASA-Goddard instrument. It identified sodium and aluminum for the first time, determined relative abundances of helium, carbon, nitrogen, oxygen, fluorine, neon, sodium, magnesium, aluminum, and silicon nuclei coming from the Sun.

During the solar storm of August 2, 1972, the highest numbers of solar high-energy particles ever measured were recorded on Pioneers 6 and 9 and on several Earth and lunar satellites. Measurements of these solar particles were not as high on Pioneer 10.

Interstellar Gas

The University of Southern California ultraviolet instrument has been able to locate concentrations of neutral hydrogen and helium and to make some tentative separations between gas originating in the Sun and that penetrating the solar system from interstellar space.

The experimenters have measured the ultraviolet light glow emitted by interplanetary neutral hydrogen and helium atoms. Preliminary findings from this data are that the interstellar neutral hydrogen appears to be entering the heliosphere in the plane of the Earth's orbit (the ecliptic) at around 100,000 km/hr (62,000 mph). The ecliptic is also roughly the plane of the orbits of all the planets.

The direction of movement of the solar system through space is about 60 degrees above the ecliptic, and the interstellar hydrogen might well have been expected to enter directly from this direction.

These findings are similar to those of three experimenters on the OGO V Earth satellite several years ago. Small differences with the OGO findings in incoming velocities of the interstellar gas might suggest some variations with time, and hence turbulence in the "interstellar wind." The ultraviolet experiment also has made the first measurements in interplanetary space of the ultraviolet glow of interplanetary helium atoms. Conclusions on properties of this interstellar helium will require further data analysis.

The Spacecraft

Pioneers 10 and G are the first spacecraft designed to travel into the outer solar system and operate there, possibly for as long as seven years and as far from the Sun as 2.4 billion kilometers (1.5 billion miles).

For these missions, the spacecraft must have extreme reliability, be of very light weight because of launch vehicle limitations, have communications systems for extreme distances, and employ non-solar power sources.

Pioneer G is identical to Pioneer 10 except that a 12th onboard experiment has been added, a fluxgate magnetometer, to measure high fields very close to Jupiter. Pioneer G is stabilized in space like a gyroscope by its five-rpm rotation, so that its scientific instruments scan a full circle about five times a minute. Designers chose spin stabilization for its simplicity and effectiveness.

Since the orbit planes of Earth and Jupiter coincide to about one degree, the spacecraft will be in or near Earth's orbit plane (the ecliptic) throughout its flight. To maintain a known orientation in this plane, controllers regularly adjust spacecraft position so that the spin axis points constantly at Earth. The spin axis coincides with the center line of the radio beam, which also points constantly at Earth.

Spacecraft navigation is handled on Earth using two-way doppler tracking and by angle-tracking.

For midcourse corrections, the Pioneer propulsion system can make changes in velocity totaling 720 km/hr (420 mph).

The spacecraft can return a maximum of 2,048 data bits per second (bps) when relatively close to the Earth. At Jupiter distance, the 64-meter (210-foot) antennas of the Deep Space Network can hear a data rate of 1,024 bps. It can store up to 49,152 data bits while other data is being transmitted.

Pioneer G is controlled largely from the Earth rather than by sequences of commands stored in onboard computers.

Launch energy requirements to reach Jupiter are far higher than for shorter missions, so the spacecraft is very light. Pioneer G weighs only 270 kilograms (570 pounds). This includes 30 kilograms (65 pounds) of scientific instruments, and 27 kilograms (60 pounds) of propellant for attitude changes and mid-course corrections.

Pioneer / Jupiter Spacecraft

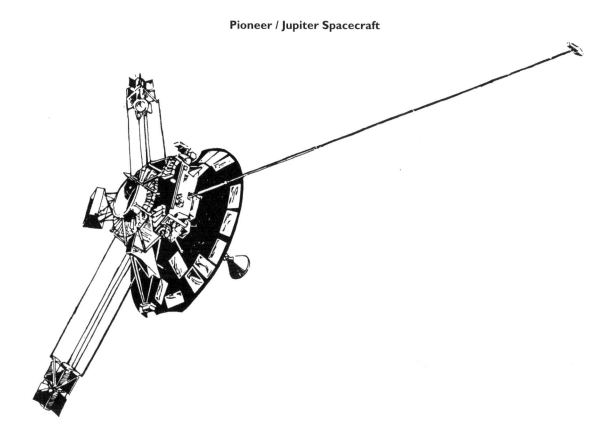

Because solar energy at Jupiter is only 4 percent of energy received at the Earth and grows steadily weaker beyond the planet, designers selected a nuclear power source over solar cells.

The mission's 14 experiments are carried out by 12 onboard scientific instruments. Two experiments use the spacecraft and its radio signal as instruments.

For reliability, spacecraft builders have employed an intensive screening and testing program for parts and materials. They have selected components designed to withstand radiation from the spacecraft's nuclear power source, and from Jupiter's radiation belts. In addition, key systems are redundant. (That is, two of the same component or subsystem are provided in case one fails.) Communications, command and data return systems, propulsion electronics, thrusters, and attitude sensors are largely redundant.

Virtually all spacecraft systems reflect the need to survive and return data for many years a long way from the Sun and the Earth.

Pioneer G Description

The spacecraft fits within the 3-meter (10-foot) diameter shroud of the Atlas-Centaur launch vehicle with booms retracted, and with its dish antenna facing forward (upward). The Earth-facing dish antenna is designated the forward end of the spacecraft. *Pioneer G* is 2.9 meters (9.5 feet) long, measuring from its farthest forward component, the medium-gain antenna horn, to its farthest rearward point, the tip of the aft-facing omnidirectional antenna. Exclusive of booms, its widest crosswise dimension is the 2.7-meter (9-foot) diameter of the dish antenna.

The axis of spacecraft rotation and the center-line of the dish antenna are parallel, and *Pioneer* spins constantly for stability.

The spacecraft equipment compartment consists of a flat box, top and bottom of which are regular hexagons. This hexagonal box is roughly 35.5 cm (14 inches) deep and each of its six sides is 71 cm (28 inches) long. One side joins to a smaller box also 35.5 cm (14 inches) deep, whose top and bottom are irregular hexagons.

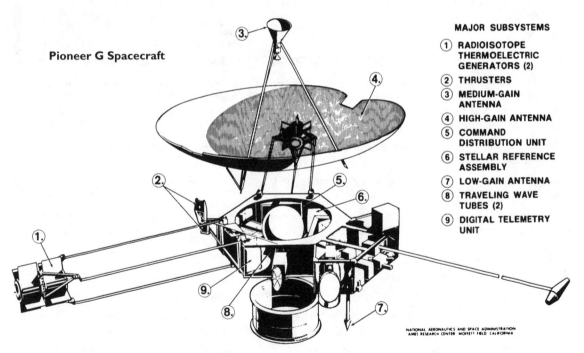

Pioneer G Spacecraft

MAJOR SUBSYSTEMS

1. RADIOISOTOPE THERMOELECTRIC GENERATORS (2)
2. THRUSTERS
3. MEDIUM-GAIN ANTENNA
4. HIGH-GAIN ANTENNA
5. COMMAND DISTRIBUTION UNIT
6. STELLAR REFERENCE ASSEMBLY
7. LOW-GAIN ANTENNA
8. TRAVELING WAVE TUBES (2)
9. DIGITAL TELEMETRY UNIT

NATIONAL AERONAUTICS AND SPACE ADMINISTRATION
AMES RESEARCH CENTER MOFFETT FIELD CALIFORNIA

This smaller compartment contains most of the 12 onboard scientific experiments. However, 12.7 kilograms (28 pounds) of the 30 kilograms (65 pounds) of scientific instruments (the plasma analyzer, cosmic ray telescope, four asteroid-meteoroid telescopes, meteoroid sensors, and the magnetometer sensors) are mounted outside the instrument compartment. The other experiments have openings cut for their sensors to look out. Together both compartments provide 1.4 square meters (16 square feet) of platform area.

Attached to the hexagonal front face of the equipment compartment is the 2.7-meter (9-foot)-diameter 46-cm (18-inch)-deep dish antenna.

The high-gain antenna feed and the medium-gain antenna horn are mounted at the focal point of the antenna dish on three struts projecting about 1.2 meters (4 feet) forward of the rim of the dish. The low-gain, omnidirectional spiral antenna extends about 0.76 meter (2.5 feet) behind the equipment compartment.

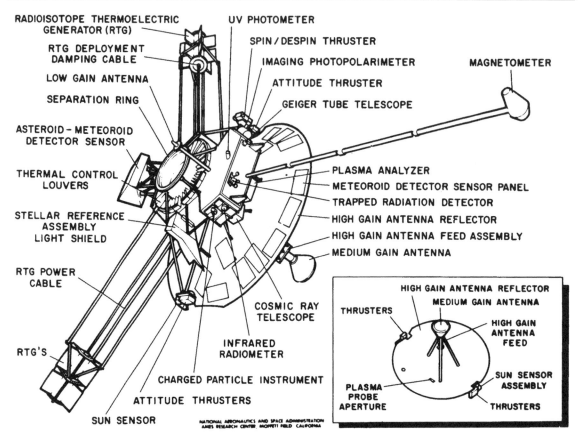

Two three-rod trusses, 120 degrees apart, project from two sides of the equipment compartment, deploying the spacecraft's nuclear electric power generator about 3 meters (10 feet) from the center of the spacecraft. A boom, 120 degrees from each of the two trusses, projects from the experiment compartment and positions the helium vector magnetometer sensor 6.6 meters (21.5 feet) from the spacecraft center. The booms are extended after launch.

At the rim of the antenna dish, two Sun sensors are mounted. A star sensor looks through an opening in the equipment compartment and is protected from sunlight by a hood.

Both compartments have aluminum frames with bottoms and side walls of aluminum honeycomb. The dish antenna is made of aluminum honeycomb.

Rigid external tubular trusswork supports the dish antenna, three pairs of thrusters located near the rim of the dish, radioisotope thermoelectric generator trusses, and launch vehicle attachment ring. The message plaque also is attached to this trusswork.

Orientation and Navigation

The spacecraft communications system also is used to orient the *Pioneer* in space.

Heart of the communications system is the spacecraft's fixed dish antenna. This antenna is as large as the diameter of the launch vehicle permits. It focuses the radio signal on Earth in a narrow beam.

The spacecraft spin axis is aligned with the center lines of its dish antenna and of the spacecraft radio beam. Except during the early part of the mission and course changes, all three are pointed toward Earth throughout the mission. *Pioneer G* maintains a known attitude in space as a result of this continuous Earth point.

For navigation, analysts will use the doppler shift in frequency of the *Pioneer* radio signal and angle-tracking by DSN antennas to calculate continuously the speed, distance, and direction of the spacecraft from Earth. Motion of the spacecraft away from Earth causes the frequency of the spacecraft radio signals to drop and wavelength to increase. This is known as the doppler shift.

Propulsion and Attitude Control

The propulsion and attitude control system provides three types of maneuvers.

It can change velocity, thus altering course to adjust the place and time of arrival at Jupiter.

Navigation

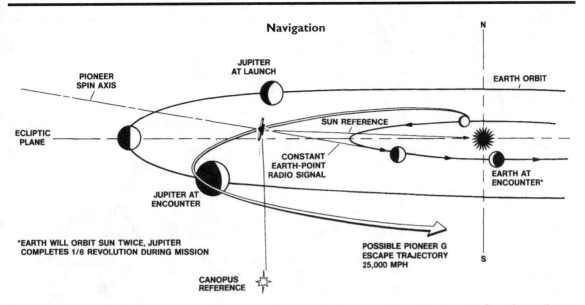

It can change the attitude of the spacecraft in space, either to point thrusters in the right direction for velocity-change thrusts or to keep the spacecraft narrow-beam antenna pointed precisely at the Earth. Controllers will command about 150 of these Earth-point adjustments as *Pioneer G* and Earth constantly change positions in space.

The system also maintains spacecraft spin at 4.8 rpm.

Over the entire mission, the propulsion and attitude control system can make changes in spacecraft velocity totaling 720 km/hr (420 mph), attitude changes totaling 1,200 degrees (almost four full rotations of the spin axis), and total spin-rate adjustments of 50 rpm. These capacities should be substantially more than needed.

The system employs six thruster nozzles which can be fired steadily or pulsed and have 0.4 to 1.4 pounds of thrust each. Electronics and attitude sensors are fully redundant.

For both attitude and velocity changes, two thruster pairs have been placed on opposite sides of the dish antenna rim. One thruster of each pair points forward, the other aft.

To change attitude, the spacecraft spin axis can be rotated in any desired direction. This is done using two nozzles, one on each side of the dish antenna. One nozzle is fired forward, one aft, in momentary thrust pulses once per spacecraft rotation. Thrusts are made at two fixed points on the circle of spacecraft rotation. Pulses are timed by a signal which originates either from the star sensor which sees the star Canopus, once per rotation, or from one of the two Sun-sensors which see the Sun once per rotation. Each pair of thrust pulses turns the spacecraft and its spin axis a few tenths of a degree, until the desired attitude is reached.

For velocity changes, the spin axis is first rotated until it points in the direction of the desired velocity change. Then two thruster nozzles, one on each side of the dish, but both facing forward or both facing aft, are fired continuously.

Flight directors can command these velocity change maneuvers directly in real time. Or they can put commands for the maneuvers into the system's attitude-control storage register. Rotation of the spin axis, velocity change thrust, and rotation back to Earth-point can then all take place automatically. This automatic sequence may be important at very long distances where, if Earth-point were lost, recovery of communication with the spacecraft might be difficult.

Operations to point the spacecraft high-gain antenna toward the Earth make use of the fact that the movable feed for the antenna dish can be offset one degree from the spin axis. This means that when the spin axis is slightly off exact Earth-point, signals received by the spacecraft from Earth vary up and down in strength.

An automatic system on the spacecraft, known as Conscan, changes attitude in a direction to reduce this variation in signal strength, returning the spin axis again to precise Earth-point. An onboard signal processor uses the varying radio signal to time attitude-change thrusts. Spatial reference for these maneuvers is provided by either Sun or star sights.

Earth-point can be trimmed by ground command if necessary, using variations in strength of the spacecraft radio signal as received by ground stations. This variation also is due to offset of the spacecraft antenna feed.

The spacecraft medium-gain horn is permanently offset 9 degrees to the spin axis and can also be used for Earth-point maneuvers.

To change spin rate, a third pair of thrusters, also set along the rim of the dish antenna, will be used. These thrusters are on a line tangent to the rim of the antenna (the circle of spacecraft rotation). One thruster points against the spin direction and is fired to increase spin rate; the other points with the direction of spin and is fired to reduce spin.

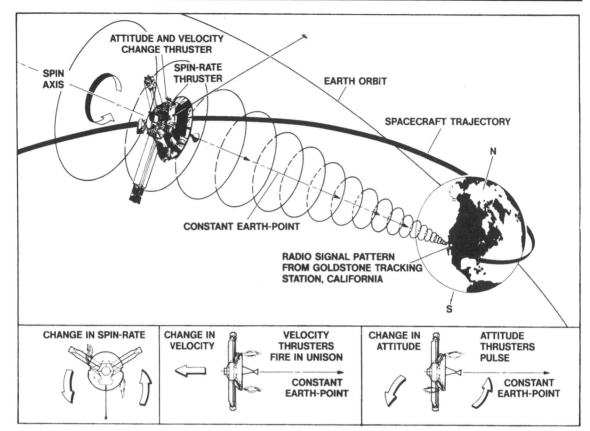

Thrust is provided by liquid hydrazine, which is decomposed into gas by a catalyst in the chamber of each thruster and then ejected from thruster nozzles. The hydrazine is stored in a single 42-cm (16.5-inch)-diameter pressurized spherical tank. Tank and connecting lines are kept from freezing by electric heaters.

Nuclear-Electric Power

Nuclear-fueled electric power for *Pioneer G* comes from four SNAP-19 radioisotope thermoelectric generators (RTGs), developed by the Atomic Energy Commission. These units turn heat from their nuclear power source into electricity.

SNAP-19 has been designed for maximum safety in case of a launch pad explosion, abort, or reentry. Before its use was authorized, a thorough review was conducted to assure the health and safety of personnel involved in the launch and of the general public. Extensive safety analyses and tests were conducted which demonstrated that the fuel would be safely contained under almost all credible accident conditions.

Two each of these power units are located at the ends of each of the two 2.7-meter (9-foot) spacecraft RTG trusses. The RTGs are located on the opposite side of the spacecraft from the experiment compartment to minimize effects of their neutron and gamma radiation on the scientific instruments. Design of experiments and spacecraft equipment has been such as to counter effects of RTG radiation.

The four RTGs are expected to provide 155 watts of power at launch, about 140 watts at Jupiter, more than 100 watts five years after launch. Spacecraft power requirements at Jupiter will be around 106 watts, of which 25 watts is for the scientific instruments. Similar units on *Pioneer 10* have performed better than expected, still providing 153 watts, ten months after launch. Power at Jupiter should be above the planned 140 watts.

The RTGs are fueled with plutonium-238. Thermoelectric converters consist of banks of thermoelectric couples surrounding the cylindrical fuel compartment, 90 for each of the four RTGs. The couples are made of lead telluride for negative elements and an alloy of tellurium, silver, germanium, and antimony for the positive elements.

Each RTG consists of a 28-cm (11-inch) cylinder with six radial fins and weighs about 14 kilograms (30 pounds).

Electric power is distributed throughout the spacecraft. It goes from each RTG through one of four inverters to form a main alternating current (AC) bus. Most of this AC power is rectified to supply the main 28-volt DC bus, with excess dumped through an external heat radiator. A battery automatically carries overloads, and is recharged when power is available. The scientific instruments and the radio transmitter's traveling wave tube amplifiers receive power from the main DC bus. Most other spacecraft systems are supplied from the central transformer, which receives power from the AC bus and provides various DC output voltages.

SNAP 19 / Pioneer Radioisotope Thermoelectric Generator

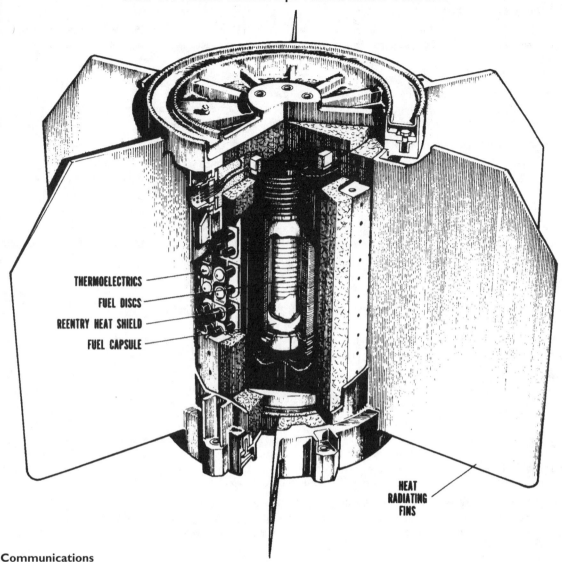

THERMOELECTRICS
FUEL DISCS
REENTRY HEAT SHIELD
FUEL CAPSULE

HEAT
RADIATING
FINS

Communications

The communications system provides for two-way communication between Earth and the spacecraft. For reliability, this system is fully redundant.

The system depends on the sensitivity of the DSN's 64-meter (210-foot) antennas and their receivers which can hear the very low amounts of energy emanating from *Pioneer* at Jupiter. When used in reverse as transmitters, the 64-meter antennas have such precision and radiated power (up to 400 kilowatts) that outgoing commands are still strong enough to be received when they reach the spacecraft.

The spacecraft system consists of high-gain, medium-gain, and low-gain antennas, used for both sending and receiving. The high-gain antenna uses the spacecraft's parabolic reflector dish; the medium-gain antenna is an Earth-facing horn mounted along with the high-gain antenna feed on struts at the focus of the dish. The low-gain antenna is a spiral, pointed to the rear, designed to provide communication at the few times when the aft end of the spacecraft is pointing toward Earth.

Each antenna is always connected to one of the two spacecraft radio receivers, and the two receivers are interchangeable by command, or automatically after a certain period of inactivity, so that if one receiver fails the other can take over.

For transmitting, the high-gain antenna produces a maximum gain of 33 decibels and has a 3.3 degree beam width; the medium-gain antenna, a gain of 12 decibels and beam width of 32 degrees. Other components for transmitting are two radio transmitters and two traveling-wave-tube power amplifiers (TWTs). The TWTs weigh only 1.75 kilograms (3.85 pounds) but produce 8 watts of power in S-band.

In addition to commands and data return, the communications system provides data for doppler tracking and maintaining spacecraft Earth-point.

Communication is on S-band frequencies, uplink to the spacecraft at 2110 and downlink at 2292 MHz. In the Asteroid Belt, the system can return 2,048 data bits per second to the DSN's 64-meter dish antennas, and at Jupiter range 1,024 bps.

Command System

Controllers use 222 different commands to operate the spacecraft. The command system consists of two command decoders and a command distribution unit. For reliability, two redundant command decoders are provided, and redundant circuits are provided throughout the logic of the command distribution unit.

Commands are transmitted to the spacecraft at a rate of one data bit per second. A single command message has 22 bits.

The command distribution unit routes commands to any one of 222 functions on the spacecraft. Seventy-three of these operate the experiments, the remaining 149, spacecraft systems. The science commands include moving the photometer telescope for pictures of the planet, calibrating instruments, and changing data types. Up to five commands can be stored for later execution.

If a command is not properly verified by the decoder, the command distribution unit does not act on the order, thus reducing the possibility of executing wrong commands. Commands also are verified by computer on the ground before transmission.

Data Handling

The spacecraft data system turns science and engineering information into an organized stream of data bits for radio transmission back to Earth.

The system also can store up to 49,152 data bits for later transmission to Earth. Storage can take place while other data are being transmitted. This allows for taking data faster than the spacecraft can transmit it.

System components include the digital telemetry unit, which prepares the data for transmission in one of nine data formats at one of eight bit rates of from 16 to 2,048 bits per second; the data storage unit; and the convolutional coder, which rearranges the data in sequential form that allows identification and correction of errors by computers on the ground. This coding allows higher bit rates to be retained for greater spacecraft distances without loss of accuracy.

The nine data formats are divided into science and engineering groups. The science group includes two basic science formats and three special-purpose science formats. The basic science formats each contain 192 bits, 144 of which are assigned to the scientific instruments; the remainder are for operating instructions of various kinds. One of the basic science formats is primarily for interplanetary flight, the other for Jupiter encounter.

The three special-purpose science formats transmit all 192 bits of data from only one or two instruments. This allows high data rates from one instrument such as the photopolarimeter during Jupiter picture taking, or the infrared radiometer whose principal purpose is to take a careful look at Jupiter's heat radiation just before closest approach.

There are four engineering formats, each of 192 bits. Each of the engineering formats specializes in one of the following areas: data handling, electrical, communications and orientation, and propulsion. These specialized data formats provide high bit rates during critical spacecraft events.

Short samples of science data can be inserted into engineering formats and vice versa. The digital telemetry unit, which puts all data together, has redundant circuits for critical functions.

Timing

The digital telemetry unit contains a stable crystal-controlled clock for all spacecraft timing signals, including those for various orientation maneuvers and experiments.

Besides redundant components and systems, many other components have redundant internal circuitry. Virtually all parts and components have been selected from those with a known and documented history of successful operation in space.

Virtually all electronic components have been "burned in" by having current applied to them before installation. For example, all transistors have been run 200 hours. Engineers have tracked all test anomalies and failures to their source via a rigid, formal control system. NASA project engineers have intensively monitored all contractor test and system integration activities.

Temperature Control

Temperature on the spacecraft is controlled at between -25 degrees and 40 degrees C (-10 degrees and 100 degrees F) inside the scientific instrument compartment, and at various other levels elsewhere for satisfactory operation of the spacecraft equipment.

The temperature control system must cope with gradually decreasing heating as the spacecraft moves away from the Sun, and with two frigid periods – one when *Pioneer G* passes through Earth's shadow at launch; the other during passage through Jupiter's shadow during flyby. The system also handles heat from the third-stage engine, atmosphere friction, spacecraft nuclear electric power generators, and other equipment.

The equipment compartments are insulated by multi-layered blankets of aluminized Mylar or Kapton. Temperature-responsive louvers at the bottom of the equipment compartment, opened by bimetallic springs, allow controlled escape of excess heat. Other equipment has individual thermal insulation and is warmed by electric heaters and 12 one-watt radioisotope heaters, fueled with plutonium-238, developed by the Atomic Energy Commission.

Magnetic Cleanliness

Pioneer G is among the most magnetically "clean" spacecraft ever built. It has been designed to provide the least interference with the magnetometer as it measures the weak interplanetary magnetic field. Since virtually any electric current flow within the spacecraft can produce a magnetic field, demagnetizing is difficult. A wide range of special components and techniques have been used in the electrical systems to achieve this low spacecraft magnetism of 0.03 gamma at the magnetometer sensor.

Reliability

In addition to reliability measures already described, the spacecraft has these added design characteristics:

It can do most mission tasks in a variety of ways. For example, several combinations of thrusters can be used for attitude and velocity changes.

The Experiments

The basic measurements made by the 14 *Pioneer G* experiments will produce important findings about Jupiter, the Asteroid Belt, and the heliosphere.

Specific mission objectives for the 14 experiments are:

For interplanetary space:

• To map the interplanetary magnetic field
• To study the radial gradient of the solar wind and its fluctuations and structure
• To study the radial and traverse gradients and arrival directions of high-energy charged particles (solar and galactic cosmic rays)
• To investigate the relationships between the solar wind, magnetic field, and cosmic rays
• To search for the boundary and shape of the heliosphere (solar atmosphere)
• To determine the density of neutral hydrogen
• To determine the properties of interplanetary dust

For the Asteroid Belt:

• To determine the size, mass, flux, velocity, and orbital characteristics of the smaller particles in the Belt
• To determine Asteroid Belt hazard to spacecraft

For Jupiter:

• To map the magnetic field
• To measure distributions of high-energy electrons and protons in the radiation belts, and look for auroras.
• To find a basis for interpreting the decimetric and decametric radio emission from Jupiter
• To detect and measure the bow shock and magnetospheric boundary and their interactions with the solar wind
• To verify the thermal balance and to determine temperature distribution of the outer atmosphere
• To measure the hydrogen / helium ratio in the atmosphere
• To measure the structure of the ionosphere and atmosphere
• To measure the brightness, color, and polarization of Jupiter's reflected light
• To perform two-color visible light imaging
• To increase the accuracy of orbit predictions and masses of Jupiter and its moons.

The instruments aboard *Pioneer G* will return data in roughly the following categories:

Magnetic Fields

Helium Vector Magnetometer (Jet Propulsion Laboratory). This sensitive instrument will measure the interplanetary magnetic field in three axes from the orbit of Earth out to the limits of spacecraft communication. It will study solar wind interaction with Jupiter, and map Jupiter's strong magnetic fields at all longitudes and many latitudes. It may be

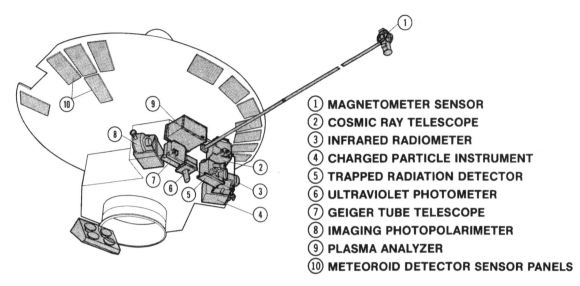

(1) **MAGNETOMETER SENSOR**
(2) **COSMIC RAY TELESCOPE**
(3) **INFRARED RADIOMETER**
(4) **CHARGED PARTICLE INSTRUMENT**
(5) **TRAPPED RADIATION DETECTOR**
(6) **ULTRAVIOLET PHOTOMETER**
(7) **GEIGER TUBE TELESCOPE**
(8) **IMAGING PHOTOPOLARIMETER**
(9) **PLASMA ANALYZER**
(10) **METEOROID DETECTOR SENSOR PANELS**

able to use these data to infer the fluid composition and other characteristics of Jupiter's interior. It will study the relationship of the fields to Jupiter's moons, and may be able to look for the heliosphere boundary. The magnetic field is carried by the solar wind, and measurements of both field and wind are needed to understand either.

The instrument is a helium vector magnetometer. Its sensor is mounted on the light-weight mast extending 6.5 meters (21.5 feet) from the center of the spacecraft to minimize interference from spacecraft fields.

The instrument operates in any one of eight different ranges, the lowest covering magnetic fields up to 2.5 gamma; the highest, fields up to 1.4 gauss (140,000 gamma). Earth's surface field is 50,000 gamma. These ranges are selected by ground command or automatically by the instrument itself. The magnetometer can measure fields as weak as 0.01 gamma.

Its sensor is a cell filled with helium, excited by radio frequencies and infrared optical pumping. Changes in the helium caused by fields the spacecraft passes through are measured. It weighs 2.6 kilograms (5.7 pounds) and uses up to 5 watts of power.

Fluxgate Magnetometer (Goddard Space Flight Center). This instrument is designed to sense strength and direction of Jupiter's magnetic field very close to the planet and up to very high values. It can measure field strength up to 10 Gauss (1 million gamma) in each of three perpendicular directions.

The experiment is designed to tell whether Jupiter's magnetic field is more like the Earth's or the Sun's field, a question which bears directly on the origin of the solar system. Jupiter's radio emissions have suggested a possible four-pole rather than a two-pole configuration for the Jovian field. If the spacecraft passes through the same field line as that occupied by Jupiter's Moon, Io, the experimenter hopes to gather data on particle and electrodynamic effects produced by Io on radio emission from Jupiter.

The instrument will be able to continuously measure the magnetic field along the spacecraft trajectory from about 12.6 Jupiter radii to the point of closest approach. It consists of two dual-axis sensors and their electronics. Each sensor is composed of a ring core, a magnetic multivibrator, a frequency doubler, and two phase-sensitive detectors. It weighs 0.6 pounds and uses approximately 0.36 watts of power.

Interplanetary Solar Wind And Heliosphere

Plasma Analyzer (Ames Research Center). The plasma instrument will map the density and energy of the solar wind (ions and electrons flowing out from the Sun). It will determine solar wind interactions with Jupiter, including the planet's bow shock wave, and will look for the boundary of the heliosphere.

The instrument consists of a high-resolution and medium-resolution analyzer. It will look toward the Sun through an opening in the spacecraft dish antenna, and the solar wind will enter like the electron beam in a TV tube. The instrument measures direction of travel, energy (speed), and numbers of ions and electrons.

The particles enter between curved metal plates and strike detectors, which count their numbers. Their energy is found by the fact that when a voltage is applied across the plates in one of 64 steps, only particles in that energy range can enter. Direction of particle travel is found from orientation of the instrument and which detector the particle struck.

In the high-resolution analyzer, the detectors are 26 continuous-channel multipliers, which measure the ion flux in energy ranges from 100 to 8,000 electron-volts. Detectors in the medium-resolution detector are five electrometers, which measure ions in ranges from 100 to 18,000 electron-volts and electrons from one to 500 electron-volts.

The plasma analyzer weighs 5.5 kilograms (12.1 pounds) and uses 4 watts of power.

Cosmic Rays, Jupiter's Radiation Belts And Radio Emissions

Charged Particle Composition Instrument (University of Chicago). This instrument has a family of four measuring systems. Two are particle telescopes primarily for interplanetary space. The other two measure trapped electrons and protons inside the Jovian magnetic field.

During interplanetary flight two telescopes will identify the nuclei of all eight chemical elements from hydrogen to oxygen and separate the isotopes deuterium, helium-3 and helium-4. Because of their differences in isotopic and chemical composition and spectra, galactic particles can be separated from solar particles as the spacecraft moves towards interstellar space. The instrument also measures the manner in which streams of high-energy particles escape the Sun and travel through interplanetary space. There is a main telescope of seven solid state detectors which measure from 1 to 500 million electron-volt (MeV) particles and a 3-element telescope which measures 0.4 to 10 MeV protons, and helium nuclei. If a key detector element is destroyed in space, diagnostic procedures and commands from Earth can bypass it.

For the magnetosphere of Jupiter, two new types of sensors were developed to cope with the extremely high intensities of trapped radiation. A solid state ion chamber operating below -40 degrees C (-40 degrees F) measures only those electrons that generate the radio waves which reach Earth. The trapped proton detector contains a foil of thorium which undergoes nuclear fission from protons above 30 million electron-volts, but is not sensitive to the presence of the intense electron radiation. The instrument weighs 3 kilograms (7.3 pounds) and uses 2.4 watts of power.

Cosmic Ray Telescope (Goddard Space Flight Center). The Cosmic Ray Telescope also will monitor solar and galactic cosmic ray particles. It will track the twisting paths of high-energy particles from the Sun and measure bending effects of the solar magnetic field on particles from the galaxy, some traveling at near light speeds. The instrument can distinguish which of the ten lightest elements make up these particles. It also will measure high-energy particles in Jupiter's radiation belts.

The instrument consists of three three-element, solid state telescopes. A high-energy telescope measures the flux of protons between 56 and 800 MeV. A medium-energy telescope measures protons with energies between three and 22 MeV, and identifies the eight elements from hydrogen to oxygen. The low-energy telescope will study the flux of electrons between 50,000 electron-volts and one MeV and protons between 50,000 electron-volts and 20 MeV.

The instrument weighs 3.2 kilograms (7 pounds) and uses 2.2 watts of power.

Jupiter's Charged Particles

Geiger Tube Telescopes (University of Iowa). This experiment will attempt to characterize Jupiter's radiation belts. The instrument employs seven Geiger-Mueller tubes to survey the intensities, energy spectra, and angular distributions of electrons and protons along *Pioneer's* path through the magnetosphere of Jupiter.

The tubes are small cylinders containing gas that generates electrical signals from charged particles.

Three tubes (considered a telescope) are parallel. Three others will form a triangular array to measure the number of multiparticle events (showers) occurring. The combination of a telescope and shower detector will enable experimenters to compare primary with secondary events in the Jovian radiation belts.

Another telescope will detect low-energy electrons (those above 40,000 electron-volts), and together with the others should shed light on Jupiter's shock front in the solar wind and the tail of its magnetosphere.

The instrument can count protons with energies above 5 million electron-volts (MeV) and electrons with energies between 2 and 50 MeV.

The instrument weighs 1.6 kilograms (3.6 pounds) and uses 0.7 watts of power.

Jovian Trapped Radiation Detector (University of California at San Diego). The nature of particles trapped by Jupiter, particle species, angular distributions, and intensities will be determined by this instrument. It will have a very broad range of energies from 0.01 to 100 MeV for electrons and from 0.15 to 350 MeV for protons (hydrogen nuclei).

Experimenters will attempt to correlate particle data with Jupiter's mysterious radio signals.

They will use five detectors to cover the planned energy range. A focused Cerenkov counter, which measures direction of particle travel by light emitted in a particular direction, will detect electrons of energy above 1 MeV and protons above 450 MeV. A second detector measures electrons at 100, 200, and 400 thousand electron-volts (KeV).

An omnidirectional counter is a solid state diode which discriminates between minimum ionizing particles at 400 KeV and high-energy protons at 1.8 MeV.

Twin DC scintillation detectors for low-energy particles distinguish roughly between protons and electrons because of different scintillation material in each. Their energy thresholds are about 10,000 eV for electrons and 150,000 eV for protons.

The instrument weighs 1.7 kilograms (3.9 1b) and uses 2.9 watts of power.

Asteroids, Meteoroids, Interplanetary Dust

Asteroid-Meteoroid Detector (General Electric Company). Samplings of the solid materials between the Earth's orbit and the spacecraft communications limit will be made by this experiment.

The instrument will measure the orbits of material in the vicinity of the spacecraft, identifying meteoroidal, asteroidal, and cometary particles and their distribution.

Four nonimaging telescopes will characterize objects, ranging from several hundred miles in diameter (asteroids) down to particles with a mass of one-millionth of a gram, by measuring sunlight reflected from them. The telescopes will measure numbers, size, velocity, and direction of individual particles. Billions of such particles strike the Earth's atmosphere each day.

The instrument can see a sand-grain sized particle out to ten meters, a small pea to 100 meters, a ping pong ball to one kilometer. Beyond a kilometer, it can see larger bodies but cannot measure their size or trajectory.

This experiment (like the photopolarimeter) will periodically measure zodiacal light (sunlight scattered by the total mass of meteoroids and dust in deep space). Comparisons of zodiacal light and particle measurements will show relationships of zodiacal cloud brightness with size distribution of particles.

Experimenters also will study effects of secondary forces such as solar radiation and planetary gravity on orbits of small particles in the solar system.

Each of its four telescopes consists of a 20-cm (8-inch) primary mirror, an 8.4-cm (3.3 inch) secondary mirror, coupling optics, and a photomultiplier tube. Each telescope has an 8 degree view cone. The four cones overlap in part, and particle distance and speed is measured by timing entry and exit of view cones. Photomultiplied pulses are counted to find particle numbers.

The instrument weighs 3.3 kilograms (7.2 pounds) and uses 2.2 watts of power.

Meteoroid Detector (Langley Research Center). To detect the distribution in space of tiny particles (masses of about one-hundred-millionth of a gram or greater), scientists will use a system of 234 pressure cells mounted on the back of the spacecraft dish antenna. The pressure cells come in 30 x 20 cm (12 x 8 inch) panels, 18 to a panel, with 13 panels. They have a total area of about one-half a square meter. In construction, the panels are something like a stainless steel air mattress.

Each of the pressure cells is filled with a gas mixture of 75 percent argon and 25 percent nitrogen. When a particle penetrates a cell it empties of gas. A transducer notes this, counting one particle impact per cell.

Laboratory tests have shown the combined mass and velocity required for a particle to penetrate a cell. With trajectory information as well, experimenters can calculate spatial distribution of the tiny meteoroids.

Cell walls are 0.002 inch stainless steel sheet, thin enough to allow penetrations of particles with a mass of about $1/100,000,000$ gram or more. On *Pioneer 10* cells with 0.001 inch walls measured particles down to one-billionth gram.

The total weight of the instrument (panels and electronics) is 1.7 kilograms (3.7 pounds) and it uses 0.7 watts of power.

Celestial Mechanics

Celestial Mechanics (Jet Propulsion Laboratory). This experiment uses the spacecraft itself as a sensitive instrument to better determine the mass of Jupiter and its satellites and to determine the harmonics and possible anomalies in Jupiter's gravity field.

Experimenters will measure the gravity effects of Jupiter and its satellites on the spacecraft flight trajectory.

Deep Space Network doppler tracking of the spacecraft determines its velocity along the Earth-spacecraft line down to a fraction of a millimeter per second, once per minute. These data are further augmented by optical and radar position measurements of the planets. Computer calculations using the spacecraft trajectory and known planet and satellite orbital characteristics should verify *Pioneer 10* findings, allowing a five-fold improvement in the accuracy of current calculations of Jupiter's mass. Masses of Jupiter's four large (Galilean) moons will be determined to an accuracy of about one percent. Experimenters expect to find the planet's dynamical polar flattening to within one half mile, and to make estimates of the mass of the planet's surface layers.

Interplanetary Hydrogen, Helium, And Dust; Jupiter's Atmosphere, Temperatures, Auroras, Moons

Ultraviolet Photometer (University of Southern California). This instrument will address some basic questions about interplanetary and interstellar space as well as about Jupiter.

During planet flyby the ultraviolet photometer will measure scattering of ultraviolet (UV) light from the Sun by Jupiter's atmosphere in two wavelengths, one for hydrogen and one for helium. Experimenters will use these data to find the amount of atomic hydrogen in Jupiter's upper atmosphere, the amount of helium and the mixing rate in Jupiter's atmosphere, and the ratio of helium to molecular hydrogen. So far helium has not been identified on Jupiter.

Virtually all theories of Jupiter's origin make assumptions about amounts of helium in the planet's makeup.

By measuring changes in intensity of ultraviolet light glow as it scans across the planet, the instrument should be able to identify Jovian auroras – the bright areas caused by an auroral oval similar to the auroral zone on Earth – if they exist.

Radio telescope measurements tell us that the solar system is immersed in an interstellar gas of cold neutral hydrogen. By measuring the scattering of the Sun's ultraviolet light the instrument can measure amounts of this neutral hydrogen within the heliosphere. Presence of neutral hydrogen (already measured near Earth) would be a result of entry of interstellar hydrogen at the heliosphere boundary.

Neutral helium may penetrate the heliosphere directly as the solar system moves through the interstellar gas at 72,000 km/hr (45,000 mph). From such helium measurements, the experimenters may be able to calculate the percent of helium in the interstellar gas. The "big bang" theory of the origin of the universe suggests that it should be about seven percent.

The instrument's measurements also should allow calculation of concentrations of hydrogen and dust in the interstellar space outside the heliosphere from measurements of the interstellar ultraviolet glow from starlight, at the hydrogen wavelength. The UV glow from scattered sunlight will diminish greatly as the spacecraft proceeds out from the Sun.

The instrument has two detectors, one which measures ultraviolet radiation at 1216 angstroms, the other at 584 angstroms – the wavelengths at which hydrogen and helium scatter solar ultraviolet.

The instrument has a fixed viewing angle and will use the spacecraft spin to scan the planet. It weighs 0.7 kilograms (1.5 pounds) and uses 0.7 watts of power.

Imaging Photopolarimeter (University of Arizona). This instrument will provide data in a number of areas, using photometry, (measurement of light intensity) and polarimetry (photometry measurements of the linear polarization of light), and imaging.

Enroute to Jupiter, the instrument will measure brightness and polarization of zodiacal light (sunlight scattered by the total mass of interplanetary dust and solid matter) several times a month to determine the amount and character of interplanetary solid material.

The instrument will take about ten images of Jupiter in the last 50 hours before closest approach to the planet (periapsis). Image scans will be alternated with polarimetry scans.

Although resolution of the images should be better than that from the best Earth telescopes, the most important fact is that the images will be taken from viewing angles impossible to obtain from Earth. At least one will show Jupiter's terminator (the line between sunlit and dark hemispheres), which is never seen from Earth. This should show texture and shapes of cloud regions.

Images will be made in both red and blue light, and it may be possible to superimpose these, providing "color pictures" of the planet in red and blue.

The instrument uses a photoelectric sensor which measures changes in light intensity, something like the light sensor for a television camera. But unlike a TV camera it will employ the five rpm spin of the spacecraft to scan the planet, in narrow strips 0.03 degrees wide.

This electronic imaging system will complete the scans for a picture in from 25 to 110 minutes, depending on distance from the planet. Scan data will be converted to digital form on the spacecraft and radioed to Earth by telemetry. Engineers then will build up the images, using various computer techniques.

One of the planet pictures may be omitted if the spacecraft trajectory allows a good view of one of Jupiter's moons.

Controllers can vary the viewing angle of the instrument's 8.64-cm (3.4-inch)-focal length telescope by about 160 degrees relative to the spacecraft's spin axis, which is fixed on the Earth-spacecraft line. The telescope can see to within 10 degrees of the spin axis, looking away from the Earth.

For the first pictures, the telescope will be looking almost down the spacecraft spin axis toward Jupiter. For the last pictures, close to the planet, it will be looking almost at right angles to the spin axis.

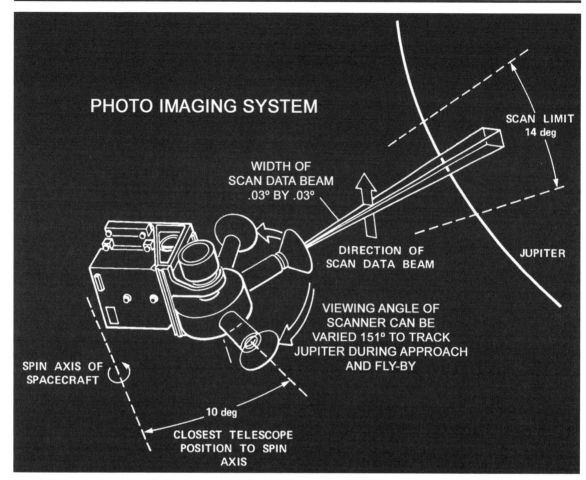

PHOTO IMAGING SYSTEM

SCAN LIMIT
14 deg

WIDTH OF
SCAN DATA BEAM
.03° BY .03°

DIRECTION OF
SCAN DATA BEAM

JUPITER

VIEWING ANGLE OF
SCANNER CAN BE
VARIED 151° TO TRACK
JUPITER DURING APPROACH
AND FLY-BY

SPIN AXIS OF
SPACECRAFT

10 deg

CLOSEST TELESCOPE
POSITION TO SPIN
AXIS

Picture reconstruction by computer on Earth will be complicated by the by the fact that picture scans will be curved as the telescope rotates with the spacecraft, while looking out at various angles to the spin axis. The only straight-line scans will be those looking straight out from the spin axis. A second factor is that Jupiter's equatorial zone rotates at 35,400 km/hr (22,000 mph). Depending on flyby trajectory, at periapsis the spacecraft may be traveling in roughly the same direction as the planet's rotation at 126,000 km/hr (78,000 mph) relative to Jupiter, or much faster.

These two factors mean that computers will have to sort out scan paths whose curvature changes steadily with spacecraft movement. They will also have to compensate for smear caused by high-speed motion of the planet and the spacecraft.

During Jupiter flyby, experimenters will use the instrument to obtain an improved understanding of the structure and composition of Jupiter's clouds, and determine the amount and nature of atmospheric gas above the clouds and aerosols in this gas. These data also will shed light on Jupiter's heat balance. The instrument will, if possible, make light scattering measurements of the Jovian satellites and any large asteroids. These data may provide understanding of the surface properties, possible atmospheres, and origins of these bodies.

The instrument includes a 2.5 cm (1-inch) aperture, 8.6-cm (3.4-inch)-focal length telescope which can be moved 160 degrees in the plane of the spacecraft spin axis by ground command or automatically. Incoming light is split by a prism according to polarization into two separate beams. Each beam is further split by going through a red filter (5800-7000 angstroms), and through a blue filter (3900-4900 angstroms). Channeltron detectors turn the light into electrical impulses, which are telemetered to Earth.

The instrument uses three viewing apertures: one 40x40 milliradians (mR) for zodiacal light measurements, a second 8x8 mR for nonimaging light measurements of Jupiter, and a third 0.5x0.5 mR for scans of the planet from which pictures will be reconstructed.

The instrument weighs 4.3 kilograms (9.5 pounds) and uses 2.2 watts of power.

Jupiter's Atmosphere, Ionosphere, Temperature

Infrared Radiometer (California Institute of Technology). Perhaps the most important mystery of Jupiter involves its heat radiation. Is the planet in fact radiating about three times more energy than it absorbs from the Sun?

experimenters will use the infrared radiometer to seek the answer by measuring Jupiter's net heat energy output. The two-channel radiometer will make measurements in the 14-25 and 29-56 micron wavelengths to study the net energy flux, its distribution over the Jovian disc, and the thermal structure and chemical composition of Jupiter's atmosphere.

Experimenters will use these data to try to find the temperature distribution in Jupiter's outer atmosphere, the existence of a polar ice cap of frozen methane, the brightness temperature of Jupiter's dark hemisphere, hot or cold spots in the atmosphere, the reason for "hot shadows" of Jupiter's moons, temperature of the Great Red Spot, and the relative abundance of hydrogen and helium in Jupiter's atmosphere. (Transparency of Jupiter's atmosphere to heat depends on the amount of helium there.)

Like the ultraviolet photometer, the radiometer uses a fixed telescope, so that its scans of the planet are made by rotation of the spacecraft. Because of this fixed viewing angle, the instrument will view the planet only during the several hours before periapsis.

The instrument has a 7.2-cm (3-inch) diameter Cassegrain telescope, and the detectors in its two channels are 88-element, thin film, bimetallic thermopiles. Its field of view is about 2,400 by 700 kilometers (1,500 by 420 miles) on Jupiter's cloud surface at closest approach of one Jupiter diameter. At this distance, it has a resolution of about 2,400 kilometers (1,500 miles).

The instrument weighs 2 kilograms (4.4 pounds) and uses 1.3 watts of power.

Occultation Experiment (Jet Propulsion Laboratory). Passage of the spacecraft radio signal through Jupiter's atmosphere as *Pioneer G* swings behind the planet will be used to measure Jupiter's ionosphere and density of the planet's atmosphere down to a pressure level of about one Earth atmosphere.

Experimenters will us; computer analysis of the incoming radio signals recorded on tape to determine the refractive index profile of Jupiter's atmosphere.

Experimenters will use computer analysis of the incoming radio signals recorded on tape to determine the refractive index profile of Jupiter's atmosphere.

This sort of analysis has been done with stars passing behind the planet, but *Pioneer* S-band telemetry is a precisely known signal source. These refraction data should allow measurements of the electron density of Jupiter's ionosphere and, used with temperature measurements, will allow inferences about the hydrogen / helium ratio in the atmosphere. Experimenters also will measure the absorption profile of the atmosphere, which should allow calculation of ammonia abundance.

If the spacecraft trajectory allows, the experimenters also will look for an atmosphere on the Jovian satellite, Io.

Experiments and Investigators

Magnetic Fields Experiment

Instrument:	Helium Vector Magnetometer
Principal Investigator:	Edward J. Smith, Jet Propulsion Laboratory, Pasadena, California
Co-Investigators:	Palmer Dyal, NASA-Ames Research Center, Mountain View, California
	David S. Colburn, NASA-Ames Research Center, Mountain View, California
	Charles P. Sonett, NASA-Ames Research Center, Mountain View, California
	Douglas E. Jones, Brigham Young University, Provo, Utah
	Paul J. Coleman, Jr., University of California at Los Angeles
	Leverett Davis, Jr. California Institute of Technology, Pasadena

Jovian Magnetic Fields Experiment

Instrument:	Flux-Gate Magnetometer
Principal Investigator:	Norman Ness, NASA-Goddard Space Flight Center, Greenbelt, Maryland
Co-Investigator:	Mario Acuna, Goddard Space Flight Center

Plasma Analyzer Experiment

Instrument:	Plasma Analyzer
Principal Investigator:	John H. Wolfe, NASA-Ames Research Center
Co-Investigators:	Louis A. Frank, University of Iowa, Iowa City
	Reimar Lust, Max-Planck-Institute fur Physik und Astrophysik
	Institute fur Extraterrestrische Physik, Munchen, Germany
	Devrie Intriligator, University of California at Los Angeles
	William C. Feldman, Los Alamos Scientific Laboratory, New Mexico

Charged Particle Composition Experiment

Instrument: Charged Particle Instrument
Principal Investigator: John A. Simpson, University of Chicago
Co-Investigators: Joseph J. O'Gallagher, University of Maryland, College Park
                       Anthony J. Tuzzolino, University of Chicago

Cosmic Ray Energy Spectra Experiment

Instrument: Cosmic Ray Telescope
Principal Investigator: Frank B. McDonald, NASA-Goddard Space Flight Center, Greenbelt, Maryland
Co-Investigators: Kenneth G. McCracken, Minerals Research Laboratory, North Ryde, Australia
                       William R. Webber, University of New Hampshire, Durham
                       Edmond C. Roelof, University of New Hampshire, Durham
                       Bonnard J. Teegarden, NASA-Goddard Space Flight Center
                       James H. Trainor, NASA-Goddard Space Flight Center

Jovian Charged Particles Experiment

Instrument: Geiger Tube Telescope
Principal Investigator: James A. Van Allen, University of Iowa, Iowa City

Jovian Trapped Radiation Experiment

Instrument: Trapped Radiation Detector
Principal Investigator: R. Walker Fillius, University of California at San Diego
Co-Investigator: Carl E. McIlwain, University of California at San Diego

Asteroid-Meteoroid Astronomy Experiment

Instrument: Asteroid-Meteoroid Detector
Principal Investigator: Robert K. Soberman General Electric Company, Philadelphia, Drexel University, Philadelphia
Co-Investigators: Herbert A. Zook, NASA-Manned Spacecraft Center, Houston, Texas

Meteoroid Detection Experiment

Instrument: Meteoroid Detector
Principal Investigator: William H. Kinard, NASA-Langley Research Center, Hampton, Virginia
Co-Investigators: Robert L. O'Neal, NASA-Langley Research Center
                       Jose M. Alvarez, NASA-Langley Research Center
                       Donald H. Humes, NASA-Langley Research Center
                       Richard E. Turner, NASA-Langley Research Center

Celestial Mechanics Experiment

Instrument: *Pioneer G* and the DSN
Principal Investigator: John D. Anderson, Jet Propulsion Laboratory
Co-Investigator: George W. Null, Jet Propulsion Laboratory

Ultraviolet Photometry Experiment

Instrument: Ultraviolet Photometer
Principal Investigator: Darrell L. Judge, University of Southern California, Los Angeles
Co-Investigator: Robert W. Carlson, University of Southern California

Imaging Photopolarimetry Experiment

Instrument: Imaging Photopolarimeter
Principal Investigator: Tom Gehrels, University of Arizona, Tucson
Co-Investigators: David L. Coffeen, University of Arizona
                       William Swindell, University of Arizona
                       Charles E. KenKnight, University of Arizona
                       Robert F. Hummer, Santa Barbara Research Center
                       Jerry Weinberg, Dudley Observatory, Albany, New York

ovian Infrared Thermal Structure Experiment

Instrument: Infrared Radiometer
Principal Investigator: Guido Munch, California Institute of Technology
Co-Investigators: Gerry Neugebauer, California Institute of Technology
                        Stillman C. Chase, Santa Barbara Research Center
                        Laurence M. Trafton, University of Texas, Austin, Texas

S-Band Occultation Experiment

Instrument: The Spacecraft Radio Transmitter and the DSN
Principal Investigator: Arvydas J. Kliore, Jet Propulsion Laboratory
Co-Investigators: Gunnar Fjeldbo, Jet Propulsion Laboratory
                        Dan L. Cain, Jet Propulsion Laboratory
                        Boris L. Seidel, Jet Propulsion Laboratory
                        S. Ichtiaque Rasool, NASA-Headquarters, Washington, D.C.

Launch Vehicle

A greatly modernized Atlas-Centaur launch vehicle, AC-30, will be used for the first time in launching the *Pioneer G* spacecraft to Jupiter.

Centaur was the nation's first high-energy, liquid hydrogen-liquid oxygen propelled launch vehicle. Developed and launched under the direction of NASA, Lewis Research Center, Cleveland, Ohio, it became operational in 1966 with the launch of *Surveyor I*, the first U.S. spacecraft to soft land on the Moon's surface.

Up to the present time Centaur has been used in combination with the Atlas booster. However, beginning next year it will also be used in combination with the Titan III to boost heavier payloads into Earth orbit and on to interplanetary trajectories. One of Titan-Centaur's first jobs will be to launch the *Viking* spacecraft to Mars in 1975.

The modernization of the Centaur vehicle, designated D-1A, primarily consists of a new integrated electronic system controlled by a 60-pound digital computer. The new "astrionics" system performs a major role in checking itself and other vehicle systems prior to launch and also maintains control of major events after liftoff. The astrionics system is mounted on a new structure called an equipment module located on the nose of the Centaur stage.

The 16,000-word-capacity computer, which is the heart of the system, replaces the original 4,800-word-capacity computer and enables it to take over many of the functions previously handled by separate mechanical and electrical systems. The new Centaur system handles navigation, guidance tasks, controls pressurization, propellant management, telemetry formats and transmission, and initiates vehicle events.

Many of the command and control functions previously performed by Atlas systems are now being handled by the Centaur equipment also. Systems which are totally integrated include guidance, flight control, telemetry and event sequence initiation.

One of the major advantages of the new Centaur D-1A system is the increased flexibility in planning new missions. In the past, hardware frequently had to be modified for each mission. Now most operational needs can be met by changing the computer software.

The AC-30 vehicle consists of an Atlas SLV-3D booster, combined with a Centaur D-1A second stage and the TE-M-364-4 third stage. The first two stages are 3.05 meters (10 feet) in diameter and are connected by an interstage adapter. Both Atlas and Centaur stages rely on internal pressurization for structural integrity.

Launch Vehicle Characteristics

The Atlas booster develops 1,919,300 newtons (431,000 pounds) of thrust at liftoff, using two 822,200-newton (184,800-pound)-thrust booster engines, one 268,100-newton (60,300-pound)-thrust sustainer engine and two vernier engines developing 3,000 newtons (675 pounds) thrust each.

Centaur carries insulation panels which are jettisoned just before the vehicle leaves the Earth's atmosphere. The insulation panels, weighing about 527 kilograms (1,162 pounds), surround the second stage propellant tanks to prevent heat or air friction from causing boil-off of liquid hydrogen during flight through the atmosphere.

To date, Centaur has successfully launched the seven unmanned Surveyor spacecraft to the Moon, *Mariners 6, 7*, and *9* and Orbiting Astronomical Observatory 2, Applications Technology Satellite 5, four Intelsat 4s and *Pioneer 10*.

The solid-fueled TE-M-364-4 third stage develops almost 66,000 newtons (15,000 pounds) of thrust. It is an uprated version of the retromotor used for the Surveyor Moon-landing vehicle.

Both third stage and spacecraft will be enclosed in an 8.7-meter (29-foot)-long, 3-meter (10-foot) diameter fiberglas shroud, which is jettisoned after leaving the atmosphere.

* Liftoff weight including spacecraft: 146,500 kilograms (323,000 pounds)

Liftoff height:	40.3 meters (132 feet)
Launch complex:	36B
Launch azimuth sector:	108 degrees

  * Measured at 5.08 centimeters (two inches) of rise.

	SLV-3D Booster	Centaur Stage	TE-M-364-4
Weight**	127,852 kilograms (284,117 pounds)	17,485 kilograms (38,856 pounds)	1,149 kilograms (2,553 pounds)
Height	22.9 meters (75 feet) (including interstage adapter)	14.6 meters (48 feet) (with payload fairing)	2.1 meters (6.75 feet) (to top interstage ring)
Thrust	1,919,300 newtons (431,000 pounds) sea level	131,200 newtons (29,500 pounds) vacuum	66,270 newtons (14,900 pounds)
Propellants	Liquid oxygen and RP-1	Liquid Hydrogen and liquid oxygen	Solid chemical fuel TP-H 3062
Propulsion	MA-5 system (two 822,200-newton (184,800-pound)-thrust engines, one 268,100-newton (60,300-pound)-sustainer engine, and two 3,000-newton (675-pound)-thrust vernier engines	Two 65,600-newton (14,750-pound)-thrust RL-10 engines. Ten small hydrogen peroxide thrusters	One solid-fuel engine
Velocity	2,438 m/s (5,454 mph) at BECO. 3,657 m/s (8,182 mph) at SECO.	10,362 m/s (23,182 mph) at MECO	14,477 m/s (32,386 mph) at spacecraft separation.
Guidance	Pre-programmed pitch rates through BECO. Switch to Centaur inertial guidance for sustainer phase.	Inertial guidance	Spin (60 rpm) imparted by spin table on top of Centaur.

** Weights are based on AC-30 configuration.

Flight Sequence

Atlas Phase

After liftoff, AC-30 will rise vertically for about 15 seconds before beginning its pitch program. Starting at two second after liftoff and continuing to T+15 seconds, the vehicle will roll to the desired flight azimuth.

After 139 seconds of flight, the booster engines are shut down (BECO) and jettisoned. BECO occurs when a acceleration of 5.7 g's is sensed by accelerometers on the Centaur and the signal is issued by the Centaur guidance system. The booster package is jettisoned 3.1 seconds after BECO. The Atlas sustainer engine continues to burn fo approximately another minute and 48 seconds propelling the vehicle to an altitude of about 146 kilometers (91 miles) attaining a speed of 13,012 km/hr (8,087 mph).

Sustainer engine cutoff (SECO) occurs at propellant depletion. Centaur insulation panels are jettisoned prior to SECO

The Atlas and Centaur stages are then separated. An explosive charge slices through the interstage adapter Retrorockets mounted on the Atlas slow the spent stage.

Centaur Phase

At four minutes, 19 seconds into the flight, the Centaur's two RL-10 engines ignite for a planned 7.5-minute burn After MECO, the Centaur stage and spacecraft are reoriented with the Centaur attitude control thrusters to place the third-stage and spacecraft on the proper trajectory after separation.

Third Stage Phase

A spin table mounted on top of the Centaur carries the third stage and the spacecraft. This is spun to 60 rpm by smal thruster rockets. Explosive bolts separate the spinning third stage and spacecraft. The TE-M-364-4 engine then ignite and burns for 56 seconds.

Separation of the *Pioneer* spacecraft is achieved by firing explosive bolts on the metal ring holding the spacecraft to the third stage. Compressed springs then push the spacecraft away from the third stage at a rate of 0.6 m/sec (2.1 feet/sec).

Retromaneuver

Requirements for the Centaur retrosystem call for a 7.6-meter (25-foot) separation distance to be achieved within 13 seconds after the third stage has separated from Centaur. This separation distance is required to prevent spacecraft damage from reflected motor exhaust products and to minimize third stage ignition transients. At one second after third stage separation Centaur retrothrust is initiated to provide the required clearance.

Atlas / Centaur / TE-M-364-4 Flight Sequence (AC-27)

	Nominal Time (seconds)	Altitude (km)	Altitude (statute miles)	Surface Range (km)	Surface Range (statute mi)	Relative Velocity (km/hour)	Relative Velocity (miles/hour)
Liftoff	0	0	0	0	0	0	0
Booster Engine Cutoff	139.1	56.8	35.3	80.6	50.14	8,957	5567
Jettison Booster	142.2	59.8	37.2	87.7	54.5	10,154	6311
Jettison Insulation Panel	184.1	97.6	60.7	192	119.37	10,265	6380
Sustainer / Vernier Cutoff	247.9	146.4	90.98	386.9	240.47	13,012	8087
Atlas / Centaur Separation	249.8	147.8	91.86	393	244.26	13,004	8082
Centaur Ignition	259.4	154.1	95.79	426.8	265.3	12,965	8058
Jettison Nose Fairing	271.4	161.4	100.3	468.7	291.3	13,171	8185
Centaur Main Engine Cutoff	709.6	164.4	102.18	3062	1902.9	37,356	23,217
Ignite Spin Rockets	779.6	180.5	112.16	3769	2342.3	37,351	23,214
Centaur / TE-M-364-4 Separation	781.6	181.5	112.82	3788	2354	37,384	23,212
TE-M-364-4 Ignition	794.6	189.2	117.6	3920	2436	37,324	23,197
TE-M-364-4 Burnout	838.4	229	142.34	4419	2746.2	50,592	31,443
TE-M-364-4 / Spacecraft Separation	939.4	472.8	293.84	5738	3566.15	50,042	31,101

Launch Operations

The Unmanned Launch Operations Directorate of the John F. Kennedy Space Center is responsible for the preparation and launch of unmanned space vehicles from its facilities at Cape Kennedy, Florida.

The *Pioneer G* spacecraft was delivered to the Cape from TRW Systems approximately six weeks before launch and placed in the spacecraft checkout area for final verification tests.

As a part of its launch operations, KSC handles scheduling of test milestones and review of data to assure that the launch vehicle has met all its test requirements and is ready for launch.

The Atlas-Centaur 30 launch vehicle was received at Cape Kennedy on January 4, 1973. The Atlas booster was erected on Complex 36B on January 9 and the Centaur second stage was mated with the booster on January 10. Launch vehicle subsystem tests were conducted from January 11 through February 15. A launch vehicle electrical test was conducted in late February.

A major milestone in launch preparations was the Terminal Countdown Demonstration (TCD) conducted approximately one month prior to launch.

The TCD main purpose is to demonstrate that all of the functions leading to the actual countdown can be performed. It is an end-to-end check of all launch vehicle systems and includes propellant loading of launch vehicle stages to verify that tanks and facilities are ready for the actual countdown.

Launch Windows

	EST	S/C Weight	Trip Time	Date Arrival	Time GMT
4/5	21:11 - 21:56	570	608	12/5/74	4:26
4/6	21:03 - 21:48	570	607	12/5/74	4:26
4/7	20:55 - 21:40	570	606	12/5/74	4:26
4/8	20:48 - 21:33	570	605	12/5/74	4:26
4/9	20:40 - 21:25	570	604	12/5/74	4:26
4/10	20:33 - 21:18	570	603	12/5/74	4:26
4/11	20:26 - 21:11	570	602	12/5/74	4:26
4/12	20:17 - 21:02	570	615	12/19/74	3:35
4/13	20:10 - 20:55	570	614	12/19/74	3:35
4/14	20:03 - 20:48	570	616	12/22/74	3:25
4/15	19:53 - 20:38	570	648	1/24/75	1:43
4/16	19:47 - 20:32	570	647	1/24/75	1:43
4/17	19:41 - 20:26	570	667	2/14/75	0:40
4/18	19:42 - 20:27	570	740	4/28/75	21:06
4/19	19:35 - 20:05	558	740	4/29/75	21:03
4/20	19:29 - 19:59	558	740	4/30/75	21:00
4/21	19:23 - 19:53	558	740	5/1/75	20:57
4/22	19:22 - 19:52	558	765	5/27/75	19:39
4/23	19:20 - 19:50	558	780	6/12/75	20:24
4/24	19:18 - 19:48	558	795	6/28/75	19:36
4/25	19:16 - 19:46	558	810	7/14/75	17:16
4/26	19:15 - 19:45	558	825	7/30/75	16:28

The *Pioneer G* spacecraft was mated with the TE-M-364-4 third stage and placed atop the Atlas-Centaur launch vehicle during the third week of March.

Following this, the Flight Events Demonstration (FED) was conducted. The purpose of the FED is to assure that the vehicle is electrically ready for final launch preparations. The FED includes running the computer and programmer through post-flight events and monitoring the data to assure correct response to all signals with the umbilical ejected.

The Composite Electrical Readiness Test (CERT) is scheduled for the space vehicle about four days prior to launch. It verifies the ability of the launch vehicle to go through post-liftoff events. Range support elements participate along with the spacecraft and launch vehicle just as during a launch.

The T-1 Day Functional Test involves final preparations in getting the entire space vehicle ready for launch, preparing ground support equipment, completing readiness procedures and installing ordnance on the launch vehicle.

The countdown is picked up at T-335 minutes. All systems are checked against readiness procedures, establishing the integrity of the vehicle and ground support equipment interface prior to tower removal at T-120 minutes. Loading of cryogenic propellants (liquid oxygen and liquid hydrogen) begins at T-80 minutes, culminating in complete vehicle readiness at T-5 minutes. The terminal count begins with the monitoring of all systems and topping off and venting of propellant and purge systems. At T-10 seconds, the automatic release sequence is initiated and the space vehicle clear for liftoff.

Mission Operations

Pioneers 10 and *G* are controlled primarily by ground commands, rather than command sequences stored in onboard computers. This means 24-hour-a-day spacecraft operation, continuous analysis of incoming science and spacecraft telemetry, and careful planning of command sequences.

The fully-ground-controlled spacecraft concept allows for great flexibility in changing plans and objectives during flight. It requires a continuously effective communications system.

Ground-commanded spacecraft probably are less expensive to build and can easily be adapted to a variety of missions.

Pioneer G control and spacecraft operations will be at the Ames-manned Pioneer Mission Area, Deep Space Network Headquarters, JPL, Pasadena, during the launch phase of the mission (first four weeks). During the interplanetary data taking and planet encounter phases of both Pioneer 10 and G missions, control is at the Pioneer Mission Operations Center (PMOC), Ames Research Center, Mountain View, California. The Pioneer Project is managed by Ames.

The PMOC has computing capability both for commanding the spacecraft and to interpret the data stream as it comes in from the DSN stations for use by flight controllers monitoring spacecraft performance.

Several Ames organizations direct and support launch, interplanetary and planet encounter operations.

The Pioneer Mission Operations Team (PMOT) consists of personnel from many government and contractor organizations, and operates under control of the Flight Director.

In addition to several assistant flight directors, the PMOT includes the following groups:

- The Spacecraft Performance Analysis Team analyzes and evaluates spacecraft performance and predicts spacecraft responses to commands.
- The Navigation and Maneuvers Team handles spacecraft navigation, and attitude control in space.
- The Science Analysis Team determines the status of the onboard scientific instruments and formulates command sequences for the instruments.

Many other support groups at Ames and other NASA facilities assist the mission operations team to insure effectiveness of computers, computer displays, computer programs, communications, physical facilities, and operation of the DSN's worldwide net of tracking stations.

Tracking and Data Retrieval

With facilities located at 120-degree intervals around the Earth, NASA's Deep Space Network will be able to support the *Pioneer G* spacecraft continuously on its two-year flight to Jupiter. As the spacecraft "sets" at one station due to the Earth's rotation, it will "rise" at the next one.

DSN stations also can provide overlapping coverage during critical events, and they will be able to track the spacecraft for the 1.5 billion miles of communications distance, and the five years of flight time beyond Jupiter, if the mission continues that long.

The DSN maintains six operational 26-meter (85-foot)-diameter dish antenna stations. These are at Madrid (Robledo de Chavela and Cebreros), Spain: Canberra, Australia; and Johannesburg, South Africa, with two stations at Goldstone,

California. In addition, the DSN has three of the extremely sensitive 64-meter (210-foot) dish antennas at Goldstone, Madrid and Canberra, able to provide continuous coverage when needed.

The 64-meter antennas can receive eight times the volume of data of the network's 26-meter antennas.

The DSN acquires data from *Pioneers 10* and *G* through the 26-meter antennas in early parts of the missions, and the 64-meter antennas later on.

A 1.2-meter (four-foot) antenna at Cape Kennedy covers pre-launch and launch phases of the flight.

All DSN stations have been redesigned and are now equipped with general purpose telemetry equipment capable of receiving, data-synchronizing, decoding and processing data at high transmission rates.

The tracking and data acquisition network is tied to the Mission Control and Computing Facility (MCCF), JPL's central data processing facility at Pasadena, by NASA's Communications Network (NASCOM).

During the critical four-week post-launch phase of the Pioneer G mission, Ames controllers and spacecraft subsystem experts at consoles in the Pioneer Mission Area at JPL will use MCCF's large data processing computers to display processed or real-time information either on television screens in their consoles or on high-speed printers.

When mission control shifts to the Pioneer Mission Operations Center at Ames, the same controllers and subsystem experts will operate the two spacecraft and analyze their performance there, using a combination of Ames and MCCF computers.

During launch, tracking will be carried out by the DSN with aid of other facilities. These are tracking antennas of Air Force Eastern Test Range and downrange elements of NASA's Spacecraft Tracking Data Network (STDN) together with the tracking ship *Vanguard* and instrumented jet aircraft. Both ship and jet are operated by STDN.

The mission is complicated by the fact that Jupiter will be half a billion miles from the Earth at the time of first encounter, and by the end of communication capability, five years after encounter, *Pioneer* will be 2.4 billion kilometers (1.5 billion miles) away. At these distances, the spacecraft radio signal becomes incredibly faint, and round-trip communications times are two to four hours.

For all of NASA's unmanned missions in deep space, such as the planet-probing *Mariners* and earlier Sun-orbiting *Pioneers*, the DSN provides tracking information on course and direction of the flight, velocity and range from Earth. It also receives engineering and science telemetry, including planetary television coverage, and sends commands for spacecraft operations. All communication links are in the S-band frequency.

DSN stations relay spacecraft doppler tracking information to the MCCF where other large computer systems calculate orbits in support of navigation and planetary target planning. High-speed data links from all stations are capable of 4,800 bps. These new data links allow real-time transmission of all data from spacecraft to Pioneer control centers at Ames or JPL. Throughout the mission, scientific data recorded on magnetic tape will be sent from DSN stations to Ames where it will be divided into individual magnetic tapes for each experimenter.

All of NASA's networks are under the direction of the Office of Tracking and Data Acquisition, NASA Headquarters, Washington, D.C. The Jet Propulsion Laboratory manages the DSN, while the STDN facilities and NASCOM are managed by NASA's Goddard Space Flight Center, Greenbelt, Maryland.

The Goldstone DSN stations are operated and maintained by JPL with the assistance of the Philco-Ford Corporation.

The Canberra station is operated by the Australian Department of Supply. The Johannesburg station is operated by the South African government through the National Institute for Telecommunications Research. The two facilities near Madrid are operated by the Spanish government's Instituto Nacional de Tecnica Aeroespacial (INTA).

Pioneer Project Management Team

Office of Space Science

Dr. John E. Naugle	Associate Administrator for Space Science
Vincent L. Johnson	Deputy Associate Administrator for Space Science
Robert S. Kraemer	Director, Planetary Programs
I. Ichtiaque Rasool	Deputy Director, Planetary Programs
Fred D. Kochendorfer	Pioneer Program Manager
Thomas P. Dallow	Pioneer Program Engineer
Dr. Albert G. Opp	Pioneer Program Scientist
Joseph B. Mahon	Director, Launch Vehicle Programs
F. Bland Norris	Manager, Medium Launch Vehicles
I. Robert Schmidt	Manager, Atlas-Centaur

Office of Tracking and Data Acquisition

Gerald M. Truszynski	Associate Administrator for Tracking and Data Analysis
Arnold C. Belcher	Network Operations
Maurice E. Binkley	Network Support

Ames Research Center, Mountain View, California

Dr. Hans Mark	Center Director
C.A. Syvertson	Deputy Director
John V. Foster	Director of Development
Charles F. Hall	Pioneer Project Manager
Dr. John H. Wolfe	Pioneer Project Scientist
Ralph W. Holtzclaw	Pioneer Spacecraft System Manager
Joseph E. Lepetich	Pioneer Experiment System Manager
Robert R. Nunamaker	Pioneer Mission Operations System Manager
J. Richard Spahr	Management Control
Robert U. Hofstetter	Mission Analysis and Launch Coordination
Norman J. Martin	Mission Control Manager
Thomas L. Bridges	Data Handling Manager
Arthur C. Wilbur	Nuclear Power
Eldon W. Kaser	Contracts
Henry Asch	Reliability and Quality Assurance
Standish R. Benbow	Reliability and Quality Assurance
Ernest J. Iufer	Magnetics

Deep Space Network, Jet Propulsion Laboratory, Pasadena, California

Dr. William H. Pickering	Director, Jet Propulsion Laboratory
Dr. Nicholas A. Renzetti	Tracking and Data System Manager
Alfred J. Siegmeth	Deep Space Network Manager for Pioneer

Lewis Research Center, Cleveland, Ohio

Bruce T. Lundin	Center Director
Edmund R. Jonash	Chief, Launch Vehicles Division
Daniel J. Shramo	Atlas-Centaur Project Manager
Edwin Muckley	Centaur Project Engineer

Kennedy Space Center, Florida

Dr. Kurt H. Debus	Center Director
John J. Neilon	Director, Unmanned Launch Operations
John D. Gossett	Chief, Centaur Operations Branch
Hugh A. Weston	Delta Project Manager (For third stage)
James W. Johnson	KSC Pioneer Project Representative
William R. Fletcher Jr.	Spacecraft Coordinator

AEC Space Nuclear Systems Division

David S. Gabriel	Division Director
Glenn A. Newby	Assistant Director for Space Electric Power
Harold Jaffe	Manager, Isotope Flight Systems Office
William C. Remini	SNAP 19 / Pioneer RTG Program Manager

TRW Systems Group

Bernard J. O'Brien	Pioneer Project Manager

Pioneer G Contractors

See *Pioneer F Contractors* in the **Pioneer 10 Press Kit**, page 44.

NATIONAL AERONAUTICS AND SPACE ADMINISTRATION
Washington, D. C. 20546
202-755-8370

FOR RELEASE:
November 19, 1974

RELEASE No: 74-292

PROJECT: Pioneer 11 Jupiter Encounter

P
R
E
S
S

K
I
T

CONTENTS

**NATIONAL AERONAUTICS AND
SPACE ADMINISTRATION**
Washington, D. C. 20546
202-755-8370

FOR RELEASE:
10 a.m. Tuesday, November 19, 1974

Nicholas Panagakos   Phone: (202) 755-3680
Headquarters, Washington, D.C.

Peter Waller   Phone: (415) 965-5091
Ames Research Center, Mountain View, California

RELEASE No: 74-292

Pioneer Encounters Jupiter December 3

Man's second spacecraft to Jupiter will reach the giant planet early on December 3, after a billion-kilometer (620-million-mile) journey that has taken nearly two years.

Hurtling through intense radiation never before encountered, *Pioneer 11* will skim Jupiter at a distance of 41,000 kilometers (26,600 miles) – three times closer than its predecessor, *Pioneer 10*.

Closest approach will be at 12:22 a.m. EST.

The 260-kilogram (570-pound) robot from Earth will sweep over the face of Jupiter at more than 171,000 kilometers (107,000 miles) an hour – the highest speed ever achieved by a manmade object.

Then, if *Pioneer* survives the radiation, it will boomerang across the solar system and head for a rendezvous with Saturn and its mysterious rings in 1979.

Pioneer 11 will be the first to use the gravity of one outer planet (Jupiter) to fly on to the next outer planet (Saturn), an essential maneuver for continued exploration of the outer solar system.

Because *Pioneer* will zoom in toward Jupiter from below, color photographs of the planet will show never-before-seen polar regions – providing an entirely different look at the colossus of the planets.

The close polar pass also will permit the first scientific measurements in high polar latitudes of both the planet and its outer environment. And it will allow the first deep probe of the inner radiation belt which girdles the planet like a doughnut.

Since radiation increases rapidly with nearness to the planet, *Pioneer*'s close trajectory will hurl it through the most intense sector of Jupiter's inner belt. Radiation levels may be ten times those encountered by *Pioneer 10*, and 40,000 times greater than Earth's belts – easily enough to destroy a spacecraft.

However, *Pioneer 11* will be moving so rapidly that no serious damage is anticipated. The sister craft, *Pioneer 10*, experienced minor malfunctions on its longer pass through the weaker outer part of the belt last December.

Controllers have aimed in *Pioneer 11* far below Jupiter's south pole. There the planet's immense gravity will suck the spacecraft almost straight up, passing it quickly through the thin disc of maximum radiation in about half an hour at a very high angle and at record speed. *Pioneer* then will loop around behind the planet and depart – still heading upward – above Jupiter's north pole.

The "corkscrew" maneuver is required to send the spacecraft to Saturn, and also to help cut down exposure in the equatorial region, the deadliest part of Jupiter's radiation zone.

The trajectory will also take the spacecraft around the planet against Jupiter's direction of rotation – permitting for the first time a look at a complete revolution of the planet's magnetic field, radiation belt and surface.

Discussing the radiation hazard, Dr. John Wolfe, Pioneer project scientist at Ames Research Center, points out:

"On this trajectory, we expect that the total radiation dose will be no more than that received by *Pioneer 10*. What we don't know is how the spacecraft will stand the short-time peak radiation as it crosses the center plane of the inner belt. At *Pioneer 10*'s closest approach, radiation striking the spacecraft was already a billion electrons per square centimeter per second and rising. Three times closer will make the peak level perhaps ten times higher for a few minutes.

"Of course, scientifically, we want to measure the intense, inner region of the belts and we need the information for future Jupiter orbiter and atmosphere probe missions."

If *Pioneer 11* survives, it has a reasonable chance of functioning long enough to reach Saturn six-and-one-half years after launch. *Pioneer 6* still performs well after ten years in space. *Pioneer 10*, well beyond Jupiter now, is operating after almost three years in space.

In the 48 hours centered on closest approach to Jupiter, *Pioneer 11* will take 22 color pictures of the brightly-banded planet, all having three to five times the resolution of pictures from Earth. All will be from angles unavailable from Earth or *Pioneer 10*, with six of them looking down on south or north polar regions. These color pictures will show a "gibbous" Jupiter (between a half and full moon) with the large semi-circle of the polar regions bordered by concentric orange and blue-gray semi-circular bands.

During the same 48 hours, *Pioneer* will take pictures of the planet-sized moons Callisto and Ganymede, and of Io, the most reflective object in the solar system. It will make the first spacecraft measurements of Amalthea, tiniest and innermost of Jupiter's 13 moons, which orbits Jupiter at a distance of 115,000 kilometers (69,000 miles).

After encounter, *Pioneer 11* will fly into an uncharted region of space about 160 million kilometers (100 million miles) above the orbit planes of most of the planets. There it can provide information on a new portion of the heliosphere. This is the region surrounding the Sun dominated by the solar wind – the gas blowing out from the Sun – and the weak magnetic fields imbedded in that wind.

Because *Pioneer 11*'s path will be through uncharted territory, scientists hope that it will flesh out the information on the giant planet returned by *Pioneer 10*. That spacecraft provided the first firm ideas of what Jupiter is really like.

The structure of the planet, from the highest levels of the atmosphere down through its center, is much more accurately known now. According to the findings, recently released, Jupiter is a whirling ball of liquid hydrogen without any detectable solid surface. In many respects, it is more like a star than a planet. It is much hotter than previously believed, with an interior that is hotter than the surface of the Sun. At the same time, at some distance below the top of the clouds, temperatures are moderate enough to conceivably support life.

Like its predecessor, *Pioneer 11* carries a plaque telling any intelligent species which might find it several million years from now, who sent it and where it came from.

Pioneer 11's radio signals, traveling at the speed of light (186,000 miles per second), will take 46 minutes to reach Earth from Jupiter.

Spacecraft operations during the encounter will be complicated by the 92-minute round-trip communications time, and the need to send 10,000 commands to the spacecraft in the two weeks centered on the closest approach – the most commands ever handled by the Deep Space Network. Operations strategy is to set most systems in one standard mode throughout the encounter, with most commands going to the spacecraft imaging system.

Pioneer 11 weighs 260 kilograms (570 pounds) and is spin-stabilized, giving its instruments a full circle scan five times a minute. It uses nuclear sources for electric power because sunlight is too weak at Jupiter and beyond for an efficient solar-powered system.

Pioneer's 2.75-meter (9-foot) dish antenna looks back at Earth throughout the mission – adjusting its view by changes in spacecraft attitude as the home planet moves in its orbit around the Sun.

Pioneer carries a 30-kilogram (65-pound) scientific payload of 12 instruments. Two other experiments use the spacecraft and its radio signal as their instruments.

Partners with the spacecraft are the three incredibly sensitive "big dish" antennas of NASA's Deep Space Network, which retrieve *Pioneer* data. *Pioneer*'s 8-watt radio signal reaches these antennas from Jupiter with a power of 1/100,000,000,000,000,000 watts. Collected for 19 million years, this energy would light a 7.5-watt Christmas tree bulb for one-thousandth of a second.

Cost of the two *Pioneer Jupiter* spacecraft, Scientific instruments, and data processing and analysis is about $100 million.

NASA's Office of Space Science has assigned management of the Pioneer Jupiter project to Ames Research Center, Mountain View, California. The spacecraft were built by TRW Systems, Redondo Beach, California. The scientific instruments were provided by NASA centers, universities, and private industry.

Tracking and data acquisition is the responsibility of NASA's Deep Space Network, operated by the Jet Propulsion Laboratory, Pasadena, California.

(End of General Release. Background information follows.)

Encounter Profile

Pioneer 11 began its journey to Jupiter and Saturn on April 5, 1973.

On its 607-day trip to Jupiter, it has followed a curving path about one billion kilometers (620 million miles) long, covering about a million miles a day.

On its flyby trajectory, *Pioneer 11* will pass through the narrow "window" (which lies in front of Jupiter as it moves on its orbit) that will allow Jupiter to throw it almost completely across the solar system to Saturn.

The spacecraft will fly a sort of corkscrew loop around the planet. As seen from above, it will cross its incoming trajectory on the way out.

Radiation Danger

Pioneer 11 will encounter its principal hazard when it flies through Jupiter's radiation belt. Most intense radiation will be encountered in a 12-hour period centered on periapsis. The belt has a total energy many millions of times that of Earth's belt.

At its 131,220 kilometers (81,000 miles) closest approach to Jupiter, *Pioneer 10* found that peak electron intensity was 10,000 times that of the Earth's belt and was increasing ten times for every three Jupiter radii closer to the planet. Proton radiation also was at very high levels. These radiation intensities were near the maximum that spacecraft systems could stand.

Since *Pioneer 11* will come three times closer to Jupiter than *Pioneer 10* – more than one Jupiter radius (55,000 miles) closer – the radiation hazard is severe.

However, peak radiation intensities were found at the equatorial plane of Jupiter's magnetic field. The field contains the radiation belt. Calculations indicate that intensity of the belt drops three times at a distance of 20 degrees above the magnetic equator, and ten times at 40 degrees above it.

To pass safely through the belt, *Pioneer 11* will fly through at a very high angle to the magnetic equatorial plane (55 degrees). Controllers will bring in the spacecraft far below Jupiter's south pole and then fly almost straight up through the plane of the magnetic equator and its intense radiation zone at a record speed of 173,340 km/hr (107,000 mph).

Project officials calculate that on this course, at this speed, total radiation experienced by *Pioneer 11* will be similar to that encountered by *Pioneer 10*, although peak radiation levels will be far higher.

If the spacecraft should be lost, so would the Saturn mission. However, substantial findings would still be available. Since *Pioneer 11* will approach Jupiter against the direction of planet rotation, large parts of the magnetic field and radiation belts would be mapped at very high latitudes, measurements of the most intense parts of the radiation belts will have been made, and incoming polar pictures of the planet will have been achieved.

The scientific instruments are susceptible to damage from Jupiter's radiation belts in the following order: The Asteroid-Meteoroid Detector is especially sensitive to radiation. Three of the four high-energy radiation counters are considered to be next most vulnerable. The solar wind, ultraviolet, and photopolarimeter counters are third most likely to be damaged. The magnetometer, the fourth high-energy radiation counter, the infrared and meteoroid instruments are least likely to suffer radiation damage.

Among spacecraft subsystems, all are equally susceptible to radiation damage.

Encounter Sequence

The two-month *Pioneer* encounter period, November 2, 1974 to January 3, 1975, breaks down as follows:

- During the three weeks, November 3 to 24, the spacecraft (still in interplanetary space) will move in from about 27 million kilometers (17 million miles) from the planet to about 10 million kilometers (6.2 million miles).
- On November 25 or 26, *Pioneer* is expected to pass through the bow shock wave in the solar wind created by Jupiter's magnetic field and enter Jupiter's magnetosphere, approaching to about 7.5 million kilometers (4.7 million miles).
- Between November 27 and December 1, *Pioneer 10* will fly to within 1.6 million kilometers (1.1 million miles) of Jupiter and will pass through the outer magnetosphere, while viewing the planet from a distance.
- During the last day and a half (38 hours), before closest approach on December 3, *Pioneer* will move into its closest distance of 42,800 kilometers (26,600 miles) above the planet's cloud tops (pericenter). During this last 38 hours, it will make its most important planet measurements and pictures, and will pass through the most intense parts of the radiation belts (between six hours before and six after pericenter). It also will traverse regions of dust concentrations near the planet.

The spacecraft will repeat this sequence as it exits the planet environment.

Pioneer 11 Jupiter Flyby

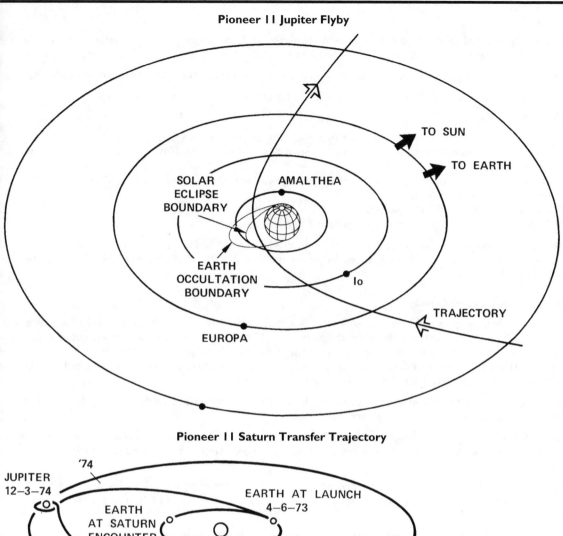

Pioneer 11 Saturn Transfer Trajectory

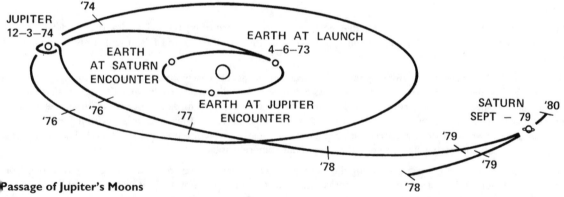

Passage of Jupiter's Moons

Early in the encounter, on November 7, about 23 million kilometers (14 million miles) out, *Pioneer* crossed the orbits of Hades and Poseidon, first two of Jupiter's four outer moons; the following day, it crossed the orbit of Pan.

On November 10, it crossed Andrastea's orbit.

On November 22, eleven days later, *Pioneer* will reach and cross the orbits of the three middle moons, Demeter, Hera, and Hestia, located 11.3 million kilometers (7 million miles) from Jupiter.

On December 2, the spacecraft will make its closest approach to all five of Jupiter's inner moons. It will come within 786,500 kilometers (488,730 miles) of Callisto, which is as large as the planet Mercury, 21 hours before pericenter. It will sweep by Ganymede at a distance of 692,300 kilometers (430,000 miles), seven hours from pericenter. Ganymede, Jupiter's largest moon, is larger than Mercury, although not nearly as dense. At 2 hours out, *Pioneer* will come within 314,000 kilometers (195,000 miles) of Io, which is larger than the Earth's moon. It will most closely approach Europa, which is about the size of our Moon, at about one hour from pericenter.

The spacecraft will come within 128,000 kilometers (80,000 miles) of Amalthea, Jupiter's tiny inner moon at an hour and a quarter after pericenter. *Pioneer*'s closest approach to Jupiter will be 67,600 kilometers (42,000 miles) inside Amalthea's orbit.

Pictures

Imaging of Jupiter and its four large moons began on November 2, 29 days from pericenter. Early images were largely for calibration. On November 25, imaging activity will go on a 23-hour-a-day basis.

At pericenter minus 48 hours, on December 1, imaging results should become equal to or better than those of Earth telescopes.

At pericenter minus about 12 hours, on December 2, the planet will overlap the spacecraft camera's view field.

Numbers of pictures:

- December 1. *Pioneer 11* is expected to take 14 images of Jupiter, and two images of Callisto.
- December 2. Plans call for 8 pictures of Jupiter, about half looking up at the South pole and one each of Callisto and Ganymede.
- December 3. After closest approach, 11 images of the planet are scheduled, many looking down at Jupiter's north pole. One picture of Io will be taken.
- December 4. There will be 22 pictures.
- December 9. Imaging will return to a schedule of 3 to 8 hours a day.

The Imaging Photopolarimeter (IPP), which takes the pictures, will use half the spacecraft's data return capacity of 1,024 bits per second during encounter, largely for picture transmission. About 60 percent of the IPP observation time will be used for pictures.

Picture Quantity and Quality

Pioneer 11's imaging system (see Experiments — Imaging Photopolarimeter) will take pictures in red and blue light. During the 96 hours centered on pericenter (beginning at 2,867,400 kilometers (1,770,000 miles) out, quantities will be: about 40 pictures of the full planet, many of portions of the planet's surface, three of the moon Callisto, and one each of Ganymede and Io. As *Pioneer 11* swings around Jupiter, Sun angles not possible from Earth (and not obtained from *Pioneer 10*) will be seen with almost every picture.

Incoming color and black and white pictures will be displayed on TV monitors as they arrive from *Pioneer*. They will then be photographed from the monitors and distributed.

Computer rectification will improve all pictures. Total time after picture receipt on Earth required for rectification (at Ames and the University of Arizona) will average 72 hours (3 days) for black and white pictures, and 96 hours (4 days) for color pictures.

In the closest pictures, looking down on either pole, general appearance of Jupiter will be dramatically different from *Pioneer 10* or Earth views. The planet will appear gibbous (between a half and full moon), showing concentric semicircular zones instead of the parallel bands.

Resolution of all pictures in the 96-hour period centered on closest approach should be better than the best views from Earth. The best pictures will be in the 24 hours before and after closest approach, when the spacecraft is within a million miles of Jupiter. These pictures should be much better than any ever made from Earth, the best ever taken of Jupiter, except those made by *Pioneer 10*. They will show cloud eddies and features invisible from Earth, and may provide deep views in to the transparent atmosphere which may exist at the poles. The several close-up pictures taken near closest approach will be somewhat distorted but comprehensible.

Average resolution of all the pictures during the close-in 96 hours may be two to three times better than that from Earth. Resolution of the last picture before pericenter (a portion of Jupiter's surface including the Red Spot) may be about five times that of the best Earth pictures.

Encounter Science Sequence

The first evidence at the spacecraft of Jupiter's presence may come on November 25 or 26. This is the earliest date for passage through the bow shock wave created as the solar wind strikes Jupiter's magnetic field.

On October 9, *Pioneer 11* made its final attitude change to assure precise pointing of the spacecraft at Earth for optimum communications. Controllers will make no further attitude changes until four days after pericenter to provide an imperturbed trajectory for the gravity-sensing experiment, which measures mass concentrations of Jupiter and orbits and masses of its moons.

Most likely dates for entering Jupiter's magnetic field are November 26 or 27.

The first ultraviolet measurements will be of Callisto and Ganymede on November 30. Because of spacecraft attitude required by the Saturn-bound flyby trajectory, there will be no ultraviolet measurements of Jupiter itself. However, there will also be ultraviolet measurements of Ganymede on December 1 and 2, and of Europa on December 2. These studies will look for hydrogen and helium and other ultraviolet glow phenomena on the three moons.

Infrared viewing to determine temperature begins the day before pericenter with measurements of Callisto. *Pioneer* will "take the temperature" of Callisto again on December 4; of Ganymede, December 2 and 3; of tiny Amalthea twice on December 2; and of Io on December 2 and 3. These will be the first temperature measurements of Amalthea, since *Pioneer 10* made none.

The most important infrared measurements will be of the planet itself, during inbound and outbound viewing periods. Inbound is a two-hour-and-45-minute period starting four hours and 21 minutes before pericenter; outbound is 2½ hours, starting one hour, 39 minutes after pericenter.

Since November 2, the imaging photopolarimeter (IPP) has been studying the nature of atmospheric gas above Jupiter's cloud – the amount of aerosols it contains, and the structure and composition of Jupiter's upper cloud layers. The IPP will provide similar information on Jupiter's moons. Nearly all of these measurements will be from view angles not available to *Pioneer 10*, making the data especially valuable.

Pericenter will occur at 9:22 p.m. PST, and some minutes later *Pioneer* will pass from the southern hemisphere through Jupiter's equatorial plane.

Twenty minutes before pericenter, the spacecraft will pass behind Jupiter itself for 42 minutes. Experimenters on Earth will make refraction studies as *Pioneer*'s radio signal passes through the atmosphere. These should tell more about electron density of Jupiter's ionosphere, and density and composition of its atmosphere – especially in the higher latitudes reached by *Pioneer 11* on its near-polar pass. Measurements will be made both going in and coming out from behind Jupiter.

Just seconds before radio blackout, the spacecraft will have entered the Jovian night for 33 and a half minutes. Returning into sunlight, *Pioneer* will begin its exit from Jupiter's environment, continuing to make measurements as it goes out.

Post-Encounter

After encounter, solar wind, particle, and magnetometer experiments will continue to look for the answers also sought by *Pioneer 10*.

The most interesting experimental questions over the next five years in the region near Jupiter's orbit and in the unexplored regions 160 million kilometers (100 million miles) above the ecliptic (plane of Earth's orbit) to be reached by *Pioneer 11* will be: What will be the solar wind and solar magnetic field expansion processes as seen over five years at various distances from the Sun and the ecliptic? What will be the flux of galactic cosmic rays and the distribution of neutral hydrogen, non-solar-wind plasmas and interstellar hydrogen and helium? What do these tell about the interstellar space beyond the boundary of the heliosphere (the Sun's atmosphere)?

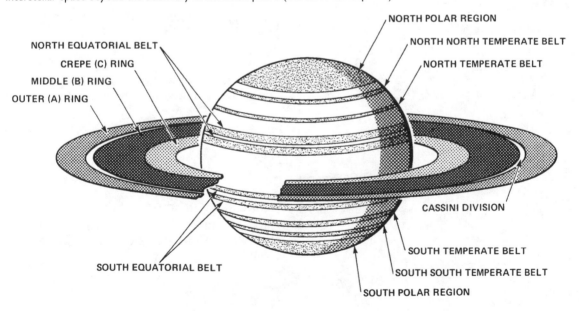

Journey to Saturn

After leaving Jupiter, *Pioneer 11* will head for Saturn. Because of its high-angle encounter trajectory, taking it far up over Jupiter's north pole, *Pioneer*'s course to Saturn will attain an overall angle to the plane of the Earth's orbit of 15.6 degrees. This will take it 161,835,000 kilometers (100,564,000 miles) above the Earth's orbit plane, by far the highest above the plane of most of the planets that any spacecraft has ever gone.

The Saturn flight will last six years. *Pioneer* will almost cross the inner solar system on the way – first passing inside Jupiter's orbit almost back into the outer fringes of the Asteroid Belt at 3.75 Astronomical Units (about 557,400,000 kilometers (348,400,000 miles) from the Sun. Then it will recross Jupiter's orbit around April 20, 1977, and fly on to Saturn, reaching the ringed planet in September 1979.

Several Saturn flyby trajectories are possible. One of these options sends the spacecraft between Saturn's rings and the planet.

Depending on which Saturn flyby course is selected, *Pioneer 11* will either escape the solar system or go into a very large oval orbit around the Sun.

Pioneer 10 And 11 Events

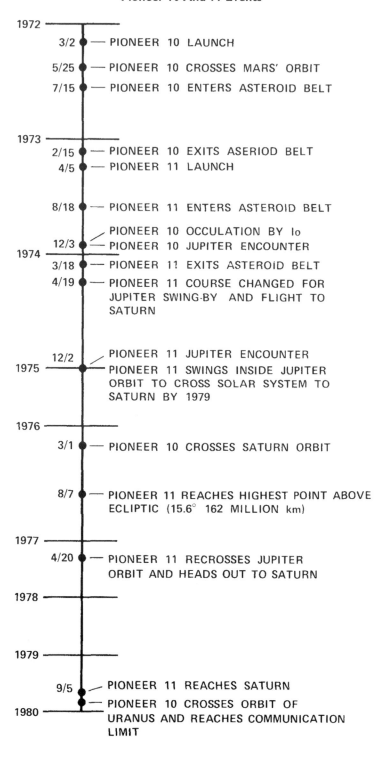

1972

3/2 — PIONEER 10 LAUNCH

5/25 — PIONEER 10 CROSSES MARS' ORBIT

7/15 — PIONEER 10 ENTERS ASTEROID BELT

1973

2/15 — PIONEER 10 EXITS ASERIOD BELT
4/5 — PIONEER 11 LAUNCH

8/18 — PIONEER 11 ENTERS ASTEROID BELT

PIONEER 10 OCCULATION BY Io
12/3 — PIONEER 10 JUPITER ENCOUNTER
1974
3/18 — PIONEER 11 EXITS ASTEROID BELT
4/19 — PIONEER 11 COURSE CHANGED FOR JUPITER SWING-BY AND FLIGHT TO SATURN

12/2   PIONEER 11 JUPITER ENCOUNTER
1975   PIONEER 11 SWINGS INSIDE JUPITER ORBIT TO CROSS SOLAR SYSTEM TO SATURN BY 1979

1976

3/1 — PIONEER 10 CROSSES SATURN ORBIT

8/7 — PIONEER 11 REACHES HIGHEST POINT ABOVE ECLIPTIC (15.6° 162 MILLION km)

1977

4/20 — PIONEER 11 RECROSSES JUPITER ORBIT AND HEADS OUT TO SATURN

1978

1979

9/5   PIONEER 11 REACHES SATURN
        PIONEER 10 CROSSES ORBIT OF
1980   URANUS AND REACHES COMMUNICATION LIMIT

Encounter Time Line

Notes:
1. All times PST.
2. One Jupiter diameter = 142,744 kilometers.

11-02-74

8:00 a.m. Begin imaging (picture taking) and polarimetry 4 to 8 hours per day through 11-24-74. Imagery primarily to support photopolarimetry measurements. Ames operations personnel will run the University of Arizona Imaging Photopolarimeter until 11-18-74 when Arizona team will join them.

9:00 a.m. Conscan measurements are made every other day throughout the Jupiter encounter to verify pointing accuracy of spacecraft antenna at the Earth. Last change in antenna pointing direction before pericenter was October 9; next change December 6, 4 days after closest approach (pericenter).

11-07-74

4:30 a.m. Cross orbit of Jupiter's outermost moon, Hades, 23,632,000 kilometers (14,683,000 miles) from Jupiter – 165 Jupiter diameters from planet.

4:49 a.m. Cross orbit of Poseidon, second of the 4 outer moons, at 23,204,000 kilometers (14,417,000 miles) – 162 Jupiter diameters from planet.

11-08-74

7:50 p.m. Cross orbit of Pan, third of the 4 outer moons, at 22,276,000 kilometers (13,841,000 miles) – 156 Jupiter diameters from planet.

11-10-74

7:29 p.m. Cross orbit of Andrastea, closest of 4 outer moons, at 20,634,000 kilometers (12,820,000 miles) – 145 Jupiter diameters from planet.

11-18-74

8:00 a.m. University of Arizona team arrives to begin intensive imaging and photopolarimetry activity. Imaging and polarimetry operations will continue up to 8 hours per day through 11-24-74.

11-21-74

10:03 a.m. Cross orbit of Hera, outermost of Jupiter's middle group of moons, at 11,667,000 kilometers (7,248,000 miles) – 82 Jupiter diameters from planet.

10:50 a.m. Cross orbit of Demeter, second of Jupiter's 3 middle moons, at 11,639,000 kilometers (7,231,000 miles) – 81.5 Jupiter diameters from planet.

5:21 p.m. Cross orbit of Hestia, closest of Jupiter's 3 middle moons, at 11,403,000 kilometers (7,084,000 miles) – 80 Jupiter diameters from planet.

11-25-74

All day Eleven images of Jupiter, polarimetry of Jupiter, Callisto, Ganymede and Europa.

4:00 p.m. Begin 23 hours-a-day imaging and polarimetry for 14 days through December 9. Both imaging and polarimetry of Jupiter will occur every day through this period; imaging a little more than half the time.

11-25-74

8:00 a.m. Earliest time for bow-shock wave crossing, inbound.

11-26-74

All day Twenty-five images of Jupiter. Polarimetry of Jupiter and Io.

8:00 a.m. Earliest time for magnetopause crossing, inbound.

12:00 noon Resolution of pictures sent back by *Pioneer* equals that of typical Earth telescope pictures.

9:00 p.m. Earliest time for magnetopause crossing, inbound. (Time period when crossing likely is 19 hours, from 11-26-74, 9:00 p.m. to 11-27-74, 4:00 p.m.)

11-27-74

All day Seventeen images of Jupiter. Polarimetry of Jupiter, Io, and Europa.

9:21 p.m. *Pioneer* 5 days from pericenter, 5,905,000 kilometers (3,669,000 miles) – 41.5 Jupiter diameters away.

9:21 p.m. Planet occupies $^1/_{10}$th (1.4 degrees) of *Pioneer*'s 14 degrees field of view. Would have a 2-inch diameter on a 21-inch TV screen.

11-28-74

All day Twenty-two images of Jupiter. Polarimetry of Jupiter and Io.

9:21 p.m. *Pioneer* 4 days from pericenter, 4,919,000 kilometers (3,050,000 miles) – 34.5 Jupiter diameters from the cloud tops.

9:21 p.m. Planet occupies 1.6 degrees of *Pioneer*'s 14 degrees view field – 2½-inch diameter on a 21-inch TV screen.

11-29-74

All day Fifteen images of Jupiter. Polarimetry of Io, Europa, Jupiter.

9:21 p.m. *Pioneer* three days from closest approach, 3,895,000 kilometers (2,420,000 miles) – 27 Jupiter diameters from cloud tops.

9:21 p.m. Planet occupies $^1/_7$th (2.1 degrees) of *Pioneer*'s 14 degrees view field – 3-inch diameter on a 21-inch TV screen.

11-30-74
 All day Twenty-four images of Jupiter. Polarimetry of Ganymede, Callisto, and Jupiter.
 9:21 p.m. All pictures from now until 48 hours after pericenter better than typical Earth telescope pictures. Average resolution in this 96-hour period two to three times better than telescope pictures. During these 96 hours *Pioneer* will return 40 pictures of the full planet, many pictures of portions of Jupiter's surface, three of Callisto, one each of Ganymede and Io.

11-30-74
 9:21 p.m. Two days from pericenter. *Pioneer* is 2,813,000 kilometers (1,748,000 miles) – 20 Jupiter diameter from the cloud tops.
 9:21 p.m. Planet occupies 2.8 degrees of *Pioneer*'s 14 degrees view field – 4¼-inch diameter on a 21-inch TV screen.
 11:28 p.m. Ultraviolet photometer measurement of Callisto.

12-01-74
 All day Fourteen images of Jupiter, two images of Callisto. Polarimetry of Callisto, Io and Jupiter.
 11:26 a.m. Ultraviolet photometer measurement of Ganymede.
 5:27 p.m. Cross orbit of Callisto, outermost Galilean moon, at 1,812,800 kilometers (1,125,295 miles).
 9:21 p.m. *Pioneer* one day from pericenter, 1,617,000 kilometers (1,005,000 miles) – 11.5 Jupiter diameters from the cloud tops.
 9:21 p.m. Planet occupies one third (4.8 degrees) of *Pioneer*'s 14 degrees view field – 7¼-inch diameter on a 21-inch TV screen.
 9:21 p.m. Begin best pictures of Jupiter. During the 24 hours before and after pericenter when *Pioneer* is within one million miles of the planet, pictures are much better than any from Earth, the best ever made of Jupiter except those taken by *Pioneer 10*.
 10:18 p.m. Infrared measurements of Callisto, inbound.
 All day Eight images of Jupiter, one image of Callisto, one image of Ganymede. Polarimetry of Jupiter.
 12:21 a.m. Closest approach to Callisto, 786,500 kilometers (488,730 miles) at 21 hours from pericenter.
 1:50 a.m. Ultraviolet measurements of Ganymede, 19 hours 31 minutes from pericenter.
 8:01 a.m. Cross orbit of Ganymede, second outermost Galilean moon, at 1,001,140 kilometers (624,953 miles) – seven Jupiter diameters from cloud tops, 13 hours 33 minutes from pericenter.
 8:21 a.m. Last full disc picture of Jupiter. Subsequent pictures will more than fill spacecraft 14 degrees view field.
 1:56 p.m. Infrared measurement Ganymede, inbound – 7 hours 25 minutes before pericenter.
 2:06 p.m. Cross orbit of Europa, second closest Galilean moon, at 601,780 kilometers (372,803 miles) – 4.2 Jupiter diameters from the cloud tops, 7.25 hours from pericenter.

12-02-74
 2:09 p.m. Closest approach to Ganymede, 692,300 kilometers (430,195 miles) – 7.2 hours from pericenter.
 2:45 p.m. Ultraviolet measurement of Europa – 6 hours 36 minutes before pericenter.
 3:21 p.m. Enter radiation belt at 3.5 Jupiter diameters from clouds, six hours from closest approach.
 4:00 p.m. Begin two-hour scan for last picture on incoming trajectory thus showing Red Spot portion of Jupiter's surface. During picture, taken from 4:00 p.m.-6:00 p.m., *Pioneer* well inside radiation belt. Range 424,000 kilometers, resolution five times Earth telescope resolution.
 5:00 p.m. to
 7:45 p.m. Infrared measurements of Jupiter – 2 hours 45 minutes of measurements starting at 4 hours 21 minutes from pericenter.
 5:13 p.m. Infrared measurement of Amalthea – 4 hours 8 minutes from pericenter.
 5:23 p.m. Cross orbit of Io, innermost Galilean moon, at 352,560 kilometers (217,945 miles) – 2.45 Jupiter diameters from cloud tops, four hours from pericenter.
 7:02 p.m. Infrared measurements of Io at 2 hours and 19 minutes before pericenter.
 7:09 p.m. Closest approach to Io, 314.000 kilometers (195,120 miles), 2 hours ten minutes pericenter.
 7:58 p.m. Crossing of *Pioneer 10*'s previous closest approach distance.
 8:10 p.m. Cross orbit of Amalthea, closest Jovian moon, at 137,260 kilometers (68,630 miles) – 0.77 Jupiter diameters from cloud tops, 1 hour nine minutes before pericenter.
 8:15 p.m. Closest approach to Europa, 586,700 kilometers (364,575 miles), 1 hour six minutes before pericenter.
 9:00:21 p.m. Enter 33 minutes 31 seconds solar occultation – starts at 20 minutes, 58 seconds before pericenter.
 9:00:42 p.m. Enter Jupiter radio occultation (blackout) – duration 42 minutes 2 seconds. Starts 20 minutes 18 seconds before pericenter.
 9:22 p.m. Pericenter – *Pioneer* is 42,828 kilometers (26,613 miles) – 0.31 Jupiter diameters from cloud tops.
 9:33:52 p.m. Exit solar occultation,12 minutes, 23 seconds after pericenter.
 9:43:03 p.m. Exit Jupiter radio occultation, 21 minutes 44 seconds after pericenter.

12-02-74
 10:30 p.m. Closest approach to Amalthea, 127,500 kilometers (79,229 miles), 1 hour 8 minutes after pericenter.

10:52 p.m. Infrared measurement of Amalthea outbound – I hour 31 minutes after pericenter.

11:00 p.m. Start 4½ hours Jupiter viewing period, outbound, for infrared instrument. Begins I hour 39 minutes after pericenter.

12-03-74

All day Eleven images of Jupiter, one image of Io, polarimetry of Jupiter, Ganymede, and Callisto.

1:30 a.m. End 4½ hours Jupiter view period for infrared instrument outbound. Ends 4 hours 9 minutes after pericenter.

3:21 a.m. Exit radiation belt at 3.5 Jupiter diameters from cloud tops, six hours after pericenter.

7:58 a.m. Infrared measurement of Io, outbound, 10 hours 37 minutes after pericenter.

9:21 p.m. One day after pericenter. *Pioneer* is 1,617,000 kilometers (1,005,000 miles) – 11.3 planet diameters from Jupiter.

11:43 p.m. Infrared measurement of Ganymede, outbound, 26 hours 22 minutes after pericenter.

12-4-74

All day Fifteen images of Jupiter. Polarimetry of Jupiter, Io, Ganymede, and Europa.

9:21 p.m. Two days after pericenter, *Pioneer* is 2,813,000 kilometers (1,748,000 miles) from Jupiter.

9:45 p.m. Infrared measurement of Callisto, outbound, 48 hours 24 minutes after pericenter.

midnight Jupiter occupies ¹/5 of 14 degrees view field.

12-5-74

All day Eighteen images of Jupiter. Polarimetry of Europa, Ganymede, Callisto, Jupiter.

9:21 p.m. Three days after pericenter. *Pioneer* is 3,895,000 kilometers (2,420,000 miles) from Jupiter.

12-6-74

All day Ten images of Jupiter. Polarimetry of Jupiter and Ganymede.

12-06-74

5:30 a.m. to

1:30 p.m. First precession maneuver after pericenter to change pointing angle at the Earth of *Pioneer* radio antenna. Maneuver lasts 6 to 8 hours and will change pointing direction about two degrees.

9:21 p.m. Four days after pericenter. *Pioneer* is 4,919,000 kilometers (3,050,000 miles) from Jupiter.

12-07-74

All day Twenty-two images of Jupiter. Polarimetry of Jupiter, Ganymede, Io, Callisto.

noon Jupiter occupies ¹/10 of *Pioneer* view field.

9:21 p.m. Five days after pericenter *Pioneer* is 5,906,000 kilometers (3,669,000 miles) from Jupiter.

12-08-74

All day Twenty-eight images of Jupiter. Polarimetry of Jupiter, Ganymede, and Callisto.

noon Earliest time for magnetopause crossing, outbound – 20-hour period when crossing expected is 12:00 noon, 12-8-74 to 4:00 a.m., 12-9-74.

12-09-74

All day Nineteen images of Jupiter. Polarimetry of Jupiter.

3:00 p.m. End of 23-hour-per-day imaging and photopolarimetry. Return to four to eight-hours-per-day operation through 1-3-75. Arizona imaging photopolarimetry team goes home, but will return for full day of operations on 12-17-74, after today (12-9-74) imaging photopolarimeter will be run by Ames operations personnel, and will be used primarily for photopolarimetry not imaging.

8:00 p.m. Latest possible time for magnetopause crossing, outbound.

12-10-74

midnight Latest time for bow shocking crossing, outbound.

12-17-74

All day Fifteen images of Jupiter by Arizona-Ames team. Arizona then returns instrument to Ames operation. Imaging and polarimetry four to eight-hours-per-day through 1-3-75.

12-20-74

4:45 a.m. to

12:45 p.m. Second precession maneuver after pericenter to change pointing angle at the Earth of *Pioneer* radio antenna. Maneuver lasts six to eight hours, changes pointing direction about two degrees.

12-30-74

6:00 a.m. to

2:00 p.m. Third precession maneuver after pericenter to change pointing angle at the Earth of *Pioneer* radio antenna. Maneuver lasts six to eight hours, changes pointing direction about two degrees.

1-3-75

8:00 a.m. End of four-to-eight-hour-per-day imaging and photopolarimetry.

4:00 p.m. End encounter period.

Jupiter – Basic Characteristics

Size, Orbit, Rotation

Galileo made the first telescopic observations of Jupiter and discovered the four large moons in 1610.

From Earth, Jupiter is the second brightest planet, and the fourth brightest object, in the sky. It is 773 million kilometers (480 million miles) from the Sun, and circles it once in just under 12 years. The planet has 13 moons, including the one discovered in September; the four outer ones are in "backward orbit" compared to other known moons. The largest, Ganymede, has a ten percent greater diameter than the planet Mercury; Callisto, the next largest, is equal in size to Mercury. Two other, Io and Europa, are similar in size to the Earth's moon.

Jupiter spins once every 10 hours, the shortest day of any of the nine planets. Because of Jupiter's size, this means that a point at the equator on its visible surface (cloud tops) races along at 35,400 km/hr (22,000 mph), compared to a speed of 1,600 km/hr (1,000 mph) for a similar point on Earth.

Jupiter's visible surface (its cloud tops) cover 62 billion square kilometers (about 24 billion square miles). The planet's gravity at cloud top is 2.36 times that of Earth.

The mass of the planet is 318 times the mass of Earth. Its volume is 1,000 times Earth's. The planet is made up of a mixture of elements similar to that in the Sun or the primordial gas cloud which formed the Sun and planets. This means at least three quarters of it is the lightest gases, hydrogen and helium. Scientists have identified hydrogen, deuterium (the heavy isotope of hydrogen), helium, methane (carbon and hydrogen), and ammonia (nitrogen and hydrogen) in Jupiter's clouds.

Picture Time Schedule

During the 52 hours centered on periapsis, Pioneer will take 25 pictures of Jupiter, one each of Callisto, Ganymede, and Io.

Earth Receipt Times (PST)		Picture
December 1	7:57 p.m. - 8:47 p.m.	Jupiter
	8:47 p.m. - 9:28 p.m.	Jupiter
	9:28 p.m. -10:14 p.m.	Jupiter
	10:56 p.m. -11:45 p.m.	Jupiter
December 2	11:45 p.m. -12:22 a.m.	Jupiter
	1:20 a.m. - 1:51 a.m.	Callisto (moon)
	2:36 a.m. - 3:35 a.m.	Jupiter
	5:25 a.m. - 6:21 a.m.	Jupiter
	6:56 a.m. - 7:55 a.m.	Jupiter
	8:30 a.m. - 9:42 a.m.	Jupiter
	10:11 a.m. -11:19 a.m.	Jupiter (small part N cut off)
	11:27 a.m. -11:55 a.m.	Ganymede (moon)
	12:54 a.m. - 2:03 p.m.	Jupiter (small part S cut off)
	2:14 a.m. - 4:05 p.m.	Jupiter (small part N cut off)
	5:02 a.m. - 6:06 p.m.	Jupiter (10% N&S cut off)
	6:10 a.m. - 7:03 p.m.	Jupiter (10% N&S cut off)
December 3	12:54 a.m. - 1:51 a.m.	Jupiter ($1/3$ S cut off)
	1:55 a.m. - 2:57 am.	Jupiter ($1/3$ S cut off)
	3:54 a.m. - 5:00 a.m.	Jupiter (5% N&S clipped)
	5:34 a.m. - 7:21 a.m.	Jupiter
	7:50 a.m. - 9:11 a.m.	Jupiter
	9:20 a.m. - 9:51 a.m.	Io
	10:57 a.m. -12:11 p.m.	Jupiter
	1:03 p.m. - 2:27 p.m.	Jupiter
	3:02 p.m. - 3:51 p.m.	Jupiter
	7:00 p.m. - 8:00 p.m.	Jupiter
	8:00 p.m. - 8:25 p.m.	Jupiter
	11:00 p.m. -11:20 p.m.	Jupiter

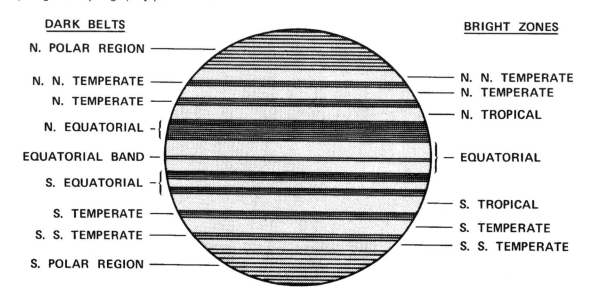

DARK BELTS   BRIGHT ZONES

N. POLAR REGION

N. N. TEMPERATE   N. N. TEMPERATE
N. TEMPERATE   N. TEMPERATE

N. TROPICAL

N. EQUATORIAL

EQUATORIAL BAND   EQUATORIAL

S. EQUATORIAL

S. TROPICAL

S. TEMPERATE

S. S. TEMPERATE   S. TEMPERATE

S. POLAR REGION   S. S. TEMPERATE

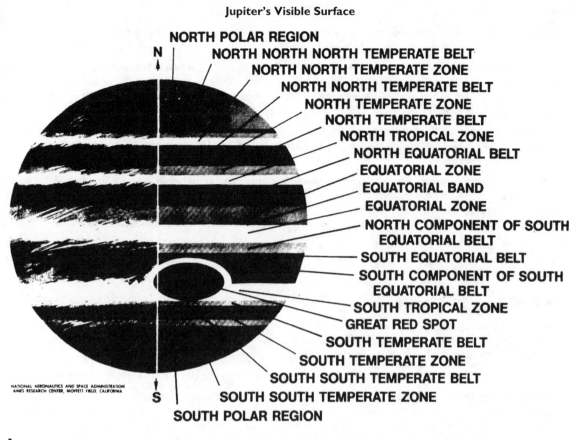

Jupiter's Visible Surface

Appearance

Jupiter has large dusky, gray regions at both poles. Between the two polar regions are five, bright salmon-colored stripes, known as zones, and four darker, slate-gray stripes, known as belts.

The Great Red Spot in the southern hemisphere is frequently bright red, and since 1665 has disappeared completely several times.

The gray polar regions may be condensed methane. The bright cloud zones have a complete range of colors from yellow and delicate gold to red and bronze. Clouds in the belts range from gray to blue-gray.

In addition to the belts and zones *Pioneer 10* found many smaller features – streaks, wisps, arches, loops, patches, lumps, and spots. Many of these are hundreds of thousands of kilometers in size.

Circulation of these features is vigorous. The entire Equatorial Zone, 20 degrees wide, sweeps around the planet 410 km/hr (225 mph) faster than the cloud regions on either side. The South Tropical current and a cloud current sweeping completely around the Great Red Spot are well-known.

Radio Signals and Heat

Earth receives more radio noise from Jupiter than from any other source except the Sun. Jupiter broadcasts three kinds of radio noise: (1) thermal – from the temperature-induced motions of the molecules in its atmosphere (wavelength, 3 centimeters); (2) decimetric (centimeter range) – from the gyrations of electrons around the lines of force of the planet's magnetic field (wavelength 3-70 centimeters); and (3) decametric (up to 10s of meters) – believed to be from huge discharges of electricity (like lightning flashes) in Jupiter's ionosphere (wavelength 70 centimeters to 60 meters).

The powerful decametric radio waves have been shown to be modulated by passages of Jupiter's close moon, Io, whose orbit is 2.5 planet diameters, 350,000 kilometers (about 217,000 miles), above Jupiter's cloud tops. Some scientists believe that Io links up magnetic lines of force in the planet's ionosphere, allowing huge discharges of built-up electrical potential.

The power of these decametric radio bursts is equal to the power of several hydrogen bombs. Their average peak value is 10,000 times greater than the power of Jupiter's decimetric signals.

Only about $1/27$th as much heat from the Sun arrives at Jupiter as reaches Earth.

Moons

Jupiter's 13 natural satellites have some odd characteristics. The second moon, Io, is the most reflective known object in the solar system. Infrared measurements show Europa and Ganymede to have water ice on their surfaces.

The inner moons include: tiny Amalthea, diameter 160 kilometers (100 miles), which orbits Jupiter twice a day at only 1.5 planet diameters, 106,000 kilometers (66,000 miles), above the cloud tops; the four large moons Io, Europa, Ganymede, and Callisto, whose orbits lie between 422,000 kilometers (262,000 miles) and 1,882,000 kilometers (1,169,000 miles) from Jupiter.

Beyond these are the seven tiny outer moons. The inner three of these, Hestia, Hera, and Demeter, have orbits which lie between 11.5 and 11.7 million kilometers (7.2 and 7.3 million miles) from Jupiter. An eighth small moon was discovered last September.

Orbits of the four outermost, Andrastea, Pan, Poseidon, and Hades, lie between 20.7 million and 23.7 million kilometers (12.9 and 14.7 million miles) from Jupiter. All are in "backward" orbits, counter to the direction of planet rotation. This suggests that they may be captured asteroids. Diameters of six of the outer moons range from 15 to 40 kilometers (9 to 24 miles), with Hestia, the seventh, having a diameter of about 130 kilometers (81 miles).

Orbital periods of the four larger inner moons range from 1.7 days (for Io) to 16.7 days. Orbital periods of the inner three of the outer seven moons are around 250 days. While the four farthest-out moons complete their retrograde orbits in around 700 days.

The View From Pioneer 10

Pioneer's target planet, Jupiter, is unique in the solar system. With more than two-thirds of all the planetary material in the solar system, it has more than 1,000 times the volume of the Earth. It has 13 moons, four around the size of the planet Mercury. By itself, the Jovian system is more than a planet. It constitutes a vast new region in space.

Scientists want to add to the spectacular discoveries made by *Pioneer 10*, just released. The findings indicate that:

- Jupiter is entirely liquid – a big, fast-spinning ball of liquid hydrogen without any detectable solid surface.
- Jupiter is very hot inside (at its core, six times hotter than the Sun's surface). This heat dates back to the formation of the planet, 4.5 billion years ago. Down-ranging densities of the four large inner moons (directly proportional to distance from the planet) seem to prove this early intense heat. The inner two, Io and Europa, seem to be rocky, with the outer two, Ganymede and Callisto, the density of water-ice and rock.
- The enormous Great Red Spot is a "permanent" 20,000-mile-wide hurricane which has been raging for at least 400 years.
- The place for a manned landing in the Jovian system (if one ever occurs) is the big moon Callisto which is outside the worst radiation and pops in and out of the radiation belts as they wobble 20 degrees up and down each ten-hour planet rotation.
- Jupiter could contain life. Its atmosphere contains ammonia, methane, hydrogen, and probably water, believed to have produced life on Earth about four billion years ago. Many scientists believe that vast regions below the clouds are at room temperature. Combined with Jupiter's abundant energy, these conditions could produce living organisms. However, rapid circulation of both atmosphere and interior of a hot, liquid planet means organisms would have to be fast-moving or short-lived.
- Jupiter's weather is fantastically different from Earth's, but has familiar "highs and lows" like Earth's. Instead of constantly moving and rotating like Earth cyclones, these features are permanent and stretch completely around the planet. They account for Jupiter's colorful bands.
- The bright zones are planet-girdling, rising columns of atmosphere, and the dark belts, descending atmosphere regions, stretched completely around the planet.
- The permanence of Jupiter weather features seems to result from its liquid character and internal heat source, which spreads heat evenly throughout the planet, not just at the equator as with solar heat on Earth.
- The big moons, Io, Ganymede, Europa, and Callisto all seem to have atmospheres. Pictures of Ganymede appear to show maria and highlands similar to the Moon's and Mars'.
- Jupiter's magnetic field is much larger than predicted. Its radiation belts are far more intense than many expected.
- Jupiter is a source of high-energy particle radiation, the only one in the solar system except the Sun.
- Jupiter apparently retains a mixture of elements similar to that in the Sun, predominantly hydrogen and helium.

Pioneer 11 hopefully will make major additions to these discoveries. Because of the polar pass, it can measure the thickness of both the radiation belts and of Jupiter's magnetosphere (space filled by the planet's magnetic field). Different interpretations of *Pioneer 10* data have protons (the heavy particles in the radiation belts) either declining inside 182,000 kilometers (114,000 miles), or else continuing to increase. This is a key question for determining the radiation hazard. Infrared and ultraviolet measurements of the big moons Callisto, Ganymede, Europa and Io should produce valuable new data.

Jupiter Atmosphere Model*
(Atmosphere Depth to Liquid Zone is 1,000 kilometers (600 miles))

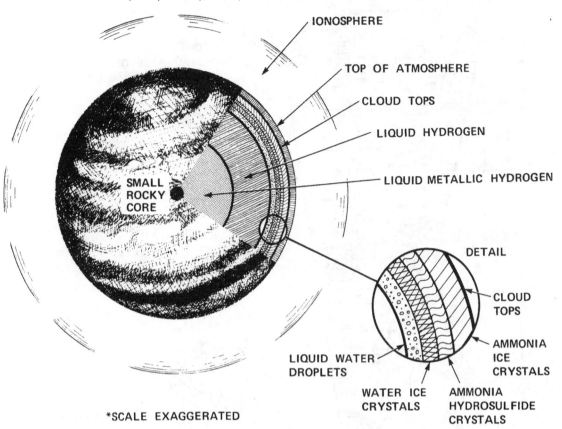

IONOSPHERE

TOP OF ATMOSPHERE

CLOUD TOPS

LIQUID HYDROGEN

LIQUID METALLIC HYDROGEN

SMALL ROCKY CORE

DETAIL

CLOUD TOPS

AMMONIA ICE CRYSTALS

LIQUID WATER DROPLETS

WATER ICE CRYSTALS

AMMONIA HYDROSULFIDE CRYSTALS

*SCALE EXAGGERATED

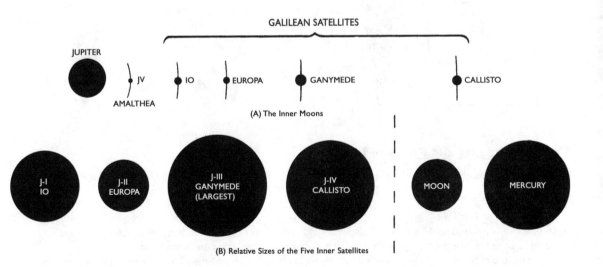

GALILEAN SATELLITES

JUPITER

JV
AMALTHEA

IO

EUROPA

GANYMEDE

CALLISTO

(A) The Inner Moons

J-I
IO

J-II
EUROPA

J-III
GANYMEDE
(LARGEST)

J-IV
CALLISTO

MOON

MERCURY

(B) Relative Sizes of the Five Inner Satellites

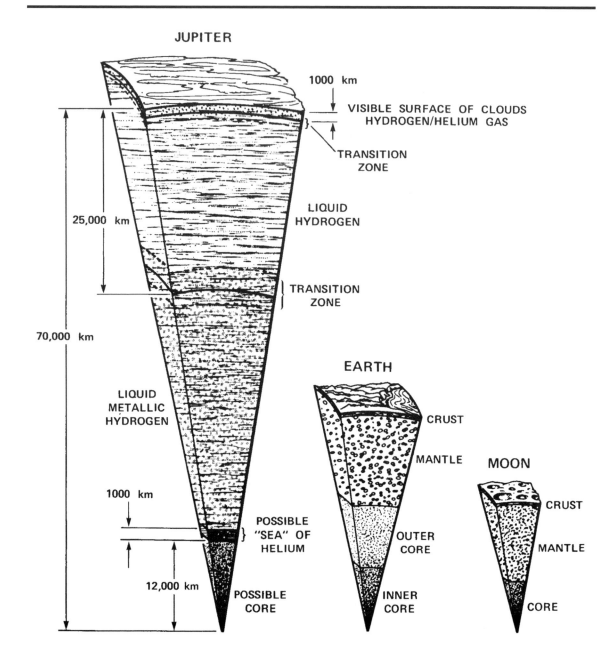

The Experiments and Projected Findings

The 14 *Pioneer 11* experiments are designed to produce findings about Jupiter and the heliosphere. Their objectives are:

For Jupiter:

- To map the magnetic field.
- To measure distributions of high-energy electrons and protons in the radiation belts, and look for auroras.
- To find a basis for interpreting the decimetric and decametric radio emissions from Jupiter.
- To detect and measure the bow shock and magnetospheric boundary and their interactions with the solar wind.
- To verify the thermal balance and to determine temperature distribution of the outer atmosphere.
- To measure the hydrogen / helium ratio in the atmosphere.
- To measure the structure of the ionosphere and atmosphere.
- To measure the brightness, color and polarization of Jupiter's reflected light.
- To perform two-color visible light imaging.
- To increase the accuracy of orbit predictions and masses of Jupiter and its moons.

For Interplanetary Space:

- To map the interplanetary magnetic field.
- To study the radial gradient of the solar wind and its fluctuations and structure.
- To study the radial and transverse gradients and arrival directions of high-energy charged particles (solar and galactic cosmic rays).
- To investigate the relationships between the solar wind, magnetic field and cosmic rays.
- To search for the boundary and extent of the heliosphere (solar atmosphere).
- To determine the density of neutral hydrogen.
- To determine the properties of interplanetary dust.

Pioneer / Jupiter Experiments

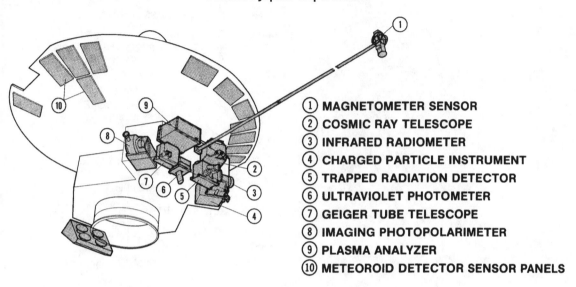

1. **MAGNETOMETER SENSOR**
2. **COSMIC RAY TELESCOPE**
3. **INFRARED RADIOMETER**
4. **CHARGED PARTICLE INSTRUMENT**
5. **TRAPPED RADIATION DETECTOR**
6. **ULTRAVIOLET PHOTOMETER**
7. **GEIGER TUBE TELESCOPE**
8. **IMAGING PHOTOPOLARIMETER**
9. **PLASMA ANALYZER**
10. **METEOROID DETECTOR SENSOR PANELS**

The instruments aboard *Pioneer 11* return data in roughly the following categories:

Magnetic Fields

Magnetometer (Jet Propulsion Laboratory). This sensitive instrument measures the interplanetary magnetic field in three axes from the orbit of Earth out to the limits of spacecraft communication. It will study solar wind interaction with Jupiter and map Jupiter's strong magnetic fields at all longitudes and many latitudes.

The instrument is a helium vector magnetometer. Its sensor is mounted on the lightweight mast extending 6.5 meters (21.5 feet) from the center of the spacecraft to minimize interference from spacecraft magnetic fields.

The instrument operates in any one of eight different ranges, the lowest covering magnetic fields up to +4 gammas; the highest, fields up to 140,000 gamma. Earth's surface field is 50,000 gamma.

These ranges are selected by ground command or automatically by the instrument itself. The magnetometer can measure fields as weak as 0.01 gamma.

Its sensor is a cell filled with helium, excited by radio frequencies and infrared optical pumping. Changes in the helium caused by fields the spacecraft passes through are measured. It weighs 2.6 kilograms (5.7 pounds) and uses up to 5 watts of power.

Fluxgate Magnetometer (Goddard Space Flight Center). This instrument is designed to sense strength and direction of Jupiter's magnetic field very close to the planet and up to very high values. It can measure field strength up to one million gamma in each of three perpendicular directions.

The instrument will be able to continuously measure the magnetic field along the spacecraft trajectory from about 12.6 Jupiter radii to the point of closest approach. It consists of two dual-axis sensors and their electronics. Each sensor is composed of a ring core, a magnetic multivibrator, a frequency doubler and two phase-sensitive detectors. The instrument weighs 0.6 pounds and uses approximately 0.36 watts of power.

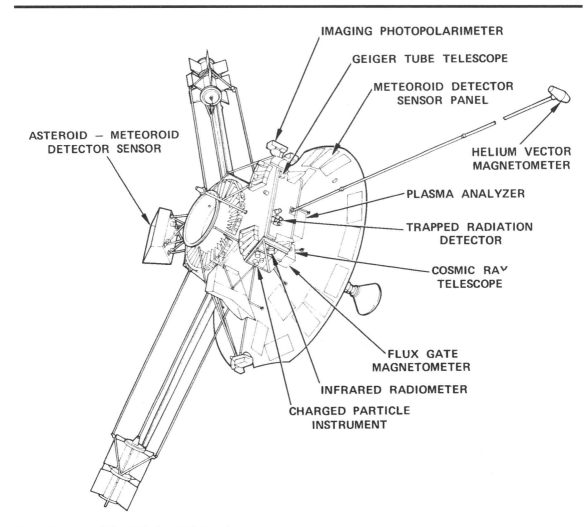

IMAGING PHOTOPOLARIMETER

GEIGER TUBE TELESCOPE

METEOROID DETECTOR
SENSOR PANEL

ASTEROID — METEOROID
DETECTOR SENSOR

HELIUM VECTOR
MAGNETOMETER

PLASMA ANALYZER

TRAPPED RADIATION
DETECTOR

COSMIC RAY
TELESCOPE

FLUX GATE
MAGNETOMETER

INFRARED RADIOMETER

CHARGED PARTICLE
INSTRUMENT

Interplanetary Solar Wind and Heliosphere

Plasma Analyzer (Ames Research Center). The plasma instrument maps the density and energy of the solar wind (ions and electrons flowing out from the Sun). It determines solar wind interactions with Jupiter, including the planet's bow shock wave.

The instrument consists of a high-resolution and medium-resolution analyzer. It will look toward the Sun through an opening in the spacecraft dish antenna, and the solar wind will enter like the electron beam in a TV tube. The instrument measures direction of travel, energy (speed) and numbers of ions and electrons.

The particles enter between curved metal plates and strike detectors, which count their numbers. Their energy is found by the fact that when a voltage is applied across the plates in one of 64 steps, only particles in that energy range can enter. Direction of particle travel is found from orientation of the instrument and which detector the particle struck.

In the high-resolution analyzer, the detectors are 26 continuous-channel multipliers, which measure the ion flux in energy ranges from 100 to 8,000 electron-volts. Detectors in the medium-resolution detector are five electrometers, which measure ions in ranges from 100 to 18,000 electron-volts and electrons from one to 500 electron-volts.

The plasma analyzer weighs 5.5 kilograms (12.1 pounds) and uses 4 watts of power.

Jupiter's Radiation Belts, Radio Emissions and Cosmic Rays

Charged Particle Composition Instrument (University of Chicago). This instrument has a family of four measuring systems. Two are particle telescopes primarily for interplanetary space. The other two measure trapped electrons and protons inside the Jovian magnetic field.

During interplanetary flight two telescopes identify the nuclei of all eight chemical elements from hydrogen to oxygen and separate the isotopes deuterium, helium-3, and helium-4. Because of their differences in isotopic and chemical composition and spectra, galactic particles can be separated from solar particles. The instrument also measures the manner in which streams of high-energy particles travel through interplanetary space. There is a main telescope of seven solid state detectors which measure from 1 to 500 million electron-volt (MeV) particles and a three-element telescope which measures 0.4 to 10 MeV protons, and helium nuclei. If a key detector element is destroyed in space, diagnostic procedures and commands from Earth can bypass it.

For the magnetosphere of Jupiter, two new types of sensors were developed to cope with the extremely high intensities of trapped radiation. A solid state electron current detector operating below -40°C (-40°F) measures only those electrons that generate the radio waves which reach Earth. The trapped proton detector contains a foil of thorium which undergoes nuclear fission from protons above 30 MeV, but is not sensitive to the presence of the intense electron radiation. The instrument weighs three kilograms (7.3 pounds) and uses 2.2 watts of power.

Cosmic Ray Telescope (Goddard Space Flight Center). The Cosmic Ray Telescope monitors solar and galactic cosmic ray particles. It tracks the twisting paths of high-energy particles from the Sun and measures bending effects of the solar magnetic field on particles from the galaxy. The instrument distinguishes which of the 10 lightest elements make up these particles. It also will measure particles in Jupiter's radiation belts.

The instrument consists of three three-element, solid state telescopes. A high-energy telescope measures the flux or protons between 56 and 800 MeV. A medium-energy telescope measures protons with energies between 3 to 22 MeV, and identifies the eight elements from helium to oxygen. The low-energy telescope studies the flux of electrons between 50,000 electron-volts and one MeV and protons between 50,000 electron-volts and 20 MeV.

The instrument weighs 3.2 kilograms (7 pounds) and uses 2.2 watts of power.

Jupiter's Charged Particles

Geiger Tube Telescopes (University of Iowa). This experiment measures Jupiter's radiation belts. The instrument employs seven Geiger-Mueller tubes to survey the intensities, energy spectra and angular distributions of electrons and protons along *Pioneer*'s path.

The tubes are small cylinders containing gas that generates electrical signals from charged particles.

Three tubes (considered a telescope) are parallel. Three others are in a triangular array to measure the number of multiparticle events (showers) which occur. The combination of a telescope and shower detector enables experimenters to compare primary with secondary events in the Jovian radiation belts.

Another telescope detects low-energy electrons (those above 40,000 electron-volts).

The instrument can count protons with energies above 5 MeV and electrons with energies greater than 50 KeV.

The instrument weighs 1.6 kilograms (3.6 pounds) and uses 0.7 watts of power.

Jovian Trapped Radiation Detector (University of California at San Diego). The nature of particles trapped by Jupiter, particle species, angular distributions and intensities is determined by this instrument. It measures a very broad range of energies from 0.01 to 100 MeV for electrons and from 0.15 to 350 MeV for protons (hydrogen nuclei).

Experimenters will attempt to correlate particle data with Jupiter's little understood radio signals.

The instrument has five detectors to cover the planned energy range. An unfocused Cerenkov counter, which measures direction of particle travel by light emitted in a particular direction, detects electrons of energy above one MeV and protons above 450 MeV. A second detector measures electrons at 100 KeV, 200 KeV and 400 KeV.

An omnidirectional counter is a solid state diode, which discriminates between minimum ionizing particles at 400 KeV and high-energy protons at 1.8 MeV.

Twin DC scintillation detectors for low-energy particles distinguish roughly between protons and electrons because of different scintillation material in each. Their energy thresholds are about 10,000 eV for electrons and 150,000 eV for protons.

The instrument weighs 1.7 kilograms (3.9 pounds) and uses 2.9 watts of power.

Jovian Dust, Meteoroids and Interplanetary Dust

Asteroid-Meteoroid Detector (General Electric Company). This instrument measures the orbits of the material in the vicinity of the spacecraft.

Four non-imaging telescopes can characterize objects, ranging from asteroids down to particles with a mass of one millionth of a gram, by measuring sunlight reflected from them. The telescopes measure particle numbers, sizes, velocity and direction.

Each of its four telescopes consists of a 20-cm (8-inch) mirror, an 8.4-cm (3.3-inch) secondary mirror, coupling optics, and a photomultiplier tube. Each telescope has an 8° view cone. The four cones overlap in part, and particle distance and speed are measured by timing entry into and exit from the cones.

The instrument is highly sensitive to hard radiation of the type expected at Jupiter, and may be damaged during encounter. It weighs 3.3 kilograms (7.2 pounds) and uses 2.2 watts of power.

Meteoroid Detector (Langley Research Center). To detect the distribution of particles (with masses of about 100-millionth of a gram or greater), scientists use a system of 234 pressure cells mounted on the back of the spacecraft dish antenna. The pressure cells come in 30 x 30-cm (12 x 12-inch) panels, 18 to a panel, with six panels. They have a total area of about one-quarter of a square meter. In construction, the panels are something like a stainless steel air mattress.

Each of the pressure cells is filled with a gas mixture of argon and nitrogen. When a particle penetrates a cell it empties of gas. A transducer notes this, counting one particle impact per cell.

Cell walls are 0.001-inch stainless steel sheet, thin enough to allow penetrates of particles with a mass of about 100-millionth of a gram or more.

The total weight of the instrument (panels and electronics) is 1.7 kilograms (3.7 pounds) and it uses 0.7 watts of power.

Celestial Mechanics

Celestial Mechanics (Jet Propulsion Laboratory). This experiment uses the spacecraft as an instrument to determine the mass of Jupiter and its satellites by measuring their effects on its trajectory.

Doppler tracking determines spacecraft velocity along the Earth-spacecraft line down to a fraction of a millimeter per second, once per minute. These data are further augmented by optical and radar position measurements of the planets. Computer calculations using the spacecraft trajectory and known planet and satellite orbital characteristics should verify *Pioneer 10* findings on Jupiter's mass, the masses of its four large moons, the planet's polar flattening and mass of its surface layers.

Jupiter's Atmosphere, Temperature, Moons, Interplanetary Hydrogen, Helium and Dust

Ultraviolet Photometer (University of Southern California). During flyby the ultraviolet photometer will measure scattering of ultraviolet (UV) light from the Sun by Jupiter's moons in two wavelengths, one for hydrogen and one for helium.

Radio telescope measurements show that the solar system is immersed in an interstellar gas of cold neutral hydrogen. By measuring the scattering of the Sun's ultraviolet light the instrument can measure amounts of this neutral hydrogen within the heliosphere.

The instrument has two detectors, one which measures ultraviolet radiation at 1216 angstroms, the other at 584 angstroms – the wavelengths at which hydrogen and helium scatter solar ultraviolet.

The instrument has a fixed viewing angle and will use the spacecraft spin to scan the planet. It weighs 0.7 kilograms (1.5 pounds) and uses 0.7 watts of power.

Pictures of Jupiter and Moons, the Planet's Atmosphere

Imaging Photopolarimeter (University of Arizona). This instrument provides data in a number of areas using photometry (measurement of light intensity) and polarimetry (photometry measurements of the linear polarization of light), and imaging.

The instrument will take pictures of Jupiter as the spacecraft approaches the planet.

Images will be made in both red and blue light, and these will be superimposed, providing "color pictures" using red and blue data plus amounts of green based on Earth telescope pictures.

The instrument uses a photoelectric sensor which measures changes in light intensity, something like the light sensor for a television camera. But unlike a TV camera it will employ the 5 rpm spin of the spacecraft to scan the planet in narrow strips 0.03° wide.

This electronic imaging system will complete the scans for a picture in from 25 to 115 minutes, depending on distance from the planet. Scan data are converted to digital form on the spacecraft and radioed to Earth by telemetry. Engineers then will build up the images using various computer techniques.

Pictures also will be taken of Jupiter's four large moons.

Controllers can vary the viewing angle of the instrument's 8.64-cm (3.4-inch)-focal length telescope by about 160° relative to the spacecraft's spin axis, which is fixed on the Earth-spacecraft line. The telescope can see to within 10 degrees of the spin axis, looking away from the Earth.

For the first pictures, the telescope will be looking almost down the spacecraft spin axis toward Jupiter. For the last pictures, close to the planet, it will be looking almost at right angles to the spin axis.

Picture reconstruction is complicated by the fact that picture scans will be curved as the telescope rotates with the spacecraft, while looking out at various angles to the spin axis. The only straight-line scans will be those looking straight out from the spin axis.

These two factors mean that computers will have to sort out scan paths whose curvature changes steadily with spacecraft movement. They will also have to compensate for smear caused by high-speed rotation of the planet and motion of the spacecraft.

Pioneer 11's record speed of nearly 110,000 mph is so fast that no pictures will be taken for the six hours of closest approach due to a severe "underlap" of spin scans.

During Jupiter flyby, experimenters also will use the instrument to study Jupiter's clouds, and to determine the nature of atmospheric gas above the clouds, and of aerosols in this gas. The instrument will measure light scattering of the Jovian satellites, to find their surface properties and atmospheres.

The instrument includes a 2.5-cm (1-inch) aperture, 8.6-cm (3.4 inch) focal length telescope which can be moved 160 degrees in the plane of the spacecraft spin axis by ground command or automatically. Incoming light is split by a prism according to polarization into two separate beams. Each beam is further split by going through a red filter (5,940-7,200 angstroms), and through a blue filter (3,900-5,000 angstroms). Channeltron detectors turn the light into electrical impulses, which are telemetered to Earth.

The instrument uses three viewing apertures: one 40 x 40 milliradians (mR) for zodiacal light measurements, a second 8 x 8 mR for non-imaging light measurements of Jupiter, and a third 0.5 x 0.5 mR for scans of the planet from which pictures will be reconstructed.

The instrument weighs 4.3 kilograms (9.5 pounds) and uses 2.2 watts of power.

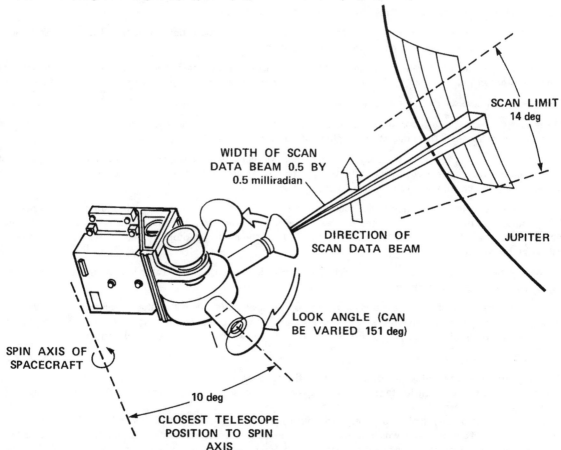

Jupiter's Atmosphere, Ionosphere, Temperature

Infrared Radiometer (California Institute of Technology). Experimenters will measure Jupiter's net heat energy output.

The two-channel radiometer will make measurements in the 14-25 and 29-56 micron wavelength regions to study the net heat energy flux, its distribution over the Jovian disc, and the thermal structure and chemical composition of Jupiter's atmosphere.

The radiometer uses a fixed telescope, so that its scans of the planet are made by rotation of the spacecraft. The instrument will view the planet during inbound and outbound periods of about two and a half hours each. It also will measure Callisto, Ganymede and Io – and make the first measurements of tiny Amalthea.

The instrument has a 7.2-cm (3-inch)-diameter Cassegrain telescope, and the detectors in its two channels are 88-element, thin film, bimetallic thermopiles. Its field of view is about 2,400 by 700 kilometers (1,500 by 420 miles) on Jupiter's cloud surface at closest approach of one Jupiter diameter. At this distance, it has a resolution of about 2,400 kilometers (1,500 miles)

The instrument weighs 2 kilograms (4.4 pounds) and uses 1.3 watts of power.

Occultation Experiment (Jet Propulsion Laboratory). Passage of the spacecraft radio signal through Jupiter's atmosphere as Pioneer swings behind the planet will measure Jupiter's ionosphere, and density of the atmosphere down to a pressure level of about one Earth atmosphere.

Experimenters will use computer analysis of the incoming radio signals recorded on tape to determine the refractive index profile of Jupiter's atmosphere.

Encounter Operations

During the two weeks of Pioneer 11's near encounter with Jupiter, the team of spacecraft controllers, analysts, and scientists at Ames Research Center, and personnel of NASA's Deep Space Network (DSN) will run a three-shift, 24-hour-a-day operation.

They will send about 10,000 commands to the spacecraft – along with the Pioneer 10 encounter, the most intensive command activity the DSN has ever handled. They will be constantly alert to analyze and recover from expected but unpredictable problems with spacecraft systems or with scientific instruments.

Their job will be complicated by the 82 minutes of round-trip communications time to Jupiter. They will wait an hour and a half between the sending of a command and verification of its execution.

The Pioneers are controlled primarily by instruction from Earth – not by operations sequences stored in onboard computers, though five commands can be stored on board. The spacecraft also can store up to 49,152 bits of science data.

Mission planners have simplified the encounter operation.

Operations strategy is to turn on, well in advance, most spacecraft subsystems and scientific instruments in a standard operating mode, and leave them this way throughout the encounter.

If all goes well, only a few routine spacecraft operations will need to be commanded. These are daily checks of pointing direction of the spin axis (spacecraft attitude), changes in telemetry data formats either twice or three times a day, and readjustments of spacecraft attitude on the third day after periapsis.

Only 8 of the 12 scientific instruments will require commands. Six of the eight will need only a few commands per day. Of the other two, the infrared radiometer will require several hundred commands near periapsis to scan Jupiter's five close moons. The bulk of the 10,000-plus commands will direct the imaging photopolarimeter in scanning the planet or its moons, for pictures of polarimetry.

Many types of problems have had to be anticipated. Six major rehearsals and training tests have been held since September, involving the entire worldwide operations team.

Controllers have tested all redundant backup systems on the spacecraft. With the large amount of DSN equipment, some failures are almost certain. Rehearsals have involved swaps of computers in case of failure, loss of high-speed data links, and coping with the 10-minute loss of command during switch-over between DSN stations.

At Jupiter, potential problems include:

- Damage to subsystems and scientific instruments by the radiation belts.
- Passage of Pioneer through Jupiter's strong magnetic field will set up eddy currents in the spinning spacecraft, creating magnetic torques and slowing spacecraft spin somewhat.
- Dust concentrations near Jupiter may mean high-velocity particle penetrations of Pioneer and damage to subsystems.
- The four scientific instruments which do not require commands include: the ultraviolet photometer, which is "on" throughout the encounter. (Its viewing periods result from the combination of the instrument's look angle and changes in spacecraft attitude as it moves around Jupiter.) Other noncommand instruments are: two of the four high-energy particle instruments and the meteoroid detector.

During encounter operations, the many other uncommanded functions on the spacecraft will be monitored in several ways. Pioneer returns complete data on its subsystems every 48 seconds. This is scanned by computer. With uncommanded changes, or functions beyond normal limits, the computer presents alarm signals on controllers' data displays.

In addition, analysts will review spacecraft data every 10 minutes.

Pioneer control and spacecraft operations are at the Pioneer Mission Operations Center (PMOC) Ames Research Center, Mountain View, California.

The PMOC has computing capability both for commanding the spacecraft and to interpret the data stream as it comes in front the DSN stations.

Several Ames organizations direct and support the encounter.

The Pioneer Mission Operations Team (PMOT) consists of personnel from many government and contractor organizations, and operates under control of the Flight Director.

In addition to several assistant flight directors, the PMOT includes the following groups:

- The Spacecraft Performance Analysis Team analyzes and evaluates spacecraft performance and predicts spacecraft responses to commands.
- The Navigation and Maneuvers Team handles spacecraft navigation and attitude control in space.
- The Science Analysis Team determines the status of the onboard scientific instruments and formulates command sequences for the instruments.

Tracking and Data Retrieval

With facilities located at 120 degree intervals around the Earth, NASA's Deep Space Network (DSN) will support *Pioneer* continuously during encounter. As the spacecraft "sets" at one station, due to the Earth's rotation, it will "rise" at the next one.

The DSN maintains three highly sensitive 64-meter (210-foot) dish antennas at Goldstone, California; Madrid, Spain; and Canberra, Australia. As a backup, DSN has six 26-meter (85-foot) dish antenna stations at Madrid (Robledo de Chavela and Cebreros); Canberra; and two stations at Goldstone.

The 26-meter stations have enough Power to command *Pioneer* at Jupiter. Where necessary they can receive data, but at the extremely low rate of 128 bits per second (bps), compared with the 1,024-bps reception rate of the 64-meter antennas.

At periapsis, there will be a 4-hour overlap between the Goldstone and Canberra 64-meter stations, with both receiving this vital data. However, Goldstone loses the spacecraft about 1 hour after closest approach. Canberra will handle the last part of periapsis operations. This will mean long-distance communications from Australia to Pioneer Control at Ames.

As *Pioneer* is drawn in by Jupiter's gravity, increases in speed will cause the doppler shift to grow very large. This will require "ramping" operations at tracking stations, complicated by communications cut-off when *Pioneer* passes behind the planet.

DSN stations will track the spacecraft on its 5-year flight to Saturn if the mission continues that long. At Saturn, round-trip communications time will be almost 3 hours.

All DSN stations have general purpose telemetry equipment capable of receiving, data-synchronizing, decoding and processing data at high transmission rates.

The tracking and data acquisition network is tied to the Mission Control and Computing Center (MCCC), JPL's central data processing facility at Pasadena, by NASA's Communications Network (NASCOM).

During encounter, some Ames controllers will be at the MCCC to provide a backup command capability in case of failure of high-speed data links between DSN stations and Ames.

For all of NASA's unmanned missions in deep space, *Pioneers* and *Mariners*, the DSN provides tracking information on course and direction of the flight, velocity, and range from Earth. It also receives engineering and science telemetry, including planetary television coverage, and sends commands for spacecraft operations. All communications links are in the S-band frequency.

DSN stations relay spacecraft doppler tracking information to the MCCC, where computers calculate orbits in support of navigation and planetary target planning. High-speed data links from all stations are capable of 4,800 bps, permitting real-time transmission of all data from spacecraft to Pioneer control centers at Ames or JPL. Throughout the mission, scientific data recorded on magnetic tape will be sent from DSN stations to Ames where it will be divided into individual magnetic tapes for each experimenter.

All of NASA's networks are under the direction of the Office of Tracking and Data Acquisition, NASA Headquarters, Washington, D.C. The Jet Propulsion Laboratory manages the DSN, while the STDN facilities and NASCOM are managed by NASA's Goddard Space Flight Center, Greenbelt, Maryland.

The Goldstone DSN stations are operated and maintained by JPL with the assistance of the Philco-Ford Corporation.

The Canberra station is operated by the Australian Department of Supply. The two facilities near Madrid are operated by the Spanish government's Instituto Nacional de Tecnica Aerospacial (INTA).

The Spacecraft

Pioneers 10 and *11* are the first spacecraft designed to travel into the outer solar system and operate there, possibly for as long as 7 years and as far from the Sun as 3.2 billion kilometers (2 billion miles).

For these missions, the spacecraft must have extreme reliability, be of very light weight, have communications systems for extreme distances, and employ non-solar power sources.

Pioneer 11 is identical to *Pioneer 10* except that a 12th onboard experiment has been added, a fluxgate magnetometer, to measure high fields very close to Jupiter. *Pioneer 11* is stabilized in space like a gyroscope by its rotation, so that its scientific instruments scan a full circle 5 times a minute. Designers chose spin stabilization for its simplicity and effectiveness.

Since the orbit planes of Earth and Jupiter coincide to about one degree, the spacecraft will be in or near Earth's orbit plane (the ecliptic) throughout its flight. To maintain a known orientation in this plane, controllers regularly adjust spacecraft position so that the spin axis points constantly at Earth. The spin axis coincides with the center line of the radio beam, which also points constantly at Earth.

Spacecraft navigation is handled on Earth using two-way doppler tracking and by angle tracking.

For course corrections, the *Pioneer* propulsion system can make changes in velocity.

The spacecraft can return a maximum of 1,024 data bits per second (bps) from Jupiter to the 64-meter (210-foot) antennas of the Deep Space Network.

Pioneer 11 is controlled largely from the Earth rather than by sequences of commands stored in onboard computers.

Launch energy requirements to reach Jupiter are far higher than for shorter missions, so the spacecraft is very light. *Pioneer 11* weighed only 270 kilograms (570 pounds) at launch. This included 30 kilograms (65 pounds) of scientific instruments.

For reliability, spacecraft builders have employed an intensive screening and testing program for parts and materials. They have selected components designed to withstand radiation from the spacecraft's nuclear power source, and from Jupiter's radiation belts. In addition, key systems are redundant. (That is, two of the same components or subsystems are provided in case one fails.) Communications, command and data return systems, propulsion electronics, thrusters and attitude sensors are largely redundant.

Virtually all spacecraft systems reflect the need to survive and return data for many years a long way from the Sun and the Earth.

Pioneer 11 Description

See *Pioneer G Description* in the **Pioneer 11 Press Kit**, page 105.

Orientation and Navigation

See *Orientation and Navigation* in the **Pioneer 11 Press Kit**, page 106.

Propulsion and Attitude Control

See *Propulsion and Attitude Control* in the **Pioneer 11 Press Kit**, page 106.

Nuclear-Electric Power

See *Nuclear-Electric Power* in the **Pioneer 11 Press Kit**, page 108.

Communications

See *Communications* in the **Pioneer 11 Press Kit**, page 109.

Command System

See *Command System* in the **Pioneer 11 Press Kit**, page 110.

Data Handling

See *Data Handling* in the **Pioneer 11 Press Kit**, page 110.

Timing

The digital telemetry unit contains a stable crystal-controlled clock for all spacecraft timing signals, including those for various orientation maneuvers and experiments.

Temperature Control

See *Temperature Control* in the **Pioneer 11 Press Kit**, page 110.

Post-Encounter Science – The Heliosphere

If its scientific instruments survive Jupiter's radiation belts, *Pioneer 11* will explore intensively a new segment of the heliosphere (the atmosphere of the Sun) after encounter. The spacecraft will survey a region of space 870 million kilometers (540 million miles) wide between the outer edge of the Asteroid Belt and the orbit of Saturn.

Pioneer also will study this region at points up to 15.6 degrees (160 million kilometers) above the ecliptic (Earth's orbit plane), far higher than measurements previously made.

The thinly diffused solar atmosphere is hundreds of times less dense than the best vacuums on Earth. Yet it is important because it contains:

• The ionized gas known as the solar wind, roughly a 50-50 mixture of protons (hydrogen nuclei) and electrons. It
  flows out from the 3,600,000 degrees F (2,000,000 degrees C) corona of the Sun in all directions at average speeds
  of 1.6 million km/hr (one million mph).
• Complex magnetic and electric fields, carried out from the Sun by the solar wind.
• Solar cosmic rays, high-energy particles thrown out by the huge explosions on the Sun's surface at up to 480 million
  km/hr (300 million mph).

The heliosphere also encompasses comets and dust. It is traversed by electromagnetic radiation from the Sun: radio waves, infrared, ultraviolet, and visible light.

The heliosphere further contains cosmic ray particles from within and beyond our galaxy, traveling at nearly the speed of light. This means enormous particle energies, up to 10^{14} million electron-volts (MeV). (10^{14} is 1 followed by 14 zeros.) There are also neutral hydrogen atoms from the interstellar gas which formed the Sun and planets.

Study of these phenomena has many applications. Storms of solar particles striking Earth interrupt radio communications and sometimes electric power transmission.

There is evidence that solar storm particles travel through the heliosphere and trigger the Earth's long-term weather cycles.

The heliosphere can be thought of as a huge laboratory where phenomena occur that cannot be simulated on Earth. For example, man cannot accelerate particles in Earth laboratories to the near-light speeds reached by galactic cosmic ray particles. These particles are observed by *Pioneer* instruments.

Solar Wind, Magnetic Field, and Solar Cosmic Rays

Near the Earth, the speed of the solar wind varies from one to three million km/hr (600,000 to 2,000,000 mph), depending on activity of the Sun. Its temperature varies from 10,000 to 1,000,000 degrees C (18,000 to 1,8000,000 degrees F). Near the Earth, collisions between streams of the solar wind use up about 25 percent of its energy. The wind also fluctuates due to features of the rotating solar corona, where it originates, and because of various wave phenomena.

The fastest solar cosmic ray particles jet out from the Sun in streams which can cover the 150 million kilometers (93 million miles) to Earth in as little as 20 minutes. Slower particle streams take one or more hours to reach Earth. Both types tend to follow the curving interplanetary magnetic field.

The positive ions from the Sun are 90 percent protons and ten percent helium nuclei, with occasional nuclei of heavier elements.

There are from none to hundreds of flare events on the Sun each year which produce high-energy solar particles, with the largest number of flares at the peak of the 11-year cycle.

The best resolution yet seen of elements making up the solar high-energy particles has been obtained by *Pioneer 10*. It identified sodium and aluminum for the first time, determined relative abundances or helium, carbon, nitrogen, oxygen, fluorine, neon, sodium, magnesium, aluminum and silicon nuclei coming from the Sun.

The solar wind, as viewed near the Earth's orbit, is most dramatically seen as containing high speed, up to 2,520,000 km/hr (1,566,000 mph) streams embedded in slower speed, 1,000,000 km/hr (624,000 mph) streams. These high-speed streams are presumably caused by high temperature regions in the solar corona. Between the Earth and Jupiter these high-speed streams speed up the gas ahead of them and they in turn are slowed down. This accounts for the large amount of turbulence in the solar wind and magnetic field still observed as far out as 510 million kilometers (420 million miles). In addition, this stream-stream interaction also produces thermal energy so that the solar wind does not cool as expected for a simple expansion model. Present estimates indicate that the solar wind will ultimately become steady (constant speed) somewhere beyond the orbit of Jupiter, perhaps as far out as the orbit of Saturn.

The Interstellar Gas

Pioneer 11 will measure neutral helium and hydrogen. Both hydrogen and helium atoms are believed to be part of the interstellar gas which has forced its way into the heliosphere as the solar system moves through interstellar space at 72,000 km/hr (45,000 mph).

The *Pioneer* ultraviolet instruments have been able to locate concentrations of neutral hydrogen and helium and to make separations between gas originating in the Sun and that penetrating the solar system from interstellar space.

The experimenters have measured the ultraviolet light glow emitted by interplanetary neutral hydrogen and helium atoms. The interstellar neutral hydrogen appears to enter the heliosphere in the plane of the Earth's orbit at around 100,000 km/hr (62,000 mph). Surprisingly, this entry point is about 60 degrees away from the direction of travel of the solar system through interstellar space, and hence the direction from which these particles ought to come.

Galactic Cosmic Rays

Galactic cosmic ray particles usually have far higher energies (velocities) than solar cosmic rays. These particles may get their tremendous energies from the explosion of stars (supernovas), the collapse of stars (pulsars), or acceleration in the colliding magnetic fields of two stars. Pioneer studies of these particles may settle questions of their origin in our galaxy and important features of the origin and evolution of the galaxy itself. These studies should answer such questions as the chemical composition of stellar sources of cosmic ray particles in the galaxy.

Galactic cosmic rays consist of protons (hydrogen nuclei), 85 percent; helium nuclei, 13 percent; nuclei of other elements, 2 percent; and high-energy electrons, 1 percent.

Because of the expected decrease in shielding by the solar wind and magnetic field, numbers of low-energy cosmic ray particles (0 to 100 million electron-volts) were expected to increase out to Jupiter. This did not occur, apparently because despite the decline of solar wind density and magnetic field strength, turbulence remained as high as near Earth. This turbulence apparently can shut out a large part of the low-energy galactic particles from the inner solar system, perhaps as far out as beyond Jupiter.

The Experimenters

See *Experiments and Investigators* in the **Pioneer 11 Press Kit**, page 117.

The Encounter Team

NASA Headquarters

Office of Associate Administrator

Dr. Rocco A. Petrone	Associate Administrator
Dr. John Naugle	Deputy Associate Administrator

Office of Space Science

Dr. Noel W. Hinners	Associate Administrator for Space Science
John Thole	Deputy Associate Administrator for Space Science
Robert S. Kraemer	Director, Planetary Programs
Fred D. Kochendorfer	Pioneer Program Manager
Dr. Albert G. Opp	Pioneer Program Scientist

Office of Tracking and Data Acquisition

Gerald M. Truszynski	Associate Administrator for Tracking and Data Analysis
Arnold C. Belcher	Network Operations
Maurice E. Binkley	Network Support

Ames Research Center

Dr. Hans Mark	Center Director
C.A. Syvertson	Deputy Director
John V. Foster	Director of Development
Charles F. Hall	Pioneer Project Manager

Dr. John H. Wolfe	Pioneer Project Scientist
Robert R. Nunamaker	Deputy Pioneer Project Manager
Ralph W. Holtzclaw	Pioneer Spacecraft System Manager
Joel Sperans	Pioneer Experiment System Manager
Richard Spahr	Management Control
Robert U. Hofstetter	Mission Operations Manager
Norman J. Martin	Chief, Flight Operations
Thomas L. Bridges	Chief, Data Systems
Arthur C. Wilbur	Nuclear Power
John J. Hurt	Contracts
Henry Asch	Reliability and Quality Assurance
John W. Dyer	Chief, Mission Analysis

Deep Space Network, Jet Propulsion Laboratory

Dr. William E. Pickering	Director, Jet Propulsion Laboratory
Dr. Nicholas A. Renzetti	Tracking and Data Systems
Alfred J. Siegmeth	Pioneer Project Support Team Manager
William E. Kirhoffer	Navigation

AEC Space Nuclear Systems Division

David S. Gabriel	Division Director
Glenn A. Newby	Associate Director, Space Nuclear Systems Division
Harold Jaffe	Manager, Isotope Flight Systems Office

TRW Systems Group

| Bernard J. O'Brien | Pioneer Project Manager |

The Bendix Corporation

| Walter L. Natzic | Pioneer Program Manager |

Pioneer 11 Contractors

See *Pioneer F* Contractors in the **Pioneer 10 Press Kit**, page 44.

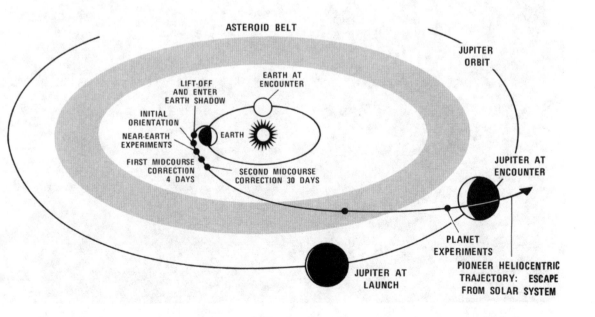

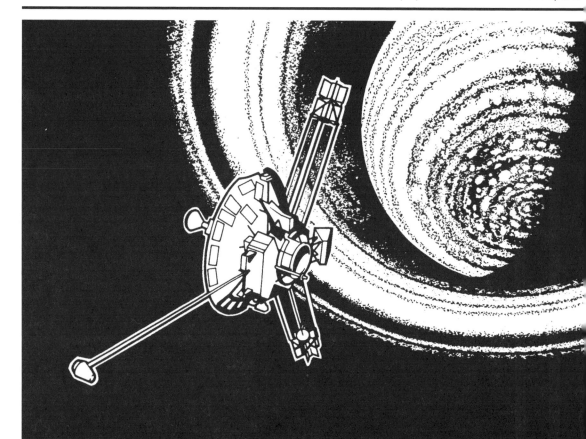

Pioneer Saturn
Encounter
Press Kit

NASA

Washington D. C. 20546

RELEASE No: 79-108

CONTENTS

First Decade...
Lunar
Landing
1969-1979

NASANews

National Aeronautics and
Space Administration

Washington, D.C. 20546
AC 202 755-8370

August 15, 1979

FOR RELEASE:
WEDNESDAY
August 22, 1979

Nicholas Panagakos
Headquarters, Washington, D.C.
Phone: (202) 755-3680

Peter Waller
Ames Research Center, Mountain View, California
Phone: (415) 965-5091

RELEASE No: 79-108

Pioneer to Encounter Saturn on September 1

NASA's *Pioneer 11* spacecraft will reach the giant ringed planet Saturn on September 1, after a six-year, three-billion kilometer (two-billion-mile) journey across the solar system.

Pioneer 11 will take the first close-up pictures and make the first close measurements of Saturn, its mysterious rings and several of its 10 satellites, including the planet-sized Titan.

Sweeping through the most intense part of Saturn's massive radiation belts, *Pioneer* will come within 21,400 kilometers (13,300 miles) of the planet and as close as 1,900 kilometers (1,200 miles) to the Saturnian rings.

Closest approach will be at 2 p.m. EDT, September 1.

The images returned by *Pioneer 11* (also called *Pioneer Saturn*) are expected to provide five to six times more detail of the planet than the best pictures taken from Earth.

Information returned by the spacecraft is expected to contribute to a better understanding of the origin and evolution of the Sun and planets. This, in turn, should provide scientists with a greater knowledge of our own Earth.

Data obtained by *Pioneer 11* will also be useful in planning the encounters of *Voyager 1* and 2 with the ringed planet in 1980 and 1981. The *Voyagers* are now Saturn-bound after encounters with Jupiter in March and July respectively.

Saturn is the second largest planet in the solar system, only slightly smaller than Jupiter. It is an immense diffuse body whose volume is 815 times that of Earth, but whose mass is only 95 times greater.

This makes Saturn the least dense planet in the solar system, lighter on the average than water. Like Jupiter, Saturn appears to be composed mainly of hydrogen and helium and to behave in many ways like a liquid planet.

Appearing to the Earth observer as a bright yellowish star of the first magnitude, Saturn is the sixth planet from the Sun, around which it revolves at an average distance of 1.4 billion kilometers (886 million miles). It completes one orbit around the Sun every 29½ years. If the years are long, the days are short – it rotates on its axis every 10 hours 14 minutes.

Difficult to observe in detail from Earth because of its great distance, Saturn is known to have a banded structure which is probably due to cloud systems similar to those of Jupiter.

Saturn's most distinctive trademark is its elaborate ring system – believed to consist of ice, ice-covered rocks or ice imbedded with rocks. The planet's rings are one of the great spectacles of the solar system. Discovered by Galileo in 1610, the rings range outward about 77,000 kilometers (48,000 miles) from Saturn's cloud tops. Total width of the three visible rings is 64,800 kilometers (40,300 miles).

The rings are very thin, with estimates of thickness ranging from several meters to as much as three kilometers (two miles). The rings are so thin that they are nearly invisible when viewed edge-on from Earth. Astronomers have determined from Earth that the orbiting ring particles are in part, and probably mostly, water ice. They may also contain metals, perhaps from the core of a broken-up moon. There is little rock. Most authorities think the particles range from snowball-sized to automobile-sized pieces with a few having a diameter of a mile or more.

The rings are believed to have originated either from the capture of a "wandering moon," which was then torn apart by Saturn's gravity, or – more likely – as a result of gravitational forces that prevented formation of a close-in satellite from the original planetary nebula.

Until the recent discovery of thin rings around Jupiter and Uranus – the two giant planets that flank Saturn – the Saturnian ring system was believed to be unique.

The Saturn encounter is the second planetary surveillance task for *Pioneer 11*. On December 2, 1974, it skimmed within 42,760 kilometers (26,725 miles) of Jupiter's cloud tops.

This close passage permitted the gravity and motion of Jupiter to act as a slingshot, increasing the spacecraft's velocity and placing it on the trajectory which is now carrying it toward its close encounter with Saturn.

The trajectory from Jupiter to Saturn is about three times as long as the first leg of *Pioneer*'s outbound journey from Earth to Jupiter. *Pioneer*'s trajectory carried it high above the plane of the ecliptic in which solar system planets revolve around the Sun, reaching a maximum height of about 160 million kilometers (100 million miles) in 1976.

As *Pioneer* approaches Saturn, its trajectory lies above the ring plane and, since the ring plane is tipped up relative to the direction of the Sun, the plane is lighted from below. The view from *Pioneer* will therefore be of a ring plane lighted from the other side and this should permit optical measurements of ring structure never before possible.

Picture resolution of Saturn will begin to exceed that from Earth on August 26 as the spacecraft approaches the planet at a distance of five million kilometers (three million miles). After August 30, *Pioneer* will be too close to Saturn to image the full planet with its rings.

At about noon on August 31, the last picture of Saturn's full disc, without the rings, will be made. Two hours before closest approach to Saturn itself, *Pioneer* will make its most detailed picture, resolving cloud features 80-100 kilometers (50-60 miles) in size.

A critical event will occur on the morning of September 1, when *Pioneer* – traveling at a speed of 85,000 km/hr (53,000 mph) – passes through the plane of Saturn's rings at a very shallow angle (4.7 degrees) on its descent toward the planet. Planetary debris could be in the area. Even impact with a small fragment could disable the spacecraft. Ring plane passage will take about 0.8 seconds.

As it rushes in toward the planet, *Pioneer Saturn* will approach from the north and cross outside the edge of Saturn's outer ring at a distance of 34,600 kilometers (21,500 miles). It will then skim in under the rings, from 2,000 to 10,000 kilometers (1,200 to 6,200 miles) below them. At the point of closest approach, on the planet's night side, *Pioneer* will come within 21,400 kilometers (13,300 miles) of Saturn's banded cloud tops.

For an hour and 20 minutes, *Pioneer* will make the historic first passage under Saturn's rings, making close-up optical measurements to determine their structure and other first-time observations.

The spacecraft will be moving too fast for good ring pictures while beneath the rings.

When *Pioneer* makes its closest approach to Saturn's cloud tops, just before it passes behind the planet, it will be traveling 114,100 km/hr (71,900 mph). Closest approach will occur at 12:34 p.m. EDT spacecraft time. Earth-received time of spacecraft signals at NASA's Ames Research Center, Mountain View, California, which manages the *Pioneers*, will be 2 p.m. EDT. Almost 1½ hours are required for radio signals to cover the Earth-Saturn distance of 1.4 billion kilometers (960 million miles). The spacecraft will make another shallow-angle crossing of Saturn's ring plane on its ascent away from the planet.

On September 2, *Pioneer* will make its closest approach to Titan, Saturn's largest moon, taking pictures and making measurements of the satellite from 356,000 kilometers (220,700 miles) away.

Titan is the largest known planetary satellite in the solar system, larger than Mercury. It appears to be wrapped in opaque orange smog and has a methane-containing atmosphere which may be as dense as Earth's at the surface, and is far denser than Mars'. Titan's hydrocarbon atmosphere is thought to be very similar to the primordial atmosphere of Earth and may well have produced organic molecules, the building blocks of life.

Like Jupiter, Saturn has white, yellow and yellow-brown banded cloud belts which appear to be the tops of planet-circling atmospheric streams, driven by the ringed planet's fast, 10-hour spin and by an internal heat source.

Pioneer 11 is expected to return approximately 150 pictures of Saturn, with the first transmission scheduled for August 20 at a distance of about one million kilometers (634,000 miles) from the planet.

About two-thirds of the 150 images expected by project scientists at the Ames Center will show more detail than is possible through observations made from Earth. These high-resolution photographs will be received at Ames from August 26 until September 8. The smallest object visible will be about 95 kilometers (50 miles) in diameter.

Travel time for spacecraft data being transmitted back to Earth at the speed of light will be 86 minutes, as compared with 52 minutes required for *Voyager 2* data to reach Earth from Jupiter in July 1979.

After passing the planet, *Pioneer Saturn* will head out of the solar system, traveling in roughly the same direction that the solar system does with respect to the local stars in our galaxy.

A sister spacecraft, *Pioneer 10*, flew past Jupiter for the first time in history in December 1973, and is now on its way out of the solar system, the first manmade object to do so. *Pioneer 10*'s solar system exit path is almost opposite to *Pioneer Saturn*'s, and it has already reached the orbit of Uranus.

On the outside chance that the *Pioneers* may be captured by an intelligent species during their nearly endless travel among the stars, each carries a plaque with a message about Earth.

Spacecraft operations during encounter will be complicated by the two-hour, 52-minute round-trip communication time and the need to send roughly 10,000 commands to the spacecraft in the two weeks centered on closest approach, most commands going to the spacecraft imaging system. Operations strategy is to set most systems in one standard mode throughout the encounter.

Pioneer Saturn weighs 260 kilograms (570 pounds) and is spin-stabilized, giving its instruments a full circle scan 7.8 time a minute. It uses nuclear sources for electric power because sunlight at Jupiter and beyond is too weak for an efficient solar-powered system.

Pioneer's 2.75-meter (9-foot) dish antenna looks back at Earth throughout the mission – adjusting its view by changes in spacecraft attitude as the spacecraft and home planet move in their orbits around the Sun.

Pioneer carries a 30-kilograms (65-pounds) scientific payload of 11 operating instruments. Two other experiments use the spacecraft and its radio signal as their instruments.

At Saturn, *Pioneer*'s instruments are expected to:

- Determine whether the planet has a magnetic field and belts of high-energy charged particles, and the effects of the rings on such belts.
- Provide a temperature and density profile from the planet's cloud tops to its core.
- Determine the presence of an internal heat source; how much heat Saturn is radiating; and effects of this heat on internal and atmospheric circulation.
- Determine the masses of Saturn's two outer rings. Are the rings made mostly of ice, or perhaps metals?
- Determine the existence and character of the hypothesized outer "E-ring," which could double total ring width, and pose a hazard to *Pioneer*, the two *Voyagers* and other future spacecraft.
- Measure light intensity and polarization from the mysterious ninth moon, Iapetus, which is six times as bright on its leading hemisphere as its trailing one; first findings of mass and density of the moon, Rhea; and confirm and improve these measurements for Titan.
- Characterize Titan's atmospheric aerosols (smog), its atmosphere structure and temperature, and take its picture.
- Measure Saturn's hydrogen / helium ratio (its two main constituents) and determine atmospheric structure, layering, and heat distribution, both horizontally and vertically.
- Study structure, temperature and flow of the atmosphere's belt and zone system, and obtain data related to atmosphere composition.
- Make at least 50 close-up color pictures of Saturn, its rings and cloud tops.

The Pioneer 10 and 11 projects are managed by Ames. The two spacecraft were built by TRW Systems, Redondo Beach, California.

(End of General Release. Background information follows.)

(Note: All times in the Background Information section of the Press Kit are Pacific Daylight Time – PDT.)

Encounter Profile

Encounter Sequence

The two-month *Pioneer 11* encounter period, August 2 to October 1, will proceed as follows:

- Beginning August 2, the spacecraft, 1.5 billion kilometers (932 million miles) from Earth, will be tracked via 64-meter (210-foot) dish antennas in Spain and Australia for about 10 hours a day. *Pioneer* will be 25 million kilometers (15 million miles) from Saturn, traveling toward it at 30,600 km/hr (19,000 mph) relative to the planet.
- On August 6, when tracking time increases to 18 hours a day and the station at Goldstone, California joins the team of tracking stations, *Pioneer* will be 1.5 million kilometers (939 million miles) from Earth. With 20 million kilometers (12 million miles) left to reach Saturn, all science instruments will be undergoing checks in preparation for the encounter. As the spacecraft moves toward Saturn at 31,000 km/hr (19,200 mph), the imaging photopolarimeter begins taking polarimetry measurements of Saturn.
- The ultraviolet photometry instrument will begin four days of Saturn observations on August 16. On the 17th, with *Pioneer* 1.53 billion kilometers (949 million miles) from Earth, tracking stations begin 24-hour tracking of the spacecraft, as it travels toward Saturn at 31,300 km/hr (19,400 mph). At this point, *Pioneer* will be 12.7 million kilometers (8 million miles) from Saturn.
- Encounter activity intensifies on August 20, when the imaging photopolarimeter begins taking images of Saturn. These first pictures are not expected to show much detail since the spacecraft will still be 10.2 million kilometers (6.3 million miles) from the planet. During the encounter, *Pioneer* will transmit more than 100 images of Saturn, its rings and the planet-sized moon Titan.
- On August 24, photopolarimetry measurements will be made of the Saturnian moon Iapetus.
- On August 25, with *Pioneer* 1.55 billion kilometers (959 million miles) from Earth, two-station tracking will begin, doubling the rate of data transmission and substantially improving image quality. At this point, 5 million kilometers (3 million miles) from Saturn, picture resolution will begin to be better than anything obtained from Earth-based telescopes.

Pioneer Saturn Voyage

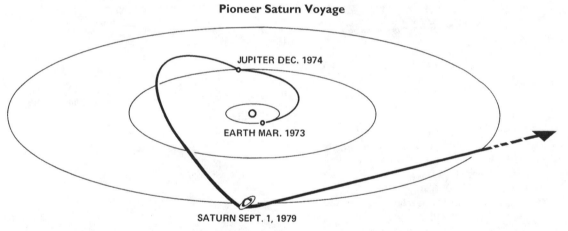

JUPITER DEC. 1974

EARTH MAR. 1973

SATURN SEPT. 1, 1979

Most of the images of the rings obtained during the mission will show the rings lighted from the other side, thus providing optical measurements not obtainable from Earth.

- On August 27, hurtling toward Saturn at 33,100 km/hr (20,500 mph), *Pioneer* will pass Phoebe, the moon farthest from Saturn, at a distance of 9,453,000 kilometers (5,860,860 miles). Scientists estimate *Pioneer* could cross the bow shock of Saturn on August 27. (The location of the bow shock at Jupiter was found to be very variable and dependent on solar wind intensity.)
- On the 28th, *Pioneer* will make the first ultraviolet photometer measurements of Saturn's large moon Titan, as well as added photopolarimetry measurements of Iapetus, Saturn's mysterious ninth moon, which is six times brighter on one side than the other. Iapetus will be encountered at a distance of 1,039,000 kilometers (644,180 miles) as *Pioneer* flies by at 33,500 km/hr (20,800 mph). At that time, *Pioneer* will be 3.3 million kilometers (2 million miles) from Saturn. More light measurements (photopolarimetry) will be made of the satellite on August 29. Also on that day, ultraviolet photometry measurements will be made of the three moons, Hyperion, Tethys and Dione.
- *Pioneer* will continue moving closer to Saturn, sending back full-disc pictures of the planet and its rings until August 30, when the spacecraft will be 2 million kilometers (1.3 million miles) from Saturn, too close to take images of the planet and rings together. The last full-planet picture of the inbound journey will be taken from 1 million kilometers (700,000 miles) away on August 31 at about 1:55 a.m. PDT and probably will be the best full-planet view.
- In the early morning hours of August 31, *Pioneer 11* will pass the eighth Saturnian satellite, Hyperion (diameter 224 kilometers or 140 miles) at a distance of 674,000 kilometers (417,880 miles), as the spacecraft speeds toward Saturn

at 37,800 km/hr (23,400 mph). At this point, *Pioneer* will be too close to the planet to get full-disc views. Photopolarimetry measurements will be made of Hyperion on August 30 and 31 and of Rhea on the 31s. Ultraviolet photometer measurements of Rhea will be made on August 30. On the 31st, the instrument will measure light from five satellites: Titan, Rhea, Dione, Tethys and Enceladus. Ultraviolet measurements of Saturn' rings also will be made on August 31. Ten hours before closest approach on September 1, the photometer will begin 14 hours of measurements of Saturn's ultraviolet light.

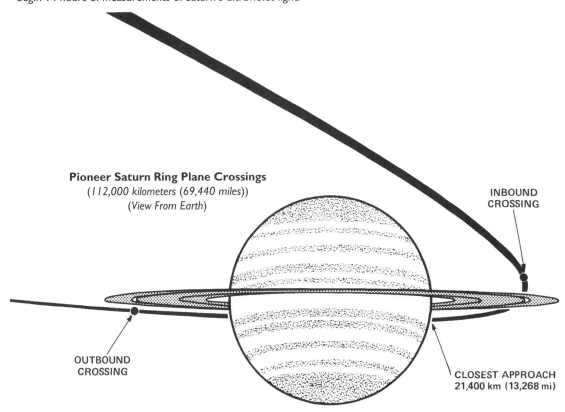

Pioneer Saturn Ring Plane Crossings
(112,000 kilometers (69,440 miles))
(View From Earth)

INBOUND
CROSSING

OUTBOUND
CROSSING

CLOSEST APPROACH
21,400 km (13,268 mi)

- Two hours before closest approach, the imaging photopolarimeter will take its most detailed pictures, showing cloud features as small as 80-100 kilometers (50-60 miles) in size.
- The most critical and dangerous event of the mission will occur at 9:05 a.m. PDT September 1, when *Pioneer* flies through the plane of Saturn's rings at a very shallow angle (4.7 degrees) on its descent toward the planet. Some planetary debris is almost certainly present in Saturn's ring plane, outside the visible rings. An impact with a fragment could destroy the spacecraft. For an hour and 20 minutes, 1.55 billion kilometers (963 million miles) away from Earth, *Pioneer 11* will make the historic first flight under Saturn's rings, making close-up optical measurements to determine their structure. (Pictures during the ring flight will be of Saturn, not of the rings.)
- At 10:30 a.m. PDT, September 1, *Pioneer* also will make its closest approach to Saturn's fifth moon, Dione, inside Dione's orbit, 291,000 kilometers (180,500 miles) away. About 20 minutes later, it will pass the second moon, Mimas, at 103,400 kilometers (64,100 miles).

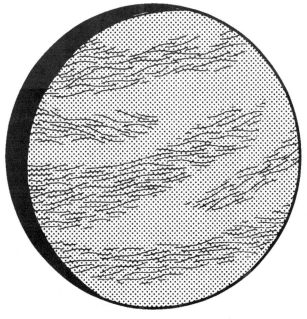

Spacecraft View of Titan
(September 2, 1979; 11:34 a.m. PDT)
(Distance – 360,000 kilometers (223,200 miles))

- At 11:05 a.m. PDT, *Pioneer* will reach its closest approach to Saturn's cloud tops at 21,400 kilometers (13,300 miles) just before it passes behind the planet. The spacecraft will be traveling at 114,100 km/hr (70,900 mph).
- One-and-a-half minutes after closest approach, *Pioneer* will pass behind Saturn and be out of radio contact with Earth for 78 minutes. After *Pioneer* emerges, it will continue to fly under the sunlit side of Saturn's rings for about 10 minutes. At 12:50 a.m. PDT, *Pioneer* will make its closest approach to the fourth moon, Tethys, at 331,700 kilometers (205,700 miles). Five minutes later, the spacecraft will make another shallow-angle crossing of Saturn's ring plane on its ascent away from the planet. As it recrosses the ring plane, it will come closest to the third moon Enceladus at 225,200 kilometers (139,600 miles). An hour after the ascent crossing of the ring plane, *Pioneer* again will have a view of the rings lighted from below.
- At 5:00 p.m. PDT, *Pioneer* will make its closest approach to the sixth moon, Rhea, at a distance of 341,900 kilometers (212,000 miles). Three hours later, from about 420,000 kilometers (260,000 miles) away, *Pioneer* will again get a full view of the planet plus its rings. *Pioneer* will be moving away from the planet at about 54,000 km/hr (33,700 mph).
- On September 2, at 12:30 p.m. PDT, *Pioneer* will make its closest approach to planet-sized Titan, Saturn's seventh moon, at a distance of 356,000 kilometers (220,700 miles). *Pioneer* will photograph and make other measurements of the 5,800-kilometer (3,600-mile)-diameter moon, the largest known satellite in the solar system. This moon dominates Saturn's satellite family both in diameter and mass, exerting a measurable gravitational force on the bodies in that system.

Pioneer Saturn Trajectory
(Satellite Positions at Titan Closest Approach)

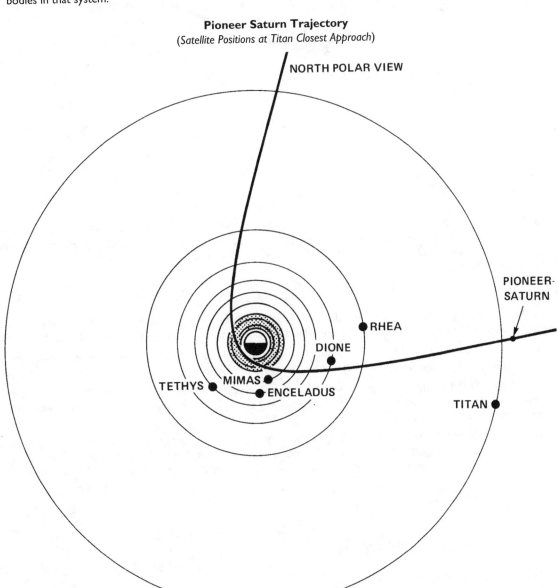

Titan Encounter

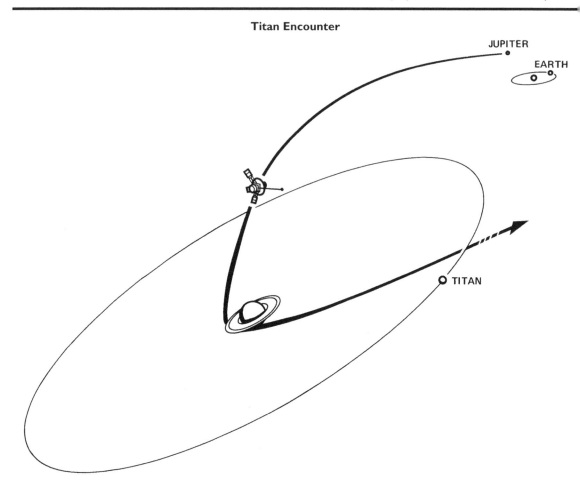

By September 3, when *Pioneer* has moved 1.4 million kilometers (900,000 miles) away from Saturn, tentative infrared and ultraviolet measurements are expected to become available. By September 6, an infrared map of Saturn may be available, as well as early calculations of the masses of the rings and of Rhea and Titan. Picture resolution will be better than Earth-based imaging through midnight, September 7.

Picture Sequences

Plans call for the imaging photopolarimeter to make about 155 images between August 20 and September 8. However, the instrument has been in space for six years, and it is possible that the total number of images could be fewer than predicted.

All of the images will not be available for distribution. Scientists hope to process for immediate distribution up to three or four images each day during the encounter period.

Between August 20 and midnight August 25, plans call for the imaging photopolarimeter to take 49 images of Saturn. The distance from the spacecraft to the planet will decrease from 10.2 million kilometers (6.3 million miles) to 6.1 million kilometers (3.8 million miles).

Between August 26 and September 8, 106 images are planned, these having better resolution than Earth-based photographs. Processing of the best images for distribution purposes will begin August 27. A maximum of seven images are planned on August 26 and another seven for August 27. By August 28, some images should be available for distribution.

Pioneer 10 and Pioneer Saturn Outbound Trajectories

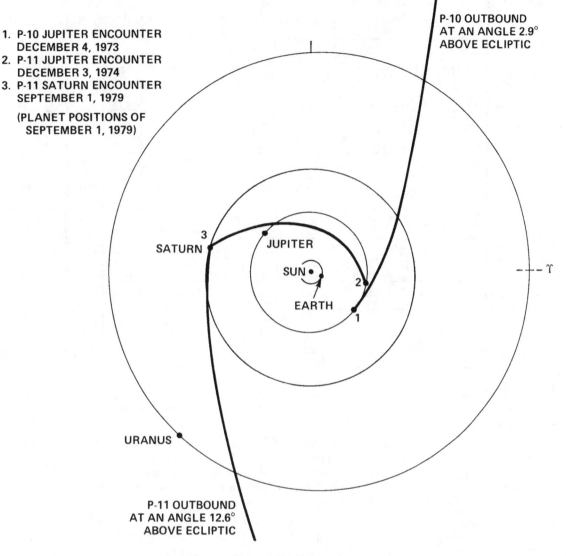

1. P-10 JUPITER ENCOUNTER
   DECEMBER 4, 1973
2. P-11 JUPITER ENCOUNTER
   DECEMBER 3, 1974
3. P-11 SATURN ENCOUNTER
   SEPTEMBER 1, 1979

   (PLANET POSITIONS OF
   SEPTEMBER 1, 1979)

P-10 OUTBOUND
AT AN ANGLE 2.9°
ABOVE ECLIPTIC

P-11 OUTBOUND
AT AN ANGLE 12.6°
ABOVE ECLIPTIC

Saturn Picture Time Schedule

Date	Time Received on Earth (PDT)	Image	Distance from Spacecraft (kilometers)	(miles)	Planned No. of Images	Pixel Size (Resolution) (km)	(miles)
August 20	3 p.m. - 8 p.m.	Saturn	10,200,000	6,300,000	2	5,100	3,200
August 21	Midnight - Midnight	Saturn	9,400,000	5,800,000	10	4,700	2,900
August 22	Midnight - Midnight	Saturn	8,600,000	5,300,000	10	4,300	2,700
August 23	Midnight - Midnight	Saturn	7,900,000	5,000,000	10	3,900	2,400
August 24	Midnight - 8 p.m.	Saturn	7,000,000	4,300,000	8	3,500	2,200
August 25	Midnight - 2 a.m.	Saturn	6,550,000	4,100,000	1	3,300	2,000
	6 a.m. - Midnight	Saturn	6,100,000	3,800,000	8	3,000	1,900
August 26	Midnight - Midnight	Saturn	5,400,000	3,350,000	7	2,700	1,700
August 27	Midnight - Noon	Saturn	4,800,000	3,000,000	4	2,400	1,500
	3 p.m. - Midnight	Saturn	4,400,000	2,700,000	3	2,200	1,350
August 28	Midnight - 4 a.m.	Saturn	4,100,000	2,500,000	1	2,100	1,300
	Noon - Midnight	Saturn	3,600,000	2,200,000	4	1,800	1,100
August 29	Midnight - 6 a.m.	Saturn	3,300,000	2,050,000	2	1,600	1,000
	8 a.m. - Midnight	Saturn	2,800,000	1,700,000	8	1,400	870
August 30	Midnight - 4 a.m.	Saturn	2,450,000	1,500,000	1	1,200	760
	6 a.m. - 8 p.m.	Saturn	2,050,000	1,300,000	6	1,000	640

Saturn Picture Time Schedule (continued)

Date	Time Received on Earth (PDT)	Image	Distance from Spacecraft (kilometers)	(miles)	Planned No. of Images	Pixel Size (Resolution) (km)	(miles)
August 31	Midnight - 3 a.m.	Saturn	1,600,000	1,000,000	1	800	500
	5 a.m. - Midnight	Saturn	1,100,000	700,000	9	540	330
September 1	Midnight - 4 a.m.	Saturn	535,800	332,200	2	270	170
	4 a.m. - 8 a.m.	Saturn	325,800	202,000	3	160	100
	8 a.m. - 9 a.m.	Saturn	94,200	58,400	1	50	30
	1 p.m. - 4 p.m.	Saturn	258,000	160,000	2	130	80
	4 p.m. - 8 p.m.	Saturn	477,600	296,100	1	240	150
	8 p.m. - Midnight	Saturn	672,600	417,000	1	340	210
September 2	Midnight - 4 a.m.	Saturn	854,400	529,700	2	430	265
	4 a.m. - 8 a.m.	Titan	423,000	262,300	1	210	130
	4 a.m. - 8 a.m.	Saturn	1,000,000	620,000	2	510	320
	8 a.m. - Noon	Titan	365,400	226,500	1	180	110
	8 a.m. - Noon	Saturn	1,200,000	740,000	2	600	370
	Noon - 4 p.m.	Titan	363,600	225,400	1	180	110
September 3	1 a.m. - 9 a.m.	Saturn	1,900,000	1,200,000	4	970	600
	10 a.m. - 1 p.m.	Saturn	2,150,000	1,300,000	1	1,080	670
	2 p.m. - 6 p.m.	Saturn	2,300,000	1,400,000	2	1,170	720
	7 p.m. - Midnight	Saturn	2,500,000	1,550,000	2	1,260	780
September 4	Midnight - 3 a.m.	Saturn	2,700,000	1,700,000	1	1,330	830
	9 a.m. - Midnight	Saturn	3,400,000	2,100,000	1	1,680	1,030
September 5	Midnight - Midnight	Saturn	3,900,000	2,400,000	10	1,900	1,200
September 6	Midnight - Midnight	Saturn	4,700,000	2,900,000	10	2,300	1,400
September 7	Midnight - Midnight	Saturn	5,500,000	3,400,000	10	2,700	1,700

Trajectory Selection

Pioneer Saturn began its journey to Jupiter and Saturn on April 5, 1973. It followed Pioneer 10, the first spacecraft to Jupiter, launched in March 1972.

When Pioneer 10 successfully completed the first flyby of Jupiter in December 1973, Pioneer 11 was retargeted to a trajectory that would provide both data that best complemented that from Pioneer 10 and included a subsequent transfer orbit to Saturn. Pioneer 11 reached Jupiter on December 3, 1974, and returned new data about the Jovian environment. It then began its 2.4 billion kilometer (1.5 billion mile) trek across the solar system toward Saturn. Preliminary maneuvers were executed in 1975 and 1976 to put the spacecraft on an interim trajectory.

From the many original targeting options, two semi-final aim points were selected. The principal difference between the two aim points, known as the "inside" and "outside" options, was their relationship to Saturn's unique ring system.

A trajectory to the inside area on Pioneer's inbound descending approach would have come very close to the planet - 3,600 kilometers (2,232 miles) inside the visible C ring. The advantage of the inside target area was hard to firmly establish, however, because of the apparent discovery of an inner D ring by Pierre Guerin in 1969. Guerin's photographs appear to show ring material within the C ring, extending close to the planet, and a new division (the Guerin division) between the C and D rings. Although the density profile for the D ring could not be measured, most assessments showed considerable risk for survival of Pioneer Saturn on the inside trajectory, even if targeted through the Guerin division at 14,400 kilometers (8,928 miles).

For an outside trajectory the inbound descending and outbound ascending flight paths can be located at any distance beyond the A ring, which extends to 77,400 kilometers (48,100 miles) from Saturn's cloud tops. However, outside targeting is again complicated by evidence of ring material in an E ring extending beyond the visible rings.

Since the existence of the E ring is critical to Voyager, NASA officials determined that the maximum return from all missions would be achieved by targeting Pioneer Saturn outside the rings.

A very crucial moment of the Pioneer Saturn mission will occur on the morning of September 1, when Pioneer passes through the ring plane at an angle of nearly 5 degrees, 112,000 kilometers (70,000 miles) from Saturn on its descent toward the planet. An impact with a fragment of the E ring could destroy the spacecraft.

Ring-plane passage takes only about 0.8 second, and is repeated four hours later when the spacecraft recrosses the ring plane on its ascent from the planet, again at 112,000 kilometers (70,000 miles) from Saturn's cloud tops.

The outside ring plane crossing at 112,000 kilometers (70,000 miles) provides a "balanced" flyby where the inbound descending and outbound ascending courses are both at that distance, with full visibility from Earth. Closest approach is 21,400 kilometers (13,300 miles). Inclination of the trajectory is 4.7 degrees to Saturn's ring plane with periapsis in the southern hemisphere. Earth and solar occultations occur shortly after periapsis.

Saturn Picture Locations

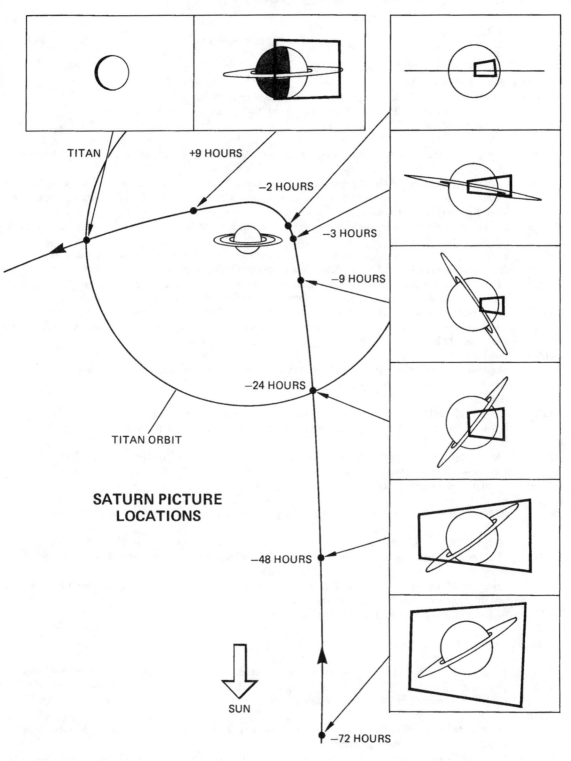

TITAN +9 HOURS

−2 HOURS

−3 HOURS

−9 HOURS

−24 HOURS

TITAN ORBIT

SATURN PICTURE LOCATIONS

−48 HOURS

SUN

−72 HOURS

Post-Saturn

Like *Pioneer 10*, *Pioneer Saturn* will escape the solar system on its post-encounter trajectory. *Pioneer Saturn* will be traveling in the opposite direction from its sister spacecraft. The escape is in the general direction of the nose of the heliosphere, which marks the boundary between the solar wind and the interstellar medium. Communications with both spacecraft should be possible through the mid-1980s. Both *Pioneers* carry a message plaque for any intelligent species that may capture the spacecraft during their years of wandering among the stars.

Encounter Timeline

Notes:

(All times are spacecraft times, PDT, except as noted.)
ERT = Earth Received Time PDT
One-way light time: August 20-22, 85 minutes; August 23-September 7, 86 minutes.
Two-way light time: August 20-22, 2 hours, 50 minutes; August 23-September 7, 2 hours, 52 minutes.
E = Closest approach to Saturn (9/1/79, 9:34 a.m.)

8/02/79	Start encounter activities. Spacecraft now tracked via 64-meter (210-foot) dishes in Spain and Australia for about 10 hours a day.
8/06/79	Tracking increases to 18 hours a day – Goldstone Station is added. Routine science data-gathering taking place. All science instruments on and being checked for encounter readiness. *Pioneer* instrument begins to make polarimetry measurements of Saturn.
8/12/79	Spacecraft is precessed to keep the high-gain antenna pointed within 0.5 degree of Earth, to maintain highest possible data rates.
8/16/79	Begin 4-day ultraviolet imaging of planet. Spacecraft turned 1.3 degrees away from Earth for ultraviolet imaging.
8/17/79	Spacecraft now tracked 24 hours a day on the 64-meter (210-foot) stations.
8/20/79	Critical encounter period begins. Photopolarimeter now begins to take images of Saturn. Distance is too great to see detail. Second precession performed to keep antenna pointing within 0.5 degree of Earth.
8/25/79	Spacecraft now covered by two stations simultaneously to increase data return. Commands sent via 26-meter (86-foot) station and data is received via 64-meter (210-foot) station. This will increase the data rate from 512 bps to 1,024 bps and substantially improve image quality.

8/26/79   (E-6 days)
            Begin picture resolution better than Earth-based. Expect at least 50 such pictures through 9/8/79. (Most will show rings, lighted from below.)
8/27/79   (E-5 days)
 1:02 a.m. Closest approach to Saturn's 10th moon, Phoebe, at 9,453,000 kilometers (5,860,860 miles).
            Possible bow shock crossing. Time or times depend on local solar wind intensity.
8/28/79   (E-4 days)
11:05 p.m. Closest approach to Saturn's ninth moon, Iapetus, (diameter 1,400 kilometers, or 900 miles) at 1,039,000 kilometers (644,180 miles). (Photometry measurements.)
8/29/79   Third precession performed to correct antenna pointing.
8/30/79
 6:34 p.m. (E-39 hours) *Pioneer* completes last image of full planet with rings, twice the best Earth resolution.
8/31/79
12:34 a.m. (E-33 hours) Last full planet picture (perhaps best full planet view); five times best Earth resolution.
 5:31 a.m. (E-28 hours) Closest approach to Saturn's eighth satellite, Hyperion, (diameter 225 kilometers, or 140 miles) at 674,000 kilometers (417,880 miles). (Ultraviolet measurements.)
10:10 a.m. Series of 21 commands, designed to keep spacecraft subsystems in proper encounter mode, are transmitted every hour for the next two days.
11:34 a.m. Begin 14 hours of ultraviolet observations of Saturn.
9/01/79
 8:00 a.m. Begin 1½ hours of infrared imaging of Saturn.

 7:34 a.m. (E-2 hours) Best picture resolution (objects 80-100 kilometers, or 50-60 miles in size), 20 times best Earth resolution.
 7:35 a.m. (9:01 a.m. ERT) *Pioneer* crosses Saturn's ring plane on its descent toward the planet. (Critical event.) (112,000 kilometers, or 70,000 miles above clouds.)
 8:09 to
11:05 a.m. Flight under Saturn's rings (lit side), 3,500 kilometers (2,200 miles) away. 1 hour, 20 minutes before occultation to 10 minutes after occultation.
 9:04 a.m. (E-½ hour) Closest approach to Saturn's fifth satellite, Dione, (diameter 1,000 kilometers, or 620 miles) at 291,100 kilometers (180,482 miles). (Photometry measurements.)

9:14 a.m.   Begin 20 minutes of radio occultation measurements of Saturn's ionosphere and atmosphere.

9:27 a.m.   Closest approach to Saturn's second moon, Mimas, (diameter 940 kilometers, or 585 miles) at 103,400 kilometers (65,108 miles).

9:34 a.m.   **(11:00 a.m. ERT) Closest approach to cloud tops of Saturn, at 21,400 kilometers (13,300 miles). Speed is 114,100 km/hr (70,900 mph). (Distance from Earth is 1.4 billion kilometers, or 963 million miles)

9:35 a.m.   Enter 78 minutes of Earth occultation.

9/01/79

9:35 a.m.   (9:35:57) Enter 79 minutes of Sun occultation – Penumbra.

9:36 a.m.   Enter Sun occultation – Umbra.

10:53 a.m. (10:53:32) Leave Earth occultation, end radio blackout and data storage. Begin 20 minutes of radio occultation measurements of Saturn's atmosphere and ionosphere.

10:54 a.m. (10:54:42) Leave Sun occultation – Umbra

10:54 a.m. (10:54:47) Leave Sun occultation – Penumbra

11:28 a.m. Closest approach to Saturn's fourth satellite, Tethys, (diameter 1,000 kilometers, or 650 miles) at 331,700 kilometers (205,654 miles). (Photometry and ultraviolet measurements.)

11:33 a.m. (E+2 hours) (12:59 p.m. ERT) *Pioneer* crosses Saturn's ring plane on its ascent away from the planet. (Critical event.) (112,000 kilometers, 70,000 miles.)

11:33 a.m. Closest approach to Saturn's third satellite, Enceladus, (diameter 1,100 kilometers, or 680 miles) at 225,200 kilometers (139,624 miles). (Ultraviolet measurements.)

12:34 p.m. (E+3 hours) Resume close-up views of rings (unlit side).

3:34 p.m.   (E+6 hours) Closest approach to Saturn's sixth satellite, Rhea, (diameter 1,600 kilometers, or 1,000 miles) at 341,900 kilometers (211,978 miles). (Photometry and ultraviolet measurements.)

6:34 p.m.   (E+9 hours) Resume half-ring plus planet view ($^1/_3$ disc).

9/02/79

4:00 a.m.   Start Titan imaging.

6:40 a.m.   (E+19 hours) First ultraviolet imaging of Titan.

8:41 a.m.   (E+24 hours) Second ultraviolet imaging of Titan.

11:05 a.m. (E+25½ hours) Closest approach to Saturn's seventh moon, Titan, (diameter 5,800 kilometers, or 3,600 miles) at 356,000 kilometers (220,720 miles).

Midnight   Titan infrared imaging.

9/03/79   (E+2 days) Tentative infrared and ultraviolet findings.

9/06/79   (E+5 days) Infrared map of Saturn; first mass calculations of Saturn's moons.

9/08/79   Start of superior conjunction (when the Sun is between the Earth and Saturn). Spacecraft passes close to Sun as viewed from Earth. Data rates will be low (16-32 bps) for the next seven days. No images will be taken during that period. From this point on, resolution ceases to surpass Earth-based resolution.

9/15/79   Superior conjunction ends. Resume normal data-gathering activity.

10/1/79   End of encounter activities.

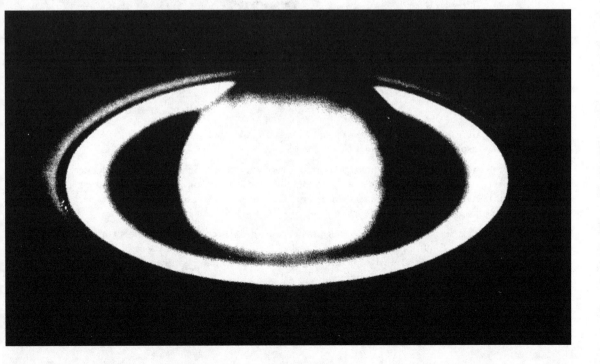

Saturn

Saturn is the only planet in the solar system less dense than water. It has a volume 815 times greater than Earth's but a mass only 95.2 times greater. It is the second largest planet. Saturn's equatorial radius is 60,000 kilometers (37,300 miles). The polar radius is considerably smaller – 53,500 kilometers (33,430 miles). The dynamic flattening, caused by Saturn's rapid rotation and increased by its low density, is the greatest of any planet yet measured.

Saturn's rings were discovered in 1610 when Galileo Galilei aimed the first astronomical telescope at Saturn. Even Galileo didn't realize what they were. He reported seeing "cup handles" in his less-than-adequate telescope. Forty-five years later, in 1655, Christiaan Huygens described the rings' true form.

A day on Saturn's equator is only 10 hours, 14 minutes – 18.5 minutes longer than a day on Jupiter. Saturn completes one orbit of the Sun in 29.46 Earth years. The average distance of Saturn from the Sun is 9.5 A.U. Saturn receives only about $1/100^{th}$ of the Sun's intensity that strikes Earth. (An astronomical unit (A.U.) is the mean distance from the Sun to the Earth – 149,600,000 kilometers (92,960,000 miles)).

Saturn, like the other outer giants, bears some resemblance to Jupiter – enough that they are often coupled together as the Jovian planets. Like Jupiter, Saturn apparently has no solid surface, but changes gradually from a thin outer atmosphere through progressively denser layers to the core, which may be a small chunk of iron and rock.

When scientists discuss the planet's atmosphere, they generally restrict their attention to a region where pressure varies from 1,000 Earth atmospheres to one 10-billionth atmosphere

Like Jupiter, the principal constituents of the Saturnian atmosphere are thought to be hydrogen and helium. Three molecules have definitely been detected in Saturn's atmosphere: hydrogen (H_2), methane (CH_4) and ethane (C_2H_6). Radio observations have shown indirect evidence for ammonia (NH_3) at atmospheric levels inaccessible to optical measurements. No other molecular or atomic species has been detected.

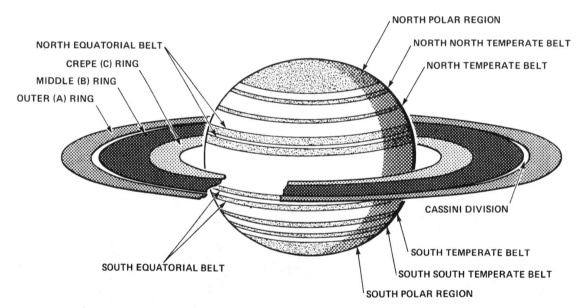

Also like Jupiter, Saturn is believed to be composed of materials in about the same ratio as the Sun, formed into the simplest molecules expected in a hydrogen-rich atmosphere.

Saturn appears to radiate nearly twice as much energy as it receives from the Sun. In the case of Jupiter, that radiation has been explained as primordial heat left over from the time, about 4.6 billion years ago, when the planet coalesced out of the solar nebula. The same may be true for Saturn. Convection is the most likely transport mechanism to carry heat from the interior of the planet to the surface.

Saturn has cloud bands similar to Jupiter's, although they are harder to see and contrast less with the planetary disc. Photographs confirm that Saturn's bland appearance is real. The blandness may be a result of lower temperatures and reduced chemical and meteorological activity compared with Jupiter or a relatively permanent and uniform high altitude haze.

The principal features of Saturn's visible surface are stripes that parallel the equator. Six dark belts and three light zones have been seen continuously over 200 years of observations.

Spots have been observed in the upper atmosphere of Saturn. Unlike the Great Red Spot of Jupiter they are not permanent nor are they easily identifiable. The spots that have been observed have lifetimes up to a few months.

Sometimes they are light, sometimes dark. They are confined to a region within 60 degrees of the equator and typically are a few thousand kilometers across. They may be comparable to hurricanes on Earth.

Saturn has 10 known satellites. The most recent discovery was Janus, found in 1966 by Audouin Dollfus. Janus has been seen in only a few photographs. It appears to travel in the plane of Saturn's rings and near them. Its low albedo and proximity to the rings make Janus difficult to observe except when the rings are edge-on to Earth and recent studies indicate that at least two separate satellites may be masquerading under the name of Janus.

The largest known satellite in the solar system is Saturn's satellite Titan. Titan has a diameter of 5,800 kilometers (3,600 miles), greater than the planet Mercury, and is known to have an atmosphere. In 1944, the late Dr. Gerard Kuiper detected a methane atmosphere on Titan. Titan's atmospheric pressure may be comparable to Earth's. Other molecules identified include ethane and probably acetylene and many scientists believe there is also a major undetected gas present. The most likely candidate is nitrogen.

Saturn Atmosphere Model*

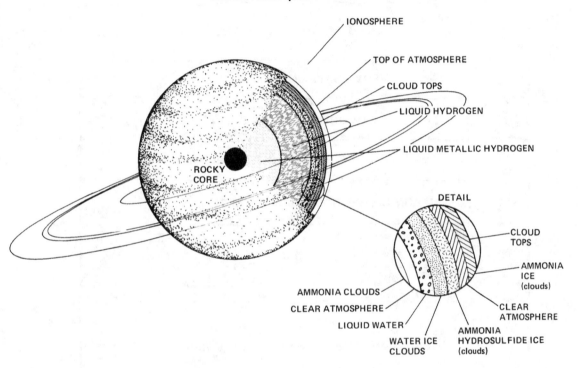

Some scientists believe organic compounds may be present on the surface of Titan, leading some to suggest it as a possible abode of some primitive life forms.

Iapetus is another Saturnian satellite that draws scientific interest. Its brightness varies by a factor of about five as it rotates on its axes, indicating that one face is bright and the other dark. The light face appears to be covered with ice but the composition of the dark face is unknown.

Saturn's rings have been a curiosity to astronomers since their discovery. Their origin is unknown but a number of hypotheses have been put forward. They might be the remains of some early satellite broken up by gravitation or remnants of the primordial material that somehow became trapped in orbit. The age of the rings is not known.

The rings lie in Saturn's equatorial plane, which is tipped 27 degrees to the orbital plane of Saturn. Although it is certain the rings are not a solid sheet, little else is known about their composition and structure. Spectroscopy shows that they are made primarily of water ice or ice-covered silicates.

The individual particles probably vary from less than a millimeter (0.04 meters) to more than a 10 meters (32.8 feet), but most are a few centimeters in size – about as big as a snowball.

Three distinct rings can be seen. The inner or "crepe" ring begins about 17,000 kilometers (11,000 miles) from the planet's visible cloud surface and extends for 15,000 kilometers (9,000 miles) to 32,000 kilometers (20,000 miles). The second ring begins at that point and extends for 26,000 kilometers (16,000 miles) to a distance of 58,000 kilometers (36,000 miles) from the planet. There, a phenomenon known as Cassini's Division breaks the rings' continuity. Cassini's Division is 2,600 kilometers (1,600 miles) wide.

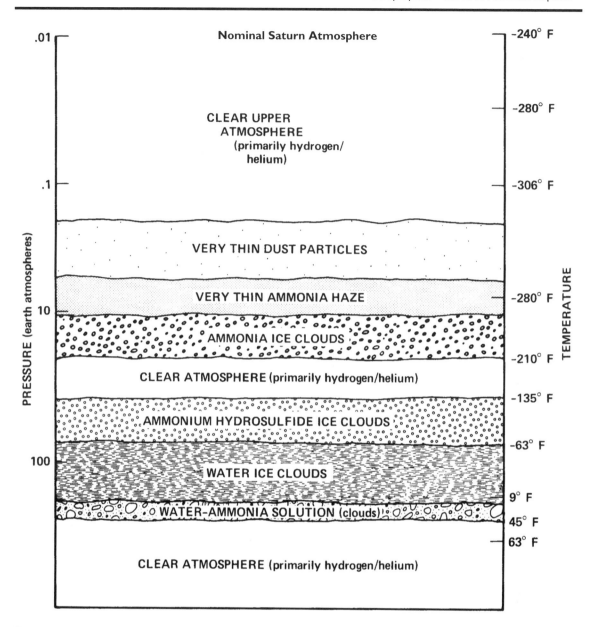

The outer ring begins 60,000 kilometers (38,000 miles) from the equator of Saturn and appears to end 16,000 kilometers (10,000 miles) farther away at a distance of 76,000 kilometers (47,000 miles).

Cassini's Division is real. It is explained by a phenomenon in celestial mechanics. Any particle at that distance would have an orbital period of 11 hours 17½ minutes, just half the period of the satellite Mimas. The particle would be nearest Mimas at the same place in its orbit every second time around. This repeated gravitational perturbation would eventually move the particle to a different distance.

Accurate measurements of the ring thickness are not possible, but limits have been placed. They appear to be somewhere between 1 and 4 kilometers (0.6 to 2.5 miles).

Until recently there was no evidence that Saturn has a magnetic field. Neither decimeter nor decametric radio emissions had been observed – the kind of "radio noise" from Jupiter that was evidence for its magnetic field. But radiometric observations from the Earth-orbiting satellite IMP-6 have provided indirect evidence for a magnetic field. If a magnetic field is present, it is probably distorted by the rings.

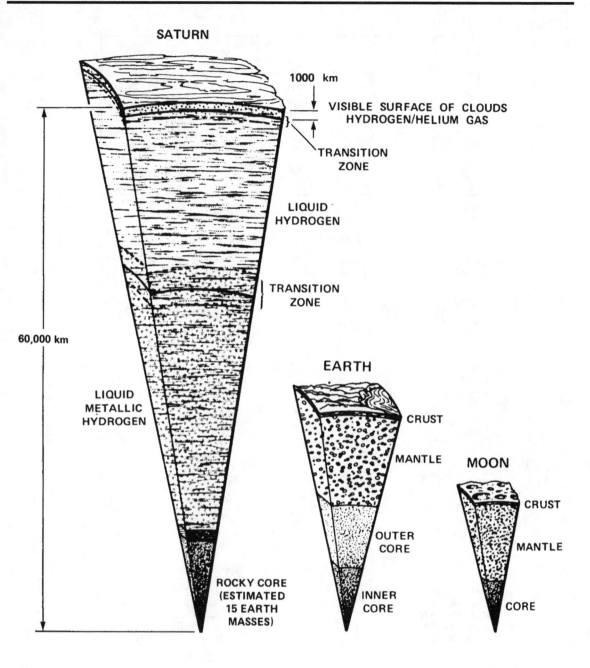

SATURN

1000 km

VISIBLE SURFACE OF CLOUDS
HYDROGEN/HELIUM GAS

TRANSITION
ZONE

LIQUID
HYDROGEN

TRANSITION
ZONE

60,000 km

LIQUID
METALLIC
HYDROGEN

EARTH

CRUST

MANTLE

MOON

CRUST

MANTLE

OUTER
CORE

ROCKY CORE
(ESTIMATED
15 EARTH
MASSES)

INNER
CORE

CORE

Earth, Jupiter, Saturn Comparison Table

Parameter	Earth	Jupiter	Saturn
Radius (Equatorial)	6,378 km (3,963 miles)	71,400 km (44,366 miles)	60,000 km (37,300 miles)
Satellites	1	13	10
Year	1	11.86	29.46
Day	23 h 56 m 04 s	9h 55m 30s	10h 14m
Mass	1	317.9	95.2
Density (water=1)	5.5	1.3	0.7
Mean distance from Sun	1 A.U.	5.2 A.U.	9.5 A.U.

The Satellites of Saturn

Name	Diameter		Semi-major Axis		Period
Janus	300 km	(187 miles) *	168,700 km	(105,000 miles)	0.815 days
Mimas	400 km	(300 miles)	185,800 km	(120,000 miles)	0.942 days
Enceladus	550 km	(390 miles)	238,300 km	(148,000 miles)	1.370 days
Tethys	1,200 km	(745 miles)	294,900 km	(183,200 miles)	1.888 days
Dione	1,150 km	(715 miles)	377,900km	(235,000 miles)	2.737 days
Rhea	1,450 km	(900 miles)	527,600 km	(328,000 miles)	4.518 days
Titan	5,800 km	(3,600 miles)	1,222,600 km	(760,000 miles)	15.945 days
Hyperion	300 km	(187 miles) *	1,484,100 km	(922,000 miles)	21.276 days
Iapetus	1,800 km	(1,120 miles)	3,562,900 km	(2,214,000 miles)	79.330 days
Phoebe**	200 km	(125 miles) *	12,960,000 km	(8,092,000 miles)	550.450 days

* Estimated. ** Phoebe's motion is retrograde.

Saturn Satellite Diameters

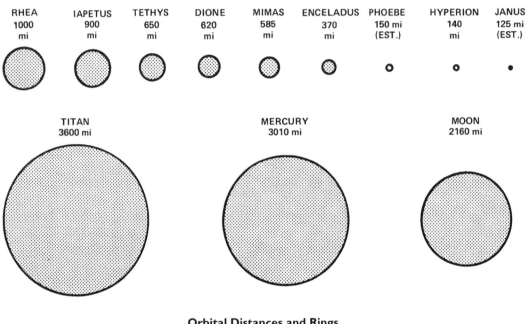

| RHEA 1000 mi | IAPETUS 900 mi | TETHYS 650 mi | DIONE 620 mi | MIMAS 585 mi | ENCELADUS 370 mi | PHOEBE 150 mi (EST.) | HYPERION 140 mi | JANUS 125 mi (EST.) |

TITAN 3600 mi

MERCURY 3010 mi

MOON 2160 mi

Orbital Distances and Rings

1. JANUS, 62,000 mi FROM SATURN
2. MIMAS, 78,000 mi
3. ENCELADUS, 110,000 mi
4. TETHYS 146,000 mi
5. DIONE, 200,000 mi

6. RHEA, 300,000 mi
7. TITAN, 720,000 mi
8. HYPERION, 880,000 mi
9. IAPETUS, 2.2 MILLION mi
10. PHOEBE, 8 MILLION mi

1/2 in. = 62,500 mi

SATURN DIAMETER	74,000 mi
SATURN TO C RING	7,812 mi
SATURN TO B RING	19,716 mi
SATURN TO CASSINI DIVISION	35,340 mi
SATURN TO A RING	38,316 mi
SATURN TO OUTER EDGE OF RINGS	48,000 mi

Pioneer Saturn Science

The Voyager and Pioneer missions to Jupiter and Saturn are addressing fundamental questions about the origin and nature of the solar system. Understanding interplanetary space and the other planets should give scientists a greater knowledge of Earth.

According to current theoretical models of the origin and evolution of the solar system, a gaseous nebula composed of solar material – gases and dust of various elements – collapsed to form the Sun. Some of the material remained behind and began to coalesce to form the planets, their satellites, the asteroids, comets and meteors. Temperature, pressure and density of the gas decreased with distance from the Sun.

Formation of the planets is believed to have resulted from accretion of the nebular material. Observed differences in the planets are accounted for in these theories by variations in material and conditions at the places where they formed. Thus, knowledge gained at each planet can be related to others and should contribute to an overall understanding of the solar system as well as our own planet Earth.

Missions to Mars, Venus, Mercury and the Moon have contributed greatly to the body of knowledge. Each planet has its own personality, significantly different from others because of its unique composition and relationship to the Sun. Individual as they are, the inner planets are related as bodies that originated near the Sun and that are composed mainly of heavier elements. They are classified as "terrestrial planets," since the Earth is approximately representative.

Scientists have known for a long time that Jupiter, Saturn and the other outer planets differ significantly from terrestrial planets. They have low average densities; only hydrogen and helium among all the elements are light enough to match observations to date. Jupiter and Saturn are sufficiently massive (318 and 95 times Earth's mass, respectively) to insure that they have retained almost all their original material. They are, however, only relatively pristine examples of the material from which the solar system formed. While almost no material has been lost, the planets have evolved over their 4.6-billion-year lifetimes and the nature and ratio of the materials may have changed. If that 4.6-billion-year evolution can be traced, scientists will obtain a clearer picture of the early state of their region of the solar system.

Scientific Objectives

Magnetic Field, Magnetosphere and Solar Wind

Determine the presence of a Saturn magnetic field. Map the field and determine its intensity, direction and structure. Find if the field's center and magnetic equator coincide with Saturn's geographical center and equator. Find if the field is dipole or multi-pole. Find the rotation rate of the deep interior of this gas planet (compared with the surface rate) by measuring the field's rotation. Measure changes in the field due to Saturn's rings. Find the shape and character of Saturn's proposed magnetosphere. Chart the magnetosphere boundary and its bow shock wave in the solar wind, and study a variety of planet interactions with electrons and protons of the solar wind. Look for energetic particles, accelerated in the magnetosphere, far out in the interplanetary medium, before *Pioneer* reaches Saturn.

Radiation Belts

Find the densities and energies of electrons and protons and other ions along the trajectory of the spacecraft through the Saturn system. Find any radiation belts in Saturn's magnetosphere. Find mini-belts between Saturn's rings and use them to determine belt origin and particle acceleration processes. Study the character of the rings, as shown by passages of high-energy particles through them.

Upper Atmosphere and Ionosphere

Determine structure, pressure, temperature, density, altitudes and some composition information on the upper atmosphere and ionosphere. Detect clouds, hazes and mixing rate in the upper atmosphere.

Saturn's Interior

Map Saturn's gravity field to an accuracy of 70 parts per million and hence find its global shape. This will allow estimates of the size of Saturn's rocky core and determination of the presence of liquid metallic hydrogen in the deep interior. Finding of a magnetic field would further confirm the presence of metallic hydrogen. Gravity measurements combined with heat data also will allow calculation of the planet's temperature and density profile from surface to core.

Planetary Heat Balance

Measure the effective temperature of Saturn's total disc, and from it determine the strength of Saturn's internal heat source by subtracting reflected solar heat from total heat radiated by the planet.

Proposed Saturn Magnetosphere
(Saturn 1 Gauss Surface Field)

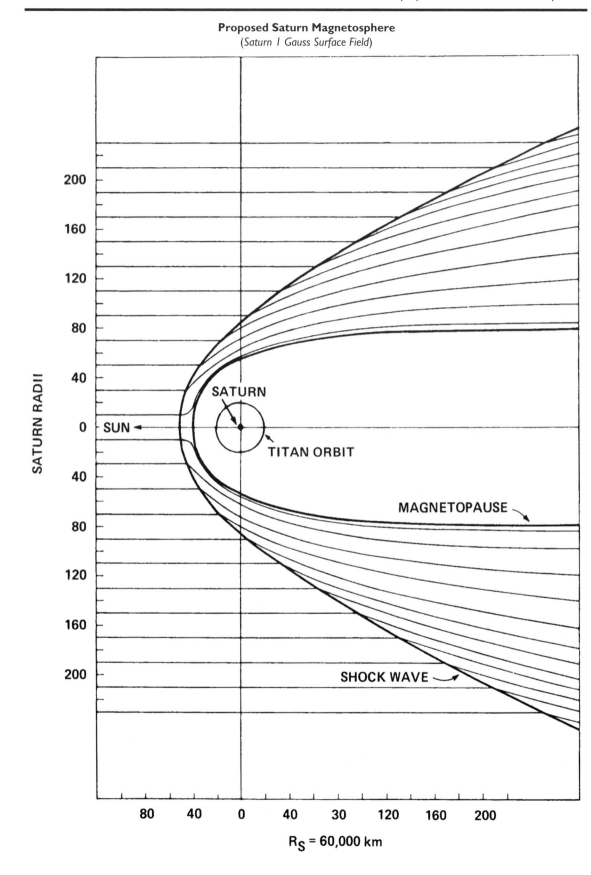

Rings

Determine individual masses of both the A ring and B ring (the outer two rings) to find whether ring material is primarily water ice or metals. (Other compositions appear to have been largely eliminated.) Study distribution, structure and character of material in all of the rings by photometry, polarimetry, two-channel ultraviolet and observation of energetic particle paths. Determine existence of the proposed E ring and density of material it contains, to find safe trajectories for *Voyager* and other spacecraft through this region. Use above observations and the ring-plane passage of *Pioneer* itself for this finding. Measure temperatures of the rings.

Satellites

Make photopolarimetry measurements of Iapetus, Dione, Tethys, Rhea and Titan and make two-channel ultraviolet observations of Titan, Hyperion, Rhea, Dione, Tethys and Enceladus to determine their surface characteristics. Make first determination of mass and density of Rhea and Titan. Improve data on satellite orbits.

Titan

Characterize Titan's atmosphere. Determine opacity and nature and layering of aerosols in its atmosphere. Look for greenhouse effect in atmosphere. Determine Titan's temperature and heat balance. Look for doughnut-shaped hydrogen region around Titan's orbit and for its magnetic wake. Improve measurements of Titan's mass to an accuracy of 0.3 percent and confirm current estimates of its density. Refine Titan orbit measurements. Make two-color images and two-channel ultraviolet observations of Titan.

Saturn System

Improve precision and confidence in determination of the total mass of the Saturn system and of Saturn's orbit. Use this to determine dynamics of the solar system and for checks on relativity theory.

Saturn's Atmosphere

Make a rough measurement of Saturn's hydrogen / helium ratio by infrared and ultraviolet observations. Make atmosphere composition calculations from radio occultation and photopolarimetry measurements. Determine atmosphere structure and layering and make observations of flow characteristics, rates of flow, vertical structure of belts and zones and polar regions by imaging and photopolarimetry.

Atmospheric Heat

Map the temperature structure of Saturn's atmosphere by infrared and radio occultation measurements providing a vertical profile. Horizontally, look for hot and cold spots and belt and zone temperature differences.

Imaging

Make 50 two-color close-up pictures of Saturn, its rings and cloud features, as well as two such pictures of Titan.

The Experiments

Instruments aboard *Pioneer Saturn* return data in the following broad categories:

Magnetic Fields

Magnetometer (Jet Propulsion Laboratory): This instrument will measure Saturn's magnetic field and magnetosphere, as well as the interplanetary magnetic field in three axes out to the limits of spacecraft communication.

The instrument is a helium vector magnetometer. Its sensor is mounted on the lightweight mast extending 6.5 meters (21.5 feet) from the center of the spacecraft to minimize interference from spacecraft magnetic fields.

The instrument operates in any one of eight different ranges, the lowest covering magnetic fields up to +4 gamma; the highest, fields up to 140,000 gamma. Earth's surface field is 50,000 gamma. These ranges are selected by ground command or automatically by the instrument itself. The magnetometer can measure fields as weak as 0.01 gamma.

Its sensor is a cell filled with helium, excited by radio frequencies and infrared optical pumping. Changes in the helium caused by fields the spacecraft passes through are measured. It weighs 2.6 kilograms (5.7 pounds) and uses up to 5 watts of power.

Fluxgate Magnetometer (Goddard Space Flight Center): This instrument is designed to measure the strength of intense planetary fields up to one million gamma in each of three perpendicular directions.

The instrument consists of two dual-axis sensors and their electronics. Each sensor is composed of a ring core, magnetic multivibrator, a frequency doubler and two phase-sensitive detectors. The instrument weighs 0.27 kilograms (0.6 pounds) and uses approximately 0.36 watts of power.

Pioneer Saturn Experiments

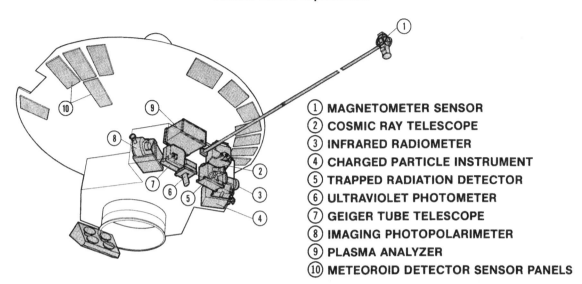

① **MAGNETOMETER SENSOR**
② **COSMIC RAY TELESCOPE**
③ **INFRARED RADIOMETER**
④ **CHARGED PARTICLE INSTRUMENT**
⑤ **TRAPPED RADIATION DETECTOR**
⑥ **ULTRAVIOLET PHOTOMETER**
⑦ **GEIGER TUBE TELESCOPE**
⑧ **IMAGING PHOTOPOLARIMETER**
⑨ **PLASMA ANALYZER**
⑩ **METEOROID DETECTOR SENSOR PANELS**

Interplanetary Solar Wind, Bow Shock Wave and Magnetosphere

Plasma Analyzer (Ames Research Center). See *Plasma Analyzer* in the *Pioneer 11 Press Kit*, page 112.

Radiation Belts and Cosmic Rays

Charged Particle Composition Instrument (University of Chicago): This instrument has a family of four measuring systems. Two are particle telescopes primarily for interplanetary space. The other two measure trapped electrons and protons in planetary radiation belts, such as those proposed for Saturn.

During interplanetary flight two telescopes identify the nuclei of all eight chemical elements from hydrogen to oxygen and separate the isotopes deuterium, helium-3 and helium-4. Because of their differences in isotopic and chemical composition and spectra, galactic particles can be separated from solar particles. The instrument also measures the manner in which streams of high-energy particles travel through interplanetary space. There is a main telescope of seven solid state detectors which measure from 1 to 500 million electron-volt particles and a three-element telescope which measures 0.4 to 10 million electron-volt protons and helium nuclei. If a key detector element is destroyed in space, diagnostic procedures and commands from Earth can bypass it.

For magnetospheres such as that proposed for Saturn, two new types of sensors were developed to cope with extremely high intensities of trapped radiation.

A solid state electron current detector operating below -40 degrees Celsius (-40 degrees Fahrenheit) measures electrons. The trapped proton detector contains a foil of thorium which undergoes nuclear fission from protons above 30 million electron-volts, but is not sensitive to electron radiation. The instrument weighs 3 kilograms (6.6 pounds) and uses 2.2 watts of power.

Cosmic Ray Telescope (Goddard Space Flight Center): The Cosmic Ray Telescope monitors solar and galactic cosmic ray particles. It tracks the twisting paths of high-energy particles from the Sun and measures bending effects of the solar magnetic field on particles from the galaxy. The instrument distinguishes which of the 10 lightest elements make up these particles. It will look for such particles in Saturn's proposed radiation belts.

The instrument consists of three three-element, solid state telescopes. A high-energy telescope measures the flux or protons between 56 and 800 million electron-volts. A medium-energy telescope measures protons with energies between 3 to 22 million electron-volts, and identifies the eight elements from helium to oxygen. The low-energy telescope studies the flux of electrons between 50,000 electron-volts and 1 million electron-volts and protons between 50,000 electron-volts and 20 million electron-volts.

The instrument weighs 3.2 kilograms (7 pounds) and uses 2.2 watts of power.

Geiger Tube Telescopes (University of Iowa): This instrument will measure Saturn's proposed radiation belts. The instrument employs seven Geiger-Mueller tubes to survey the intensities, energy spectra and angular distributions of electrons and protons along *Pioneer*'s flight path by Saturn.

The tubes are small cylinders containing gas that generates electrical signals from charged particles.

Three tubes (considered a telescope) are parallel. Three others are in a triangular array to measure the number of multiparticle events (showers) which occur. The combination of a telescope and shower detector will enable experimenters to compare primary with secondary events in Saturn's proposed radiation belts.

Another telescope detects low-energy electrons (those above 40,000 electron-volts). The instrument can count protons with energies above 5 million electron-volts and electrons with energies greater than 50,000 electron-volts.

The instrument weighs 1.6 kilograms (3.6 pounds) and uses 0.7 watts of power.

Trapped Radiation Detector (University of California at San Diego): The nature of particles believed trapped by Saturn, particle species, angular distributions and intensities is determined by this instrument. It measures a very broad range of energies from 0.01 to 100 million electron-volts for electrons and from 0.15 to 350 million electron-volts for protons (Hydrogen nuclei).

The instrument has five detectors to cover the planned energy range. An unfocused Cerenkov counter, which measures direction of particle travel by light emitted in a particular direction, detects electrons of energy above 1 million electron-volts and protons above 450 million electron-volts. A second detector measures electrons at 100,000 electron-volts, 200,000 electron-volts and 400,000 electron-volts.

An omnidirectional counter is a solid state diode which discriminates between minimum ionizing particles at 400,000 electron-volts and high-energy protons at 1.8 million electron-volts.

Twin direct current scintillation detectors for low-energy particles distinguish roughly between protons and electrons because of different scintillation material in each. Their energy thresholds are about 10,000 electron-volts for electrons and 150,000 electron-volts for protons.

The instrument weighs 1.7 kilograms (3.9 pounds) and uses 2.9 watts of power.

Dust, Meteoroids and Interplanetary Dust

Asteroid-Meteoroid Detector (General Electric Co.): This instrument was disabled in Jupiter's intense radiation belts.

Meteoroid Detector (Langley Research Center): To detect the distribution of particles (with masses of about 100 millionth of a gram or greater), scientists use a system of 234 pressure cells mounted on the back of the spacecraft dish antenna. The pressure cells come in 30 x 30-cm (12 x 12-inch) panels, 18 to a panel, with six panels. They have a total area of about one-quarter of a square meter, or one square yard.

Each pressure cell is filled with a gas mixture of argon and nitrogen. When a particle penetrates a cell it empties of gas. A transducer notes this, counting one particle impact per cell. Cell walls are 0.001-inch stainless steel sheet, thin enough to allow penetration of particles with a mass of about 100-millionth of a gram or more.

The total weight of the instrument (panels and electronics) is 1.7 kilograms (3.7 pounds) and it uses 0.7 watts of power.

Celestial Mechanics

Celestial Mechanics (Jet Propulsion Laboratory): This experiment uses the spacecraft as an instrument to determine the mass of Saturn and its satellites by measuring their effects on its trajectory.

Doppler tracking determines spacecraft velocity along the Earth-spacecraft line down to a fraction of a millimeter per second, once per minute. These data are further augmented by optical and radar position measurements of the planets. Computer calculations using the spacecraft trajectory and known planet and satellite orbital characteristics will provide findings on: Saturn's mass, the masses of two of its moons, the rings, the planet's polar flattening and mass of its surface layers and deep interior.

Atmosphere, Temperature, Moon, Interplanetary Hydrogen, Helium and Dust

Ultraviolet Photometer (University of Southern California): During flyby, the ultraviolet photometer will measure scattering of ultraviolet light from the Sun by Saturn, Titan, other moons and the rings in two wavelengths, one for hydrogen and one for helium. It will look for other properties.

Radio telescope measurements show that the solar system is immersed in an interstellar gas of cold neutral hydrogen. By measuring the scattering of the Sun's ultraviolet light the instrument can measure amounts of this neutral hydrogen within the heliosphere.

The instrument has two detectors, one which measures ultraviolet radiation at 1216 angstroms, the other at 584 angstroms — the wavelengths at which hydrogen and helium scatter solar ultraviolet.

The instrument has a fixed viewing angle and will use the spacecraft spin to scan the planet. It weighs 0.7 kilograms (1.5 pounds) and uses 0.7 watts of power.

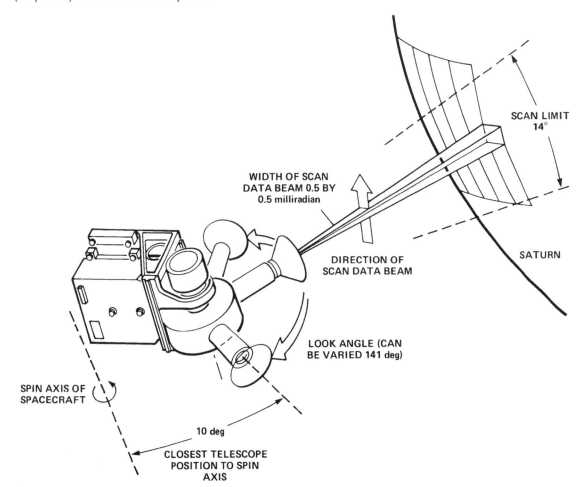

Pictures, Atmospheres and Moon Surfaces

Imaging Photopolarimeter (University of Arizona): This instrument provides data in a number of areas using photometry (measurement of light intensity), polarimetry (photometry measurements of the linear polarization of light) and imaging.

The instrument will take pictures of Saturn, its rings and Titan, as the spacecraft passes the planet.

Images will be made in both red and blue light, and these will be superimposed, providing "color pictures" using red and blue data plus amounts of green based on Earth telescope pictures.

The instrument uses a photoelectric sensor which measures changes in light intensity, something like the light sensor for a television camera. But unlike a TV camera it will employ the 5-rpm spin of the spacecraft to scan the planet, in narrow strips 0.03 degrees wide.

This electronic imaging system will complete the scans for a picture in from 25 to 110 minutes, depending on distance from the planet. Scan data are converted to digital form on the spacecraft and radioed to Earth by telemetry. Engineers then will build up the images, using various computer techniques.

Controllers can vary the viewing angle of the instrument's 8.64-cm (3.4-inch)-focal length telescope by about 160 degrees relative to the spacecraft's spin axis, which is fixed on the Earth-spacecraft line. The telescope can see to within 10 degrees of the spin axis, looking away from the Earth.

Picture reconstruction is complicated by the fact that picture scans will be curved as the telescope rotates with the spacecraft, while looking out at various angles to the spin axis. The only straight-line scans will be those looking straight out from the spin axis.

This factor means that computers will have to sort out scan paths whose curvature changes steadily with spacecraft movement. They will also have to compensate for smear caused by rotation of the planet and motion of the spacecraft.

During flight under the rings, the spacecraft will take no pictures due to a 20 to one "underlap" of picture scans.

During Saturn flyby, experimenters also will use the instrument to study Saturn's clouds, and to determine the nature of atmospheric gas above the clouds, and of aerosols in this gas as well as various other atmosphere properties. The instrument will measure light scattering of the Saturn satellites, to find their surface properties and atmospheres. It will make similar measurements of the rings.

The instrument includes a 2.5-cm (1-inch)-aperture, 8.6-cm (3.4-inch)-focal length telescope which can be moved 160 degrees in the plane of the spacecraft spin axis by ground command or automatically. Incoming light is split by a prism according to polarization into two separate beams. Each beam is further split by going through a red filter (5,940-7,200 angstroms), and through a blue filter (3,900-5,000 angstroms). Channeltron detectors turn the light into electrical impulses, which are telemetered to Earth.

The instrument uses three viewing apertures: one 40 x 40 milliradians for zodiacal light measurements, a second 8 x 8 milliradians for non-imaging light measurements and a third 0.5 x 0.5 milliradians for scans of the planet from which pictures will be reconstructed.

The instrument weighs 4.3 kilograms (9.5 pounds) and uses 2.2 watts of power.

Atmosphere, Ionosphere, Temperature

Infrared Radiometer (California Institute of Technology): Experimenters will measure the net heat energy emitted by Saturn and Titan, as well as vertical and horizontal temperature measurements in Saturn's atmosphere and a rough measurement of the atmosphere's hydrogen / helium ratio.

The two-channel radiometer will make measurements in the 14-25 and 29-56 micron wavelength regions to study the net heat energy flux, its distribution and the thermal structure and chemical composition of Saturn's atmosphere.

The radiometer uses a fixed telescope, so that its scans of the planet are made by rotation of the spacecraft. The instrument will view the planet during the two and a half hours just before closest approach.

The instrument has a 7.2-cm (3-inch)-diameter Cassegrain telescope, and the detectors in its two channels are 88-element thin film, bimetallic thermopiles.

The instrument weighs 2 kilograms (4.4 pounds) and uses 1.3 watts of power.

Occultation Experiment (Jet Propulsion Laboratory): The passage of spacecraft radio signals through Saturn's atmosphere and ionosphere as *Pioneer* swings behind the planet will be used to measure temperature, density and other characteristics of Saturn's ionosphere and Saturn's atmosphere down to a pressure level of about one Earth atmosphere.

Experimenters will use computer analysis of the incoming radio signals recorded on tape to determine the refractive index profile of Saturn's atmosphere.

Encounter Operations

During encounter operations, the many functions on the spacecraft will be monitored in several ways. *Pioneer* returns complete engineering data on its subsystems every 48 seconds. This is scanned by computer. With uncommanded changes, or functions beyond normal limits, the computer presents alarm signals on controllers' data displays. In addition, analysts will review spacecraft data every 10 minutes.

Pioneer control and spacecraft operations are at the Pioneer Mission Operations Center, NASA Ames Research Center, Mountain View, California. The operations center has computing capability both for commanding the spacecraft and to interpret the data stream as it comes in from the Deep Space Network stations.

Several Ames organizations direct and support the encounter.

The Pioneer Mission Operations Team consists of personnel from many government and contractor organizations and operates under control of the flight director. In addition to several assistant flight directors, the operations team includes the following groups:

- The spacecraft Performance Analysis Team analyzes and evaluates spacecraft performance and predicts spacecraft responses to commands.
- The Navigation and Maneuvers Team handles spacecraft navigation and attitude control.
- The Science Analysis Team determines the status of the onboard scientific instruments and formulates command sequences for the instruments.

Tracking and Data Retrieval

With facilities located at 120 degree intervals around the Earth, NASA's Deep Space Network will support *Pioneer* continuously during encounter. As the spacecraft "sets" at one station, due to the Earth's rotation, it will "rise" at the next one.

The network maintains three highly sensitive 64-meter (210-foot) dish antennas at Goldstone, California; Madrid, Spain; and Canberra, Australia. As a backup, it has six 26-meter (85-foot) dish antenna stations at Madrid (Robledo de Chavela and Cebreros), Canberra and two stations at Goldstone.

The 26-meter (85-foot) stations have enough power to command *Pioneer* at Saturn. Where necessary they can receive data, but at the extremely low rate of 128 bits per second, compared with the 1,024 bits-per-second rate of the 64-meter (210-foot) antennas.

At periapsis, there will be a four-hour overlap between the Goldstone and Canberra 64-meter (210-foot) stations, with both receiving this vital data. However, Goldstone loses the spacecraft about one hour after closest approach. Canberra will handle the last part of periapsis operations. This will mean long-distance communications from Australia to Pioneer Control at Ames.

As *Pioneer* is drawn in by Saturn's gravity, increases in speed will cause the doppler shift to grow very large. This will require "ramping" operations at tracking stations, complicated by communications cut-off when *Pioneer* passes behind the planet. Also, at Saturn, round-trip communications time will be almost three hours.

All stations have general-purpose telemetry equipment capable of receiving, data-synchronizing, decoding and processing data at high transmission rates.

The tracking and data acquisition network is tied to the Jet Propulsion Laboratory's Mission Control and Computing Center at Pasadena, California, by NASA's Communications Network.

The Deep Space Network provides tracking information on course and direction of the flight, velocity and range from Earth. It also receives engineering and science telemetry, including planetary television coverage, and sends commands for spacecraft operations. All communications links are in the S-band frequency.

The stations relay spacecraft doppler tracking information to the control center, where computers calculate orbits in support of navigation and planetary target planning. High-speed data links from all stations are capable of 7,200 bits per second, permitting real-time transmission of all data from spacecraft to Pioneer control centers at Ames or JPL.

All networks are under the direction of the Office of Space Tracking and Data Systems, NASA Headquarters, Washington, D.C. The Jet Propulsion Laboratory manages the Deep Space Network, while the Goddard Space Flight Center, Greenbelt, Maryland, manages all the others.

The Goldstone stations are operated and maintained by the Jet Propulsion Laboratory with the assistance of Bendix Field Engineering. The Canberra station is operated by the Australian Department of Supply. The two facilities near Madrid are operated by the Spanish government's Instituto Nacional de Tecnica Aerospacial.

The Spacecraft

Pioneer 11 and *Pioneer 10* are the first spacecraft designed to travel into the outer solar system and operate there for many years. They must have extreme reliability, have communications systems for distances of billions of miles, and employ non-solar power sources. Designers chose spin stabilization for its simplicity and effectiveness.

Launch energy requirements to the outer planets are far greater than for short missions, so the spacecraft is very light. *Pioneer Saturn* weighed only 270 kilograms (570 pounds) at launch. This included 30 kilograms (65 pounds) of scientific instruments.

For reliability, spacecraft builders have employed an intensive screening and testing program for parts and materials. They have selected components designed to withstand radiation from the spacecraft's nuclear power source and from planetary radiation belts. In addition, key systems are redundant. (That is, two of the same components or subsystems are provided in case one fails.) Communications, command and data return systems, propulsion electronics, thrusters and attitude sensors are largely redundant.

The Earth-facing dish antenna is in effect the forward end of the spacecraft. *Pioneer Saturn* is 2.9 meters (9.5 feet) long, measuring from its farthest forward component, the medium-gain antenna horn, to its farthest rearward point, the tip of the aft-facing omnidirectional antenna. Exclusive of booms, its widest crosswise dimension is the 2.7-meter (9-foot) diameter of the dish antenna.

The axis of spacecraft rotation and the centerline of the dish antenna are parallel, and *Pioneer* spins constantly for stability.

The spacecraft equipment compartment is a hexagonal box, roughly 35.5 cm (14 inches) deep. Each of its six sides is 71 cm (28 inches) long. One side joins to a smaller box also 35.5 cm (14 inches) deep, whose top and bottom are irregular hexagons.

This smaller compartment contains most the 12 onboard scientific experiments. However, 12.7 kilograms (28 pounds) of the 30 kilograms (65 pounds) of scientific instruments (the plasma analyzer, cosmic ray telescope, four asteroid-meteoroid telescopes, meteoroid sensors and the magnetometer sensors) are mounted outside the instrument compartment. The other experiments have openings for their sensors. Together both compartments provide 1.4 square meters (16 square feet) of platform area.

Attached to the hexagonal front face of the equipment compartment is the 2.7-meter (9-foot)-diameter, 46-cm (18-inch)-deep dish antenna.

The high-gain antenna feed and the medium-gain antenna horn are mounted at the focal point of the antenna dish on three struts projecting about 1.2 meters (4 feet) forward of the rim of the dish. The low-gain, omnidirectional spiral antenna extends about 0.76 meters (2.5 feet) behind the equipment compartment.

Two three-rod trusses, 120 degrees apart, project from two sides of the equipment compartment, supporting the spacecraft's nuclear electric power generators about 3 meters (10 feet) from the center of the spacecraft. A boom, 120 degrees from each of the two trusses, projects from the experiment compartment and positions the helium vector magnetometer sensor 6.6 meters (21.5 feet) from the spacecraft center. The booms are extended after launch.

At the rim of the antenna dish, two Sun sensors are mounted. A star sensor looks through an opening in the equipment compartment and is protected from sunlight by a hood.

Both compartments have aluminum frames with bottoms and side walls of aluminum honeycomb. The dish antenna is made of aluminum honeycomb.

Rigid external tubular trusswork supports the dish antenna, three pairs of thrusters located near the rim of the dish, radioisotope thermoelectric generator trusses and launch vehicle attachment ring. The message plaque also is attached to this trusswork.

The Heliosphere

On its trip to Saturn from Jupiter, *Pioneer 11* explored intensively a new segment of the heliosphere (the atmosphere of the Sun). The spacecraft has surveyed a region of space 870 million kilometers (540 million miles) wide between the outer edge of the Asteroid Belt and the orbit of Saturn.

Pioneer also studied this region at points up to 15.6 degrees (160 million kilometers) above the ecliptic (Earth's orbit plane), far higher than measurements previously have been made. A major result of this flight was discovery of the basic dipole structure of the Sun's magnetic field.

After leaving Saturn, *Pioneer* will study the outer heliosphere until spacecraft communication ceases, perhaps around 1990 near the orbit of Neptune.

The thinly diffused solar atmosphere is hundreds of times less dense than the best vacuums on Earth. Yet it is important because it contains:

- The ionized gas known as the solar wind, roughly a 50-50 mixture of protons (hydrogen nuclei) and electrons. It flows out from the 3,600,000 degrees F (2,000,000 degrees C) corona of the Sun in all directions at average speeds of 1.6 million km/hr (one million mph).
- Complex magnetic and electric fields, carried out from the Sun by the solar wind.
- Solar cosmic rays, high-energy particles thrown out by the huge explosions on the Sun's surface at up to 480 million km/hr (300 million mph).

The heliosphere also encompasses comets and dust. It is traversed by electromagnetic radiation from the Sun: radio waves, infrared, ultraviolet, and visible light.

The heliosphere further contains cosmic ray particles from within and beyond our galaxy, traveling at nearly the speed of light. This means enormous particle energies, up to 10^{14} million electron-volts. (10^{14} is 1 followed by 14 zeros.) There are also neutral hydrogen atoms from the interstellar gas which formed the Sun and planets.

Study of these phenomena has many applications. Storms of solar particles striking Earth interrupt radio communications and sometimes electric power transmission.

There is evidence that solar storm particles travel through the heliosphere and trigger the Earth's long-term weather cycles.

The heliosphere can be thought of as a huge laboratory where phenomena occur that cannot be simulated on Earth. For example, man cannot accelerate particles in Earth laboratories to the near-light speeds reached by galactic cosmic ray particles. These particles are observed by *Pioneer* instruments.

Solar Wind, Magnetic Field, and Solar Cosmic Rays

Near the Earth, the speed of the solar wind varies from one to three million km/hr (600,000 to 2,000,000 mph), depending on activity of the Sun. Its temperature varies from 10,000 to 1,000,000 degrees C (18,000 to 1,800,000 degrees F). The wind also fluctuates due to features of the rotating solar corona, where it originates, and because of various wave phenomena.

The fastest solar cosmic ray particles jet out from the Sun in streams which can cover the 150 million kilometers (93 million miles) to Earth in as little as 20 minutes. Slower particle streams take one or more hours to reach Earth.

The positive ions from the Sun are 90 percent protons and 10 percent helium nuclei, with occasional nuclei of heavier elements.

There are from none to hundreds of flare events on the Sun each year which produce high-energy solar particles, with the largest number of flares at the peak of the 11-year cycle.

The best resolution yet seen of elements making up the solar high-energy particles has been obtained by *Pioneer 10* from the August 1972 solar flare. It identified sodium and aluminum for the first time, determined relative abundances of helium, carbon, nitrogen, oxygen, fluorine, neon, sodium, magnesium, aluminum, and silicon nuclei coming from the Sun.

The solar wind, as viewed near the Earth's orbit, is most dramatically seen as containing high-speed, up to 2,520,000 km/hr (1,566,000 mph) streams embedded in slower-speed, 1,000,000 km/hr (624,000 mph) streams. These high-speed streams are presumably caused by high-temperature regions in the solar corona.

Between the Earth and Jupiter these high-speed streams speed up the gas ahead of them and they in turn are slowed down. This accounts for the large amount of turbulence in the solar wind and magnetic field still observed as far out as 510 million kilometers (420 million miles). In addition, this stream interaction also produces thermal energy so that the solar wind does not cool as expected for a simple expansion model. Solar wind streams become much more regular beyond the orbit of Saturn.

The Interstellar Gas

Pioneer Saturn measures neutral helium and hydrogen. Both hydrogen and helium atoms are believed to be part of the interstellar gas which has forced its way into the heliosphere as the solar system moves through interstellar space at 72,000 km/hr (45,000 mph).

The *Pioneer* ultraviolet instruments have been able to locate concentrations of neutral hydrogen and helium and to make separations between gas originating in the Sun and that penetrating the solar system from interstellar space.

The experimenters have measured the ultraviolet light glow emitted by interplanetary neutral hydrogen and helium atoms. The interstellar neutral hydrogen appears to enter the heliosphere in the plane of the Earth's orbit at around 100,000 km/hr (62,000 mph). Surprisingly, this entry point is about 60 degrees away from the direction of travel of the solar system through interstellar space, and hence the direction from which these particles ought to come.

Galactic Cosmic Rays

Galactic cosmic ray particles usually have far higher energies (velocities) than solar cosmic rays. These particles may get their tremendous energies from the explosion of stars (supernovas), the collapse of stars (pulsars), or acceleration in the colliding magnetic fields of two stars. Pioneer studies of these particles may settle questions of their origin in our galaxy and important features of the origin and evolution of the galaxy itself. These studies should answer such questions as the chemical composition of stellar sources of cosmic ray particles in the galaxy.

Galactic cosmic rays consist of protons (hydrogen nuclei), 85 percent; helium nuclei, 23 percent; nuclei of other elements, 2 percent; and high-energy electrons, 1 percent.

Because of the expected decrease in shielding by the solar wind and magnetic field, numbers of low-energy cosmic ray particles (0 to 100 million electron-volts) were expected to increase by Saturn distance. This did not occur, apparently because despite the decline of solar wind density and magnetic field strength, turbulence remained as high as near Earth. This turbulence shuts out a large part of the low-energy galactic particles from the inner solar system, as far out as Saturn and beyond.

Pioneer Experimenters

See *Experiments and Investigators* in the **Pioneer 11 Press Kit**, page 117.

The Encounter Team

NASA Headquarters
Office of Space Science

Dr. Thomas A. Mutch	Associate Administrator for Space Science
Andrew J. Stofan	Deputy Associate Administrator for Space Science
Dr. Adrienne F. Timothy	Assistant Associate Administrator for Space Science
Angelo Guastaferro	Director, Planetary Programs
Fred D. Kochendorfer	Pioneer Program Manager
Dr. Albert G. Opp	Pioneer Program Scientist

Office of Space Tracking and Data Systems

Dr. William C. Schneider	Associate Administrator for Space Tracking and Data Systems
Charles A. Taylor	Director, Network Operations

Ames Research Center

Clarence A. Syvertson	Director
A. Thomas Young	Deputy Director
Dr. Dean Chapman	Director of Astronautics
Charles F. Hall	Pioneer Project Manager
Dr. John H. Wolfe	Pioneer Project Scientist
Robert P. Hogan	Chief, Flight Operations
Henry Asch	Reliability and Quality Assurance
John W. Dyer	Chief, Mission Analysis

NASA Jet Propulsion Laboratory

Dr. Bruce C. Murray	Director
W.P. Spaulding	Tracking and Data Systems Manager
Richard Miller	Pioneer Tracking and Data Systems Manager
Robert Ryan	Pioneer Project Support Team Manager
William E. Kirhofer	Navigation

Department of Energy
Advanced Nuclear Systems and Projects Division

David S. Gabriel	Director
Glenn A. Newby	Associate Director, Space Nuclear Systems Division
Harold Jaffe	Manager, Isotope Flight Systems Office

TRW Systems Group

Bernard J. O'Brien	Pioneer Project Manager

Bendix Field Engineering Corporation

Thomas F. Groves	Pioneer Program Manager
Pat Barclay	Deputy Pioneer Program Manager

Pioneer Saturn Contractors

See *Pioneer F Contractors* in the **Pioneer 10 Press Kit**, page 44.

National Aeronautics and Space Administration, Ames Research Center, Moffett Field, CA

March 2003

Astrogram

Communication for the Information Technology Age

Pioneer 10 Finally Sends Last Signal To Earth

By Michael Mewhinney

After more than 30 years, it appears the venerable *Pioneer 10* spacecraft has sent its last signal to Earth. *Pioneer*'s last, very weak signal was received on January 22, 2003.

NASA engineers reported that *Pioneer 10*'s radioisotope power source has decayed, and it may not have enough power to send additional transmissions to Earth. NASA's Deep Space Network (DSN) did not detect a signal during the last contact attempt on February 7. The previous three contacts, including the January 22 signal, were very faint, with no telemetry received. The last time a *Pioneer 10* contact returned telemetry data was April 27, 2002. NASA has no additional contact attempts planned for *Pioneer 10*.

"*Pioneer 10* was a pioneer in the true sense of the word. After it passed Mars on its long journey into deep space, it was venturing into places where nothing built by humanity had ever gone before," said Dr. Colleen Hartman, director of NASA's Solar System Exploration Division, NASA Headquarters. "It ranks among the most historic, as well as the most scientifically rich, exploration missions ever undertaken," she said.

Originally designed for a 21-month mission, *Pioneer 10* exceeded all expectations and lasted more than 30 years. "It was a workhorse that far exceeded its warranty and I guess you could say we got our money's worth," said *Pioneer 10* Project Manager Larry Lasher of Code SFS.

Pioneer 10 was built by TRW Inc., Redondo Beach, California, and was launched on March 2, 1972 on a three-stage Atlas-Centaur rocket. *Pioneer 10* reached a speed of 32,400 mph needed for the flight to Jupiter, making it the fastest human-made object to leave Earth; fast enough to pass the Moon in 11 hours and to cross Mars' orbit, about 50 million miles away, in just 12 weeks.

On July 15, 1972, *Pioneer 10* entered the asteroid belt, a doughnut-shaped area that measures some 175 million miles wide and 50 million miles thick. The material in the belt travels at speeds up to 45,000 mph and ranges in size from dust particles to rock chunks as big as Alaska.

Artist rendition of the Pioneer 10 spacecraft shown in orbit around Jupiter.

Pioneer 10 was the first spacecraft to pass through the asteroid belt, considered a spectacular achievement, and then headed toward Jupiter. Accelerating to a speed of 82,000 mph, *Pioneer 10* passed by Jupiter on December 3, 1973.

The spacecraft was the first to make direct observations and obtain close-up images of Jupiter. *Pioneer 10* also charted the gas giant's intense radiation belts, located the planet's magnetic field and established that Jupiter is predominantly a liquid planet. In 1983, *Pioneer 10* became the first human-made object to pass the orbit of Pluto, the most distant planet from the sun.

ollowing its encounter with Jupiter, *Pioneer 10* explored the outer regions of the solar system, studying energetic articles from the sun (solar wind) and cosmic rays entering our portion of the Milky Way. The spacecraft continued o make valuable scientific investigations in the outer regions of the solar system until its science mission ended on 1arch 31, 1997.

ince that time, *Pioneer 10*'s weak signal has been tracked by the DSN as part of a new advanced concept study of ommunication technology in support of NASA's future Interstellar Probe mission. At last contact, *Pioneer 10* was 7.6 illion miles from Earth, or 82 times the nominal distance between the sun and the Earth. At that distance, it takes 1ore than 11 hours and 20 minutes for the radio signal, traveling at the speed of light, to reach the Earth.

From Ames Research Center and the Pioneer Project, we send our thanks to the many people at the Deep Space 1etwork and the Jet Propulsion Laboratory (JPL) who made it possible to hear the spacecraft signal for this long," said *ioneer 10* Flight Director David Lozier, also of Code SFS at Ames.

ioneer 10 explored Jupiter, traveled twice as far as the most distant planet in our solar system, and as Earth's first missary into space, is carrying a gold plaque that describes what we look like, where we are and the date when the 1ission began. *Pioneer 10* will continue to coast silently as a ghost ship through deep space into interstellar space, eading generally for the red star Aldebaran, which forms the eye of the constellation Taurus (the bull). Aldebaran is bout 68 light years away. It will take *Pioneer 10* more than 2 million years to reach it. Its sister ship, *Pioneer 11*, ended s mission on September 30, 1995, when the last transmission from the spacecraft was received.

Pioneer spacecraft in assembly jig

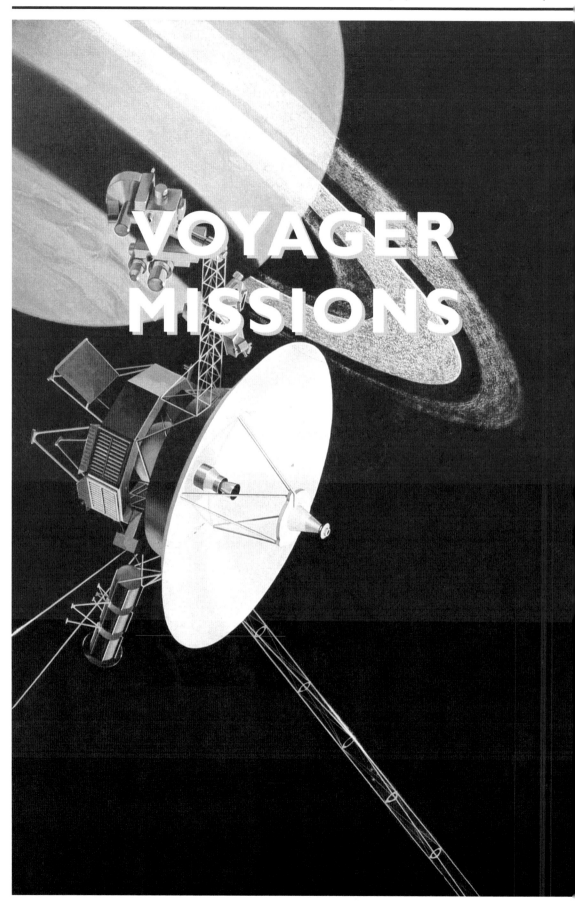

National Aeronautics and
Space Administration

Prepared by Betty Shultz

Jet Propulsion Laboratory
California Institute of Technology
Pasadena, California 91109

Office of Public Information
Telephone (213) 354-5011

Voyager Mission Data Summary

(Times are Pacific Daylight Time)

	Voyager 1	Voyager 2
1. Objective	Fly past and study Jupiter and five of its major satellites (Io, Europa, Ganymede, Callisto, and Amalthea), and Saturn and four of its satellites (Titan, Tethys, Enceladus, and Rhea).	Fly past and study Jupiter and five of its major satellites (Io, Europa, Ganymede, Callisto, and Amalthea); Saturn and nine of its satellites (Iapetus, Hyperion, Titan, Dione, Mimas, Enceladus, Tethys, Rhea, Phoebe); and possibly on to Uranus and Neptune.
2. Project Managers	Raymond Heacock (Jupiter 1 & 2, Saturn 1) Esker Davis (Saturn 2) Richard Laeser (Post-encounter)	Same
Project Scientist	Edward C. Stone, Caltech	Same
3. Launch Date	September 5, 1977	August 20, 1977
4. Encounter Dates	Jupiter: March 5, 1979 Io: March 5, 1979 Ganymede: March 5, 1979 Callisto: March 6, 1979 Europa: March 5, 1979 Amalthea: March 5, 1979 Saturn: November 12, 1980 (3:46 pm) Titan: November 11, 1980 Tethys: November 12, 1980 Enceladus: November 12, 1980 Rhea: November 12, 1980	Jupiter: July 9, 1979 Io: July 9, 1979 Ganymede: July 9, 1979 Callisto: July 8, 1979 Europa: July 9, 1979 Amalthea: July 9, 1979 Saturn: August 25, 1981 (8:25 pm) Iapetus: August 22, 1981 Hyperion: August 24, 1981 Titan: August 25, 1981 Dione: August 25, 1981 Mimas: August 25, 1981 Enceladus: August 25, 1981 Tethys: August 25, 1981 Rhea: August 25, 1981 Phoebe: September 4, 1981 Uranus: January, 1986 Neptune: September, 1989
5. Spacecraft-to-Earth Distance At Arrival	Jupiter: 428 million miles Saturn: 948 million miles	Jupiter: 576 million miles Saturn: 967 million miles
6. Flyby Distance	Jupiter: 172,750 miles Io: 11,800 miles Ganymede: 69,600 miles Callisto: 77,000 miles Europa: 455,000 miles Amalthea: 261,000 miles Saturn: 77,340 miles Titan: 2,500 miles Tethys: 258,000 miles Enceladus: 125,800 miles Rhea: 44,700 miles Dione: 100,000 miles Hyperion: 546,000 miles Ipaetus: 1,537,000 miles	Jupiter: 400,000 miles Io: 702,000 miles Ganymede: 37,000 miles Callisto: 136,000 miles Europa: 125,000 miles Amalthea: 347,076 miles Saturn: 63,000 miles Iapetus: 560,000 miles Titan: 413,000 miles Hyperion: 300,000 miles Tethys: 58,000 miles Enceladus: 54,000 miles Rhea: 401,000 miles Mimas: 193,000 miles Dione: 312,000 miles Phoebe: 1,290,000 miles Uranus: Neptune:
7. Arc Distance	Jupiter: 595 million miles Saturn: 1.349 billion miles	Jupiter: 577 million miles Saturn: 1.422 billion miles

Voyager Mission Data Summary

(continued)

	Voyager 1	*Voyager 2*
8. Built By	JPL	Same
9. Weight of Spacecraft	1,797 lbs; Instruments: 254 lbs	Same
10. Instruments on Spacecraft	a) Wide-angle and narrow-angle television cameras (P.I.: Dr. Bradford A. Smith, University of Arizona) b) Cosmic-ray detectors (P.I.: Dr. Rochus E. Vogt, Caltech) c) Infrared spectrometers and radiometers (P.I.: Dr. Rudolf A. Hanel, Goddard Space Flight Center) d) Low-energy charged-particle detectors (P.I.: Dr. S. M. Krimigis, Johns Hopkins University) e) Magnetometers (P.I.: Dr. Norman Ness, Goddard Space Flight Center) f) Photopolarimeter (P.I. Dr. Arthur L. Lane, JPL) g) Planetary radio astronomy instruments (P.I.: Dr. James W. Warwick, Department of Astro-Geophysics, University of Colorado) h) Plasma experiment (P.I.: Dr. Herbert Bridge, Massachusetts Institute of Technology) i) Plasma wave experiment (P.I.: Dr. Frederick L. Scarf, TRW) j) Ultraviolet spectrometers (P.I.: Dr. A. Lyle Broadfoot, Kitt Peak National Observatory)	
11. Cost of Project	Cost of Voyager 1 & 2: $338 million	
12. Pictures Returned By Spacecraft	More than 35,000 photos of Jupiter were returned by both Voyager 1 and 2; more than 33,000 of Saturn.	
13. Responsible Managing Center	JPL	Same
14. Written Material Available	a) JPL / PIO Voyager Background Kit b) JPL / PIO Voyager Press Kits c) JPL Status Bulletins d) JPL / PIO Voyager Fact Sheet e) Voyager color brochure f) National Geographic Magazine (January 1980) g) Science Magazine Reprints (June 1979, November 1979) h) Nature Magazine (August 1979) i) Sky & Telescope (September 1979)	

NASA Facts

An Educational Publication
of the
National Aeronautics and
Space Administration

NF-89

The Voyager Mission

Jupiter, The Giant of the Solar System

CONTENTS

In orbit 780 million kilometers (485 million miles) from the Sun, Jupiter is four and a half times as distant from the Sun as is the Earth. The planet is unusual by Earthly standards; it is only slightly denser than water, but 318 times more massive than the Earth. Because of this great mass, which is more than that of all the other planets combined, Jupiter dominates the Solar System after the Sun. Its powerful gravity pulls the orbits of the other planets out of shape, and may have prevented the asteroids from forming a planet. The orbits of comets are also affected by Jupiter; some comets have even been captured by the big planet and are under its control.

The giant Jupiter differs greatly from Earth and the terrestrial rocky planets, Mercury, Venus and Mars, in that Jupiter does not appear to have any solid surface, but is a great ball of liquid topped by a thin shell of gas. It consists mainly of hydrogen, with a small quantity of helium and traces of ammonia, methane, water, ethane, and acetylene. The interior of the planet is believed to change gradually with depth from a gaseous mixture of hydrogen and helium – about one atom of helium for every ten molecules of hydrogen – to an interior of liquid metallic hydrogen. There may also be a small rocky core.

After the Sun, Jupiter is also the noisiest source of radio signals in the solar system. The radiation we receive at Earth from Jupiter consists of three types. Thermal radio waves are produced by molecules moving about in the atmosphere of the giant planet. Radio waves in the decameter range are produced by electrical discharges in the atmosphere and are somehow influenced by the movement of one of the satellites, Io, along its orbit. Radio waves in the decimeter range are produced by electrons in the magnetosphere of Jupiter. It was these decimetric waves that led scientists to conclude that Jupiter has a magnetic field like that of the Earth because the particles responsible for generating them must be trapped in such a field. This magnetic field and magnetosphere of Jupiter was confirmed by NASA's *Pioneer* spacecraft flying by the planet in 1973 and 1974.

The visible surface of Jupiter is the top of turbulent clouds. Estimates of the depth of the atmosphere beneath these clouds vary from 100 to 5,800 kilometers (60 to 3,600 miles). Below the atmospheric shell is a deep ocean of liquid hydrogen. Astronomers agree generally about the internal structure of Jupiter (Figure 1), but they differ about the details. At the high pressure deep within the planet, hydrogen

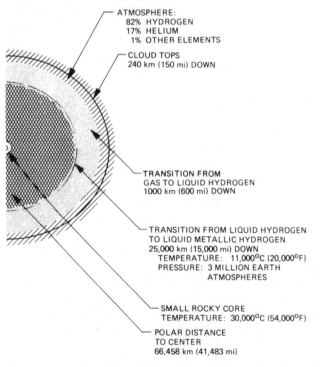

ATMOSPHERE:
82% HYDROGEN
17% HELIUM
1% OTHER ELEMENTS

CLOUD TOPS
240 km (150 mi) DOWN

TRANSITION FROM
GAS TO LIQUID HYDROGEN
1000 km (600 mi) DOWN

TRANSITION FROM LIQUID HYDROGEN
TO LIQUID METALLIC HYDROGEN
25,000 km (15,000 mi) DOWN
TEMPERATURE: 11,000°C (20,000°F)
PRESSURE: 3 MILLION EARTH
ATMOSPHERES

SMALL ROCKY CORE
TEMPERATURE: 30,000°C (54,000°F)

POLAR DISTANCE
TO CENTER
66,458 km (41,483 mi)

Figure 1. Diagram of the interior of the planet Jupiter according to a commonly accepted model.

becomes metallic in form and readily conducts heat and electricity. There is a shell of this liquid metallic hydrogen surrounding a rocky core, which also probably contains heavier metals. The helium of the planet might be mixed with the hydrogen in solution, or might have condensed into a shell of liquid helium surrounding the core. At the core temperatures might be as high as 30,000°C (54,000°F). Such a hot core could account for Jupiter radiating into space some two and a half times as much heat as it receives from the Sun.

At one time astronomers thought that Jupiter generates heat by the pressure of its own gravity compressing its interior and causing it to contract in size. Now some astronomers favor the explanation that Jupiter is hot today because it is still cooling down from its formation. As it cools it continues to contract. Jupiter is too small for the weight of its massive bulk to compress the material at its center sufficiently to ignite nuclear reactions there. Had Jupiter been larger it might have formed a second sun, so that the Solar System would have become a binary star system.

Origin of Jupiter

Scientists think that the cloud of gas from which Jupiter was formed gave birth to the giant planet about 4½ billion years ago by a 70,000 year convulsive contraction that was followed by a headlong collapse to form the planet in a brief three months. The initial cloud from which Jupiter originated was probably 650 million kilometers (400 million miles) in diameter, but within the 70,000 years of initial formation it rapidly shrank and heated from a temperature of -218°C (-360°F) to 2,200°C (4,000°F). The hydrogen molecules ionized and broke up, and the protoplanet collapsed within a few months to about five times its present diameter. As it did, enormous quantities of heat were developed by the tremendous pressures. The interior of Jupiter reached a temperature of 50,000°C (90,000°F), and the planet glowed red hot at its surface. Over the billions of years to the present, the planet has gradually cooled. Evidence for this red hot phase is thought to be the decreasing density of the big satellites of Jupiter with increasing distance from the planet. The innermost moons were heated so much by the hot planet that most lightweight substances (volatiles) were driven into space. Water was prevented from condensing or remaining in any substantial quantities on them.

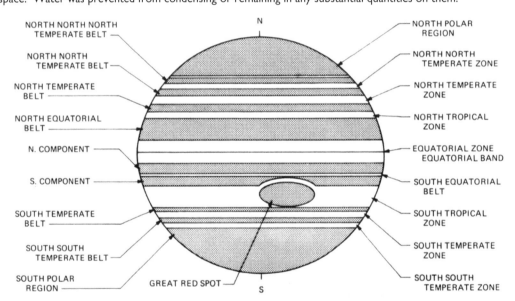

Figure 2. The visible surface of Jupiter consists of a banded structure of clouds that are consistent enough to have been given the names shown in this drawing.

Jupiter as Seen from Earth

Jupiter completes an orbit about the Sun in 11.86 years. When Jupiter appears opposite to the Sun as seen from Earth, astronomers say that the planet is in opposition. At this time the planet is closest to the Earth and shines its brightest in the night sky, directly south at midnight. During NASA's Voyager mission to Jupiter, oppositions of the planet will occur during the winter months December 1977 in the constellation of Taurus, January 1979 in Gemini, and February 1980 in Cancer. Even small telescopes will reveal bands of clouds and show the four big satellites as star-like objects moving in their orbits about Jupiter.

Jupiter completes a rotation on its axis in 9 hours, 55 minutes, and 30 seconds. This rapid rotation has flattened the poles and bulged the equator of Jupiter. The polar and equatorial diameters are 133,516 kilometers and 142,796 kilometers (82,966 miles and 88,734 miles) respectively. Seen by telescope from Earth, Jupiter appears as a striped banded disc (Figure 2), with all the markings parallel to the equator. Dusky gray regions cover each pole. The dark late gray stripes are called belts. They are believed to be regions of descending air masses. The lighter, salmon colored bands, called zones, are believed to be rising cloudy air masses. A transparent atmosphere extends high above the visible cloud tops of the belts and zones.

Among the belts and zones are many smaller features – streaks, wisps, arches, plumes and spots – some of which may be knots of clouds. They change form rapidly even though they are thousands of kilometers in size. This indicates a very turbulent atmosphere.

The cloud features move around Jupiter more slowly at increasing distance from the equator. A great equatorial current sweeps around the planet some 400 km/hr (250 miles/hr) faster than the regions on either side of it. Telescopic observations show that air masses move parallel to Jupiter's equator rather than from the equator toward the poles, as on Earth. Cloud motions have been measured in the zones that indicate speeds of 320 km/hr (200 miles/hr) relative to the whole planet.

The bands around Jupiter are explained as convective cells in the atmosphere that are stretched out because of the planet's rapid rotation. At the top of these light zones, the cooler material moves toward the equator and toward the pole and is deflected in an east-west direction by the rotation of the planet before it sinks back to lower altitude in the dark belts. Above 50 degrees north and below 50 degrees south latitudes, the banded structure ends and the atmosphere is turbulently disorganized into many convective cells (Figure 3).

In the southern hemisphere is a 24,000-kilometer (15,000-mile)-diameter oval feature known as the Great Red Spot (Figure 4), which has intrigued astronomers and theorists since it was first observed centuries ago. The spot has varied in intensity, color, and size over the years, and several times it has almost completely disappeared. There are also smaller red spots and white spots within the turbulent cloud systems of Jupiter.

Scientists now think they have discovered some of the main features of Jupiter's atmosphere and weather. The weather probably occurs in a thin skin of the atmosphere as it does on Earth. The bottom of the weather layer is at the base of the lowest clouds of water, below which is a relatively uniformly heated deep atmosphere of hydrogen and helium. Instead of moving from the equator to the poles and back, as on Earth, Jupiter's weather circulation seems to consist of a relatively local turbulence in the polar regions, and a flow parallel to the equator in the temperate and equatorial regions. This means that weather circulation on the whole planet is somewhat like weather in the Earth's tropical regions.

The heat rising from the interior of Jupiter appears to be shunted away internally from the equatorial regions, which are warmed by incoming solar radiation. The result is that the planet behaves as though controlled by a giant thermostat so that its atmosphere has a temperature at the polar regions similar to that at the equator, and in both day and night hemispheres. If there is, indeed, no flow of heat in the atmosphere from the equator toward the poles, the weather on Jupiter must be driven by the internal heat, not by solar heat as on Earth.

Zones are believed to be similar to cyclones on Earth, but on Jupiter they are stretched out around the planet (Figure 5). Heat released from the condensation of water within rising masses of atmosphere called cells could drive the rising cells even higher. More water-laden atmosphere would be sucked into the rising columns and reinforce this effect.

Figure 3. This enlargement of a Pioneer 11 picture covers part of Jupiter's north temperate zone and its north polar region. It shows the breakup of the regular banded structure of Jupiter's clouds northward toward the pole.

The zones may thus be planet-girdling, cyclone-type features consisting of long strips of rising atmosphere powered endlessly by the condensation of water vapor. There are, however, other explanations for Jupiter's weather systems, such as a slight equator to pole heat flow, and the effects of the huge size of the planet on the way fluids flow within it. Also, the liquid interior of Jupiter may affect convective flow in a way that gives rise to zones in the atmosphere of differing heat, which causes them to rise and fall. The NASA *Voyager* spacecraft will gather new information that shoul‹ allow scientists to choose between the various current theories. It seems clear, however, that the basic difference⸱ between weather of the Earth and of Jupiter arise from the lack of a solid surface on Jupiter and from internal heatin‹ of the giant planet as opposed to the Earth's heating from the Sun. Clouds and atmospheric features can go unchange‹ for decades or for centuries on Jupiter because there is very little friction between the thin cloud layers and th‹ compressed atmospheric gas below them. There is no solid surface like that of the Earth to affect the flow ‹ atmospheric gases.

Figure 4. In the southern hemisphere of Jupiter there is a 25,000 kilometer long oval feature called the Great Red Spot, which has been seen on the planet for hundreds of years.

Figure 5. Convection patterns proposed for flow of Jupiter's liquid interior are shown in this diagram. Near the poles, the circular flow of liquid hydrogen by convection (rising of heated material) is conventional. In the equatorial regions, however, there may be a series of turning convective rolls, parallel to Jupiter's rotation axis. Convective flow would occur in the zone of molecular hydrogen and be unable to cross the boundary into the region of liquid metallic hydrogen. Convective flow in a rapidly rotating liquid sphere is different from that in a static liquid.

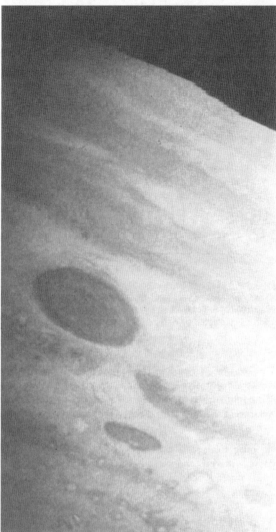

Figure 6. The Great Red Spot, seen in close up by Pioneer 11, displays a mottled appearance within its interior, which may be turbulent cloud systems. Other smaller spots can be seen nearby.

Permanent features seen in the atmosphere, such as the Great Red Spot (Figure 6) could be "freewheeling" phenomena that require relatively small inputs of energy to keep them going. Jupiter's hydrogen atmosphere retains energy more easily than does Earth's atmosphere of nitrogen and oxygen. Consequently, once a free wheeling phenomenon has received energy, a long time can elapse before the energy is dissipated. The Great Red Spot turns like a wheel between the north and south halves of the South Tropical Zone, which is split by the spot. Over the 300 years since it was first observed, the spot has drifted westward at an irregular rate. There are smaller red spots in other zones, but these have not lasted as long as the big spot. The spot has been simulated on computer models by assuming that it is a rising column of gas spun between the halves of the tropical zone in which it is located. Some scientists speculate that this column rises to a height in the atmosphere at which traces of phosphorus can condense and thereby create its deep red color. The top of the red cloud is some 8 kilometers (5 miles) above the surrounding clouds.

The highest clouds on Jupiter are the deepest colored, and the red color may be due to phosphorus. The next highest clouds are white, believed to be ammonia crystal clouds. Below these are gray or brown clouds that may consist of ammonia and sulfur compounds or complex organic molecules. The water clouds, where water condenses and releases latent heat, may form a transition region between the uniformly mixed lower atmosphere and the upper region of weather. (See Figure 7.)

However, we are still quite ignorant of the true nature and origin of the material that makes up the colors of the clouds of Jupiter. The colored material might originate in high temperature chemistry taking place beneath the cloud layers and be swept to high altitudes by the upwelling storm systems. Full disc color movies of Jupiter will be obtained

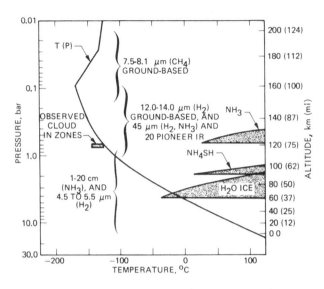

Figure 7. Diagram of the atmosphere of Jupiter showing the position of the main cloud layers and the temperatures and pressures at their levels as calculated by A. Ingersoll of Caltech.

by the *Voyager* spacecraft and may show the interactions between cloud motions and color or the planet.

Atmospheric features on Jupiter will be tracked by *Voyager's* cameras in an attempt to understand the meteorology of Jupiter better. Such understanding will help us also understand the complex terrestrial patterns of weather systems. *Voyager* will be able to show us how the Jupiter weather features change over a period of time and confirm whether or not they are connective storms like huge hurricanes. The Great Red Spot and other smaller spots may be dynamic meteorological phenomena i.e., huge storm systems. The mottled appearance within the Great Red Spot as seen in the *Pioneer* pictures suggests such connective motion. If the Great Red Spot can be photographed at the edge of the planet, we might be able to obtain its profile and find out exactly how high it pokes above the rest of the cloud systems. Differences in color revealed by high magnification imaging may provide us with clues on the composition of the clouds. On the dark side of the planet, we can also look for lightning discharges in an attempt to discover if these are causes of the decametric radio waves.

To explain the Great Red Spot, we need to find out how the smaller red spots form and later fade by looking at them with the high magnification possible with the *Voyager* cameras. Then we can see all the details of their structure and find out how these details change as the spots evolve. At the same time, instruments carried by the *Voyager* spacecraft will look at the thermal and other properties of the atmosphere of Jupiter.

The temperature of the upper atmosphere of Jupiter appears to be higher than would be expected from heating by incoming solar radiation and outgoing heat from the planet itself. Some unknown mechanism could be the cause. Possibly there is an interaction between particles in Jupiter's magnetosphere and the gases of the upper atmosphere. The *Voyager* spacecraft will try to resolve the question by observing ultraviolet radiation from the planet's upper atmosphere. This will provide information about the nature of the processes taking place there.

The Magnetosphere

The magnetosphere of Jupiter (Figure 8) in which protons and electrons are controlled or trapped by the magnetic field of the planet has a volume one million times that of the Earth. Jupiter's magnetosphere is large because relatively low-energy particles spinning off from the top of the ionosphere stretch the magnetic field lines. Out-flowing low-energy particles, however, can produce only a very weak magnetic field at great distances from the planet. High-speed gusts of the solar wind periodically push back this weak field and force the magnetosphere to collapse. As it does, energy is imparted to electrons and they are accelerated almost to the speed of light. They squirt across the solar system to Earth and even to Mercury. One of the unexplained effects is that electrons surge out from Jupiter once every ten hours.

The entire magnetosphere of Jupiter behaves like an enormous, fast rotating, leaky bag, the outer skin of which moves about 2½ times as fast as the 1.6 million km/hr (1 million mph) solar wind, the blizzard of electrons and protons streaming outward from the Sun. The continuous buffeting and shaking of this huge spinning bag of particles as the magnetosphere of Jupiter rubs and bumps against the pressure of the solar wind, interrupts the regular movement of particles from pole to pole within the magnetic field of Jupiter along the field lines, and throws them inward toward the planet, thereby recirculating and energizing them. The magnetosphere contains radiation belts that are 5,000 to 10,000 times as intense as those of Earth. This tremendous intensity may originate from recirculation of particles with increase of their energy each time around.

Particles from the radiation belts around Jupiter move to the turbulent outer magnetosphere along lines connected to the magnetic poles of the planet, which are offset from the rotation poles. There they are redirected by snarled magnetic fields and some of the particles move back to the inner radiation belt. The inward motion increases the energy of these particles. Each time they repeat the cycle they obtain more energy. Although many of the particles are absorbed by the Galilean satellites, some eventually acquire enormous energies and squirt out from Jupiter across the Solar System.

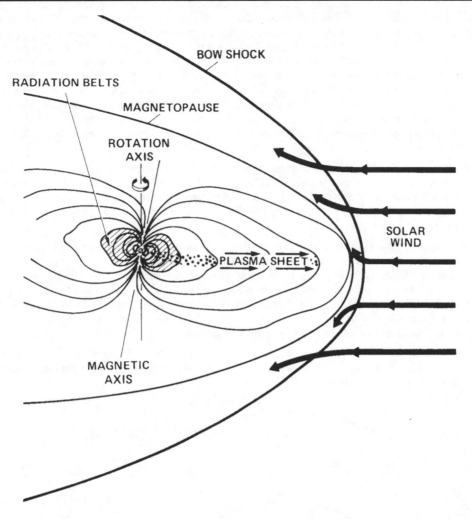

Figure 8. Jupiter's magnetosphere is created by the planet's powerful magnetic field holding ionized particles of the solar wind away from the planet. The magnetic field is at least doubled in size by ionized particles from the top of Jupiter's ionosphere. As these particles are spun out by centrifugal force, they drag the magnetic field with them. Jupiter's magnetosphere has an average diameter of nine million miles, big enough, if it could be seen from Earth, half a billion miles away, to occupy 2 degrees of sky, compared to the 0.5 degrees occupied by the Sun and the Moon.

The radiation belts of Jupiter would probably be 100 times more intense without the sweeping effects of the Galilean satellites and the squirting of particles into interplanetary space. The particles in the radiation belts probably come from the top of Jupiter's ionosphere rather than from the solar wind. By contrast, Earth's radiation belts are made up of particles from the solar wind trapped by Earth's magnetic field.

The magnetosphere of Jupiter has three distinct regions. The inner magnetosphere, like that of the Earth, is doughnut shaped with the planet in the hole. It contains the intense, multi-shelled radiation belts in which three of Jupiter's big moons orbit and constantly sweep out particles. This inner magnetosphere extends about ten diameters from Jupiter; i.e., to about 1,400,000 kilometers (900,000 miles).

A middle magnetosphere contains a current sheet of electrified particles moving out from Jupiter under centrifugal force, thereby stretching magnetic field lines with them. This disc extends between 10 and 30 diameters of Jupiter, to about 4,300,000 kilometers (2,700,000 miles). These particles, which in the middle magnetosphere are mostly electrons, tend to be confined to a thin current sheet like a ring around the planet.

An outer magnetosphere lies between 30 and 45 Jupiter diameters (up to 6,400,000 kilometers or 4,000,000 miles), but buffeting by the solar wind causes the magnetic fields to be irregular and the size of this region to fluctuate widely at different times.

The spacecraft *Pioneers 10* and *11* provided much information about this magnetic environment of Jupiter, which contains the plasma of electrons and protons surrounding the planet, and also how the solar wind interacts with it. The *Voyager* spacecraft in turn are expected to provide even better and more complete information about the region, including the position and shape of the bow shock. This shock is created where the magnetic field of the planet overrides the field carried by the solar wind. The two *Pioneer* spacecraft observed tremendous variations in the location of the magnetopause region between the solar wind and the field of the planet. These variations are still not thoroughly understood. The pressure of the solar wind varies a great deal at different times and it could be that these changes cause the fluctuations in the magnetopause observed by the *Pioneers*. To this end, the *Voyagers'* instruments will attempt to determine the extent and the physical characteristics of the electrons and ions forming the plasma within these three regions of the magnetosphere, particularly where the electrons change from being part of the inner magnetosphere, which rotates with the planet, and the region which is influenced by outward current flow to become a magnetotail streaming behind the planet in a direction away from the Sun. The tail has been detected as far away as the orbit of Saturn.

An important discovery of the *Pioneer* spacecraft was the strongly distorted nature of the outer magnetosphere. The general three-dimensional shape of the magnetosphere is still not thoroughly understood, because it is complicated by the rapid spin of the planet and the effects of the solar wind buffeting the magnetosphere. The *Voyager* mission will provide two more tracks of spacecraft through the complex magnetosphere and thus provide a better basis for our understanding of Jupiter's interaction with the solar wind.

There is a concentration of lines of magnetic force that extends from the surface of Jupiter to the satellite Io. This is referred to as a flux tube and it is somehow connected with the emission of radio waves from Jupiter. The origin of these decametric radiation bursts from Jupiter is still unknown, but it is believed that they may be connected in some way with instabilities of plasma within the ionosphere of Jupiter which are, in turn, caused by a flow of energy down the flux tube from Io. Scientists speculate that these enormous surges of energy might be detected by *Voyager* as ultraviolet emissions on the night side of Jupiter. Also, one of the *Voyager* spacecraft will fly through the flux tube and provide information about its true nature and the energy flowing along it.

This is the third in a series of publications on the Voyager mission to Jupiter and Saturn. The first publication in the series was entitled *Voyager, Mission to the Outer Planets*; the second was entitled *Satellites of the Outer Planets*: a future issue will be devoted to the planet Saturn.

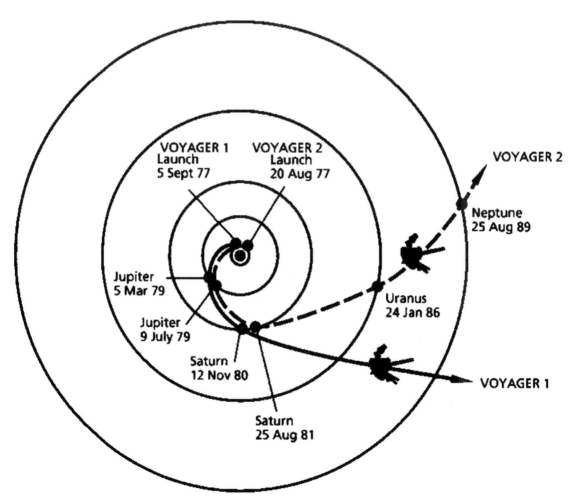

Voyager Bulletin

MISSION STATUS REPORT No. 36   **FEBRUARY 23, 1979**

In transit – *Voyager I* took this photo of Jupiter, Io, and Europa on February 13, 1979. Io is about 350,000 kilometers (220,000 miles) above Jupiter's Great Red Spot, while Europa is about 600,000 kilometers (375,000 miles) above Jupiter's clouds. Although both satellites have about the same brightness, Io's color is very different from Europa's. Io's equatorial region shows two types of material – dark orange, broken by several bright spots – producing a mottled appearance. The poles are darker and reddish. Preliminary evidence suggests color variations within and between the polar regions. Io's surface composition is unknown, but it may be a mixture of salts and sulfur. Europa is less strongly colored, although still relatively dark at short wavelengths. Markings on Europa are less evident than on the other satellites, although this picture shows darker regions toward the trailing half of the visible disk. Jupiter is about 20 million kilometers (12.4 million miles) from the spacecraft at the time of this photo. At this resolution (about 400 kilometers or 250 miles) there is evidence of circular motion in Jupiter's atmosphere. While the dominant large-scale motions are west-to-east, small-scale movement includes eddy-like circulation within and between the bands.

NASA

National Aeronautics and
Space Administration

Jet Propulsion Laboratory
California Institute of Technology
Pasadena California

Encounter Minus 10 Days

Recorded Mission Status (213) 354-7237
Status Bulletin Editor (213) 354-4438
Public Information Office (213) 354-5011

CONTENTS

Mission Highlights

Voyager 1 is no longer "going" to Jupiter – it is there! Every instrument's data are showing strong indications of the planet's presence – and its presence is overwhelming.

Auroral-type activity around Io and Jupiter has been observed by the ultraviolet spectrometer (UVS). The gaseous torus cloud associated with Io does not seem to be composed of readily identifiable neutral atoms or singly ionized atoms.

The plasma wave instrument (PWS) is detecting very low frequency emissions (about 10 to 60 kilohertz) that are not directly related to decametric emissions from Jupiter that have long been observed at Earth. The signals probably originate near or beyond the orbit of Io, and *Voyager 1* should fly through this source area, obtaining direct information on the signals. The calculated power of the signals is about 1 billion watts, about the same as Earth's total radiated power. Jupiter's decametric radiation is about 100 billion watts. (These measurements assume isotropy, that is, that the same values can be measured in any direction.)

In related studies, the planetary radio astronomy (PRA) instrument is seeing arc structures in its radio spectrum data from about 30 megahertz to less than 1 megahertz. Jupiter's magnetic axis is offset from its spin axis by about 11°, so that the north magnetic pole is sometimes tilted toward Earth and the spacecraft. The PRA data shows arcs curving to the left before the magnetic pole tips toward *Voyager*, and to the right after tipping. Jupiter appears to be the source of the signals, but they are being affected in an area between the planet and the spacecraft – an area through which *Voyager 1* is expected to fly.

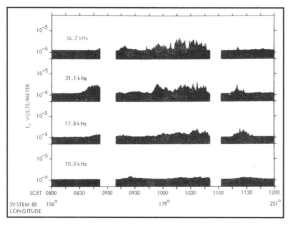

The magnetometers, low-energy charged particles (LECP) instrument, and plasma instrument are giving strong indications of nearing the bow shock, the interface between the solar wind and the planet's magnetosphere. Bow shock crossing may come sooner than predicted (February 26).

Images taken during January and early February have been processed into a recently released color movie. The rotation movie was compiled from intensive imaging on January 30 through February 3. Atmospheric changes (including current flows) through several rotations of Jupiter are visible, as well as satellites in transit around the planet. Counterclockwise rotation in the Great Red Spot is clearly visible.

PWS – Voyager 1's plasma wave instrument is recording very low frequency radio emissions from Jupiter which appear to be related to the north and south magnetic poles. This spectrum was recorded on January 20 at about 44 million miles from the planet. The north magnetic pole passage occurs at about 210° longitude.

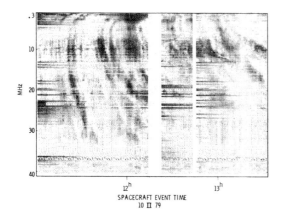

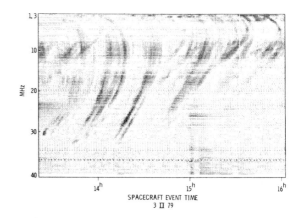

ARCS – These radio spectra, collected by Voyager 1's PRA instrument on two different days, clearly show the arc structures associated with Jupiter's north magnetic pole each time it points toward the spacecraft.

A Miniature Solar System

With its thirteen, possibly fourteen, satellites, Jupiter forms what many liken to a miniature solar system. All of the inner satellites are denser and more massive than the other satellites. (Also true for the planets of the solar system – Mercury, Venus, Earth and Mars are all far more dense than Jupiter, Saturn, Uranus, Neptune, or Pluto.

Voyager I will observe the five satellites closest to the planet: Amalthea, Io, Europa, Ganymede, and Callisto. Each is unique and intriguing.

Tiny Amalthea, about 120 to 240 kilometers (75 to 150 miles) in diameter, is the innermost satellite, and orbits the planet once every 12 hours (approximately). In the past, it was speculated to be a captured asteroid, because of its small size and its reflectivity characteristics. Its average distance from the planet is 181,500 kilometers (70,077 miles).

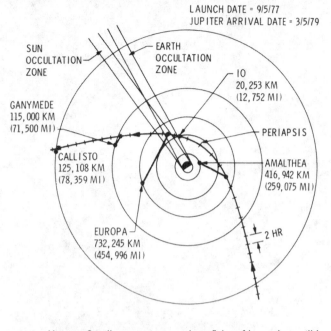

LAUNCH DATE = 9/5/77
JUPITER ARRIVAL DATE = 3/5/79

Voyager I is most interested in Io, so much so that it will risk Jupiter's intense radiation to get close to it. *Voyager 2* will not attempt a close flyby of Io, and so will be exposed to less radiation. With reddish polar caps and a tenuous atmosphere, Io is also surrounded by a yellow glow – thought to be a cloud of sodium sputtered off the satellite's surface by particle bombardment.

At about 5.9 R_J (Jupiter radii), Io is very much within the Jovian magnetosphere, and, indeed, seems to influence the pattern of Jupiter's decametric radio bursts. In addition, a region known as the Io "flux tube" is a magnetic link between the satellite's surface and the planet. *Voyager I* will spend about 4½ minutes in this area as it flies beneath the satellite's south pole. *Voyager I*'s closest approach to Io will be from 20,253 kilometers (12,752 miles) three hours after closest approach to Jupiter on March 5. Io is about the size of our Moon (3,636 kilometers or 2,259 miles), and is the third largest of the four Galilean satellites. It rotates about the planet in about 42½ hours at a distance of 422,000 kilometers.

Like Io, Europa appears to be a rocky body and recent pictures have shown a dark equatorial band. With a diameter of about 3,066 kilometers (1,905 miles), it is slightly smaller than our Moon and circles Jupiter in about 3½ days from about 671,400 kilometers. *Voyager I* will pass its closest to this satellite, 732,243 kilometers (454,996 miles), on March 5.

Ganymede and Callisto are the second and third largest planetary satellites in the solar system (Saturn's satellite Titan is the largest). Both are larger than the planet Mercury. Ganymede is thought to be mostly liquid water with a mud-core and a crust of ice. With a diameter of about 5,216 kilometers (3,241 miles), Ganymede circles Jupiter in about 7 days at an average distance of 1 million kilometers (62 million miles). *Voyager I* will pass 115,000 kilometers (71,500 miles) from Ganymede on March 5.

Callisto is thought to be half water, although its dark reflectivity indicates a rocky surface. Over 1.8 million miles from Jupiter, Ganymede makes one rotation in about 16 days 16½ hours. Its diameter is about 4,890 kilometers (3,039 miles). *Voyager I*'s closest look at Ganymede will be March 6 from about 125,108 kilometers (78,359 miles).

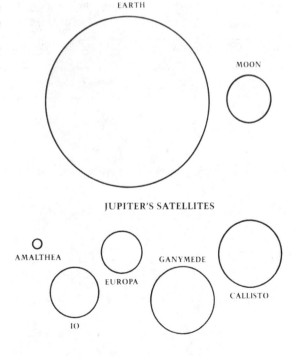

JUPITER'S SATELLITES

One Plus Three — Jupiter, its Great Red Spot and three of its four largest satellites are visible in this photo taken February 5, 1979, by Voyager 1. The spacecraft was 28.4 million kilometers (17.5 million miles) from the planet at the time. The innermost large satellite, Io, can be seen against Jupiter's disk. Io is distinguished by its bright, brown-yellow surface. To the right of Jupiter is the satellite Europa, also very bright but with fainter surface markings. The darkest satellite, Callisto (still nearly twice as bright as Earth's Moon), is barely visibly at the bottom left of the picture. Callisto shows a bright patch in its northern hemisphere. All three orbit Jupiter in the equatorial plane, and appear in their present position because Voyager is above the plane. All three satellites always show the same face to Jupiter — just as Earth's Moon always shows us the same face. This photo shows the sides of the satellites that always face away from the planet.

Closest Approaches

Body	Day*	Time*	Range
Amalthea	3/4	10:21 p.m.	416,942 km (259,075 miles)
Jupiter	3/5	4:42 a.m.	280,000 km (174,000 miles)
Io	3/5	7:50 a.m.	20,253 km (12,752 miles)
Europa	3/5	11:19 a.m.	732,245 km (454,996 miles)
Ganymede	3/5	6:52p.m.	115,000 km (71,500 miles)
Callisto	3/6	9:46 a.m.	125,108 km (78,359 miles)

* Earth-received time; all times are Pacific Standard Time.

Voyager *Bulletin*

MISSION STATUS REPORT No. 43   **JUNE 5, 1979**

VOYAGER 1 – January 24, 1979
40 million kilometers (25 million miles)

VOYAGER 2 – May 9, 1979
46 million kilometers (29 million miles)

Jupiter is sporting quite a different face than it did just four months ago, as these photos by the two *Voyager* spacecraft clearly show. Although individual features in the Jovian atmosphere are long-lived, the winds blow at greatly different speeds at various latitudes, causing the clouds to move independently of each other and to change longitudes.

Important changes appear in the region of the Great Red Spot: One of the white ovals has drifted from a position southwest of the Great Red Spot in late January to its present position 60 degrees eastward. Its movement has allowed another feature to move in behind it, from a January position just west of the white oval to a May position directly beneath the GRS. The white oval is drifting east at a rate of about 0.35 degree a day (1.57 mph), while the GRS itself is drifting west at about 0.26 degree a day (1.48 mph).

The bright "tongue" extending upward from the red spot is interacting with a thin, bright cloud above it that has traveled twice around Jupiter in four months. Turbulent wave patterns to the west of the GRS, which have been observed since 1975, appear to be breaking up. This area has undergone three major periods of activity in the last 15 years.

The *Voyager 2* photo shows a dark spot which has developed along the northern edge of the dark equatorial region. A similar feature was observed by *Pioneer 10* in December, 1973. Dark spots in the northern latitudes in the *Voyager 1* photo are still present. These spots are thought to be holes in the upper cloud decks penetrating into the warmer lower cloud layers.

Ganymede is visible at the lower left of the *Voyager 1* photo.

NASA

National Aeronautics and
Space Administration

Jet Propulsion Laboratory
California Institute of Technology
Pasadena California

VOYAGER 2: JUPITER MINUS 34 DAYS
VOYAGER 1: SATURN MINUS 525 DAYS

Recorded Mission Status (213) 354-7237
Status Bulletin Editor (213) 354-4438
Public Information Office (213) 354-5011

CONTENTS

Mercator projections – Preliminary shaded relief maps of the Galilean satellites have been prepared from images taken during Voyager 1's flight through the Jovian system. The scale here is 1:100 million. Io shows a complex system of calderas, flows, and volcanic features. Featureless, smooth plains lie between the volcanic regions. The plumes of eight active volcanoes identified to date are indicated by arrows.

Mission Highlights

Thirty-four days from its closest approach to Jupiter, *Voyager 2* is operating smoothly with only a few special concerns. Heating in one section of the buss (the spacecraft's main body) causes frequency drifts in the ship's remaining radio receiver (the primary receiver and a tracking loop capacitor on the remaining receiver failed in April, 1978). The frequency drifts limit routine commanding. Such heating occurs primarily when the spacecraft is maneuvered off the Sun line (as in special tests or calibrations, or maneuvers) or when the spacecraft power consumption changes. Events likely to cause temperature increases have been identified and plans for commanding during these periods have been revised.

Final target selection for the imaging, ultraviolet, and infrared experiments has been completed and the computer sequences are being finished to be relayed to the spacecraft before the near encounter period (July 8-9). Some planned observations have had to be simplified due to the space occupied in the processor by the backup mission load (BML). The BML is designed to automatically operate the spacecraft (although at a reduced activity level) through a Saturn encounter in August, 1981 should the remaining receiver fail. In that event, the spacecraft would not be able to receive radio signals from Earth, but could still transmit data to Earth.)

The photopolarimeter instrument will not operate the polarization wheel due to sticking problems similar to those on *Voyager 1*'s instrument. Near-ultraviolet photometry of Jupiter and Ganymede will be obtained, however, as well as a multi-color study of Io's ion torus.

The infrared interferometer spectrometer and radiometer (IRIS) is in a long heating cycle scheduled to end about June 20. The heating was required to preserve instrument performance due to degradation of a bonding material in the motor. The degradation could cause the mirror alignments to be off just fractions of a wavelength of light, but even 0.001 cm (0.00005 inch) would be enough to affect the quality of the data.

Imaging

Voyager 2 is now mosaicing the disk of Jupiter, taking a set of 12 narrow angle and 2 wide angle frames once an hour to ensure full coverage of the planet.

An "approach zoom" movie showing changes at the Great Red Spot since *Voyager 1*'s flyby is now being assembled from images taken during the past four weeks as the spacecraft "zoomed" closer to the planet.

Fifty-hours of nearly continuous picture taking from May 27 to May 29 covered five rotations of Jupiter and will supplement *Voyager 1*'s color movie of ten rotations made last January. The ultraviolet spectrometer experiment will also use the frames to search for auroral activity on the planet.

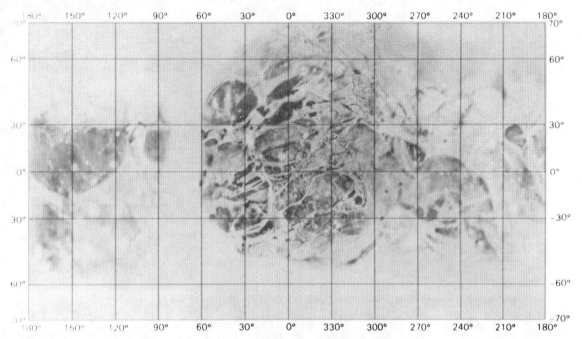

Ganymede – An older, cratered terrain is criss-crossed by a younger system of grooves which divide Ganymede's surface into features as large as 1,000 kilometers across. The grooves may have resulted from faulting or surface expansion, while the absence of larger craters or mountains on Ganymede suggests an icy crust which collapses under the weight of heavier features.

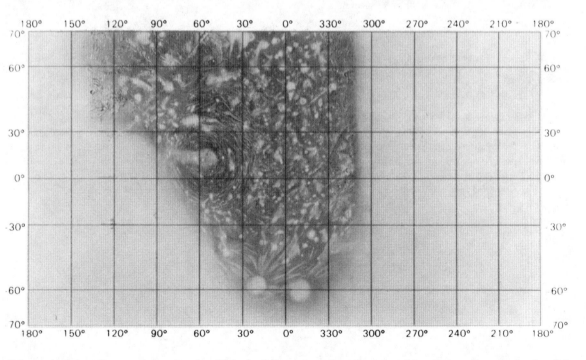

Callisto – More heavily cratered than the other Galileans, Callisto has no grooved terrain as does Ganymede, suggesting that their crusts have evolved very differently. The large mullet-ring structure centered at about +10° latitude, 60° longitude, has no central basin, ring mountains, or radial ejects.

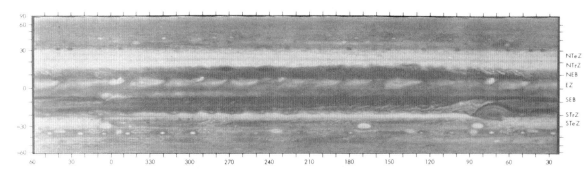

Cylindrical Jupiter – This computer generated map was made from 10 color images of Jupiter taken February 1, 1979, by Voyager 1, during a single, 10 hour rotation of the planet. Computers at the Jet Propulsion Laboratory's Image Processing Lab then turned the photos into this cylindrical projection. Such a projection is invaluable as an instantaneous view of the entire planet. Along the northern edge of the north equatorial belt (NEB) are four dark brown, oblong regions believed by some scientists to be openings in the more colorful upper cloud decks, allowing the darker clouds beneath to be seen. The broad equatorial zone (EZ) is dominated by a series of plumes, possibly regions of intense convective activity, encircling the entire planet. In the southern hemisphere the Great Red Spot is located at about 75 degrees longitude. South of the Great Red Spot in the south temperate zone (STeZ) three large white ovals, seen from Earth-based observatories for the past few decades, are located at 5 degrees, 85 degrees and 170 degrees longitude. Resolution in this map is 375 miles (600 kilometers). Since Jupiter's atmospheric features drift around the planet, longitude is based on the orientation of the planet's magnetic field. Symbols at right edge of photo denote major atmospheric features (dark belts and light zones): NTeZ – north temperature zone; NTrZ – north tropical zone; NEB – north equatorial belt; EZ – equatorial zone; SEB – south equatorial belt; STrZ – south tropical zone; and STeZ – south temperate zone.

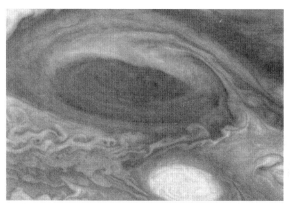 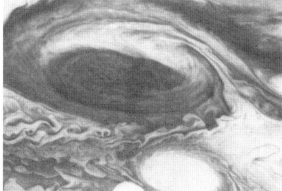

Color enhancement – Voyager's color photographs are actually composites of at least three exposures taken through different filters. The photo at left shows "normal" color, while the photo at right has been "color enhanced" emphasizing red and blue to make some features stand out. Some bizarre and beautiful pictures result from color enhancement techniques. In this view, the Great Red Spot and a white oval with a wake of counter rotating vortices extend about 24,000 kilometers (15,000 miles) from the top to the bottom of the frame. Puffy features inside the GRS and "reverse-S" spirals inside both the GRS and white oval are visible. The Great Red Spot appears to be one of the coldest areas on the planet, while the white ovals are also cold.

Voyager  Bulletin

MISSION STATUS REPORT No. 45   *JULY 5, 1979*

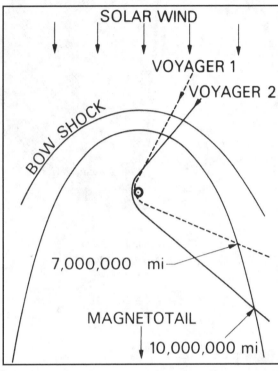

Magnetotail passage – Voyager 2 will spend a longer period taking measurements in Jupiter's magnetosphere, perhaps as long as 30 days, in comparison to nearly 13 days for Voyager 1. On its outbound journey, Voyager 2 may cross the magnetopause as far as 10 million miles from Jupiter.

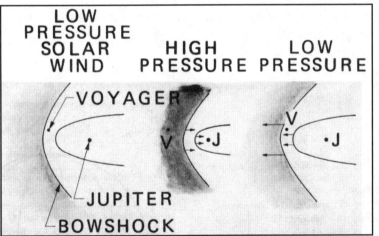

Bow shock crossings – Voyager 2's first crossing of Jupiter's bow shock came July 2 at a distance of about 7 million kilometers (4.4 million miles) from the planet. At least eleven crossings have been noted by the plasma instrument, magnetometers, and plasma wave instrument as of noon on July 5, as the solar pressure ebbed and flowed, sometimes causing the bow shock to overtake tine spacecraft again. The bow shock is a surface separating the essentially undisturbed supersonic solar wind from the deflected subsonic solar wind outside the magnetosphere where particles are trapped by the planet's magnetic field.

VOYAGER 2: JUPITER MINUS 4 DAYS

NASA

National Aeronautics and
Space Administration

Jet Propulsion Laboratory
California Institute of Technology
Pasadena California

Recorded Mission Status (213) 354-7237
Status Bulletin Editor (213) 354-4438
Public Information Office (213) 354-5011

CONTENTS

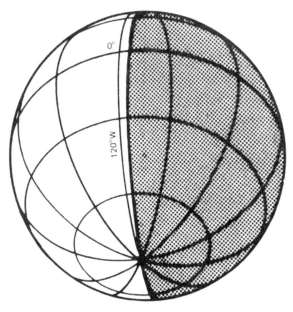

Ganymede
Closest Approach J-15.5 hr
From 63,000 kilometers

EUROPA
Closest Approach J-4.5 hr
From 207,000 kilometers

Opposite faces – These computer generated plots show Voyager 2's view of Europa and Ganymede at the times of closest approach. Flying over 500 thousand kilometers closer to Europa than Voyager 1 did, Voyager 2 will learn more about the linear features seen on the Moon-sized satellite in March.

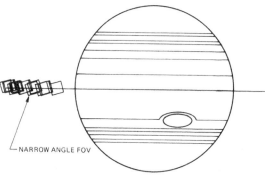

NARROW ANGLE FOV

INBOUND RING PLANE CROSSING
JULY 8   5:03 pm PDT

Ring crossing – Voyager 2 will cross the plane of Jupiter's thin ring twice, once inbound and once outbound. Both wide and narrow angle frames, some through color filters, will be taken to learn more about the thickness and composition.

Radio occultation – Voyager 2 will pass nearer the south pole of Jupiter, in contrast to Voyager 1's nearly equatorial pass. Only three hours after Jupiter closest approach in March, Voyager 1 moved into the planet's shadow, all radio signals blocked by the planet for nearly two hours. All data during this period, including the closest approach to Io, had to be tape recorded for later playback. Voyager 2 will enter Earth occultation nearly 22 hours after its closest approach to Jupiter.

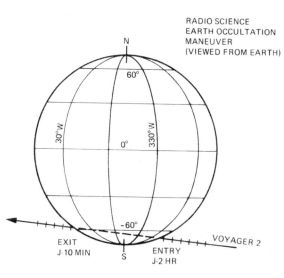

RADIO SCIENCE
EARTH OCCULTATION
MANEUVER
(VIEWED FROM EARTH)

EXIT
J-10 MIN

ENTRY
J-2 HR

VOYAGER 2

TICKS ARE AT 10-MIN INTERVALS

Voyager 2 Flyby of Jupiter – July 7-11, 1979

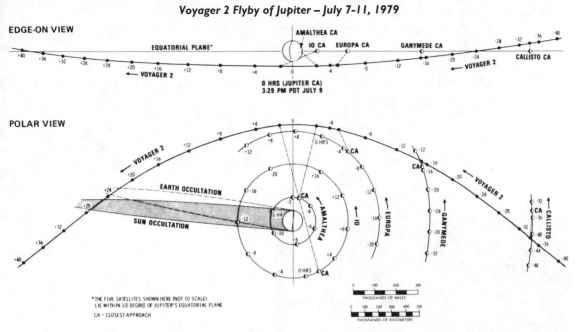

Close encounters — Voyager 2 will make its closest approaches to Callisto, Ganymede, Europa, and Amalthea before its nearest pass to Jupiter on July 9. Steering clear of Jupiter's intense radiation, Voyager 2 will not repeat Voyager 1's close flyby of Io, Instead flying outside the orbit of Europa.

Sampling of Voyager 2 Encounter Activities

(Continuous observations by radio science, magnetometers, low-energy charged particle, plasma, and cosmic ray investigations, as well as mapping by imaging cameras, photopolarimeter, ultraviolet and infrared spectrometers.)

(All times are Earth-received event start times, Pacific Daylight Time)

June 25		First targeted Callisto images
June 30		First targeted Ganymede images
July 2-3		Expected bow shock crossing
July 3		First targeted Europa images
July 7		First Callisto mosaic; First targeted Io images
July 8	2:39 a to 2:51 a.m.	Ring observations
	6:13 a.m.	Callisto – closest approach 1214,886 kilometers)
	7:32 a.m.	Final Callisto mosaic
	5:04 p to 5:29 p.m.	Ring observations
	5:32 p.m.	First Ganymede mosaic
July 9	1:06 a.m.	Ganymede – closest approach (62,297 kilometers)
	1:12 a.m.	Final Ganymede mosaic
	10:02 a.m.	Final Europa mosaic
	11:45 a.m.	Europa – closest approach (205,848 kilometers)
	1:53 p.m.	Amalthea – closest approach (558,565 kilometers)
	4:21 p.m.	Jupiter – closest approach (721,750 kilometers)
	4:34 p.m.	Begin 10 hour to "volcano watch"
	5:09 p.m.	Io – closest approach (1,129,850 kilometers)
July 10	2:30 a.m.	Conclude to "volcano watch"
	2:21 p to 4:08 p.m.	Earth occultation
	5:10 p to 7:48 p.m.	Sun occultation
	6:48 p to 7:11 p.m.	Ring observations
July 11	8:43 a to 9:27 a.m.	Ring observations

Summary of Voyager Close Approaches

Body	Voyager 1 Distance	Best Imaging Resolution (km)	Voyager 2 Distance
Jupiter	280,000 km (174,000 miles)	6	721,750 km (448,470 miles)
Amalthea	417,100 km (259,173 miles)	7.8	558,600 km (347,100 miles)
Io	20,500 km (12,738 miles)	1*	1,129,850 km (702,056 miles)
Europa	733,800 km (454,962 miles)	33**	205,850 km (127,900 miles)
Ganymede	114,600 km (71,209 miles)	2	62,300 km (38,700 miles)
Callisto	126,100 km (78,355 miles)	2.3	214,900 km (136,500 miles)

* Best Io resolution was limited by image smear due to timing offset caused by radiation
   levels, not distance of closest approach.

** Final Europa images were obtained at a range of 18 million kilometers, well before closest
   approach to the satellite.

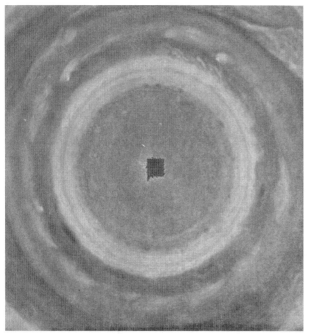

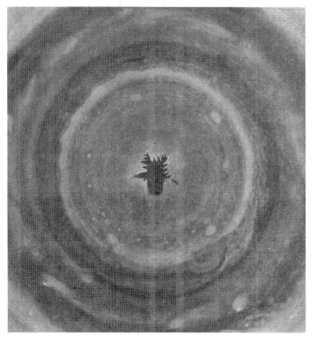

Polar projections — Jupiter's northern and southern hemispheres as they might be seen from directly above the poles are shown in these polar stereographic projections constructed by JPL's Image Processing Lab from Voyager 1 photos. The resolution is 600 kilometers (375 miles). The dark objects in the centers are areas where no pictures were available at that resolution.

In the northern hemisphere, the northward extent of the belt zone structure is clearly shown to at least 50 degrees north latitude. At the northern edge of the equatorial region, the plumes are evenly space around the planet. Positions of the active cloud plumes, marked by bright nuclei, are not symmetrical. At about 32 degrees north, dark cloud vortices that move in westerly currents at about 30 meters per second (67 miles an hour) can be seen. The spacings of those features vary, and cloud systems have been seen to roll over one another in the region. The broad white region is divided by the North Temperate Belt's high-speed jet, seen as a thin brown line.

The southern hemisphere image shows three white ovals and a large region of the same zone without any discrete feature. Smaller scale spots, almost equally spaced. cover almost 270 degrees of longitude, while the disturbances trailing from the Great Red Spot extend about 180 degrees in longitude.

Voyager 1

Encounter with Saturn
Press Kit

NASA

National Aeronautics and
Space Administration

RELEASE No: 80-159

Jet Propulsion Laboratory
4800 Oak Grove Drive
Pasadena, California 91103
AC 213 354-4321

CONTENTS

NASA News

National Aeronautics and
Space Administration

Washington, D.C. 20546
AC 202 755-8370

Nicholas Panagakos   Phone: (202) 755-3680
Headquarters, Washington, D.C.

FOR RELEASE:

Frank Bristow   Phone: (213) 354-5011
NASA Jet Propulsion Laboratory, Pasadena, California

10 A.M. EST, TUESDAY
October 28, 1980

October 23, 1980

RELEASE No: 80-159

Voyager to Take a Close Look at Saturn on November 12

America's *Voyager 1* will arrive at Saturn early next month, marking the third episode in a decade long, multibillion-mile space odyssey to the outer planets and beyond.

Voyager 1 will encounter the ringed planet on November 12, making its closest approach at approximately 6:45 p.m. EST at a distance of 124,200 kilometers (77,174 miles) from the visible cloud tops.

The flyby of Saturn comes more than 20 months after *Voyager*'s close approach to Jupiter in March 1979, and 16 months after a sister craft, *Voyager 2*, made its Jovian encounter in July 1979.

Voyager 1's detailed scientific examination of Saturn, its rings and moons, is the NASA spacecraft's final planetary encounter before it leaves the solar system about 1990. *Voyager 2* will encounter Saturn later next year, followed by possible Uranus and Neptune encounters in 1986 and 1989 respectively.

The close-up looks at the outer planets are the most current steps in the United States program of systematic planetary exploration, in which the solar system is used as a natural laboratory. By being able to compare the similarities and differences of the various planets, scientists hope to learn more about the history and future of the solar system and particularly our own planet Earth. The outer planets, evolving at a different rate and under different conditions than the inner planets – Mercury, Venus and Mars – play a particularly important role in this effort.

A prime target for study by *Voyager 1* will be Saturn's satellite Titan, the largest moon in the solar system and the only moon known to have retained a substantial atmosphere. The spacecraft will approach Titan at a distance of 4,000 kilometers (2,500 miles) from the surface, the closest approach to any body in the Voyager mission, on November 12 at 12:41 a.m. EST (spacecraft time).*

The same day, *Voyager* also will make its closest approach to Saturn and the satellites Tethys (415,320 kilometers or 258,067 miles) at 5:16 p.m. EST; Mimas (88,820 kilometers or 55,190 miles) at 8:42 p.m.; Enceladus (202,521 kilometers or 125,840 miles) at 8:50 p.m.; and Dione (161,131 kilometers or 100,122 miles) at 10:39 p.m. Closest approach to Rhea (72,000 kilometers or 44,739 miles) will be at 1:21 a.m. EST on November 13.

When the two *Voyager* spacecraft have completed their television experiments they will have sent more than 70,000 pictures of Jupiter, Saturn and its rings, 13 satellites and black space near the two planets. Black-space imaging at Jupiter resulted in discovery of one its three newly found moons.

At Saturn, as at Jupiter, *Voyager*'s radio signals will be used extensively to measure the atmospheres of the planet and satellites and assess the size and density of the ring particles.

The *Voyager 1* trajectory includes occultations of the Earth and Sun by Titan, Saturn and the rings. During these three occultations, Earth will be eclipsed by Titan, Saturn and the rings from the spacecraft's point of view. As *Voyager* passes behind each body, its radio signals will pass through the atmospheres of Saturn and Titan and through the rings toward Earth; resultant changes in the radio signals will provide information on characteristics of the planet's and satellite's atmospheres, ionospheres and the size and density of ring particles.

---

* If one were able to watch the event from Earth, the eye would see the close encounter one hour, 25 minutes later – the time it takes light to travel the 1.5 billion kilometers (930 million miles) back to Earth. The same velocity is true using the *Voyager* radio to observe the event. Scientists call this the "Earth-received time." Times listed in this Press Kit are spacecraft times.

Each *Voyager* uses 10 instruments and the spacecraft radio system to study the planets, their principal satellites, Saturn's rings, the magnetic and radiation regions surrounding the planets, and the interplanetary medium. (The photopolarimeter on *Voyager 1* has failed and will not be used at Saturn; the one on *Voyager 2* is still operating.)

The *Voyagers* are carrying telescope equipped television cameras, cosmic ray detectors, infrared spectrometers and radiometers, low-energy charged particle detectors, magnetometers, photopolarimeters, planetary radio astronomy receivers, plasma detectors, plasma wave instruments and ultraviolet spectrometers.

The spacecraft began the observatory phase of its encounter on August 22 and will conclude the 117-day surveillance of the Saturnian system on December 15.

Voyager 2 will follow its sister spacecraft to Saturn in August 1981. Its possible four-planet itinerary includes encounters with Uranus in January 1986 and Neptune in August 1989.

Saturn is the sixth planet from the Sun, second largest in the solar system. Like Jupiter, it is a giant sphere of gas, mostly hydrogen and helium, with a small core of rocky material.

The planet takes 29.46 years to complete one orbit around the Sun, which is approximately 1.42 billion kilometers (886 million miles) away. A day on Saturn lasts 10 hours, 39 minutes, 24 seconds.

Until recently, Saturn was believed to be the only planet encircled by rings. But both Jupiter and Uranus were discovered to have thin, barely visible rings. (The Jovian ring was discovered by *Voyager*.) Saturn's rings, however, are much richer in material, probably chunks of ice and dirt, and are bright and highly visible from Earth-based telescopes.

At least six rings surround Saturn. From the planet outward, they are designated D, C, B, A, F and E. Divisions between the rings are believed to be caused by the three innermost satellites, Mimas, Enceladus and Tethys. The Cassini Division, a space between the B and the A ring, is the only division clearly visible with a small telescope from Earth.

The Encke Division is a small gap in the rings. The *Pioneer* spacecraft, which flew past Saturn in September 1979, detected a gap between the A and F rings. *Pioneer* also reported a third division between the C and B rings.

Saturn's face, in comparison to Jupiter's, is bland and tranquil in appearance. This may be due to thick, high-altitude haze that gives Saturn its opaque appearance, hiding weather activity in the planet's atmosphere.

Saturn has at least 13 satellites, though many more not yet discovered may orbit the planet. The existence of 10 has been known for some time and these objects have been named. Of the three recently discovered satellites, two appear to share the orbit of Dione and one of Enceladus.

From the planet outward, the 10 named satellites are Janus (whose existence is now doubtful), Mimas, Enceladus, Tethys, Dione, Rhea, Titan, Hyperion, Iapetus and Phoebe.

In its excursion though the Jovian system, *Voyager 1* discovered a wispy ring of particles around Jupiter and towering volcanoes on Io. Super-bolts of lightning and immense auroras were found in the planet's violently churning atmosphere.

More than 18,000 pictures of Jupiter, the four Galilean satellites (Io, Europa, Ganymede and Callisto), and tiny Amalthea were obtained with *Voyager 1*'s two-camera imaging system during the 98-day encounter.

Voyager 2's visit to the Jovian system yielded an additional 15,000 pictures of the giant planet and its satellites.

Using Jupiter's enormous gravity, the two spacecraft were hurled onto trajectories toward Saturn.

When *Pioneer 10* and *11* passed through Jupiter's intense radiation in their 1973 and 1974 encounters, the two *Voyagers* were still being assembled. Based on data from the *Pioneers*, some *Voyager* parts were exchanged for more radiation resistant components, electronic circuits were modified, and additional radiation shielding was added to each spacecraft. As a result, the *Voyagers* suffered little damage in their flybys of Jupiter.

Saturn's radiation is much gentler than Jupiter's. When *Pioneer 11* flew beneath Saturn's rings, radiation intensity dropped dramatically, showing that charged particles are interrupted and absorbed by particles in the rings.

Voyager 2 was launched August 20, 1977, from Cape Canaveral, Florida, aboard a Titan-Centaur rocket. Two weeks later, on September 5, *Voyager 1* was launched on a faster, shorter trajectory and sped past its twin before the end of the year. By the time *Voyager 1* reaches Saturn, *Voyager 2* will be nine months behind.

When *Voyager 1* encounters Saturn it will have traveled more than 2 billion kilometers (1.24 billion miles) since launch.

After completing their planetary missions, each spacecraft will search for the outer limit of the solar wind – that presumed boundary somewhere in our part of the Milky Way where the influence of the Sun gives way to that of other stars of the galaxy.

The Voyager Project is managed for NASA by the Jet Propulsion Laboratory, Pasadena, California, a government owned facility, operated for the space agency by the California Institute of Technology.

NASA program manager is Frank A. Carr, and Dr. Milton A. Mitz is NASA program scientist. Voyager project manager is Raymond L. Heacock, JPL, and Dr. Edward C. Stone of Caltech is project scientist.

Estimated cost of the Voyager Project, exclusive of launch vehicles, tracking and data acquisition and flight support activities is $338 million.

(End of General Release. Background information follows)

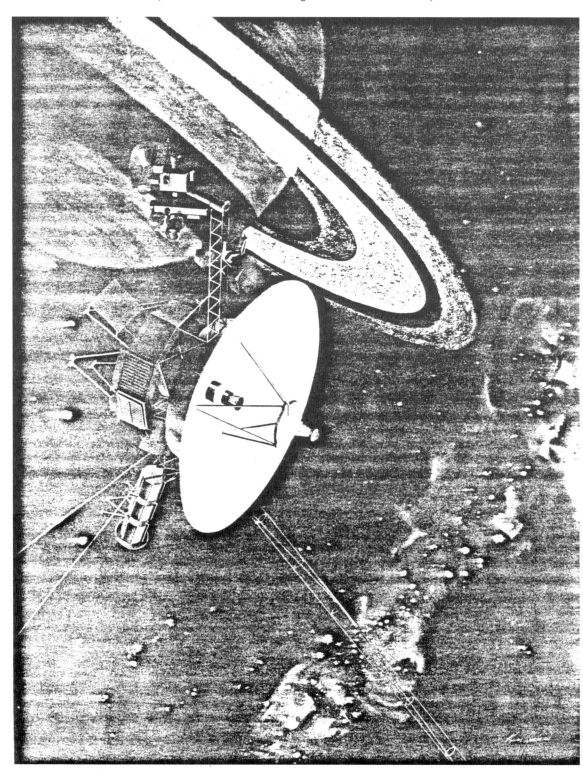

Voyager I Saturn Encounter Events

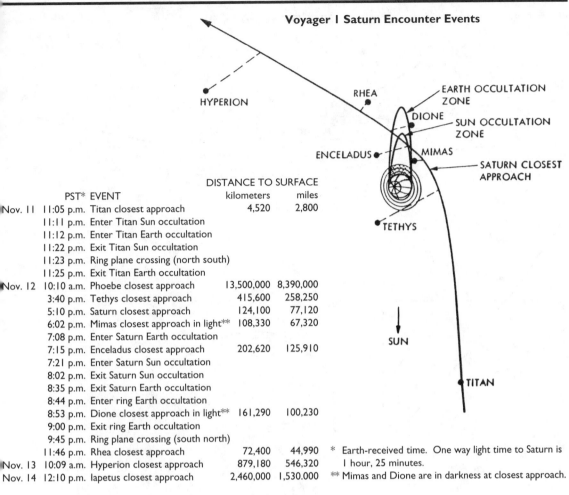

| | PST* EVENT | DISTANCE TO SURFACE | |
		kilometers	miles
Nov. 11	11:05 p.m. Titan closest approach	4,520	2,800
	11:11 p.m. Enter Titan Sun occultation		
	11:12 p.m. Enter Titan Earth occultation		
	11:22 p.m. Exit Titan Sun occultation		
	11:23 p.m. Ring plane crossing (north south)		
	11:25 p.m. Exit Titan Earth occultation		
Nov. 12	10:10 a.m. Phoebe closest approach	13,500,000	8,390,000
	3:40 p.m. Tethys closest approach	415,600	258,250
	5:10 p.m. Saturn closest approach	124,100	77,120
	6:02 p.m. Mimas closest approach in light**	108,330	67,320
	7:08 p.m. Enter Saturn Earth occultation		
	7:15 p.m. Enceladus closest approach	202,620	125,910
	7:21 p.m. Enter Saturn Sun occultation		
	8:02 p.m. Exit Saturn Sun occultation		
	8:35 p.m. Exit Saturn Earth occultation		
	8:44 p.m. Enter ring Earth occultation		
	8:53 p.m. Dione closest approach in light**	161,290	100,230
	9:00 p.m. Exit ring Earth occultation		
	9:45 p.m. Ring plane crossing (south north)		
	11:46 p.m. Rhea closest approach	72,400	44,990
Nov. 13	10:09 a.m. Hyperion closest approach	879,180	546,320
Nov. 14	12:10 p.m. Iapetus closest approach	2,460,000	1,530.000

* Earth-received time. One way light time to Saturn is
1 hour, 25 minutes.

** Mimas and Dione are in darkness at closest approach.

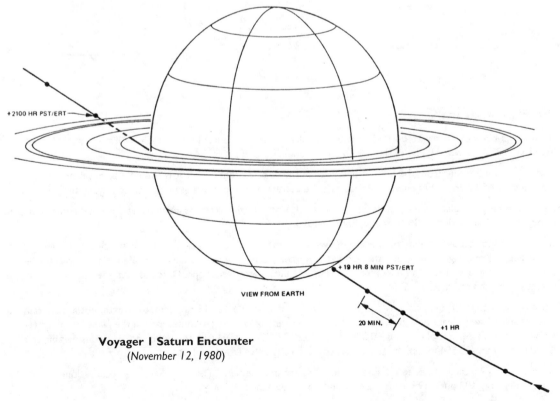

VIEW FROM EARTH

Voyager I Saturn Encounter
(November 12, 1980)

View From Earth

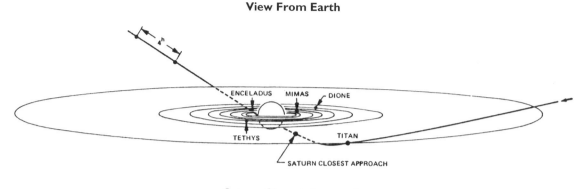

Saturn Closest Approach

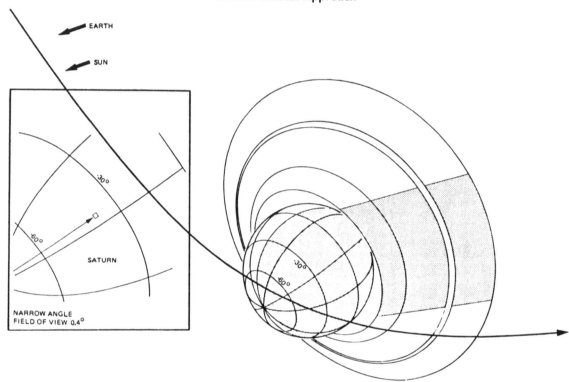

The Planet Saturn

The ancients believed Saturn to be the most distant planet from the Sun. Not until Sir William Herschel discovered Uranus in 1781 did anyone know of the existence of planets beyond Saturn.

Saturn is the sixth planet from the Sun. It is unique in the solar system in that it is the only planet lighter than water, with a density of 0.7 grams per cubic centimeter. (The density of water is one gram per cubic centimeter.)

Saturn's rings are the planet's most distinctive feature. It was thought to be the only planet encircled by rings until the discoveries of Uranus' rings in 1977 and Jupiter's in 1979.

The first telescopic observations of the planet were made by Galileo Galilei in 1610. His telescope, which gave a magnification of only eight power, led the early astronomers to believe he had seen not one but three planets because of the confusing image of Saturn's encompassing rings. Using a better telescope, Dutch astronomer Christiaan Huygens correctly identified the rings in 1655.

Saturn is the second largest planet in our solar system. It has a volume 815 times that of Earth, but a mass only 95.2 times greater. Like its giant neighbor, Jupiter, Saturn's extremely rapid rotation has caused the planet to be flattened at its poles. Saturn's equatorial radius is 60,330 kilometers (37,490 miles), while the polar radius is considerably smaller – 54,000 kilometers (33,554 miles). Its surface gravity is 1.15 (Earth's gravity is 1.0).

Saturn takes 29.46 Earth years to complete one orbit around the Sun. Though a Saturnian year is long, its days are short, lasting only 10 hours, 39 minutes, 24 seconds (as determined by *Voyager*).

Two Views of Voyager I Flyby of Saturn
(November 11-13, 1980)

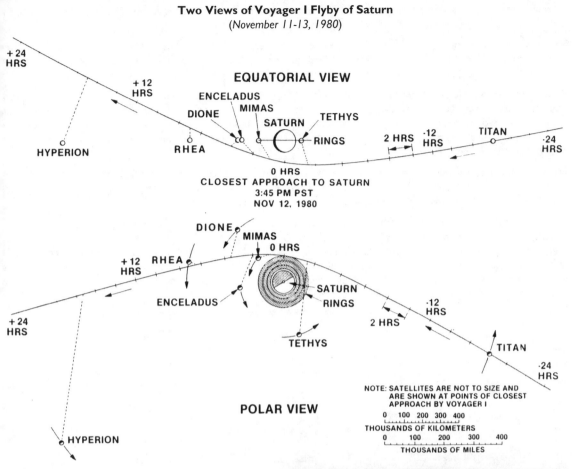

In its slow orbit around the Sun, Saturn is perturbed by the other planets – especially Jupiter – so its orbital path is not strictly elliptical. The planet wavers in its distance from the Sun in a region of between 9.0 AU and 10.1 AU. (An astronomical unit, or AU, is the mean distance from the Sun to Earth – 149,597,870 kilometers or 92,955,806.8 miles). Saturn receives only about one one-hundredth as much sunlight as that which reaches Earth.

Like the three other gas giants – Jupiter, Neptune and Uranus – Saturn has no solid surface, but is a huge, multilayered globe of gas with a small core of iron and rocky material.

Because Saturn is farther from the Sun, it is colder than Jupiter. Material in its atmosphere freezes at greater depths than on Jupiter. Ammonia, for example, freezes and forms clouds on Saturn at a depth of two or three Earth atmospheres (two to three times the surface pressure of Earth, which is 1,000 millibars), compared to six-tenths of an atmosphere on Jupiter.

Gravity field analysis and temperature profile measurements made by *Pioneer 11* suggest that Saturn's core, extending out about 13,800 kilometers (8,575 miles) from the center, is twice the size of Earth, but is so compressed that its iron and rocky core contains 15 to 20 times the mass of Earth.

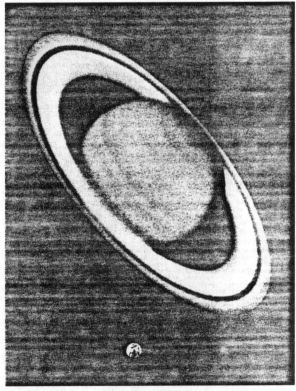

Comparative Sizes of Saturn and Earth

Surrounding the core is a layer of electrically conductive metallic hydrogen. This form of hydrogen has not been observed on Earth, since immense pressure is required for its production.

The outermost layer of the planet is an envelope of hydrogen and helium. The interaction between these two elements may explain Saturn's emission of heat. Both Jupiter and Saturn release about twice the amount of energy they receive from the Sun. But scientists believe the two planets produce their energy in entirely different ways.

Jupiter emits energy left over from the gravitational contraction that occurred when the planet was formed 4.6 billion years ago. But because Saturn is smaller, scientists conclude that it cannot emit energy in the same manner – any heat left over from Saturn's gravitational contraction would have dissipated long ago. Instead, Saturn's heat production may be the result of the separation of hydrogen and helium in the outer layer, with heavier helium sinking through the planet's liquid hydrogen interior.

From observations, the velocity of winds at Saturn's cloud tops appears to be about 1,400 kilometers (900 miles) an hour – twice that on Jupiter. The calculations are based on observations from Earth of spots in Saturn's clouds, combined with *Voyager's* radio measurements of the rotation of the planet's interior.

Saturn has at least 13 satellites, though the family may finally number many more. Titan is the largest satellite in the solar system; with a diameter of 4,345 to 5,632 kilometers (2,700 to 3,500 miles), its size approximates that of Mercury. It is the only satellite in the solar system known to have retained a substantial atmosphere. Titan's atmospheric density is now calculated to be between 20 millibars and 2,000 millibars at the surface. (Earth's atmospheric density is about 1,000 millibars; Mars atmospheric density is about 10 millibars.) Its prevailing temperature is low – about -198 degrees Celsius (-324 degrees Fahrenheit).

Major Features of Saturn

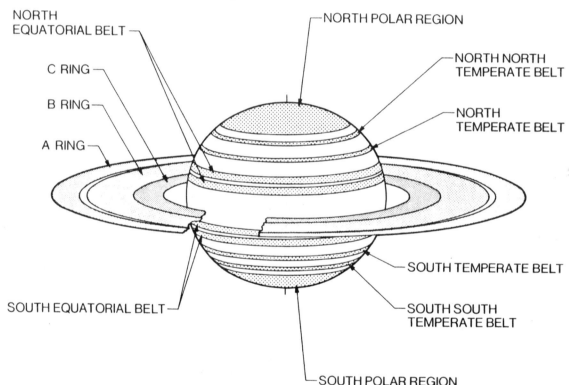

Saturn's other satellites are like no others we have ever seen. Most of them are larger than Mars' tiny satellites, Phobos and Deimos, but they are not as large as the four Galilean satellites of Jupiter. They are mid-sized bodies, expected to consist of mixtures of rock and ice, with rocky material averaging 30 to 36 percent, or they could be all ice. It now appears that *Voyager* may discover a number of new, small satellites outside the A ring.

The rings' designations were assigned in the order of discovery; the letters have nothing to do with their relative positions. Beginning with the outermost ring, the rings are:

• E ring; extending to about eight Saturn radii (480,000 kilometers or 298,260 miles) from the planet. It has been photographed from Earth.
• F ring; identified in images taken by *Pioneer 11*. It is a very narrow ring just outside the A ring, but distinct and separate from the E ring.

- A ring; outermost ring visible with small telescopes.
- B ring; the brightest ring, lies inside the A ring, and is separated from it by Cassini's Division. (The division is not completely clear of material.) From Earth the B ring appears completely filled with material. However, a few *Pioneer 11* images show light leaking through, leading to speculation there may be holes in the ring.
- C ring (or "crepe ring"); barely visible in small telescopes; lies inside the B ring; 17,000 kilometers (11,000 miles) from cloud tops.

In addition, a D ring has been claimed between the C ring and the cloud tops. Some scientists doubt its existence.

The Cassini Division between the B and A ring is the most prominent gap between the rings and easily visible with Earth-based telescopes. A thin space within the outermost edge of the A ring is the Encke Division. *Pioneer* identified a new, 3,600-kilometer (2,200-mile) division within the A ring.

The composition of the rings is unknown. The rings appear to be entirely ice or ice-covered material about 1 to 30 centimeters (0.3 to 12 inches) in diameter. Scientists want to determine the particle sizes and densities and discover if other material is present. How and why the rings formed is also a major question.

Jupiter's thin ring is made up of dust particles 0.05 cm (0.0002 inches) in diameter, as perceived by *Voyager* from forward scattering of sunlight.

But Saturn's rings contain much larger particles, so *Voyager* will use radio signals aimed at the rings and toward Earth at occultation to determine the presence of particles the size of 30 to 90 cm (1 to 3 feet) in diameter.

The B ring contains the greatest amount of material and is so dense that little light can penetrate.

Scientists want to determine why density apparently differs from ring to ring, and if there are waves within the rings. Large objects in the rings themselves may also cause local variations in density.

Saturn's magnetic field is 1,000 times stronger than Earth's and 20 times weaker than Jupiter's. Saturn's magnetosphere is unique in that its north-south axis is within one degree or less of the planet's rotation axis, unlike the 10 degree tilt to the rotation axis of Earth, Jupiter and the Sun.

The magnetosphere, a magnetic bubble in the solar wind that surrounds the planet, is larger than Earth's but smaller than Jupiter's. It is a teardrop shaped magnetic envelope about 8 million kilometers (5 million miles) wide, that extends about 2.5 million kilometers (1.5 million miles) from the planet.

Saturn possesses radiation belts of high-energy electrons and protons comparable in intensity to Earth's, though the region they occupy is about 10 times larger. The intensity of Saturn's radiation was unknown until *Pioneer 11* flew past the planet and determined that the radiation belts are several hundred times weaker than Jupiter's. When the spacecraft passed below the rings, the radiation intensity dropped dramatically, indicating that ring material interrupts and absorbs the charged particles that bounce back and forth between the planet's poles.

The absorption of charged particles makes the area around the rings the most radiation free sector of space yet found in the solar system. The innermost satellites also absorb radiation particles.

Pioneer's infrared instrumentation showed the equatorial zone is cooler than adjoining higher-latitude regions. The difference in temperature indicates varying heights of belts and zones.

Ultraviolet radiation detected at Saturn's poles is possibly the result of auroral activity like that which *Voyager* discovered at Jupiter.

Each of the *Voyager* experiments – 10 instruments and the spacecraft radio – gather specific data individually, but it is through the synergistic combination of data from several instruments that the science teams will assemble a coherent picture of the entire Saturn system.

Saturn Science Experiments

Voyager 1's science investigations at Saturn fall into four broad classifications:

1. Saturn's atmosphere, studied by:
   - Imaging
   - Infrared Interferometer Spectrometer and Radiometer
   - Ultraviolet Spectrometer
   - Radio Science.

2. Seven of the satellites of Saturn – Titan, Rhea, Dione, Mimas, Iapetus, Hyperion and Enceladus. A study will be conducted of recently discovered small satellites, and a search for undiscovered bodies. The studies will be conducted by:
   - Imaging
   - Infrared Interferometer Spectrometer and Radiometer

- Ultraviolet Spectrometer
- Radio Science.

3. Saturn's magnetic field and its interaction with the solar wind and the satellites, studied by:
- Plasma
- Low-energy charged particles
- Cosmic ray
- Magnetometers
- Planetary radio astronomy
- Plasma wave
- Radio science.

4. Saturn's rings, studied by:
- Imaging
- Infrared Interferometer Spectrometer and Radiometer
- Ultraviolet Spectrometer
- Radio Science.

Magnetic Fields Investigation

The magnetic field of a planet is an externally measurable indication of conditions deep within its interior. Four magnetometers aboard *Voyager 1* gather data on the planetary magnetic fields at Saturn, the satellites, solar wind and satellite interactions with the planetary field, and the interplanetary (solar) magnetic field.

The magnetometers effectively probe the interplanetary medium – the tenuous, ionized and magnetized gas called the solar wind.

To form the solar wind, the Sun constantly emits electrically charged particles (protons and electrons) from the ionization of hydrogen. Those particles are in the fourth state of matter, called plasma (the other three states are solid, liquid and neutral gas). The plasma is of extremely low density and it fills all interplanetary space. Because it is ionized the solar wind is an electrically conducting medium.

The solar wind is deflected by planetary magnetic fields (such as Earth's, Jupiter's and Saturn's) and streams around and past the obstacles, confining the planetary magnetic fields to regions called the magnetosphere.

The shape of Saturn's magnetic field is not clearly understood. But because *Voyager 1* and *Voyager 2* arrive nine months apart, scientists can link long-term measurements of the solar wind to the magnetosphere as the latter changes size and shape in response to variations in the solar wind.

The outer regions of Jupiter's, and probably Saturn's, magnetic fields are shaped much differently from Earth's. The rapid rotation of Jupiter and Saturn may be one explanation as their magnetic fields co-rotate with the planets. At great distances, the planets' magnetic field lines appear to form a spiral structure, which could be explained by outward flowing plasma. A striking difference between the magnetic fields of the three planets is that Earth's and Jupiter's north magnetic poles are offset at least 10 degrees from their geographic poles, while Saturn's is offset 1 degree or less, according to data from *Pioneer 11*, which studied Saturn in September 1979.

Cosmic Ray Investigation

Cosmic rays are the most energetic particles in nature and are atomic nuclei (primarily protons) and electrons. They comprise all known natural elements. Over certain energy ranges and at certain periods of time, the content of cosmic rays is similar in proportion to that of all the matter in the solar system.

Generally, however, their composition varies significantly with energy, indicating to scientists that a variety of astrophysical sources and processes contributes to their numbers.

Cosmic rays may, as we search for their origins, tell much about the solar system's origins and processes. Cosmic rays are material samples of the galaxy, and can tell us much about how stars synthesize the various elements in their interiors.

Voyager's cosmic ray instrument will study the energy content, origin and acceleration process, and life history of cosmic rays.

Planetary Radio Astronomy

The planetary radio astronomy will measure radio emissions from Saturn in the low-frequency range from 20 kilohertz to 40.5 megahertz. (AM radio stations broadcast at frequencies between 550 kilohertz and 1.6 megahertz.) Scientists say emissions ranging in wavelength from less than 10 meters (33 feet) to thousands of meters can result from wave particle plasma interactions in the magnetosphere and ionosphere of Saturn.

Saturn's weak, kilometer-wavelength radio bursts have been used by the scientists to determine the planet's rotation rate with high precision. That rotation rate, determined with data from this instrument, is 10 hours, 39 minutes, 24 seconds.

Another goal of the planetary radio astronomy experiment is to search for lightning in Saturn's atmosphere, since lightning was discovered in the atmosphere of Jupiter, and it has been postulated as a catalyst for the formation of life. Together with the plasma wave experiment and several optical instruments, planetary radio astronomy may be able to demonstrate the existence of lightning on Saturn.

One current theory says that lightning in an atmosphere of hydrogen, methane, ammonia and water could set off reactions that eventually form complex organic molecules.

Infrared Interferometer Spectrometer and Radiometer

Saturn presents a more subtle appearance to Earth-based observers than Jupiter, with its well defined and colorful light zones, dark belts and large storm systems.

Voyager's infrared interferometer spectrometer and radiometer — called, simply, IRIS — will try to determine why the differences are so marked between the two apparently similar planets.

Each chemical compound has a unique spectrum. Measuring the infrared radiation emitted and reflected by an object provides data on atmospheric gas composition, abundance, clouds, haze, temperatures, circulation and heat balance.

Hydrogen, deuterium (heavy hydrogen), methane, ammonia, ethane and phosphine have been identified in Saturn's atmosphere. There is also evidence for the presence of helium, although the amount is uncertain.

Knowledge of the absorption properties of the atmosphere's major constituents can be used to measure the temperature at various depths as it changes with pressure. The presence of a considerable quantity of dust in the atmosphere is deduced from the sharp drop in overall reflectivity from 5,000 angstroms to 3,000 angstroms.

Satellite temperature maps will be constructed using the distinctive spectral signatures of ices and minerals found on the surfaces. Maps and pictures of the satellites can be used to study the geology and evolution of the bodies, and how they differ with distance from Saturn.

Radio Science

The radio that provides tracking and communication with Voyager also explores the planets and space.

Measurements of Voyager's radio signals provide information on gravity fields and atmospheres of Saturn and its satellites, the solar corona and general relativity.

Changes in frequency, phase, delay, intensity and polarization of radio signals between spacecraft and Earth provide information about the space between the two, and about gravitational forces that affect the spacecraft and alter its path.

When the spacecraft moves behind a body as viewed from Earth (called occultation), radio waves coming from the spacecraft pass through the ionosphere and atmosphere on their way Earthward. Changes in signal characteristics during those events give information about the vertical structure of the ionosphere, atmosphere, clouds and turbulence.

Imaging

Voyager 1's two television cameras will take about 18,000 pictures during the Saturn encounter. By themselves and coupled with data from other instruments, the cameras can provide vast amounts of scientific data about the objects to be studied — Saturn and at least seven of its satellites.

On a gross scale, one would expect Saturn to be very much like Jupiter — a vast accumulation of hydrogen and helium surrounding a small, dense, rocky core.

But there are also a variety of differences between the two; while both have a layer of high altitude atmospheric haze, Saturn's appears from Earth to be much thicker than Jupiter's. That, plus Saturn's greater distance from Earth, makes motions within the planet's atmosphere difficult to study from Earth. As a result, imaging scientists expect to make the first extensive study of global wind systems in the atmosphere of Saturn.

Though the gross cloud structure in Saturn's atmosphere is dominated by the familiar belt zone pattern as seen on Jupiter, there is a marked absence of any long-lived features analogous to Jupiter's Great Red Spot. The few spots or features that have been seen from Earth indicate that tropical latitudes on Saturn rotate more rapidly than do the temperate and polar regions (recall that this is analogous to Jupiter). One explanation for the sparsity of spots in Saturn's atmosphere may be the presence of a thick haze layer that diminishes atmospheric contrast. Scientists hope that high-resolution, high-contrast Voyager images will reveal the elusive features.

Most scientists believe Saturn's atmosphere must be as active as Jupiter's. The bland appearance of the cloud tops, they feel, is most likely due to obscuration by the high altitude haze.

Atmospheric features will be affected by such phenomena as high wind speeds on Saturn, atmospheric convection and a rotation rate similar to Jupiter, radiation of internal heat that is more than 2.5 times greater than the radiation received from the Sun, the coriolis force created by such rapid rotation, and the effects of the ring shadow on portions of the cloud tops.

Determining the structure of the atmosphere of Saturn is considered one of the classic problems to be solved by *Voyager* photography. Constant monitoring of the planet and motion picture sequences will help provide data to satisfy that requirement.

Since particles in Saturn's rings are thought to be extremely small (from a few centimeters to a few meters in size), *Voyager*'s cameras will not resolve the individual ring particles. But waves or clumps of particles, as well as large, satellite-sized objects orbiting within or near the rings, should be identifiable. (The radio science experiments are intended to resolve many of the scientific questions about particle size.)

The satellites of Saturn (along with the rings) are expected by imaging scientists to provide the most surprises. Little is known about these satellites besides rough estimates of the sizes and densities.

Their densities indicate they could be solid ice or some combination of ice and rock up to about 30 to 40 percent rock. Infrared reflection spectra from Earth telescopes have also identified water ice or frost on many of the satellite surfaces.

It is known that Titan's clouds are reddish-colored; through a telescope Titan looks as red as Mars. But even the surface pressure of Titan's atmosphere is unknown; current estimates range from 2 bars, twice the surface atmospheric pressure of Earth, to about 20 millibars, about twice the surface pressure at Mars, or 2 percent of Earth's surface pressure. But sizes, surface characteristics, densities and composition of the other satellites are not known.

Low-Energy Charged Particles

The low-energy charged particle instrument is a strong coupling factor in *Voyager*'s complement of fields and particles investigations, contributing to many areas of interest, including the solar wind, solar flares, particle accelerations, magnetic fields and cosmic rays.

Two detectors allow measurements during the long interplanetary cruise and the encounters. The wide dynamic range, combined with wide coverage in energy and species, allow characterization of a wide range of energetic particle environments, including the intense trapped radiation environment around the planets. The low-energy charged particle measures particles traveling 2,400 to more than 150,000 kilometers (1,500 to more than 93,000 miles) a second. (High-energy particles travel at or near the speed of light 299,792 kilometers or 186,282 miles a second.)

Observations of particle acceleration provide data for understanding solar flare processes, cosmic ray acceleration and processes in Earth's magnetosphere.

Plasma

Voyager's plasma instrument measures plasma properties, including velocity, density and temperature, for a wide range of flow directions in the solar wind and planetary magnetospheres.

Traveling at supersonic speed (averaging 400 kilometers or 250 miles a second), plasma streams in all directions radially from the Sun, forming the solar wind. Solar wind interaction with Earth's magnetic field results in the northern lights and geomagnetic storms. Similar events have been observed at other planets.

At Saturn, the plasma team will study the interaction of the solar wind with Saturn; the sources, properties, forms and structure of Saturn's magnetospheric plasma; and the interaction of the magnetospheric plasma with the satellites and rings.

Although the plasma instrument cannot directly observe the neutral atoms in the magnetosphere, neutral gas is eventually ionized and becomes part of the Saturnian magnetosphere plasma.

It is possible that Titan, largest satellite in the solar system and the only one known to have a substantial atmosphere, has a torus, or donut-shaped cloud, of ionized material associated with it. If that is the case, the plasma instrument should detect some of those ions when *Voyager* nears the orbit of Titan.

Saturn's magnetosphere probably extends into space about 40 times the radius of Saturn (2.4 million kilometers or 1.49 million miles). That distance appears typical for a quiet magnetosphere and quiet solar wind. (When *Pioneer 11* arrived at Saturn late in August 1979, the Saturnian magnetosphere was strongly affected by violent solar wind activity.)

During *Voyager 1*'s encounter with Saturn, the pressure of the solar wind at Saturn and the size of Saturn's magnetosphere can be predicted using data from *Voyager 2* – farther from Saturn and closer to the Sun. Comparing data from both spacecraft during the *Voyager 1* encounter will reveal the reaction of the Saturnian magnetosphere to changes in the incoming solar wind. The leading edge of the magnetosphere will pulse in and out depending on changes in solar wind pressure.

Voyager's first encounter with Saturn's magnetosphere will be detected when the spacecraft crosses the bow shock wave, a region of demarcation between the solar wind and the Saturn environment. Voyager I is expected to cross the bow shock about November 10, two days before closest approach.

Immediately behind the bow shock is a transition region called the magnetosheath that separates the solar wind from the magnetosphere. The inner boundary of the magnetosheath, the magnetopause, separates the modified solar wind plasma in the magnetosheath from the plasma in the magnetosphere proper. Plasma in the magnetosphere comes from several sources – Saturn's ionosphere, ions from satellites and atmospheres, and the solar wind.

In the inner magnetosphere, plasma trapped by the magnetic field is forced to rotate with the planet. This region of co-rotation may extend as far as the magnetopause; the farther from the planet, the more the centrifugal force causes stretching of the magnetic field lines, more or less parallel to Saturn's equator.

Plasma Wave

Voyager I is surrounded by a low-density, ionized gas called a plasma. That plasma, composed entirely of atoms that are broken apart into electrons and positively charged ions, is a good electrical conductor with properties that are strongly affected by magnetic fields.

Plasma sources include the Sun, the planets and some of their satellites. Low-density plasmas are unusual; ordinary collisions between ions are unimportant, and individual ions and electrons interact with the rest of the plasma by means of emission and absorption of waves.

Localized interactions between waves and particles strongly control the dynamics of the entire plasma medium, and Voyager's plasma wave instruments are providing the first measurements of those phenomena at the outer planets.

Plasma waves are low-frequency oscillations that have their origins in instabilities within the plasma. They are of two types – electrostatic oscillations (similar to sound waves) or electromagnetic waves of very low frequency.

The plasma wave instrument measures the electric field component between 10 and 56,000 hertz. By way of comparison, Voyager's magnetometers measure the magnetic components of electromagnetic plasma waves below 10 hertz, while the planetary radio astronomy instrument measures waves with frequencies above 20 kilohertz.

Plasma ions and electrons collectively emit and absorb plasma waves. While the resulting particle / wave interactions affect the magnetospheric dynamics of the outer planets and the properties of the distant interplanetary medium, they have never been directly observed in those regions, since plasma waves cannot be observed far from their source and since there have been no previous wave studies at the outer planets.

Some effects to be studied include heating of solar wind particles at the outer planet bow shocks, acceleration of solar wind particles that produce high-energy trapped radiation, and the maintenance of boundaries between the rotating inner magnetospheres and the solar wind streaming around the planets.

Another objective of the plasma wave experiment is to study the influence of wave particle effects on interactions between inner satellites and the planet's rapidly rotating magnetosphere.

Detection of lightning bolts in the atmosphere of Saturn, as discussed earlier, would also be significant in searches for energy sources that contribute to the planet's structure. The plasma wave instrument searches for whistler signals that escape into the magnetosphere from lightning discharges.

The descending-scale whistle that is characteristic of lightning is caused by different velocities for plasma waves traveling along the magnetic field; higher frequencies arrive at the receiver sooner than lower frequencies.

Using the high-rate telemetry usually reserved for transmission of imaging data, the plasma wave instrument will play to Earth the entire audible signal of space plasma waves, spacecraft power, thruster firing and other instruments.

Ultraviolet Spectrometer

Voyager's ultraviolet spectrometer will study the composition and structure of Saturn's atmosphere and the material surrounding the satellites. Two techniques, airglow measurements and occultation, have been developed to probe a planet's atmosphere without entering that atmosphere:

- Airglow observations require coverage of a large area for maximum sensitivity to the weak emissions high in the atmosphere where collisions between atoms and molecules are rare.
- Occultation measurements require an instrument that reads ultraviolet radiation from the Sun as it is absorbed and scattered by a planet's atmosphere as the spacecraft moves into the shadow with the planet between it and the Sun.

Airglow observations measure atomic hydrogen and helium in the upper atmosphere by recording the resonance scattering of sunlight. Resonance scattering occurs when atoms and molecules absorb solar ultraviolet at specific wavelengths and reradiate at the same wavelengths. That differs from fluorescence, in which the activating wavelength is absorbed and energy is re-emitted at different wavelengths. It is also possible that auroral emissions will be observed at Saturn, since they were observed at Jupiter.

As the spacecraft disappears behind Saturn, the planet's atmosphere passes between the Sun and the ultraviolet spectrometer. Since the gases that make up an atmosphere have identifiable absorption characteristics at short wavelengths (they absorb various wavelengths differently), the ultraviolet spectrometer can measure how much of each gas is present at what temperature.

The important point is not how much sunlight enters the atmosphere, but what happens to it as it enters how it is absorbed and scattered.

Experiment	Principal Investigator	Instrument and Functions
Imaging Science	Bradford A. Smith, University of Arizona	Two TV cameras with 1,500 mm. f/8.5 and 200 mm, f/3 optics; multiple filters, variable shutter speeds and scan rates. Wide angle field of view = 56 x 55 milliradian (about 3 degrees square); on scan platform.
Infrared Interferometer Spectrometer	Rudolf H. Hanel, Goddard Space Flight Center	Spectrometer radiometer measuring temperature and molecular gas composition, with narrow, ¼ degree field of view, producing measurements every 48 seconds; on scan platform.
Ultraviolet Spectrometer	A. Lyle Hroadfoot, University of Southern California	Grating spectrometer measuring ion, atomic and small molecular gas abundances; spectral range 500-1,700 Angstroms; on scan platform.
Photopolarimeter	Arthur L. Lane, Jet Propulsion Laboratory	(Instrument not operating on Voyager 1.)
Plasma	Herbert S. Bridge, Massachusetts Institute of Technology	Dual plasma detectors, one aligned toward Earth and Sun and one perpendicular, with detection ranges from 10 eV to more than 6 KeV per nucleon.
Low-Energy Charged Particles	S. M. Krimigis, Johns Hopkins Applied Physics Laboratory	Dual rotating solid state detector sets, covering various ranges from 10 KeV to more than 30 MeV per nulceon.
Cosmic Ray	Rochus E. Vogt, California Institute of Technology	High-energy, low-energy and electron telescope systems using arrays of solid state detectors, several ranges from 0.5 to 500 MeV per nucleon.
Magnetometer	Norman F. Ness, Goddard Space Flight Center	Two low-field triaxial fluxgate magnetometers located on a 13-meter (43-foot) boom; two high-field (about 20 gauss) instruments on spacecraft body.
Planetary Radio Astronomy	James W. Warwick, Radiophysics, Inc.	Two 10-meter (33-foot) whip antennas and two-band receiver 1.2 KHz to 40.5 MHz detecting planetary radio emissions and bursts and solar stellar bursts.
Plasma Wave	Frederick L. Scarf, TRW Defense and Space Systems	Uses 10-meter (33-foot) planetary radio astronomy antennas with step frequency detector and waveform analyzer to measure plasma waves, thermal plasma density profiles at Jupiter and Saturn, satellite / magnetosphere interactions and wave particle interactions.
Radio Science	G. Len Tyler, Stanford University	Uses spacecraft S-band / X-band links in occultations of planets, satellites and rings to perceive changes in refractivity and absorption; celestial mechanics information calculated from tracking data.

Voyager Jupiter Science Results

Voyager 1 began its encounter with Jupiter in January 1979, completing it in early April. Voyager 2 began its encounter with Jupiter a few weeks later, in April 1979, and continued into August. The two spacecraft took more than 33,000 pictures of Jupiter and its five major satellites.

Although astronomers had studied Jupiter from Earth for more than three centuries, scientists were astonished by the findings from Voyager 1 and 2. They found new physical, geological and atmospheric processes in the planet, its satellites and magnetosphere.

Discovery of active volcanism on the satellite Io was probably the greatest surprise and may turn out to have the most far reaching effects. It is likely that activity on Io may be affecting the entire Jovian system. Io appears to be the source of matter that pervades the Jovian magnetosphere, the region of space surrounding the planet that is primarily influenced by the planet's strong magnetic field. Sulfur, oxygen and sodium, apparently erupted by Io's many volcanoes, were detected as far as the outer edge of the magnetosphere. Particles of the same material were detected inside Io's orbit, where they are accelerated to more than 10 percent of the speed of light.

The following is a summary of the more important science results from the Voyager encounters with Jupiter:

Jupiter's Atmosphere

• Atmospheric features of broadly different sizes appear to move with uniform velocities. That suggests that mass motion (movement of material) and not wave motion (movement of energy through a relatively stationary mass) is being observed.

- Rapid brightening of features in the atmosphere was observed, followed by spreading of cloud material. That is probably the result of disturbances that trigger convective (upwelling and downwelling) activity.
- A belt zone pattern of east to west winds was seen in the polar regions, roughly similar to the pattern seen in the more temperate areas. Previous investigations led scientists to believe the polar regions are dominated by convective upwelling and downwelling. Information from the two *Voyagers* shows that this was not the case.
- Material associated with the Great Red Spot, the planet's most prominent atmospheric feature, moves in a counterclockwise (anticyclonic) direction. The material appears to have a rotation period of about six days.
- Smaller spots appear to interact with the Great Red Spot and with each other.
- Auroral emissions (similar to Earth's northern lights) were observed in Jupiter's polar regions. They have been seen in both ultraviolet and visible light. The ultraviolet auroral emissions were not present during the *Pioneer 10* encounter in 1973. They appear to be associated with material from Io that spirals along magnetic field lines to fall into Jupiter's atmosphere.
- Cloud-top lightning bolts, similar to super-bolts in Earth's high atmosphere, were detected.
- The atmospheric temperature at 5 to 10 millibars is about 160 degrees Kelvin (-171 degrees Fahrenheit), part of an inversion layer — a warm region above a cold layer, similar to the phenomenon that traps smog in the Los Angeles Basin. (Earth's surface pressure is about 1,000 millibars.)
- The *Voyagers* observed an ionospheric temperature that changed with altitude, reaching about 1,100 Kelvin (1,520 degrees Fahrenheit). That was also not observed by *Pioneer 10*, and Voyager scientists believe they are witnessing large temporal or spatial changes in the ionosphere of Jupiter.
- Atmospheric helium was measured; its percentage compared to hydrogen is important to understand the composition and history of the atmosphere and indirectly, the primordial cloud out of which the Sun and planets formed. The relative abundance of helium is about 11 percent that of hydrogen.
- The atmospheric temperature above the Great Red Spot is somewhat colder (3 degrees Celsius or 5.4 degrees Fahrenheit) than in surrounding regions.

Satellites and Rings

- Eight currently active (erupting) volcanoes, probably driven by tidal heating, were positively identified on Io by *Voyager 1*. Many more are suspected. *Voyager 2* saw six more still active, but the largest had shut down. An eighth was out of range of *Voyager 2*'s instruments. Plumes from the volcanoes extend up to 250 kilometers (155 miles) above the surface. The material is being ejected at rates up to 1 kilometer per second (2,200 mph). By comparison, ejection velocities have been measured at Mount Etna, one of Earth's most explosive volcanoes, at 50 meters per second (112 mph). The volcanism on Io is apparently associated with heating of the satellite by tidal pumping. Io's orbit is perturbed by Europa and Ganymede, two other large satellites nearby, and Io is drawn back again by Jupiter. That action causes tidal bulging as great as 100 meters (330 feet).
- *Voyager 1* measured the temperature of two hot spots on Io. Both are associated with volcanic features. While the surrounding terrain has a temperature of -138° C (-216° F), the hot spot's temperature is about 20° C (68° F). That hot spot may, scientists believe, be a lava lake, although the temperature indicates that the surface of the spot is not molten; it is, at least, reminiscent of lava lakes on Earth. The other hot spot is associated with Plume No. 1, called Pele. That hot spot, though smaller, registered temperatures as high as 300° C (572° F).
- Europa displayed a large number of intersecting linear features in the distant, low-resolution photos from *Voyager 1*. Scientists at first believed the features might be deep cracks, caused by crustal rifting or tectonic processes. The closer, high-resolution photos from *Voyager 2*, however, left scientists puzzled: the features were so lacking in topographic relief that they "might have been painted on with a felt marker," one scientist commented. There is a possibility that Europa may be internally heated due to tidal forces, although at a much lower level than Io.
- Ganymede showed two distinct types of terrain — cratered and grooved. That suggests to scientists that Ganymede's entire, ice-rich crust has been under tension from global tectonic processes.
- Callisto has an ancient, heavily-cratered crust, with remnant rings of enormous impact basins. The basins themselves have been erased by the flow of the ice-laden crust — to the degree that little topographic relief is apparent in the high-resolution pictures taken by the *Voyagers*.
- Amalthea has an elliptical shape. It is 265 kilometers (164 miles) by 140 kilometers (87 miles). Amalthea is about 10 times bigger than Mars' larger satellite, Phobos.
- A ring of material was discovered around Jupiter. The outer edge is 128,000 kilometers (79,500 miles) from the center of the planet, and is no more than 30 kilometers (18.6 miles) thick. There appears to be more than one ring, and the material appears to gradually fall inward to the planet itself. Studies of the scattering of light by the ring should yield a good indication of the size of the particles.

Magnetosphere

- An electric current of about 2.5 million amperes was detected in the flux tube connecting Jupiter and Io. That was 2½ times stronger than the current predicted before *Voyager 1*'s arrival. The spacecraft did not fly through the flux tube, as has been planned, since the flux tube had been displaced 7,000 kilometers (4,300 miles) from the expected location.

- The *Voyagers* detected ultraviolet emissions from doubly- and triply-ionized sulfur and from doubly-ionized oxygen. Since *Pioneer 10* and *11* did not detect those emissions, that indicates a hot plasma was not present in 1973 and 1974 when the *Pioneer* encounters took place. The sulfur apparently originates in Io's volcanoes.
- Plasma electron densities in some regions of the Io torus (a tube-shaped ring of matter in the region of Io's orbit) exceeded 4,500 per cubic centimeter.
- A cold plasma, rotating with Jupiter, was discovered inside six Jupiter radii (428,000 kilometers or 266,000 miles) from the planet. Ions of sulfur and oxygen were detected.
- High-energy trapped particles were also detected in the same region near Jupiter. They had significantly enhanced abundances of oxygen, sodium and sulfur.
- A hot plasma was measured near the Jovian magnetopause (the outer edge of the magnetosphere), composed mostly of protons, oxygen and sulfur.
- Kilometric radio emissions were detected coming from Jupiter. The emissions, in the frequency range from 10 kilohertz to 1 megahertz, might be associated with the Jovian auroral processes.
- Plasma flows were detected in the day-side middle magnetosphere; they rotate with the planet at a 10 hour period.
- *Voyager 1* saw evidence of a transition from closed magnetic field lines to a magnetotail on the antisolar side of Jupiter. Although such a magnetotail was never in serious question, its existence had not been detected before.
- *Voyager* also measured radio spectral arcs (from about 1 megahertz to more than 30 megahertz) in patterns that correlate with Jovian longitude.

Voyager Science Teams

Cosmic Ray

- Rochus E. Vogt,   California Institute of Technology, Principal Investigator
- J. Randy Jokipii,   University of Arizona
- Frank B. McDonald,   Goddard Space Flight Center
- A. W. Schardt,   Goddard Space Flight Center
- Edward C. Stone,   California Institute of Technology
- James H. Trainor,   Goddard Space Flight Center
- William R. Webber,   University of New Hampshire

Infrared Radiometry and Spectrometry

- Rudolf A. Hanel,   Goddard Space Flight Center, Principal Investigator
- Barney Conrath,   Goddard Space Flight Center
- Dale Cruikshank,   University of Hawaii
- F. Michael Flasar,   Goddard Space Flight Center
- Daniel Gautier,   Observatoire de Paris, France
- Peter Gierasch,   Cornell University
- Shailendra Kumar,   University of Southern California
- Virgil Kunde,   Goddard Space Flight Center
- William Maguire,   Goddard Space Flight Center
- John Pearl,   Goddard Space Flight Center
- Joseph Pirraglia,   Goddard Space Flight Center
- Cyril Ponnamperuma,   University of Maryland
- Robert Samuelson,   Goddard Space Flight Center

Imaging Science

- Bradford A. Smith,   University of Arizona, Team Leader
- Geoffrey Briggs,   NASA Headquarters
- Allan F. Cook II,   Center for Astrophysics
- G. Edward Danielson,   California Institute of Technology
- Merton E. Davies,   Rand Corp.
- Garry E. Hunt,   University College London
- Torrence V. Johnson,   Jet Propulsion Laboratory
- Harold Masursky,   U.S. Geological Survey
- Tobias Owen,   State University of New York
- Carl Sagan,   Cornell University
- Laurence Soderblom,   U.S. Geological Survey
- Verner E. Suomi,   University of Wisconsin

Low-Energy Charged Particles

- S.M. (Tom) Krimigis,   Johns Hopkins University, Principal Investigator
- Thomas P. Armstrong,   University of Kansas
- W. Ian Axford,   Max Planck Institut fur Aeronomie
- Carl O. Bostrom,   Johns Hopkins University
- Chang Yun Fan,   University of Arizona
- George Gloeckler,   University of Maryland
- Ed Reath,   Johns Hopkins University
- Louis J. Lanzerotti,   Bell Laboratories

Magnetic Fields

- Norman F. Ness,   Goddard Space Flight Center, Principal Investigator
- Mario F. Acuna,   Goddard Space Flight Center
- Ken W. Behannon,   Goddard Space Flight Center
- Len F. Burlaga,   Goddard Space Flight Center
- Ron P. Lepping,   Goddard Space Flight Center
- Fritz M. Neubauer,   Der Technischen Universat Braunschweig

Plasma Science

- Herbert S. Bridge,   Massachusetts Institute of Technology, Principal Investigator
- John W. Belcher,   Massachusetts Institute of Technology
- Len F. Burlaga,   Goddard Space Flight Center
- Christoph R. Goertz,   Max Planck Institut fur Aeronomie
- Richard E. Hartle,   Goddard Space Flight Center
- Art J. Hundausen,   High Altitude Observatory
- Alan J. Lazarus,   Massachusetts Institute of Technology
- Keith Ogilvie,   Goddard Space Flight Center
- Stanislaw Olbert,   Massachusetts Institute of Technology
- Jack D. Scudder,   Goddard Space Flight Center
- George L. Siscoe,   University of California, Los Angeles
- James D. Sullivan,   Massachusetts Institute of Technology
- Vytenis M. Vasyliunas,   Max Planck Institut fur Aeronomie

Photopolarimetry

This instrument, on *Voyager 1*, is not operating correctly and has been turned off.
The photopolarimeter on *Voyager 2* is operating and will be used at Saturn.)

- Arthur L. Lane,   Jet Propulsion Laboratory, Principal Investigator
- David Coffeen,   Goddard Institute for Space Studies
- Larry Esposito,   University of Colorado
- James E. Hansen,   Goddard Institute for Space Studies
- Charles W. Hord,   University of Colorado
- Kevin Pang,   Science Applications, Inc.
- Makiko Sato,   Goddard Institute for Space Studies
- Robert West,   University of Colorado

Planetary Radio Astronomy

- James W. Warwick,   Radiophysics, Inc., Principal Investigator.
- Joseph K. Alexander,   Goddard Space Flight Center
- Andre Boischot,   Observatoire de Paris, France
- Walter E. Brown, Jr.,   Jet Propulsion Laboratory
- Thomas D. Carr,   University of Florida
- Samuel Gulkis,   Jet Propulsion Laboratory
- Fred T. Haddock,   University of Michigan
- Christopher C. Harvey,   Observatoire de Paris, France
- Michael L. Raiser,   Goddard Space Flight Center
- Yolande LeBlanc,   Observatoire de Paris
- Jeffrey B. Pearce,   Radiophysics, Inc.
- Robert G. Peltzer,   Martin Marietta Corp.
- Roger Phillips,   Jet Propulsion Laboratory
- Anthony C. Riddle,   University of Colorado
- David H. Staelin,   Massachusetts Institute of Technology

Plasma Wave

- Frederick L. Scarf,   TRW Defense and Space Systems, Principal Investigator
- Donald A. Gurnett,   State University of Iowa

Radio Science

- G. Len Tyler,   Stanford University, Team Leader
- John D. Anderson,   Jet Propulsion Laboratory
- Thomas L. Croft,   SRI International
- Von R. Eshelman,   Stanford University
- Gerald S. Levy,   Jet Propulsion Laboratory
- Gunnar F. Lindal,   Jet Propulsion Laboratory
- G.E. Wood,   Jet Propulsion Laboratory

Ultraviolet Spectroscopy

- A. Lyle Broadfoot,   University of Southern California, Principal Investigator
- Sushil K. Atreya,   University of Michigan
- Michael J.S. Belton,   Kitt Peak National Observatory
- Jean L. Bertaux,   Service d'Aeronomie du CNRS
- Jacques E. Blamont,   Jet Propulsion Laboratory
- Alexander Dalgarno,   Harvard College Observatory
- Thomas M. Donahue,   University of Michigan
- Richard Goody,   Harvard University
- John C. McConnell,   York University
- Michael B. McElroy,   Harvard University
- H. Warren Moos,   Johns Hopkins University
- William R. Sandel,   University of Southern California
- Donald E. Shemansky,   University of Southern California
- Darrell F. Strobel,   Naval Research Laboratory

Voyager Mission Summary

August 1977 to September 1980

Voyager 2, first of the two spacecraft to begin the journey to the outer planets, was launched from Cape Canaveral Florida, aboard a Titan-Centaur launch vehicle, at 10:29 a.m. EDT, August 20, 1977.

Voyager 1 was launched 16 days later, at 8:46 a.m. EDT, September 5.

On December 10, 1977, both spacecraft entered the asteroid belt, a band of rock and dust 360 million kilometers (223 million miles) wide that circles the Sun between the orbits of Mars and Jupiter. *Voyager 1* flew out of the asteroid belt September 8, 1978. *Voyager 2* left the belt behind on October 21.

Earlier – on December 15, 1977 – *Voyager 1*, flying a shorter and faster course to Jupiter, had overtaken *Voyager 2*. At the time, they were 170 million kilometers (105 million miles) from Earth.

In the early months of 1978, two major problems occurred, one on each spacecraft.

During a calibration of *Voyager 1*'s scan platform, the movable housing for the cameras and some other instruments became jammed. Engineers determined that a tiny piece of soft debris had found its way into the gears. By maneuvering the platform, engineers were able to free the platform.

Voyager 2's primary radio receiver failed on April 5, 1978, and the spacecraft's computer command subsystem automatically switched to the backup receiver. The backup receiver, however, developed a faulty tracking loop capacitor. But engineers have provided a work-around solution to the tracking loop capacitor. The spacecraft now cannot lock on a signal from Earth and track it as the frequency varies because of the spacecraft's velocity and the Earth's rotation (doppler shift). But ground-based computers have been programmed to precisely control the frequencies transmitted so they can arrive at the spacecraft at the frequency at which *Voyager 2* can receive them.

And an automated sequence for Saturn has been stored aboard *Voyager 2*'s computer. In the event that the second receiver should fail, and the spacecraft can no longer be commanded, *Voyager 2* will perform its Saturn encounter experiments and return useful data to Earth.

Both *Voyagers* conducted fields and particles experiments during the long cruise to Jupiter, and the period between Jupiter and Saturn and Saturn encounters.

To achieve the highest-ever velocity of a spacecraft (over 51,500 km (32,000 miles) per hour), *Pioneer 10* was launched on its trip to Jupiter on an Atlas-Centaur booster with an added third stage.

Communication with the *Pioneer* spacecraft at unprecedented distances relied upon the large antennas of the Deep Space Network, such as this one at Goldstone in California's Mojave Desert.

This spectacular series of consecutive pictures was taken over a period of four hours, between 44½ and 40½ hours before periapsis. The Great Red Spot is prominent, and the shadow of Io traverses the disc of the planet. Cosmetic enhancement was applied to all these images.

Only 11 hours after launch *Pioneer 10* passed the orbit of the Moon, shown conceptually in this artist's sketch. The Moon was actually at a more distant part of its orbit than shown here. By comparison, the Apollo spacecraft took four days to reach the Moon.

NASA's *Pioneer 11* image of Saturn and its moon Titan (upper left). The irregularities in ring silhouette and shadow are due to technical anomalies in the preliminary data. Looking at the rings from left to right, the ring area begins with the outer A ring; the Encke Division; the inner A Ring; Cassini Division; the B Ring; the C Ring; and the innermost area where the D Ring would be. *Pioneer 11* was 2,846,000 kilometers (1,768,422 miles) from Saturn when his image was made.

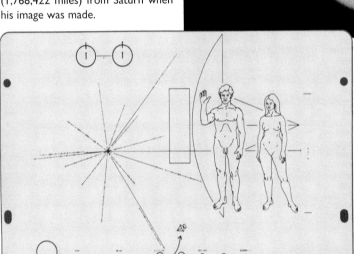

Both *Pioneer* spacecraft carried a pictorial message intended for other intelligent species, if any exist, who might find the spacecraft thousands of years from now in some other star system. The plaque tells when the *Pioneers* were launched, from where, and by whom. The plaque was designed by Dr. Carl Sagan, Director of the Laboratory for Planetary Studies, Cornell University.

The *Voyager* spacecraft weighed 795 kg (1,753 pounds) including 113 kg (249 pounds) of scientific instruments. The large antenna at the top is 3.66 meters (12 feet) in diameter. A gold-plated copper record of Earth greetings, sights, and sounds is inside a gold plated aluminum canister, which has instruction symbols on its face and is attached to one of the ten sides of the spacecraft (center). On the boom at the lower left are the three nuclear power generators.

Pioneer 11 followed *Pioneer 10* toward Jupiter on April 5, 1973. Launching took place within a few seconds of the opening of the launch window.

The *Voyager* spacecraft were among the most sophisticated, automatic, and independent robots of their time ever sent to explore the planets. Here *Voyager* undergoes final tests in a space simulator chamber.

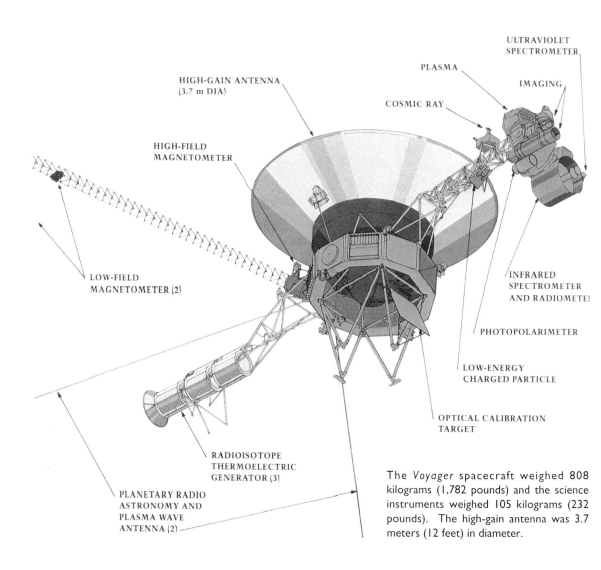

ULTRAVIOLET
SPECTROMETER

PLASMA

IMAGING

HIGH-GAIN ANTENNA
(3.7 m DIA)

COSMIC RAY

HIGH-FIELD
MAGNETOMETER

LOW-FIELD
MAGNETOMETER (2)

INFRARED
SPECTROMETER
AND RADIOMETE

PHOTOPOLARIMETER

LOW-ENERGY
CHARGED PARTICLE

OPTICAL CALIBRATION
TARGET

RADIOISOTOPE
THERMOELECTRIC
GENERATOR (3)

PLANETARY RADIO
ASTRONOMY AND
PLASMA WAVE
ANTENNA (2)

The *Voyager* spacecraft weighed 808 kilograms (1,782 pounds) and the science instruments weighed 105 kilograms (232 pounds). The high-gain antenna was 3.7 meters (12 feet) in diameter.

The main elements of NASA's Deep Space Network (DSN) are the Deep Space Stations based around the globe, the network control center, and the ground communications facility that links the stations and control center together. For the *Voyager* 2 encounters with Uranus and Neptune, the DSN's the Australian tracking complex was augmented by the Parkes Radio Astronomy antenna (inset), a 64-meter dish located approximately 200 miles from the main complex, near Canberra in New South Wales.

A Titan IVB-Centaur rocket served as the launch vehicle for the *Voyager* spacecraft, which was the last planned used of this type of launch vehicle prior to the era of the Space Transportation System (Space Shuttle).

In this artist's concept of the *Voyager* spacecraft the 13-meter (43-foot) magnetometer boom, and the two planetary radio astronomy and plasma wave antennas can be seen.

Jupiter is more massive than all of the other planets combined, plus their satellites, the asteroids and all the comets. The *Voyager* spacecraft produced higher-resolution pictures of Jupiter's colorful bands, zones and storm features than either Earth-based telescopes or the earlier Pioneer missions.

Jupiter and its four planet-sized Galilean satellites were photographed in early March 1979 by *Voyager 1* and assembled into this collage. They are not to scale but are in their relative positions. Reddish Io (upper left) is nearest Jupiter; then Europa (center); Ganymede and Callisto (lower right). Nine other much smaller satellites circle Jupiter, one inside Io's orbit and the others millions of miles from the planet.

Voyager 1 took this picture of Ganymede (center-left) from a distance of 1.6 million miles. Ganymede is the seventh and largest of Jupiter's known moons. Voyager scientists discovered that Ganymede has its own magnetosphere embedded inside Jupiter's large one.

As *Voyager 1* flew by Jupiter, it captured this photo of the Great Red Spot, an anti-cyclonic (high-pressure) storm that can be likened to the worst hurricanes on Earth. It is so large that three Earths could fit inside it. The Great Red Spot had been observed from Earth for hundreds of years, yet never before with this clarity and closeness (objects as small as six hundred kilometers can be seen).

This montage of images of the Saturnian system was prepared from an assemblage of images taken by the *Voyager 1* spacecraft during its Saturn encounter in November 1980. This artist's view shows Dione in the forefront, Saturn rising behind, Tethys and Mimas fading in the distance to the right, Enceladus and Rhea off Saturn's rings to the left, and Titan in its distant orbit at the top.

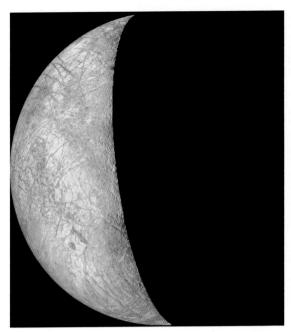

This mosaic of Europa, the smallest Galilean satellite, was taken by *Voyager 2*. The bright areas are probably ice deposits, whereas the darkened areas may be the rocky surface or areas with a more patchy distribution of ice.

This *Voyager 2* view, focusing on Saturn's C-ring (blue) was compiled from three separate images, taken through ultraviolet, clear and green filters, at a distance of 2.7 million kilometers (1.7 million miles) from the planet.

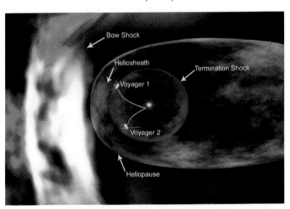

At left is an artist's concept of the positions of the *Voyager* spacecraft in relation to structures formed around our Sun by the solar wind. Also illustrated is the termination shock, a violent region the spacecraft must pass through before reaching the outer limits of the solar system.

This computer generated montage (below-left) shows Neptune as it would appear from a spacecraft approaching Triton, Neptune's largest moon at 2,706 km (1,683 miles) in diameter.

This picture of Neptune was produced from images taken through the green and orange filters on the *Voyager 2* narrow-angle camera at a range of 4.4 million miles, 4 days and 20 hours before closest approach.

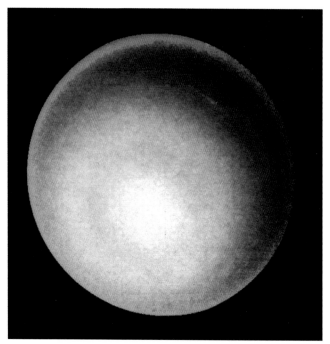

This computer enhancement of a *Voyager 2* image emphasizes the high-level haze in Uranus' upper atmosphere. Clouds are obscured by the overlying atmosphere.

At the Kennedy Space Center's Spacecraft and Assembly Encapsulation Facility 2 (SAEF-2), the planetary spacecraft checkout facility, clean-suited technicians work on the *Galileo* spacecraft prior to moving it to the Vehicle Processing Facility (VPF) for mating with the inertial upper stage (IUS).

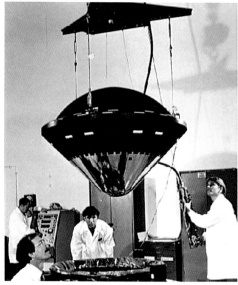

At top is an artist's concept of the *Galileo* spacecraft, a system that includes a Jupiter-orbiting observatory and an entry probe. The atmospheric probe descent module is suspended above the deceleration module aeroshell (center) prior to mating. At the Kennedy Space Center's Spacecraft and Assembly Encapsulation Facility 2 (below), clean-suited technicians lower the *Galileo* probe descent module into test stand via overhead crane.

In the Vertical Processing Facility (VPF), the spacecraft *Galileo* is prepared for mating with the Inertial Upper Stage (IUS) booster prior to its more than six-year journey to Jupiter.

The *Galileo* spacecraft was launched into orbit and deployed from the the Space Shuttle *Atlantis* as the main objective of mission STS-34, launched on October 18, 1989.

The *Galileo* spacecraft and its Inertial Upper Stage (IUS) booster were deployed from the cargo bay of STS-34 *Atlantis* on October 18, 1989, less than 7½ hours after launch.

This mosaic of asteroid Ida consists of five images acquired by the *Galileo* spacecraft's solid state imaging camera at distances ranging from 3,057 to 3,821 kilometers on August 28, 1993.

Probing Jupiter's Atmosphere – the red-hot nose cone separates from the probe portion of the *Galileo* spacecraft as it "hangs on the shrouds" and samples the atmosphere of Jupiter.

Cassini Mission photographic artwork of the *Cassini* Saturn Orbiter and Titan *Huygens* Probe Spacecraft.

NASA's *Galileo* spacecraft acquired its highest resolution images of Jupiter's moon Io on July 3, 1999 during its closest pass to Io since orbit insertion in late 1995. This color mosaic uses the near-infrared, green and violet filters (slightly more than the visible range) of the spacecraft's camera and approximates what the human eye would see.

JPL technicians clean and prepare the upper equipment module for mating with the nuclear propulsion module subsystem of the *Cassini* orbiter in the Payload Hazardous Servicing Facility at KSC.

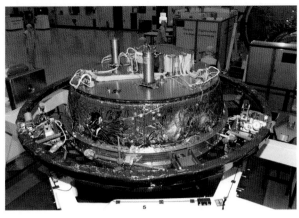

The descent module of the *Huygens* probe undergoes preflight processing in the Payload Hazardous Servicing Facility. The cylinders on the top of the probe contain antennas; the small square box has a parachute.

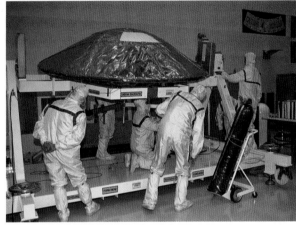

The *Huygens* probe on a custom handling fixture in the Payload Hazardous Servicing Facility (PHSF) being readied for mating with the *Cassini* orbiter.

The *Huygens* probe being installed into the *Cassini* orbiter in the Payload Hazardous Servicing Facility (PHSF) at Kennedy Space Center

The *Huygens* probe on its custom handling fixture, prepared for mating with the *Cassini* orbiter by the Dornier Satelliten Systeme workers.

Dornier Satelliten Systeme workers lift the heat shield of the *Huygens* probe (top) in the Payload Hazardous Servicing Facility. The spacecraft was returned to the PHSF after damage to thermal insulation was discovered inside *Huygens* from an abnormally high flow of conditioned air. Internal inspection, insulation repair and a cleaning of the probe were required. The *Cassini* spacecraft in a JPL assembly room (right). The large spacecraft (orbiter) is shown with ESA's *Huygens* Titan probe attached. The orbiter mass at launch will be nearly 5,300 kg, over half of which is propellant for trajectory control. The mass of the 2.7-meter-diameter Titan probe is roughly 350 kg. Below, JPL workers examine the *Huygens* probe after removal from the *Cassini* spacecraft in the Payload Hazardous Servicing Facility (PHSF) at KSC.

The *Cassini* spacecraft undergoes vibration and thermal testing at JPL facilities, subjecting it to weeks of "shake and bake" tests that imitate the forces and extreme temperatures that the spacecraft will experience.

Below, an environmental health specialist from EG&G Florida Inc., KSC's base operations contractor, uses an ion chamber dose rate meter to measure radiation levels in one of the three radioisotope thermoelectric generators that will power the *Cassini* spacecraft.

A Titan IVB-Centaur booster carried the *Cassini* orbiter and its attached *Huygens* probe into space on October 15, 1997. A 2.2-billion mile, seven-year journey brought *Cassini* to Saturn in July 2004.

This artist's conception shows the *Huygens* Probe approaching Titan's surface. Saturn appears dimly in the background through Titan's thick atmosphere. The *Cassini* spacecraft flies overhead with its high gain antenna pointed at the *Huygens* probe.

Below is an artist's concept of the *Deep Space 1* spacecraft, ion engine thrusting, approaching the comet Borelly. The large solar panels convert sunlight into electricity to power the xenon ion engine.

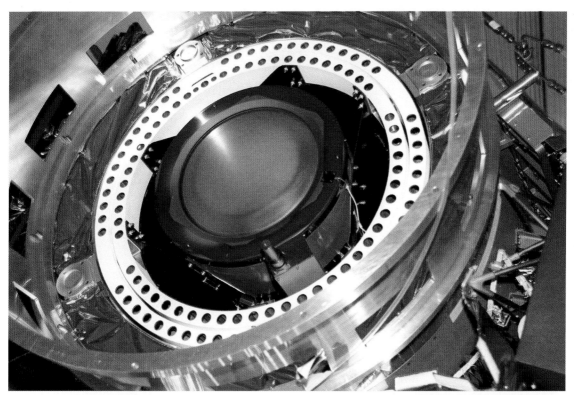

Deep Space 1 is lifted from its work platform, giving a closeup view of the experimental solar-powered ion propulsion engine. This engine is the first non-chemical propulsion to be used as the primary means of propelling a spacecraft. *Deep Space 1* is designed to validate 12 new technologies for space missions.

In the Payload Hazardous Servicing Facility, KSC workers place insulating blankets on *Deep Space 1* to prepare it for launch. *Deep Space 1*'s onboard experiments include the ion propulsion engine and software that tracks celestial bodies so that the spacecraft can make its own navigation decisions without the intervention of ground controllers.

Deep Space 1 was launched in October 1998 aboard a Boeing Delta 7326 rocket from Launch Pad 17A, Cape Canaveral Air Station, Florida. Delta II rockets are medium capacity expendable launch vehicles derived from the Delta family of rockets built and launched since 1960. *Deep Space 1* is part of NASA's New Millennium Program.

The Deep Space Operations Center, also known as the Dark Room, is the heart of the Space Flight Operations Facility at JPL and links over 800 independent workstations all over the world. These state-of-the-art engineering workstations run the UNIX operating system and some commercial off-the-shelf software. Most of the software has been uniquely developed or adapted at JPL to perform the specialized telemetry processing and spacecraft commanding tasks required for operating multiple missions.

The sun rises behind Launch Pad 17-B at Cape Canaveral Air Force Station, Florida, where the Boeing Delta II rocket carrying the *Deep Impact* spacecraft waits for launch. Gray clouds above the horizon belie the favorable weather forecast for the afternoon launch. Liftoff was at 1:47 p.m. EST January 12, 2005. *Deep Impact* will rendezvous with Comet Tempel 1 when the comet is 83 million miles from Earth.

In the clean room at Astrotech Space Operations near Kennedy Space Center, the plastic protective cover is lifted from the *Deep Impact* spacecraft prior to it undergoing functional testing to verify its state of health after the over-the-road journey from Ball Aerospace and Technologies Corporation in Boulder, Colorado. This was followed by loading updated flight software and beginning a series of Mission Readiness Tests. The spacecraft was launched December 30, 2004, aboard a Boeing Delta II rocket from Launch Complex 17 at Cape Canaveral Air Force Station, Florida.

Voyager I began the observatory phase of its Jupiter encounter January 4, 1979, while it was 61 million kilometers (38 million miles) from the planet. Immediately obvious to the scientists on Voyager's imaging team was the change in Jupiter's cloud features in the four years since *Pioneer II* had flown past Jupiter (in December 1974).

Voyager I made its closest approach to Jupiter at 4:05 a.m. PST, spacecraft time, on March 5, 1979, at an altitude of 278,000 kilometers (173,000 miles). (See Jupiter science results section of Press Kit.) It passed the Galilean satellites after passing Jupiter, then began its 20-month journey to Saturn, boosted on its way by Jupiter's orbital motion and immense gravity.

Despite the intense radiation dose dealt *Voyager I* by particles in Jupiter's magnetosphere, the spacecraft experienced only one failure: circuitry in the photopolarimeter failed about six hours before closest approach. It has not recovered and will not be used during the Saturn encounter.

Voyager 2's encounter with Jupiter began April 25, just 19 days after *Voyager I*'s encounter ended. The same general sequence was followed with the *Voyager 2* encounter, with some important differences:

• *Voyager 2* passed farther from Jupiter than *Voyager I*, 650,000 kilometers (404,000 miles) above the cloud tops.

• *Voyager 2*'s science sequences were designed from knowledge gained by *Voyager I*.

• *Voyager 2* observed the Galilean satellites at close range on the inbound leg of the trajectory. Those satellites always present the same face to Jupiter, so between the two spacecraft, photos of both faces of the satellites were obtained.

The closest approach point at Jupiter was chosen to provide a trajectory to Saturn (for *Voyager 2*) that in turn would yield a trajectory to Uranus. At Jupiter and Saturn, the flight path and velocity of the spacecraft are changed by gravitational attraction of the planets and their orbital motion.

Voyager I's encounter with Saturn began August 22, 1980. It will make its closest approach to Saturn on November 12. *Voyager 2*'s encounter will begin in May 1981; closest approach occurs August 26. Then *Voyager 2* begins the cruise toward Uranus.

The Satellites of Saturn

Name	Radius (km)	Radius (mi.)	Distance (km)	Distance (mi.)	Orbital Time (Earth Days)
Janus*	150	93	168,700	104,825	0.815
Mimas	175	109	185,500	115,264	0.942
Enceladus	260	162	238,000	147,886	1.370
Tethys	510	317	294,700	183,118	1.888
Dione	550	342	377,400	234,506	2.737
Rhea	750	466	527,000	327,463	4.518
Titan	2,915	1,811	1,222,000	759,316	15.945
Hyperion	150	93	1,484,000	922,115	21.276
Iapetus	800	497	3,562,000	2,213,324	79.330
Phoebe	40	25	12,960,000	8,052,971	550.450

* Janus' existence is in doubt.

A number of other satellites have been discovered during the past 14 months, both by the *Pioneer* spacecraft and from Earth; *Voyager I* will conduct several sequences to verify those new satellites and their orbits and to search for others as yet undiscovered.)

Voyager Team

NASA Office of Space Science

Dr. Thomas A. Mutch	Associate Administrator for Space Science
Andrew J. Stofan	Deputy Associate Administrator
Dr. Adrienne F. Timothy	Assistant Associate Administrator for Science
Angelo Guastaferro	Director, Planetary Division
Dr. Geoffrey A. Briggs	Deputy Director, Planetary Division
Dr. C. Howard Robins	Manager, Operations
Frank A. Carr	Program Manager, Acting
Dr. Milton A. Mitz	Program Scientist

NASA Office of Tracking and Data Systems

Robert E. Smylie	Associate Administrator for Tracking and Data Systems
Charles A. Taylor	Director, Network Systems Division
Dr. Richard Green	Manager, Deep Space Network Operations
Harold G. Kimball	Director, Communications and Data Systems Division

Jet Propulsion Laboratory

Dr. Bruce C. Murray	Director
Gen. Charles H. Terhune, Jr.	Deputy Director
Robert J. Parks	Assistant Director for Flight Projects
Raymond L. Heacock	Project Manager
Esker K. Davis	Deputy Project Manager
Richard Laeser	Mission Director
George P. Textor	Deputy Mission Director
Richard P. Rudd	Deputy Mission Director
Charles E. Kohlhase	Manager, Mission Planning Office
Robert G. Polansky	Manager, Ground Data Systems
Marvin R. Traxler	Manager, Tracking and Data System
Kurt Heftman	Manager, Mission Computing System
Dr. Charles H. Stembridge	Manager, Flight Science Office
Dr. Ellis D. Miner	Assistant Project Scientist
Edward L. McKinley	Manager, Flight Engineering Office
Raymond J. Amorose	Manager, Flight Operations Office

California Institute of Technology

Dr. Edward C. Stone	Project Scientist

Conversion Table

Multiply	By	To Get
Inches	2.54	Centimeters
Centimeters	0.3937	Inches
Feet	30.48	Centimeters
Centimeters	4.7244	Feet
Feet	0.3048	Meters
Meters	3.2808	Feet
Yards	0.9144	Meters
Meters	1.0936	Yards
Statute Miles	1.6093	Kilometers
Kilometers	0.6214	Miles
Feet/Second	0.3048	Meters/Second
Meters/Second	3.281	Feet/Second
Meters/Second	2.237	Statute Miles/Hour
Feet/Second	0.6818	Miles/Hour
Miles/Hour	1.6093	Kilometers/Hour
Kilometers/Hour	0.6214	Miles/Hour
Pounds	0.4563	Kilograms
Kilograms	2.2046	Pounds

- To convert Fahrenheit to Celsius (Centigrade), subtract 32 and multiply by $5/9$.
- To convert Celsius to Fahrenheit, multiply by $9/5$ and add 32.
- To convert Celsius to Kelvin, add 273.
- To convert Kelvin to Celsius, subtract 273.

NASA
National Aeronautics and
Space Administration

Press Kit
January 1986

RELEASE No: 85-165   January 1986

CONTACTS

James F. Kukowski	Headquarters, Washington, D.C.	Phone: (202) 453-1548
Frank E. Bristow	Jet Propulsion Laboratory, Pasadena, California	Phone: (818) 354-5011
Mary Beth Murrill	Jet Propulsion Laboratory, Pasadena, California	Phone: (818) 354-5011

RELEASE No: 85-165

January 1986

CONTENTS

National Aeronautics and
Space Administration

Washington, D.C. 20546
AC 202-453-8400

FOR RELEASE:
IMMEDIATE

Voyager Nears Uranus

NASA's *Voyager 2* will make its closest approach to Uranus, flying 81,500 kilometers (50,600 miles) above the cloud tops of the seventh planet, at 1 p.m. EST, January 24, 1986. It will be the first spacecraft to reach the planet, providing our first close look at this system.

The *Voyager 2* Uranus encounter, which began November 4, 1985, continues through February 25, 1986. During that period, the spacecraft's 11 instruments will perform close-range studies of the planet, its five known satellites and nine rings. The spacecraft also will search for a planetary magnetic field, new satellites and new rings.

Encounter activity peaks during a 6-hour period on January 24, 1986, when the highest priority observations will take place. In about a quarter of a day, scientists will obtain more information about Uranus, its satellites and rings, than has been learner since Sir William Herschel discovered the planet March 13, 1781. Much of the data collected during the spacecraft's closest approach, however, will be recorded on the spacecraft for playback to Earth on following days.

Uranus will be the third planet visited by *Voyager 2* since the spacecraft was launched from Earth on August 20, 1977. *Voyager 2* flew past Jupiter on July 9, 1979, and then Saturn on August 25, 1981.

An identical spacecraft, *Voyager 1*, was launched September 5, 1977, and flew past Jupiter on March 5, 1979, and Saturn on November 12, 1980. *Voyagers 1* and 2 returned to scientists an unprecedented amount of information on the two planets, their rings, moons and the interplanetary medium during the spacecraft cruise phases. More than 70,000 photos were taken of Jupiter and Saturn by *Voyager 1* and 2.

Among the major discoveries at Jupiter were active volcanism on the satellite Io, thin rings of dust and ice encircling Jupiter, and three new satellites orbiting our solar system's largest planet.

The spacecraft found that Titan's (one of Saturn's moons) atmosphere is composed primarily of nitrogen, that it contains simple organic compounds, and has one and one half times the pressure of Earth's. Several new Saturnian satellites were discovered. Most surprising, perhaps, was the discovery of unexpected phenomena in the planet's rings, including thousands of tiny wave like features, some apparently caused by the gravitational influence of some of the satellites, and spoke-like features that may be electrically charged dust particles levitated above the main ring plane.

The Jupiter and Saturn data are still being studied by scientists.

Scientists and engineers at the Jet Propulsion Laboratory have been focusing on the Uranus encounter since *Voyager 2* left Saturn behind in September 1981. All of *Voyager's* 11 science instruments are functioning and all will make observations of Uranus and its environment.

Because Uranus is about twice as far from Earth as Saturn, the rate at which *Voyager* will be able to transmit data to Earth is slower.

Normally, this would have seriously limited the number of photographic and other data that could be sent back to Earth, but engineers and scientists have programmed one of the spacecraft's computers to compress and encode the imaging data in order to return about 200 images a day. In addition, several antennas at each of NASA's Deep Space Network (DSN) sites will be electronically linked to increase their receiving power, allowing more of *Voyager's* faint radio signal to be captured. This technique, called arraying, greatly enhances the overall strength and quality of the signal received. Antenna arraying will be used at the Australia, Spain and California complexes.

Most of the key data obtained during the Uranus encounter and all of that during the closest approach will be received by the DSN's antenna complex in Canberra, Australia.

The Canberra complex, which also will be electronically linked with the Australian government's 64-meter (210-foot) Parker Radio Astronomy Observatory, is critical to the encounter for several reasons. The spacecraft track will be almost directly above the Australian complex during the encounter closest approach, allowing up to 12 hours of coverage of *Voyager 2* daily. As a result of this geometric relationship, the spacecraft's signal quality will be enhanced because it will pass through a thinner slice of Earth's atmosphere than it will at the lower elevation at the antennas in California and Spain.

The Australian continent provides the added benefit of having large distances between antennas. The Parkes and Tidbinbilla antennas, for example, are 320 kilometers (200 miles) apart. This reduces the risk that data at both stations might be subject to any degradation resulting from potentially simultaneous rain showers.

The Australian complex can accommodate a higher data rate than the other two DSN complexes both because of its advantageous viewing geometry in relation to the spacecraft, and because *Voyager's* signal quality will be improved by combining or arraying the output of several large antennas in Australia.

Antenna arraying also will be used at the California and Spain complexes.

Uranus is the third largest of the solar system's nine planets. Its polar axis lies nearly in the ecliptic plane rather than perpendicular to it, as most of the other solar system planets. Scientists do not know why Uranus is tipped 95 degrees from its vertical axis, but some speculate that early in its formation, Uranus was struck and the axis of rotation tipped by an object about the size of Earth.

Uranus orbits the sun once in 84 years, with one pole in sunlight for 42 years while the other pole is in darkness. When *Voyager 2* approaches Uranus, the planet's south pole will be pointing toward the sun.

Because of the unique orientation of the Uranian system, the planet's polar region will dominate *Voyager's* view as the spacecraft approaches. The entire Uranian system will present a "bulls-eye" appearance to the spacecraft. Although Uranus will loom progressively larger in *Voyager's* field-of-view, the spacecraft's perspective on the planetary system essentially won't change until just hours before closest approach.

Uranus has five known satellites. They are considerably smaller than the Galilean moons of Jupiter, Saturn's Titan and Earth's Moon, but they are still among the largest satellites in the solar system. Closest to the planet and smallest is Miranda, about 500 kilometers (300 miles) in diameter. Next is Ariel, whose diameter is about 1,330 kilometers (825 miles). Umbriel is the third satellite, with a diameter of about 1,110 kilometers (690 miles).

Titania is fourth from Uranus and has a diameter of 1,600 kilometers (995 miles). Outermost of the five is Oberon which, with a diameter of 1,630 kilometers (1,010 miles), is the largest satellite of Uranus.

Very little is known about these moons. They could range in composition from being mostly rock to mostly ice. Scientists believe Uranus' satellites are probably similar to some of Saturn's.

They are about the same in size, but have water ice on their surfaces. They differ, however, in that they are darker. There is no evidence that they are uniformly gray – they could display mottled dark and light surfaces like Jupiter's Callisto, or even show a surface as extreme as the black and white surface of Saturn's moon Iapetus.

Uranus is circled by at least nine thin rings that are among the darkest objects in the solar system – as dark as charcoal. The outermost ring varies in size from 20 to 100 kilometers (12 to 60 miles); two rings are about 10 kilometers (6 miles) and two are about 3 kilometers (2 miles). The widths of the other four are smaller, but have not been determined. Most are elliptical in shape. The rings' composition is not known, but some scientists believe that the ring particles may have contained methane which has decomposed into darker carbon materials.

Uranus is about 20 times farther from the sun than the Earth, four times farther than Jupiter and twice as far as Saturn. Uranus receives only one four-hundredth of the sunlight that Earth receives, one-sixteenth that of Jupiter and one-quarter that of Saturn.

Voyager's cameras therefore must take extremely long exposures in order to register images of the planet and its satellites.

To cope with these constraints on Voyager's photography, a special technique called image motion compensation will be used to prevent smearing of images at these low light levels. The technique was successfully tested at Saturn.

It involves rotating the spacecraft (which is traveling more than 45,000 mph) while the camera's shutter is open, in much the same way a photographer moves his camera while taking a picture of a speeding object.

Without image motion compensation, the best detail visible at Miranda would be 56 kilometers (35 miles). Using the technique, the cameras can record detail of 0.6 kilometers (one-third mile). The technique will improve resolution of details on Oberon's surface from 48 kilometers (30 miles) to 12.5 kilometers (7.75 miles). Detail visible in images of Ariel will be improved from 50 kilometers (30 miles) to 2.3 kilometers (1.4 miles).

Voyager 2's photopolarimeter will observe two ring occultations similar to observations made at Saturn. Three of Uranus' rings will pass in front of the star Sigma Sagitarrii (Nunki) and all the rings will pass in front of Beta Persei (Algol). Based on how much starlight passes through the rings, the photopolarimeter will provide information on the width and thickness of the rings and the material they contain.

Radio science measurements will complement and extend this data using information from the spacecraft signal passing through the ring material. Radio science data on Uranus atmospheres, and ionospheres will be obtained as the spacecraft enters and exits occultation periods.

In addition to two cameras, photopolarimeter and a spacecraft radio *Voyager* carries an infrared interferometer / spectrometer and radiometer, an ultraviolet spectrometer, a cosmic ray detector, a plasma instrument, a low-energy charged particle detector, magnetometers, planetary radio astronomy receiver, and a plasma-wave instrument.

The *Voyager* spacecraft are based on the highly successful *Mariner* design. Each *Voyager* weighs 825 kilograms (1,819 pounds) and is dominated by its 3.66-meter (12-foot) antenna. Electric power is provided by three radioisotope thermoelectric generators (RTGs).

This system is required because solar cells could not receive sufficient solar energy for the power needed at the great distances that the spacecraft travel from the Sun.

When the Uranus encounter ends in February 1986, *Voyager* 2 will be on a course that will take it to Neptune, its last planetary visit on August 25, 1989 (GMT).

The Jet Propulsion Laboratory in Pasadena, California, manages the Voyager project for NASA's Office of Space Science and Applications. Earl Montoya is Voyager Program Manager at NASA Headquarters, and Dr. William Brunk, NASA Headquarters, is Voyager Program Scientist. Richard P. Laeser of JPL is Voyager Project Manager and Dr. Edward C. Stone of the California Institute of Technology is Voyager Project Scientist.

(End of General Release. Background information follows)

Uranus and Earth

Uranus

Uranus is one of the four giant gaseous planets of the solar system. It was the first planet to be discovered with a telescope, and was found by amateur astronomer Sir William Herschel on March 13, 1781. The planet has never before been encountered by a spacecraft.

The most distinctive feature of Uranus is its unusual rotational position, tipped over on its axis. Scientists theorize that early in Uranus' history, a collision with another, smaller body might have tilted the planet from a vertical or near vertical axis to its present orientation. In the planet's 84-year-long solar orbit, one pole is in sunlight for 42 years while the other pole is in darkness. These factors are expected to give Uranus unusual weather patterns. The length of a Uranian day is uncertain, but direct measurements indicate it is either 16 or 24 hours. Theoretical models based on the rate of precession of the planet's rings set the day closer to 16 hours.

Uranus and Neptune are about the same size. Uranus is about 51,000 kilometers (31,800 miles) in diameter; Neptune is about 49,600 kilometers (30,500 miles). Based on their similar compositions and sizes, Uranus and Neptune can be considered as a set of twin planets (though the two are dissimilar in many ways), and Jupiter and Saturn as another set of twins.

The bulk of Jupiter and Saturn's composition is gas hydrogen and helium surrounding relatively small cores of mostly molten rock and various ices.

Uranus' and Neptune's cores are about the same size as Jupiter's and Saturn's. The atmospheres of Uranus and Neptune, however, are much smaller. The two may represent how Jupiter and Saturn might appear if stripped of much of their massive gas envelopes.

Uranus and Neptune's modest atmospheres consist mostly of gaseous compounds of hydrogen, carbon, nitrogen, oxygen and perhaps helium. In Uranus' case, scientists believe an ocean of melted water ice with dissolved ammonia could exist beneath its atmosphere and hazy cloud layer.

Each planet employs some mechanism to distribute the heat it absorbs and the energy it emits. Jupiter, Saturn and Neptune each have significant internal heat sources – they emit more energy than they absorb from the Sun, while Uranus, like Earth, emits very little heat of its own making. On Earth, sunlight, is absorbed mostly at the equator. From there, the oceans distribute the heat north and south to the poles to maintain global temperature equilibrium.

Uranus' southern hemisphere currently receives all the sunlight incident on the planet, so there may be significant meridianal flows – atmospheric motion that crosses the latitudes of the planet like the seams in a beach ball – carrying heat from one pole to the other.

Rings

There are nine thin, black rings known to surround Uranus. They were found in 1977 when Uranus passed in front of a bright star, affording astronomers the opportunity to detect the rings as they blocked out the light of the star. The outermost ring, called the epsilon ring, reflects only about 5 percent of the light it receives. It is about as dark as the black side of Saturn's moon Iapetus, or as dark as charcoal. The other eight rings are expected to be equally black. Inward from the epsilon ring they are: delta, gamma, eta, beta, alpha, and rings 4, 5 and 6. (The discrepancy in the nomenclature of the rings is due to their simultaneous discovery by two independent groups of observers. One group decided to call the rings by Greek letters, ordering them from the inside out. The other group numbered the rings they found and ordered them from the outside in).

Three of the rings, eta, gamma, and delta, are very nearly circular. The rest are somewhat eccentric. The epsilon ring is most unusual, being both quite eccentric and varying in width by tens of miles. It is not known if the rings formed with Uranus when the solar system originated 4.6 billion years ago, or if they are a more recent development, perhaps the remnants of a broken-up moon, meteoroids or due to a combination of these processes.

The darkness of the epsilon ring (it is the only ring whose brightness has been measured) implies that most of the particles lack bright water coatings, but they could be coated with a residue of carbon compounds leftover from methane decomposed by sunlight or by energetic particle radiation.

The Uranian rings are expected to share some of the physical characteristics unveiled in Saturn's rings by *Voyagers 1* and 2. Scientists expect to find small shepherd moons, like those found at Saturn, which probably herd the rings into their unusual shapes.

Moons

Little is known of the five moons orbiting Uranus; even their sizes and masses are not well defined. They have icy surfaces but the amount of rock inside is not known.

Some scientists have suggested that some of the moons could have formed from the debris left over after a collision between Uranus and another body. In this scenario, a body one or two times the size of Earth smashed into Uranus, tipped it on its side,

	Diameter		Dist. from center of Uranus	
Moon	(km)	(mi)	(km)	(mi)
Miranda	500±220	310±35	130,000	80,000
Ariel	1,330±130	825±80	192,000	119,000
Umbriel	1,110±100	690±60	267,000	166,000
Titania	1,600±120	995±70	438,000	272,000
Oberon	3,630±340	1,010±85	586,000	364,000

The Moons of Uranus

and splashed part of the planet's atmosphere into space. The resulting mix of rock and gases would have formed a disk around the planet, out of which the moons could have formed.

The moons are about the same size as the intermediate satellites of Saturn, such as Dione or Enceladus. The darkness of their surfaces suggests that they, like the rings, may have been darkened by the effect of radiation on methane. They could exhibit tectonic features, such as cracked surfaces or frozen flows of icy magma.

Perturbations of the Uranian rings detected from ground-based observations have been interpreted by some scientists as evidence for the possible existence of a small, sixth satellite orbiting between Miranda and the epsilon ring. If it (or other new moons) exists, it should be easily detectable by *Voyager's* instruments.

Magnetosphere

Scientists assumed that because evidence of auroral activity has been detected at Uranus, that it might possess a magnetic field. Auroras on Earth and Jupiter are caused by the interaction of their magnetic fields with the stream of atomic particles emitted by the Sun, called the solar wind.

The magnetic field of Uranus, if one exists, will be unique in the solar system due to the planet's odd polar axis orientation. (A planet's magnetic field is thought to be generated by fluid motion in the planet's interior, and rotates with the interior as well).

In those planets with a magnetosphere, the axis of the field is roughly aligned with the body's rotational axis. Since Uranus' rotational axis currently faces the sun, its magnetic field would meet the solar wind nearly pole on. The magnetic field in the polar region would be funnel shaped, dipping inward at the pole. This may allow the solar wind to penetrate closer to the planet than would otherwise be possible.

As with other planets, the solar wind would deform the other regions of the magnetic field to produce, in this case, a long tail extending directly away from the planet's northern pole. The magnetotail may also be twisted into a spiral by the rotation of the planet. In Uranus' case, the magnetosphere might extend 1 million kilometers (620,00 miles) from the planet. *Voyager*'s instruments could detect such a structure if it exists.

There may be energetic particles trapped within the Uranian magnetosphere (like those at Saturn). The resulting radiation effects could explain the moderately dark surfaces of the moons as well as the extremely dark rings.

Data from the planetary radio astronomy instrument on *Voyager 2* are being analyzed daily for signals that would show evidence of a Uranian magnetic field. By early December 1985, however, the spacecraft already was much closer to Uranus than it was to either Jupiter or Saturn when their radio signals were first detected. No Uranian signals had yet been heard. Although the existence of a Uranian magnetic field cannot be ruled out, the presence of aurora-like emissions may not be an indication of a magnetic field, but may instead be airglow in the planet's atmosphere. (Airglow is a phenomenon associated with photochemical reactions of atmospheric gases).

If Uranus has no magnetic field, the planet may interact with the solar wind in much the way Venus does. Venus has an almost non-existent magnetic field. Thus its electrically conductive ionosphere diverts the flow of the solar wind around the planet. The same situation could exist at Uranus, with the Uranian ionosphere meeting the solar wind in lieu of a magnetic field. The bow shock – the area where the solar wind first responds to the presence of a planet – would be close to Uranus on the sunward side and taper off into a tear-shaped wake behind the planet.

Uranus might also interact with the solar wind in much the same way as a comet does, resulting in a very weak bow shock and an ionized plasma tail behind the planet.

Science Objectives

Voyager 2's complement of 11 instruments will be dedicated to more than two dozen major scientific objectives during the Uranus encounter.

In January 1986, as it did at Jupiter in 1979 and Saturn in 1981, *Voyager 2* will encounter a broad range of planetary phenomena. At Uranus, the spacecraft will find an atmosphere and weather system of potentially great complexity; a dark nine-banded ring system unique in orientation and composition; and a collection of at least five icy and/or rocky moons.

Scientists met in February 1984* to establish a scientific framework for the *Voyager 2* encounters with Uranus and Neptune. Working groups compiled a list of high priority Uranus science objectives, and *Voyager 2* has been programmed to perform observations designed to meet these goals. Twenty-seven of these measurements will occur in the 96-hour near-encounter period of January 22-26, 1986 – most of them within 6 hours of the spacecraft's closest approach to Uranus.

Each *Voyager* carries 11 scientific instruments – actually 10 plus the spacecraft radio. They can be divided into two general classes: those that require pointing (target-body sensors) and those that don't (fields and particles sensors).

There are five pointable sensors: the imaging science subsystem (consisting of wide- and narrow-angle television cameras), infrared interferometer spectrometer and radiometer, photopolarimeter subsystem, radio science subsystem, and ultraviolet spectrometer. All but the radio ride on the spacecraft's steerable scan platform.

The other six instruments measure energetic particles, radio emissions and magnetic fields, in space and near planets. They are the magnetic fields experiment (consisting of four magnetometers) plasma subsystem, low-energy charged particle detector, cosmic ray subsystem, plasma wave subsystem and planetary radio astronomy experiment.

Voyager 2 observations with these instruments can be divided into four groups at Uranus: atmosphere, rings, satellites and magnetosphere.

Atmosphere

Voyager scientists hope to observe and define the global circulation and meteorology of the upper, visible clouds of Uranus, as well as the horizontal and vertical distribution of clouds and hazes.

The ultraviolet spectrometer will examine the upper atmosphere, while the infrared and radio science experiments will provide information on the composition, pressures and temperatures deeper in the atmosphere.

---

* Proceedings published as *Uranus and Neptune*, NASA Conference Publication 2330, Jay T. Bergstralh, editor, National Technical Information Service, 1984.

Various instruments, notably the infrared spectrometer, will be used to determine the heat balance at Uranus – the ratio of internal energy emitted to solar energy absorbed. Any measurable excess of energy – as was found at Jupiter and Saturn – would have important implications for theories on Uranus' formation and weather mechanisms.

Imaging of the sunlit southern hemisphere will help characterize wind speeds at different latitudes. Ultraviolet measurements building on those of the International Ultraviolet Explorer (IUE) from Earth orbit will search for auroral activity at the poles or find a more diffuse airglow emission.

Rings

The nine known rings of Uranus are so dark that many of the most important observations will occur after closest approach, when *Voyager 2* can look at the rings as they are backlit by the Sun.

Voyager's instruments will be used to determine the size, distribution and reflective properties of ring particles. Toward this end, *Voyager* will conduct both stellar occultation studies (measurements of starlight passing through the rings) and radio occultation studies, obtained when the spacecraft is passing behind the rings as viewed from Earth.

Photopolarimeter and radio science will pinpoint the locations of the known rings and perhaps reveal others while providing information on their structures. *Voyager* will search for tiny satellites embedded in the rings and for satellites that serve as "shepherds," herding material between them.

Satellites

The satellites are expected to be airless bodies, so most of what can be learned will come from imaging of their surfaces. *Voyager 2*'s cameras will acquire images at the closest approaches to each of the five known moons; longer range photography will produce full-disk color pictures.

Voyager imagery will provide information on sizes, shapes, surface markings and surface relief. The color images will provide information on the distribution of different materials on the surface, and about any processes that have modified the moons' surfaces. Precise radio tracking of the spacecraft – especially during the close pass of Miranda – could provide improved mass estimates for some or all of the moons.

The cameras may also reveal the existence of previously unknown satellites, including ring shepherds.

Magnetosphere

The excess radiation revealed by the IUE spacecraft had been interpreted as auroral emission – an indication that Uranus may possess a magnetic field. More recently, however, the ultraviolet emissions have been interpreted as being atmospheric airglow, which would not necessarily be indicative of a planetary magnetic field. No additional evidence of a magnetic field has been observed.

If the planet does have a magnetosphere, measurements by the planetary radio astronomy experiment of the rotation of the magnetic field would, by inference, give the rotation rate of Uranus' interior.

Because of its rotational orientation, Uranus could offer the first example of a "pole-on" magnetosphere. In such a configuration, the solar wind might flow deep down into the polar regions, rather than arriving and being strongly deflected in the equatorial regions as at Earth.

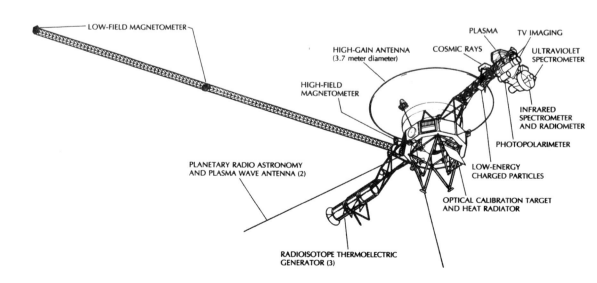

Science Experiments

Imaging Science Subsystem

The imaging science subsystem (ISS) will observe and record the visible characteristics of Uranus, its atmosphere, satellites and rings. In addition, groups of pictures returned by the ISS wide- and narrow-angle television cameras will be used to map satellite surfaces. Images of the satellites against the background stars will help Voyager engineers navigate the spacecraft.

Specific ISS goals at Uranus are to obtain high-resolution photography of atmospheric motions, colors and unusual features (analogous to Jupiter's Great Red Spot and similar smaller spots), and to characterize the vertical structure of the atmosphere, comparative and detailed geology of the satellites, satellite size and rotation, and details of the rings.

The ISS consists of two television-type cameras mounted on the scan platform. The wide-angle camera has a focal length of 200 mm and is sensitive in the range from 4,000 to 6,200 angstroms. The narrow-angle camera has a focal length of 1,500 mm and can image in the range from 3,200 to 6,200 angstroms.

For several days before and after closest approach, *Voyager* will have several simultaneous imaging opportunities: high-resolution photography of Uranus as it grows larger than the field of view; a close pass by Miranda of just 29,000 kilometers (18,000 miles); more distant photography of the other satellites; and high-resolution imaging of the rings.

To exploit such a variety of opportunities, it is necessary for the spacecraft to return large quantities of imaging data over a wide range of telemetry rates in real time. Data also can be recorded on board the spacecraft, as during the spacecraft's occultation by Uranus (when the spacecraft will be behind Uranus and out of radio touch with Earth) and later played back to Earth.

Each camera is equipped with a filter wheel whose individual filters have a wide variety of uses. These filters permit specified types or wavelengths of light to pass through and block all other types from reaching the camera detectors.

The wide-angle camera carries one clear filter; one filter each in blue, green and orange wavelengths; a sodium-D filter, and two filters for study of the distribution of atmospheric methane. The narrow-angle camera carries two clear filters, two green filters and one filter each of violet, blue, orange and ultraviolet.

The design of the *Voyager* imaging system was based on those of previous *Mariner* spacecraft, with advances and changes dictated by the specific requirements of the Jupiter and Saturn encounters. The Uranus encounter takes the spacecraft beyond the environment and lifetime for which the cameras were designed, but no problems are anticipated.

Image-smear conditions will be much more severe at Uranus due to lower light levels. The relatively low reflectivity of the rings and satellites, will necessitate longer exposures. (Uranus receives one-sixteenth the sunlight received at Jupiter, about one-quarter that at Saturn.) A technique called image-motion compensation (see "Customizing *Voyager 2* for Uranus") will be used to prevent image smearing.

The increasingly remote distance of *Voyager 2* also affects the rate at which the spacecraft can transmit pictures to Earth. Each picture consists of more than 5 million bits of information. At Jupiter, *Voyager* could send back as many as 75 pictures per hour. At Uranus, *Voyager* will be able to send back a maximum of 12-17 pictures per hour. Even this rate of return (about 200 pictures per day) can be accomplished only through special techniques of data compression and encoding, in which the 5 million bits per picture will be compressed to 2 million bits and fed into the telemetry stream at a lower rate than at Jupiter.

On the approach to Uranus, the narrow-angle camera will photograph the planet at planned regular intervals in an attempt to discern cloud motions. Resolution will improve steadily from about 1,600 kilometers (990 miles) to a few hundred kilometers.

About one week from closest approach, the disk of Uranus will exceed the field of view of the narrow-angle camera. At that time, the detail visible will be about 130 kilometers (81 miles). The wide-angle camera will begin its work, and the narrow-angle camera will shift its focus to portions of the planet that warrant special scientific interest.

The ISS weighs 38.2 kilograms (84.2 pounds) and uses 41.9 watts of power.

Dr. Bradford A. Smith of the University of Arizona is imaging team leader.

Infrared Interferometer Spectrometer and Radiometer

The infrared interferometer spectrometer and radiometer (IRIS) will focus primarily on Uranus' atmosphere. The instrument will gather data on:

- Uranus' heat balance – the amount of energy the planet emits versus the amount it receives from the sun;
- The temperature, structure, dynamics and composition of the atmosphere, in particular the ratio of helium to hydrogen; and
- The characteristics of clouds and aerosols in the atmosphere.

Uranus will be especially interesting because of its unusual polar orientation. Comparison of the sunlit south pole and dark north pole is an important IRIS task.

The telescope-based IRIS system provides broad spectral coverage in the infrared from 2.5 to 50 microns and visible radiometry from 0.3 to 2 microns.

The instrument weighs 18.4 kilograms (40.6 pounds) and dissipates 14 watts average power.

Dr. Rudolf A. Hanel of NASA's Goddard Space Flight Center is principal investigator.

Photopolarimeter Subsystem

The photopolarimeter subsystem (PPS) will study aerosol particles in Uranus' atmosphere, and the textures and compositions of the surfaces of the satellites. It also will measure the size, albedo and spatial distribution of particles in the rings, as well as the rings' optical and geometric thickness.

The PPS consists of a telescope fitted with filters and polarization analyzers. The system measures the way its targets reflect light, and hence, determines their structures, as the reflected light is polarized by chemicals and aerosols (in the case of Uranus' atmosphere) or by small particles (in the rings and on the solid satellite surfaces).

A special high-speed ultraviolet photometry mode will be used for two stellar occultations of Uranus' rings. By measuring the light from two stars – Algol (Beta Persei) and Nunki (Sigma Sagittarii) – seen through the rings, the PPS will help determine ring structure with high resolution.

The experiment weighs 4.41 kilograms (9.72 pounds) and uses 2.4 watts average power.

Dr. Arthur L. Lane of the Jet Propulsion Laboratory is principal investigator.

Radio Science Subsystem

Voyager 2's two-way radio communications link with Earth also will be used to conduct scientific investigations at Uranus. Precise measurements of the phase and amplitude of the radio signal can be analyzed to detect minute variations due to the passage near or through the planetary atmosphere and rings. At Uranus, the radio science subsystem (RSS) will:

- Study the structure of the atmosphere, including temperature, pressure, density and turbulence as the spacecraft passes behind the planet;
- Determine the optical depth, structure and particle size distribution of the Uranian rings, again during occultation passage;
- Determine the mass and gravity field of Uranus and its satellites; and
- Conduct experimental tests of Einstein's theory of relativity when the radio signal passes close to the sun to determine the influence of the sun's gravity on the radio signal.

Dr. G. Len Tyler of the Center for Radar Astronomy at Stanford University is team leader.

Ultraviolet Spectrometer

The ultraviolet spectrometer (UVS) will study Uranus' atmosphere, gathering data on its composition by means of the techniques of atomic emission and atomic absorption. UVS data will be used at Uranus to:

- Determine distributions of major constituents of the upper atmosphere as a function of altitude;
- Measure the absorption of solar ultraviolet radiation by the upper atmosphere as the sun is occulted by Uranus;
- Measure ultraviolet airglow emissions of the atmosphere from the bright disk of the planet, its bright limb (the outer edge of the disk), terminator (the dividing shadow between night and day) and dark side; and
- Determine the auroral morphology at Uranus, and the results of an orientation that places one pole toward the Sun and the other in darkness for years at a time.

In addition, the UVS will continue to study the distribution and ratios of hydrogen and helium in interplanetary and interstellar space.

The UVS employs a grating spectrometer that is sensitive to ultraviolet radiation in the range from 500 to 1,700 angstroms. The experiment weighs 4.49 kilograms (9.90 pounds) and uses 2 watts of power.

Dr. A. Lyle Broadfoot of the University of Arizona is principal investigator.

Cosmic Ray Subsystem

The primary function of the cosmic ray subsystem (CRS) is to measure the energy spectrum of electrons and cosmic ray nuclei. The seven energetic particle telescopes in the CRS are also designed to:

- Determine the elemental and isotopic composition of cosmic ray nuclei and solar energetic particles;
- Determine the distribution and composition of high-energy particles trapped in the Uranian magnetic field; and

- Determine the intensity and directional characteristics of energetic particles as a function of radial distance from the Sun, and determine the location of the modulation boundary where the influence of the heliosphere ends and true interstellar space begins.

The CRS uses seven independent solid state detector telescopes. Working together, they cover the energy range from 0.5 million to 500 million electron-volts.

The experiment weighs 7.52 kilograms (16.6 pounds) and uses 5.2 watts of power.

Dr. Edward C. Stone of the California Institute of Technology is principal investigator.

Low-Energy Charged Particle Detector

The-low energy charged particle detector (LECP) is designed to characterize the composition, energies and angular distributions of charged particles. In addition to studying Uranus' magnetosphere, the LECP will investigate:

- The composition of low-energy charged particles trapped in the Uranian magnetic field;
- Interactions of charged particles with the satellites and rings;
- The propagation of solar particles in the vicinity of Uranus;
- The quasi-steady interplanetary fluxes and high-energy components of the solar wind; and
- The origins and interstellar propagation of galactic cosmic rays (those originating outside the solar system).

Two solid state particle detector systems are mounted on a rotating platform. Their sensitivity to charged particles ranges from 15,000 to more than 160 million electron-volts.

The LECP weighs 7.47 kilograms (16.5 pounds) and draws 4.2 watts of power during encounter.

Dr. S.M. (Tom) Krimigis of the Applied Physics Laboratory at Johns Hopkins University is principal investigator.

Magnetic Fields Experiment

The magnetic fields experiment (MAG) will determine the existence and character of a Uranian magnetosphere. The four magnetometers making up the experiment will study the interaction of the magnetic field with satellites orbiting within it, and study the interplanetary interstellar magnetic fields in the vicinity of Uranus.

Two low-field magnetometers are mounted on a 13-meter (43-foot) boom away from the magnetic field of the spacecraft itself. Two high-field magnetometers are mounted on the spacecraft body. The low-field sensors can measure fields as weak as 0.002 gamma (or about one ten-millionth that of the Earth's equatorial field); the high-field sensors can measure fields more than 30 times stronger than that at Earth's surface.

Total MAG experiment weight is 5.5 kilograms (12 pounds). The experiment uses 3.2 watts of power.

Dr. Norman Ness of NASA's Goddard Space Flight Center is principal investigator.

Planetary Radio Astronomy Experiment

Voyager 2's planetary radio astronomy experiment (PRA) will search for and study a variety of radio signals emitted by Uranus. The PRA will determine the relationship of these emissions to the satellites, the magnetic field, atmospheric lightning and plasma environment.

The detector also measures planetary and solar radio bursts from new directions in space and relates them to measurements made from Earth.

Using two 10-meter (33-foot) electric antennas as detectors, which it shares with the plasma wave subsystem, the PRA receiver provides coverage from 20 kilohertz to 40.5 megahertz in the radio frequency band.

The instrument weighs 7.66 kilograms (16.9 pounds) and uses 6.8 watts of power.

Dr. James W. Warwick of Radiophysics Inc., Boulder, Colorado, is principal investigator.

Plasma Subsystem

The plasma subsystem (PLS) studies the very hot ionized gases, or plasmas, that exist in interplanetary regions and within planetary magnetospheres. About 99 percent of the matter in the universe is in the plasma state, mostly at temperatures in excess of 10,000 degrees Kelvin (17,500 degrees Fahrenheit). Among its several scientific objectives, the PLS will:

- Study the overall extent and configuration of Uranus' magnetosphere and the nature and sources of the internal plasma;
- Measure the properties of the solar wind (density, temperature and velocity) as it flows into the outer solar system and beyond;

- Study the interaction of the solar wind with Uranus and the other outer planets; and
- Determine the extent of the solar atmosphere (solar wind) and the nature of the boundary between the Sun's atmosphere and the interstellar medium.

The PLS consists of two plasma detectors that are sensitive to solar and planetary plasmas – both the positive ions and electrons – with energies between 10 and 6,000 electron-volts.

The experiment weighs 9.89 kilograms (21.8 pounds) and draws 8.3 watts of power.

Dr. Herbert S. Bridge of the Massachusetts Institute of Technology is principal investigator.

Plasma-Wave Subsystem

Voyager's plasma-wave subsystem (PWS) is designed to measure the electric field components of local plasma waves. At Uranus, the PWS will measure the density and distribution of plasma, interactions of plasma waves with energetic particles, and the interactions of the Uranian satellites with the planet's magnetosphere.

The PWS will provide key information on phenomena related to the interaction between plasma waves and particles that control the dynamics of the magnetosphere. The satellites of Uranus may provide important localized sources of plasma and field-aligned currents, which could significantly affect the trapped particle populations.

Two extendible electric antennas, shared with the planetary radio astronomy experiment, serve as plasma-wave detectors. The PWS system covers the frequency range of 10 hertz to 56 kilohertz. In normal mode, the PWS acts as a scanner, stepping from one frequency to another. In a second mode at selected times during encounter, the system can record electric field wave forms across all frequencies in a broad band (50 hertz to 10 kilohertz).

The experiment weighs 1.37 kilograms (3.02 pounds). It uses 1.4 watts of power in normal step-frequency mode and 1.6 watts in the step-frequency-plus-waveform-analyzer mode.

Dr. Frederick L. Scarf of TRW Defense and Space Systems Group, is principal investigator.

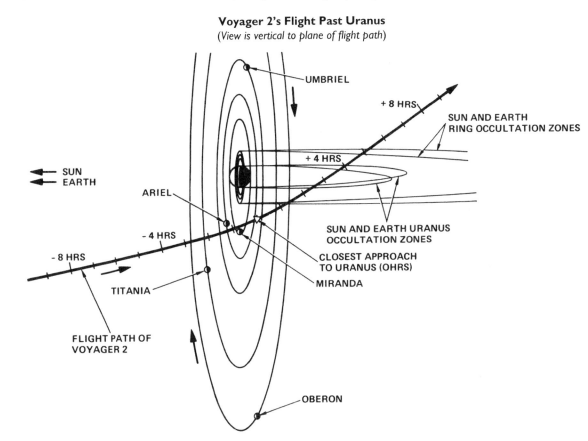

Voyager 2's Flight Past Uranus
(*View is vertical to plane of flight path*)

Customizing Voyager 2 for Uranus

Voyager 2 is a relatively old spacecraft. When launched in 1977, the two *Voyagers* were funded to conduct Jupiter and Saturn encounters only, so the spacecraft were designed to operate at their peak performance levels for about 5 years.

Since then, Voyager engineers and science planners have designed new techniques for collecting, processing and sending data to allow *Voyager 2* to carry out a full menu of experiments and observations at Uranus and Neptune. To this end, *Voyager 2* has been heavily reprogrammed using Earth-based commands during its flight between Saturn and Uranus. Its six onboard computers now incorporate newly developed and more optimum methods of processing data, such as data compression. Large savings in transmitted signal power can be effected by sending only changes in the data, rather than the complete data set. The increasing distance between the spacecraft and tracking stations on Earth decreases the signal strength.

At Jupiter, data rates of 115,200 bits per second (bps) were possible; at Saturn, the rate had dropped to 44,800 bps. At Uranus, the spacecraft signal will be considerably weaker, but the same signal, received at two or more antennas (different groupings of 34-meter (112-foot) and 64-meter (210-foot) antennas) at each of NASA's three Deep Space Network (DSN) complexes, will be combined using a technique called arraying. This technique reinforces the strength of the signal received by electronically combining the same telemetry recorded at more than one station. Arraying will make possible a data rate of 21,600 bps at Uranus. Without arraying, fewer than half the observations planned could be performed and the data returned to Earth. The Australian government has provided NASA with use of its Parkes Radio Astronomy Observatory 64-meter antenna to be specially instrumented and arrayed with the Canberra DSN complex during critical phases of the Uranus and Neptune encounters.

While the data rates will be much lower at Uranus, the data compression technique makes it possible to squeeze as many as 200 images a day using the relatively weak data stream signal strength expected during the encounter. This has been made possible in part by reprogramming one of six computers on the spacecraft to preprocess all imaging data prior to its transmission to Earth. Instead of transmitting the full eight bits (containing 256 levels of gray) for each picture element, or pixel, only the difference between the brightness of successive pixels is transmitted. The result of this image data compression will be at least a 60 percent reduction in the number of bits needed per image.

Computer processing of the imaging data at JPL will restore the correct brightness to each pixel to produce complete black-and-white and color images.

From Uranus, the minimum amount of time one picture will take up in *Voyager's* radioed data stream will be four minutes, compared to 2.4 minutes at Saturn and 0.8 minutes at Jupiter. Without the new image data compression technique, each picture would take up nearly 13 minutes in the data stream, severely limiting the number of images that could be returned.

Voyager 2 will be flying past Uranus and its moons at a velocity of more than 45,000 miles an hour. This poses a problem for the cameras on board – to get an unsmeared photo of an object, the camera has to track the image while the shutter is open. In addition, Uranus is twice as far from the Sun as Saturn; light levels are four times lower and the targets have inherently darker surfaces, so longer exposure times will be required. All of these factors make the potential of image smearing more of a problem.

Image-motion compensation, a technique conceived since launch, involves turning the spacecraft itself. This technique worked successfully in minimizing image smear in close up photography of Saturn's moons. It will be used for most of *Voyager's* closest approaches to the moons and rings of Uranus.

The most challenging photo opportunity exists at the innermost moon, Miranda. Photographing this small, dark object at relatively close range, 29,000 kilometers (18,000 miles), requires that the spacecraft rotate at a rate of more than $1/30^{th}$ of a degree per second. This rate of rotation is in excess of that allowed by the onboard computer that maintains the spacecraft attitude. If such a roll were normally attempted, the onboard computer would direct *Voyager* to override the command and return to a stable position to prevent it from losing contact with Earth. Voyager engineers have designed and tested a method to override this safeguard and to safely turn the spacecraft faster in order to allow clear close ups of Miranda.

Voyager 2 Spacecraft Health

On April 5, 1978, the spacecraft's computer command subsystem automatically switched to the back-up receiver. The back-up receiver, however, had hitherto concealed a problem of its own – a faulty tracking loop capacitor – meaning that the receiver could not lock strongly on to the frequency of the transmitted signal. This meant that the ground transmitter had to send the precise frequency, which, after undergoing change (doppler shift) while traveling the distance between the Earth and spacecraft, had to match the frequency that the receiver on the spacecraft was expecting. That frequency depends on a number of factors, including the receiver's temperature. When the prime receiver was turned back on, it failed almost immediately – requiring that the rest of the mission be flown on the malfunctioning back-up receiver.

Voyager engineers have determined how the tuning depends on temperature, and how the operation of different subsystems on board affects the temperature of the receiver. Even so, there is a period after any change in the spacecraft's configuration when it is impossible to know the receiver's temperature with adequate precision. As a result, commands cannot be routinely transmitted to *Voyager* after a change in the spacecraft configuration until the receiver temperature has had time to stabilize.

If need be, controllers can send commands to the spacecraft at different frequencies in rapid succession to ensure that one will be picked up by the receiver. This, and other techniques to work around the crippled receiver, were successfully employed at Jupiter, and have been further refined in ensuing years.

There is a chance that the back-up receiver could fail or lose contact with Earth permanently. The Voyager team has planned against this possibility by programming the spacecraft computer with simplified encounter routines for execution at Uranus. The spacecraft has been instructed to send data back to Earth even in the event that it loses uplink contact.

In 1981, *Voyager 2*'s scan platform jammed in one axis just after its Saturn encounter. The jamming prevented further pointing of the instruments for the duration of the encounter.

After 2 days, the platform was again movable. Three years of analysis and testing showed that the problem was due to a loss of lubricant and consequent damage to a bearing in the high-speed gear train of the platform. The lubricant apparently migrated back into the gear train after a short period of rest.

Voyager engineers have determined that slow-rate motion of the platform can be safely accomplished during the Uranus encounter, and a prohibition against moving the platform at a high rate will help ensure that the platform is fully usable when the spacecraft reaches Neptune.

In addition, tests designed to detect the onset of another similar failure of the scan platform have been strategically placed in the command sequences that will be executed during the encounter. In the event *Voyager* senses another such problem approaching, it will execute a back-up near-encounter command sequence which avoids use of critical components that could fail.

Mission Ground Operations

Commands for controlling all of *Voyager 2*'s systems and operations are sent to the spacecraft in a single beam of radio signals from an antenna at one of NASA's Deep Space Network (DSN) complexes. One command load of up to 2,500 18-bit words can contain instructions for sequences of activities for the spacecraft to carry out over a period ranging from 2 days to 6 months.

The Voyager science teams determine the observations needed to be made to accomplish mission objectives. In turn, the Voyager sequence team designs blocks of time in which the spacecraft will make the science observations as it concurrently performs engineering and navigation tasks. The team tests the sequences with computer-based simulations of the spacecraft to ensure that they are consistent with the spacecraft's hardware, software and operational constraints.

The Voyager project has maintained nearly the same operating procedures that were employed during the Jupiter and Saturn encounters, but with about one third the staff. This has resulted in fewer but longer command transmission loads during the cruise between Saturn and Uranus, and limited the number of calibrations performed.

The DSN is comprised of large antennas at three communications complexes, in Spain, Australia and California.

Commands are sent at a rate of 16 bits per second through the antenna dishes at any one of the stations. Traveling at the speed of light, they will reach the spacecraft (at Uranus) in about 2 hours and 45 minutes. With this long delay, engineers on the ground are unable to respond quickly if a spacecraft problem develops. For this reason, the spacecraft's master computer, called the computer command subsystem (CCS), has been programmed with a set of stored responses to anticipated problems.

The computer allows the spacecraft to act autonomously and quickly to protect itself from situations that could threaten spacecraft communications or operations. The CCS also contains the back-up mission load (BML), which carries basic commands that would allow *Voyager 2* to conduct rudimentary investigations of Uranus if the spacecraft's radio were to fail.

Occasionally the need arises to change the state of the spacecraft or one of its instruments beyond the scope of the commands already in the spacecraft computer. Commands of this type are called real-time commands, and are usually sent for immediate execution by the spacecraft.

The telemetry received at the complexes is transmitted to JPL via wide-band and/or high-speed, telephone-quality data lines. The wide-band lines are primarily used for high-bit-rate science telemetry, while high-speed lines are used for engineering telemetry and low-bit-rate science collected during the cruise phase.

Overseas lines are routed through NASA's Goddard Space Flight Center via satellite links. The Goldstone transmissions are sent directly to JPL through ground microwave stations.

Both wide-band and high-speed transmissions are received at JPL by the Network Operations Center (NOC), where they are logged on tape and also routed in real-time to the Mission Control and Computing Center for further processing.

All *Voyager* telemetry received and processed by the NOC is routed to the Mission Control and Computing Center (MCCC). The MCCC is responsible for display, control, decoding and routing of real-time telemetry to the Test and Telemetry System (TTS) and the JPL's Multimission Image Processing Laboratory (MIPL). The TTS displays engineering telemetry in real-time for the spacecraft team and the mission control team, and processes and displays science data for each of the science teams.

Imaging data are transferred to the MIPL for processing and analysis. Here, the imaging data are decompressed. During this process, the images can be enhanced to bring out subtle features, and in some cases, corrected for errors.

All imaging and other data are collected and processed into Experiment Data Records (EDR), which contain all available science and engineering data from a given instrument. The EDRs are the final data product forwarded to investigators for analysis. A companion record called the Supplementary Experiment Data Record (SEDR) accompanies the EDRs and contains the best estimate of the conditions under which the observations were taken.

Deep Space Network Support

The Uranus encounter presents an unprecedented challenge in deep space communication. The Voyager X-band radio signal, for example, will be less than one-sixteenth as strong as it was at Jupiter in 1979. NASA's Deep Space Network (DSN) has just undergone a major upgrade adding, among other capabilities, arraying, new 34-meter antennas and automatic network monitor and control, which will complement improvements in the *Voyager* flight data system program, and significantly increase the potential data return.

The DSN has carried out all tracking and communication with the *Voyager*s since injection onto their interplanetary trajectories after launch.

DSN stations are located around the world, in multi-antenna complexes at Goldstone, in California's Mojave Desert; near Madrid, Spain; and near Canberra, Australia. The three complexes are spaced at widely separated longitudes so that spacecraft can be in continuous view as the Earth rotates.

Each location is equipped with a 64-meter (210-foot) antenna, a 34-meter (112-foot) antenna and a 26-meter (85-foot) antenna. In addition, a high-efficiency 34-meter antenna has been recently added at Goldstone and Canberra. A third such antenna is scheduled to begin operations at Madrid in March 1987, in anticipation of *Voyager* 2's 1989 encounter at Neptune.

In addition to the giant antennas, each of the complex's signal processing centers house equipment for transmission, receiving, data handling and interstation communication. The downlink radio frequency system includes cryogenically cooled, low-noise amplifiers.

Uplink

The uplink operates at S-band radio frequency (2,113 megahertz), carrying commands and ranging signals from ground stations to the spacecraft. The 64-meter antenna stations have 400-kilowatt transmitters (normally operated at 60 kw for Voyager); transmitter power at the standard 34-meter stations is 20 kw. The high-efficiency 34-meter antennas at Goldstone and Canberra are presently used only for reception of spacecraft signals.

Downlink

The downlink is transmitted from the spacecraft at S-band (approximately 2,295 MHz) and X-band (approximately 8,415 MHz) frequencies. All the standard 34-meter antennas and the 64-meter antennas can receive the S- and X-band signals simultaneously; the high-efficiency 34-meter antennas receive only X-band transmissions.

Arraying

During low-telemetry-rates cruise phase, a combination of 34-meter or 64-meter support for 16 hours per day is satisfactory. During critical mission high-telemetry-rate phases, *Voyager* requires continuous 34-meter and 64-meter coverage of the spacecraft and the arraying of antennas. At the Uranus encounter, all DSN antennas at each longitude will be arrayed, so that their combined collecting areas will increase the amount of signal captured and thus improve the potential for high-rate, low-error data return. In the case of Canberra, the three DSN antennas – one 64-meter and two 34-meter – will be arrayed for *Voyager* encounter using the 64-meter Parkes Radio Astronomy Observatory which is operated by the Australian Central Science and Industry Research Organization.

At Saturn, 10 astronomical units (AU) from Earth, the signal strength supported a maximum data rate of 44,800 bits per second (bps). At Uranus, 19 AU away, the maximum supportable rate will be 21,600 bps; under special conditions, 29,900 bps may be used.

Arraying of the DSN tracking antennas will allow project engineers to approximately double the collecting area and increase the expected signal strength to one-half instead of one-fourth the Saturn level when *Voyager 2* reaches Uranus.

This is accomplished in the following manner by the new 34-meter high-efficiency antennas at Goldstone and Canberra, which will each effect a 25-percent increase in potential data. The 64-meter Parkes antenna in Australia (arrayed with the Canberra DSN stations by means of a 320-kilometer (200-mile) microwave link will effect at least a further 50-percent increase in collection area.

Australian activities will be critical to encounter support because the high southern (-23 degree) declination of *Voyager 2* will result in long, 12-hour spacecraft view periods at Canberra and 9-hour viewing at Parkes. (The shorter time at Parkes results from antenna pointing constraints). The quality of data received at the Australian facility is also likely to be higher than that received in California and Spain because of the large distances between antennas, which decreases the risk of data being lost due to local weather conditions.

The combined Canberra-Parkes facilities will obtain the critical closest approach imaging and science data for Uranus and all its satellites on encounter day, in addition to data recorded on the spacecraft and played back the next day. This Canberra-Parkes array is also expected to obtain various telemetry data during closest approach, and all radio science data during the critical Uranus and ring occultation periods on encounter day. In all, the Parkes Radio Astronomy Observatory will provide 61 passes of array support, including daily array support for three weeks in January 1986.

Voyager Mission Summary

Voyager 2 was the first of the two *Voyagers* to begin the journey to the outer planets. It was launched from Cape Canaveral, Florida, aboard a Titan-Centaur launch vehicle at 10:29 a.m. EDT on August 20, 1977.

Voyager 1 was launched 16 days later at 8:46 a.m. EDT, September 5. Flying a shorter and faster route to Jupiter, it overtook *Voyager 2* on December 15, 1977. At the time, they were 170 million kilometers (105 million miles) from Earth.

Both spacecraft entered the asteroid belt, a band of rock and dust 360 million kilometers (223 million miles) wide that circles the sun between the orbits of Mars and Jupiter. Both spacecraft left the asteroid belt without incident. *Voyager 1* flew out on September 8, 1978, and *Voyager 2* on October 21, 1978.

Voyager 1 made its closest approach to Jupiter at 7:05 a.m. EST on March 5, 1979, at an altitude of 278,000 kilometers (173,000 miles). It encountered the four Galilean satellites after passing Jupiter, and began its 20-month cruise to Saturn, boosted on its way by Jupiter's gravitational assist.

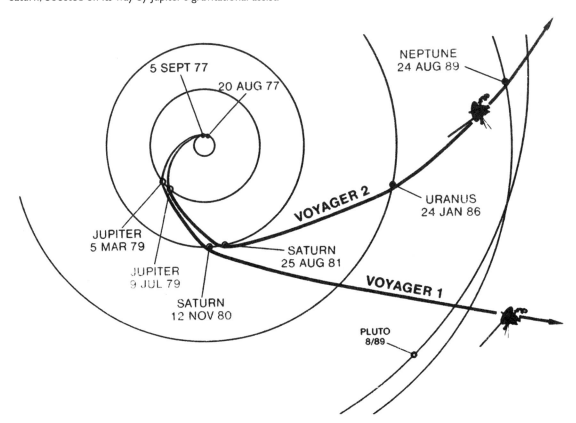

Voyager 2's closest approach to Jupiter occurred July 9, 1979. The same general encounter sequence was followed, with some important differences: *Voyager 2* passed farther from Jupiter than *Voyager 1* – 650,000 kilometers (404,000 miles) above the cloud tops, and *Voyager 2* observed the Galilean satellites as it entered the Jovian system, so that between the two spacecraft, photos of both faces of the satellites were obtained.

The point of closest approach to Jupiter for both spacecraft was selected to provide the proper gravity assist to target each *Voyager* toward Saturn. *Voyager 1* flew past the ringed planet November 12, 1980, and *Voyager 2* followed on August 25. 1981.

Voyager 1 has completed its planetary encounters and is now climbing through unexplored space on a path upward from the ecliptic plane (the broad disk in which Earth and most of the other planets orbit the sun). It is already farther above the ecliptic plane than any other spacecraft, and is returning information on the fields and particles it encounters.

Both spacecraft are expected to cross the heliopause, the unknown boundary separating the solar system from interstellar space, in the late 1990s.

Voyager 1 will travel in the direction of the star Raselhaugue (Alpha Ophiuchus); *Voyager 2* toward Sirius (Alpha Canis Majoris).

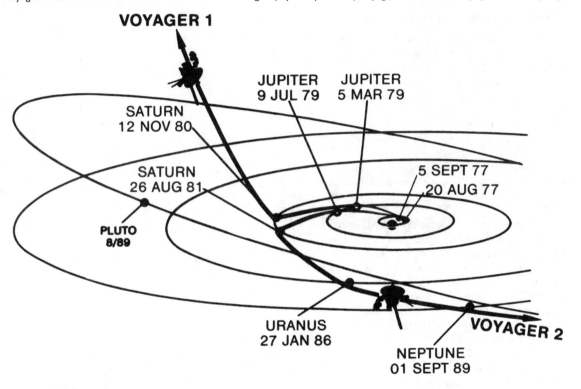

Quick Look Facts

Close approach distances and times January 24, 1986 (PST / EST):

Uranus:	81,500 kilometers	(50,600 miles)	10 a.m.	1 p.m.	(from cloud tops of Uranus)
Uranus:	107,000 kilometers	(64,500 miles)			(from center of Uranus)
Titania:	365,200 kilometers	(227.000 miles)	7:10 a.m.	10:10 a.m.	
Oberon:	470,600 kilometers	(300,000 miles)	8:13 a.m.	11:13 a.m.	
Ariel:	127.000 kilometers	(79,000 miles)	8:22 a.m.	11:22 a.m.	
Miranda:	29,000 kilometers	(18,000 miles)	9:04 a.m.	12:04 p.m.	
Umbriel:	325,000 kilometers	(202,000 miles)	12:53 p.m.	3:53 p.m.	

One way light time, Earth to Uranus: 2 hours, 44 minutes, 50 seconds.
Distance of *Voyager 2* from Earth on January 24: 2,965,400,000 kilometers (1,842,610,000 miles).

Velocity of *Voyager 2* on January 24:

Geocentric:	36 km/sec	(80,200 mph)
Heliocentric:	22 km/sec	(49,000 mph)

Launch dates:

Voyager 1	September 5, 1977
Voyager 2	August 20, 1977

Voyager 2 Neptune encounter: August 25, 1989

Neptune close approach distance: 32,000 kilometers   (20,000 miles)
From cloud tops of Neptune: 1300 kilometers   (800 miles)
From center of Neptune: 26,000 kilometers   (16,160 miles)

Cost of Voyager Project (excluding launch and tracking): $600 million (through Uranus encounter)

Voyager Science Teams

Cosmic Ray

- Edward C. Stone, California Institute of Technology, Principal Investigator
- J. Randy Jokipii, University of Arizona
- Frank B. McDonald, NASA Headquarters
- James H. Trainor, Goddard Space Flight Center
- William R. Webber, University of New Hampshire

Infrared Radiometry and Spectrometry

- Rudolf A. Hanel, Goddard Space Flight Center, Principal Investigator
- Barney Conrath, Goddard Space Flight Center
- Dale Cruikshank, University of Hawaii
- F. Michael Flasar, Goddard Space Flight Center
- Daniel Gautier, Observatoire de Paris, France
- Peter Gierasch, Cornell University
- Virgil Kunde, Goddard Space Flight Center
- William Maguire, Goddard Space Flight Center
- John Pearl, Goddard Space Flight Center
- Joseph Pirraglia, Goddard Space Flight Center
- Robert Samuelson, Goddard Space Flight Center

Imaging Science

- Bradford A. Smith, University of Arizona, Team Leader
- Geoffrey Briggs. NASA Headquarters
- Allan F. Cook II, Center for Astrophysics
- G. Edward Danielson, California Institute of Technology
- Merton E. Davies, Rand Corp.
- Garry E. Hunt, Imperial College, London
- Torrence V. Johnson, Jet Propulsion Laboratory
- Harold Masursky, U.S. Geological Survey
- Tobias Owen, State University of New York
- Carl Sagan, Cornell University
- Laurence Soderblom, U.S. Geological Survey
- Verner E. Suomi, University of Wisconsin

Low Energy Charged Particles

- S.M. (Tom) Krimigis, Johns Hopkins University, Principal Investigator
- Thomas P. Armstrong, University of Kansas
- W. Ian Axford, University of Wellington, New Zealand
- Carl O. Bostrom, Johns Hopkins University
- George Gloeckler, University of Maryland
- Ed Keath, Johns Hopkins University
- Louis J. Lanzerotti, Bell Laboratories

Magnetic Fields

- Norman F. Ness, Goddard Space Flight Center, Principal Investigator
- Mario F. Acuna, Goddard Space Flight Center
- Ken W. Behannon, Goddard Space Flight Center
- Len F. Burlaga, Goddard Space Flight Center
- Jack Connerney, Goddard Space Flight Center
- Ron P. Lepping, Goddard Space Flight Center
- Fritz Neubauer, Universitat zu Koln, Federal Republic of Germany

Plasma Science

- Herbert S. Bridge, Massachusetts Institute of Technology, Principal Investigator
- John W. Belcher, Massachusetts Institute of Technology
- Len F. Burlaga, Goddard Space Flight Center
- Christoph K. Goertz, University of Iowa
- Richard E. Hartle, Goddard Space Flight Center
- Art J. Hundhausen, High-Altitude Observatory
- Alan J. Lazarus, Massachusetts Institute of technology
- Keith Ogilvie, Goddard Space Flight Center
- Stanislaw Olbert, Massachusetts Institute of Technology
- Jack D. Scudder, Goddard Space Flight Center
- George L. Siscoe, University of California, Los Angeles
- James D. Sullivan, Massachusetts Institute of Technology
- Vytenis M. Vasyliunas, Max Planck Institut fur Aeronomie

Photopolarimetry

- Arthur L. Lane, Jet Propulsion Laboratory, Principal Investigator
- David Coffeen, Goddard Institute of Space Studies
- Larry Esposito, University of Colorado
- James E. Hansen, Goddard Institute for Space Stud
- Charles W. Hord, University of Colorado
- Makiko Sato, Goddard Institute for Space Studies
- Robert West, Jet Propulsion Laboratory

Planetary Radio Astronomy

- James W. Warwick, Radiophysics Inc., Principal Investigator
- Joseph K. Alexander, Goddard Space Flight Center
- Andre Boischot, Observatoire de Paris, France
- Walter E. Brown, Jr., Jet Propulsion Laboratory
- Thomas D. Carr, University of Florida
- Samuel Gulkis, Jet Propulsion Laboratory
- Fred T. Haddock, University of Michigan
- Christopher C. Harvey, Observatoire de Paris, France
- Michael L. Kaiser, Goddard Space Flight Center
- Yolande LeBlanc, Observatoire de Paris, France
- Robert G. Peltzer, Martin Marietta Corp.
- Roger Phillips, Southern Methodist University
- Anthony C. Riddle, University of Colorado
- David H. Staelin, Massachusetts Institute of Technology

Plasma Wave

- Frederick L. Scarf, TRW Defense and Space Systems, Principal Investigator
- Donald A. Gurnett, University of Iowa
- William Kurth, University of Iowa

Radio Science

- G. Len Tyler, Stanford University, Team Leader
- John D. Anderson, Jet Propulsion Laboratory
- Von R. Eshleman, Stanford University
- Gerald S. Levy, Jet Propulsion Laboratory
- Gunnar F. Lindal, Jet Propulsion Laboratory
- Gordon E. Wood, Jet Propulsion Laboratory

Ultraviolet Spectroscopy

- A. Lyle Broadfoot, University of Arizona, Principal Investigator
- Sushil K. Atreya, University of Michigan
- Michael J. S. Belton, Kitt Peak National Observatory
- Jean L. Hertaux, Service d'Aeronomie du CNRS
- Jacques E. Blamont, Jet Propulsion Laboratory
- Alexander Dalgarno, Center for Astrophysics

- Thomas M. Donahue, University of Michigan
- Richard Goody, Harvard University
- Jay B. Holberg, University of Arizona
- John C. McConnell, York University, Canada
- Michael B. McElroy, Harvard University
- H. Warren Moos, Johns Hopkins University
- Bill R. Sandel, University of Arizona
- Donald E. Shemansky, University of Arizona
- Darrell F, Strobel, Johns Hopkins University

Voyager Management Team

NASA Office of Space Science

Dr. Burton I. Edelson	Associate Administrator for Space Science
Samuel W. Keller	Deputy Associate Administrator
Dr. Geoffrey A. Briggs	Director, Solar System Exploration Division
Al V. Diaz	Deputy Director, Solar System Exploration Division
Earl J. Montoya	Program Manager
Dr. William E. Brunk	Program Scientist

NASA Office of Tracking and Data Systems

Robert O. Aller	Associate Administrator for Space Tracking and Data Systems
Norman Pozinsky	Deputy Associate Administrator for Space Tracking and Data Systems
Charles T. Force	Director, Ground Networks Division

Jet Propulsion Laboratory

Lew Allen	Director
Robert J. Parks	Deputy Director
Walker E. Giberson	Assistant Laboratory Director for Flight Projects
Peter T. Lyman	Assistant Laboratory Director Tracking and Data Acquisition
Richard P. Laeser	Project Manager
George P. Textor	Mission Director / Deputy Project Manager
Richard P. Rudd	Deputy Mission Director
Charles E. Kohlhase	Manager, Mission Planning Office
Allan L. Sacks	Manager, Ground Data Systems
J. Pieter deVries	Manager, Flight Science Office
Dr. William I. McLaughlin	Manager, Flight Engineering Office
Douglas G. Griffith	Manager, Flight Operations Office
Marvin R. Traxler	Manager, Tracking and Data Systems
Dr. Ellis D. Miner	Assistant Project Scientist

California Institute of Technology

Dr. Edward C. Stone	Project Scientist

1986...
A Year for
Space Science

NASA Facts

National Aeronautics and
Space Administration

Jet Propulsion Laboratory
California Institute of Technology
Pasadena, CA 91109

Voyager Mission to the Outer Planets

CONTENTS

The twin spacecraft *Voyager 1* and *Voyager 2* were launched by NASA in separate months in the summer of 1977 from Cape Canaveral, Florida. As originally designed, the *Voyagers* were to conduct close-up studies of Jupiter and Saturn, Saturn's rings, and the larger moons of the two planets.

To accomplish their two-planet mission, the spacecraft were built to last five years. But as the mission went on, and with the successful achievement of all its objectives, the additional fly-bys of the two outermost giant planets, Uranus and Neptune, proved possible – and irresistible to mission scientists and engineers at the Voyagers' home at the Jet Propulsion Laboratory in Pasadena, California.

As the spacecraft flew across the solar system, remote-control reprogramming was used to endow the *Voyagers* with greater capabilities than they possessed when they left the Earth. Their two-planet mission became four. Their five-year lifetimes stretched to 12 and more.

Eventually, between them, *Voyager 1* and *2* would explore all the giant outer planets of our solar system, 48 of their moons, and the unique systems of rings and magnetic fields those planets possess.

Had the Voyager mission ended after the Jupiter and Saturn flybys alone, it still would have provided the material to rewrite astronomy textbooks. But having doubled their already ambitious itineraries, the *Voyagers* returned to Earth information over the years that has revolutionized the science of planetary astronomy, helping to resolve key questions while raising intriguing new ones about the origin and evolution of the planets in our solar system.

Mission Concept

The Voyager mission was designed to take advantage of a rare geometric arrangement of the outer planets in the late 1970s and the 1980s. This layout of Jupiter, Saturn, Uranus and Neptune, which occurs about every 175 years, allows a spacecraft on a particular flight path to swing from one planet to the next without the need for large onboard propulsion systems. The flyby of each planet bends the spacecraft's flight path and increases its velocity enough to deliver it to the next destination. Using this "gravity assist" technique, the flight time to Neptune can be reduced from 30 years to 12.

While the four-planet mission was known to be possible, it was deemed to be too expensive to build a spacecraft that could go the distance, carry the instruments needed and last long enough to accomplish such a long mission. Thus, the *Voyagers* were funded to conduct intensive flyby studies of Jupiter and Saturn only. More than 10,000 trajectories were studied before choosing the two that would allow close flybys of Jupiter and its large moon Io, and Saturn and its large moon Titan; the chosen flight path for *Voyager 2* also preserved the option to continue on to Uranus and Neptune.

From the NASA Kennedy Space Center at Cape Canaveral, Florida, *Voyager 2* was launched first, on August 20, 1977. *Voyager 1* was launched on a faster, shorter trajectory on September 5, 1977. Both spacecraft were delivered to space aboard Titan-Centaur expendable rockets.

The prime Voyager mission to Jupiter and Saturn brought *Voyager 1* to Jupiter in 1979 and Saturn in 1980, while *Voyager 2* flew by Jupiter in 1979 and Saturn in 1981.

Voyager 1's trajectory, designed to send the spacecraft closely past the large moon Titan and behind Saturn's rings, bent the spacecraft's path inexorably northward out of the ecliptic plane – the plane in which most of the planets orbit the Sun. *Voyager 2* was aimed to fly by Saturn at a point that would automatically send the spacecraft in the direction of Uranus.

After *Voyager 2*'s successful Saturn encounter, it was shown that *Voyager 2* would likely be able to fly on to Uranus with all instruments operating. NASA provided additional funding to continue operating the two spacecraft and authorized JPL to conduct a Uranus flyby. Subsequently, NASA also authorized the Neptune leg of the mission, which was renamed the Voyager Neptune Interstellar Mission.

Voyager 2 encountered Uranus on January 24, 1986, returning detailed photos and other data on the planet, its moons, magnetic field and dark rings. *Voyager 1*, meanwhile, continues to press outward, conducting studies of interplanetary space. Eventually, its instruments may be the first of any spacecraft to sense the heliopause – the boundary between the end of the Sun's magnetic influence and the beginning of interstellar space.

Following *Voyager 2*'s closest approach to Neptune on August 25, 1989, the spacecraft flew southward, below the ecliptic plane and onto a course that will take it, too, to interstellar space. Reflecting the *Voyagers*' new transplanetary destinations, the project is now known as the Voyager Interstellar Mission.

Voyager 1 is now leaving the solar system, rising above the ecliptic plane at an angle of about 35 degrees at a rate of about 520 million kilometers (about 320 million miles) a year. *Voyager 2* is also headed out of the solar system, diving below the ecliptic plane at an angle of about 48 degrees and a rate of about 470 million kilometers (about 290 million miles) a year.

Both spacecraft will continue to study ultraviolet sources among the stars, and the fields and particles instruments aboard the Voyagers will continue to search for the boundary between the Sun's influence and interstellar space. The *Voyagers* are expected to return valuable data for two or three more decades. Communications will be maintained until the Voyagers' nuclear power sources can no longer supply enough electrical energy to power critical subsystems.

The cost of the *Voyager 1* and 2 missions – including launch, mission operations from launch through the Neptune encounter and the spacecraft's nuclear batteries (provided by the Department of Energy) – is $865 million. NASA budgeted an additional $30 million to fund the Voyager Interstellar Mission for two years following the Neptune encounter.

Operations

Voyagers 1 and 2 are identical spacecraft. Each is equipped with instruments to conduct 10 different experiments. The instruments include television cameras, infrared and ultraviolet sensors, magnetometers, plasma detectors, and cosmic ray and charged-particle sensors. In addition, the spacecraft radio is used to conduct experiments.

The *Voyagers* travel too far from the Sun to use solar panels; instead, they were equipped with power sources called radioisotope thermoelectric generators (RTGs). These devices, used on other deep space missions, convert the heat roduced from the natural radioactive decay of plutonium into electricity to power the spacecraft instruments, omputers, radio and other systems.

he spacecraft are controlled and their data returned through the Deep Space Network (DSN), a global spacecraft acking system operated by JPL for NASA. DSN antenna complexes are located in California's Mojave Desert; near 1adrid, Spain; and in Tidbinbilla, near Canberra, Australia.

upiter

oyager 1 made its closest approach to Jupiter on March 5, 1979, and Voyager 2 followed with its closest approach ccurring on July 9, 1979. The first spacecraft flew within 206,700 kilometers (128,400 miles) of the planet's cloud ops, and Voyager 2 came within 570,000 kilometers (350,000 miles).

apiter is the largest planet in the solar system, composed mainly of hydrogen and helium, with small amounts of 1ethane, ammonia, water vapor, traces of other compounds and a core of melted rock and ice. Colorful latitudinal ands and atmospheric clouds and storms illustrate Jupiter's dynamic weather system. The giant planet is now known ɔ possess 16 moons. The planet completes one orbit of the Sun each 11.8 years and its day is 9 hours, 55 minutes.

lthough astronomers had studied Jupiter through telescopes on Earth for centuries, scientists were surprised by many f the Voyager findings.

he Great Red Spot was revealed as a complex storm moving in a counter-clockwise direction. An array of other naller storms and eddies were found throughout the banded clouds.

iscovery of active volcanism on the satellite Io was easily the greatest unexpected discovery at Jupiter. It was the first me active volcanoes had been seen on another body in the solar system. Together, the Voyagers observed the eruption f nine volcanoes on Io, and there is evidence that other eruptions occurred between the Voyager encounters.

lumes from the volcanoes extend to more than 300 kilometers (190 miles) above the surface. The Voyagers observed aaterial ejected at velocities up to a kilometer per second.

's volcanoes are apparently due to heating of the satellite by tidal pumping. Io is perturbed in its orbit by Europa and anymede, two other large satellites nearby, then pulled back again into its regular orbit by Jupiter. This tug-of-war esults in tidal bulging as great as 100 meters (330 feet) on Io's surface, compared with typical tidal bulges on Earth of ne meter (three feet).

appears that volcanism on Io affects the entire Jovian system, in that it is the primary source of matter that pervades apiter's magnetosphere – the region of space surrounding the planet influenced by the Jovian magnetic field. Sulfur, xygen and sodium, apparently erupted by Io's many volcanoes and sputtered off the surface by impact of high-energy articles, were detected as far away as the outer edge of the magnetosphere, millions of miles from the planet itself.

uropa displayed a large number of intersecting linear features in the low-resolution photos from Voyager 1. At first, :ientists believed the features might be deep cracks, caused by crustal rifting or tectonic processes. The closer high-esolution photos from Voyager 2, however, left scientists puzzled. The features were so lacking in topographic relief ıat as one scientist described them, they "might have been painted on with a felt marker." There is a possibility that uropa may be internally active due to tidal heating at a level one-tenth or less than that of Io. Europa is thought to ave a thin crust (less than 30 kilometers or 18 miles thick) of water ice, possibly floating on a 50-kilometer (30-mile)-eep ocean.

anymede turned out to be the largest moon in the solar system, with a diameter measuring 5,276 kilometers (3,280 iiles). It showed two distinct types of terrain – cratered and grooved – suggesting to scientists that Ganymede's ntire icy crust has been under tension from global tectonic processes.

allisto has a very old, heavily cratered crust showing remnant rings of enormous impact craters. The largest craters ıve apparently been erased by the flow of the icy crust over geologic time. Almost no topographic relief is apparent the ghost remnants of the immense impact basins, identifiable only by their light color and the surrounding subdued ngs of concentric ridges.

faint, dusty ring of material was found around Jupiter. Its outer edge is 129,000 kilometers (80,000 miles) from the enter of the planet, and it extends inward about 30,000 kilometers (18,000 miles). Two new, small satellites, Adrastea ıd Metis, were found orbiting just outside the ring. A third new satellite, Thebe, was discovered between the orbits f Amalthea and Io.

piter's rings and moons exist within an intense radiation belt of electrons and ions trapped in the planet's magnetic ıld. These particles and fields comprise the Jovian magnetosphere, or magnetic environment, which extends three to even million kilometers toward the Sun, and stretches in a windsock shape at least as far as Saturn's orbit – a distance f 750 million kilometers (460 million miles).

As the magnetosphere rotates with Jupiter, it sweeps past Io and strips away about 1,000 kilograms (one ton) of material per second. The material forms a torus, a doughnut-shaped cloud of ions that glow in the ultraviolet. The torus' heavy ions migrate outward, and their pressure inflates the Jovian magnetosphere to more than twice its expected size. Some of the more energetic sulfur and oxygen ions fall along the magnetic field into the planet's atmosphere, resulting in auroras.

Io acts as an electrical generator as it moves through Jupiter's magnetic field, developing 400,000 volts across its diameter and generating an electric current of 3 million amperes that flows along the magnetic field to the planet's ionosphere.

Saturn

The Voyager 1 and 2 Saturn flybys occurred nine months apart, with the closest approaches falling on November 12 and August 25, 1981. Voyager 1 flew within 64,200 kilometers (40,000 miles) of the cloud tops, while Voyager 2 came within 41,000 kilometers (26,000 miles).

Saturn is the second largest planet in the solar system. It takes 29.5 Earth years to complete one orbit of the Sun, and its day was clocked at 10 hours, 39 minutes. Saturn is known to have at least 17 moons and a complex ring system. Like Jupiter, Saturn is mostly hydrogen and helium. Its hazy yellow hue was found to be marked by broad atmospheric banding similar to but much fainter than that found on Jupiter. Close scrutiny by Voyager's imaging systems revealed long-lived ovals and other atmospheric features generally smaller than those on Jupiter.

Perhaps the greatest surprises, and the most puzzles, were found by the Voyagers in Saturn's rings. It is thought that the rings formed from larger moons that were shattered by impacts of comets and meteoroids. The resulting dust and boulder- to house-size particles have accumulated in a broad plane around the planet varying in density.

The irregular shapes of Saturn's eight smallest moons indicates that they too are fragments of larger bodies. Unexpected structure, such as kinks and spokes, was found in addition to thin rings and broad, diffuse rings not observed from Earth. Much of the elaborate structure of some of the rings is due to the gravitational effects of nearby satellites. This phenomenon is most obviously demonstrated by the relationship between the F-ring and two small moons that "shepherd" the ring material. The variation in the separation of the moons from the ring may explain the ring's kinked appearance. Shepherding moons were also found by Voyager 2 at Uranus.

Radial, spoke-like features in the broad B-ring were found by the Voyagers. The features are believed to be composed of fine, dust-size particles. The spokes were observed to form and dissipate in time-lapse images taken by the Voyagers. While electrostatic charging may create spokes by levitating dust particles above the ring, the exact cause of the formation of the spokes is not well understood.

Winds blow at extremely high speeds on Saturn – up to 1,800 km/hr (1,100 mph). Their primarily easterly direction indicates that the winds are not confined to the top cloud layer but must extend at least 2,000 kilometers (1,200 miles) downward into the atmosphere. The characteristic temperature of the atmosphere is 95° Kelvin.

Saturn holds a wide assortment of satellites in its orbit, ranging from Phoebe, a small moon that travels in a retrograde orbit and is probably a captured asteroid, to Titan, the planet-sized moon with a thick nitrogen-methane atmosphere. Titan's surface temperature and pressure are 94° Kelvin (-292° Fahrenheit) and 1.5 atmospheres. Photochemistry converts some atmospheric methane to other organic molecules, such as ethane, that is thought to accumulate in lakes or oceans. Other more complex hydrocarbons form the haze particles that eventually fall to the surface, coating it with a thick layer of organic matter. The chemistry in Titan's atmosphere may strongly resemble that which occurred on Earth before life evolved.

The most active surface of any moon seen in the Saturn system was that of Enceladus. The bright surface of this moon, marked by faults and valleys, showed evidence of tectonically induced change. Voyager 1 found the moon Mimas scarred with a crater so huge that the impact that caused it nearly broke the satellite apart.

Saturn's magnetic field is smaller than Jupiter's, extending only one or two million kilometers. The axis of the field is almost perfectly aligned with the rotation axis of the planet.

Uranus

In its first solo planetary flyby, Voyager 2 made its closest approach to Uranus on January 24, 1986, coming within 81,500 kilometers (50,600 miles) of the planet's cloud tops.

Uranus is the third largest planet in the solar system. It orbits the Sun at a distance of about 2.8 billion kilometers (1.8 billion miles) and completes one orbit every 84 years. The length of a day on Uranus as measured by Voyager 2 is 17 hours, 14 minutes.

Uranus is distinguished by the fact that it is tipped on its side. Its unusual position is thought to be the result of a collision with a planet-sized body early in the solar system's history. Given its odd orientation, with its polar regions exposed to sunlight or darkness for long periods, scientists were not sure what to expect at Uranus.

Voyager 2 found that one of the most striking influences of this sideways position is its effect on the tail of the magnetic field, which is itself tilted 60 degrees from the planet's axis of rotation. The magnetotail was shown to be twisted by the planet's rotation into a long corkscrew shape behind the planet.

The presence of a magnetic field at Uranus was not known until Voyager's arrival. The intensity of the field is roughly comparable to that of Earth's, though it varies much more from point to point because of its large offset from the center of Uranus. The peculiar orientation of the magnetic field suggests that the field is generated at an intermediate depth in the interior where the pressure is high enough for water to become electrically conducting.

Radiation belts at Uranus were found to be of an intensity similar to those at Saturn. The intensity of radiation within the belts is such that irradiation would quickly darken (within 100,000 years) any methane trapped in the icy surfaces of the inner moons and ring particles. This may have contributed to the darkened surfaces of the moons and ring particles, which are almost uniformly gray in color.

A high layer of haze was detected around the sunlit pole, which also was found to radiate large amounts of ultraviolet light, a phenomenon dubbed "dayglow." The average temperature is about 60° Kelvin (-350° degrees Fahrenheit). Surprisingly, the illuminated and dark poles, and most of the planet, show nearly the same temperature at the cloud tops.

Voyager found 10 new moons, bringing the total number to 15. Most of the new moons are small, with the largest measuring about 150 kilometers (about 90 miles) in diameter.

The moon Miranda, innermost of the five large moons, was revealed to be one of the strangest bodies yet seen in the solar system. Detailed images from Voyager's flyby of the moon showed huge fault canyons as deep as 20 kilometers (12 miles), terraced layers, and a mixture of old and young surfaces. One theory holds that Miranda may be a reaggregration of material from an earlier time when the moon was fractured by a violent impact.

The five large moons appear to be ice-rock conglomerates like the satellites of Saturn. Titania is marked by huge fault systems and canyons indicating some degree of geologic, probably tectonic, activity in its history. Ariel has the brightest and possibly youngest surface of all the Uranian moons and also appears to have undergone geologic activity that led to many fault valleys and what seem to be extensive flows of icy material. Little geologic activity has occurred on Umbriel or Oberon, judging by their old and dark surfaces.

All nine previously known rings were studied by the spacecraft and showed the Uranian rings to be distinctly different from those at Jupiter and Saturn. The ring system may be relatively young and did not form at the same time as Uranus. Particles that make up the rings may be remnants of a moon that was broken by a high-velocity impact or torn up by gravitational effects.

Neptune

When Voyager flew within 5,000 kilometers (3,000 miles) of Neptune on August 25, 1989, the planet was the most distant member of the solar system from the Sun. (Pluto once again will become most distant in 1999.)

Neptune orbits the Sun every 165 years. It is the smallest of our solar system's gas giants. Neptune is now known to have eight moons, six of which were found by Voyager. The length of a Neptunian day has been determined to be 16 hours, 6.7 minutes.

Even though Neptune receives only three percent as much sunlight as Jupiter does, it is a dynamic planet and surprisingly showed several large, dark spots reminiscent of Jupiter's hurricane-like storms. The largest spot, dubbed the Great Dark Spot, is about the size of Earth and is similar to the Great Red Spot on Jupiter. A small, irregularly shaped, eastward-moving cloud was observed "scooting" around Neptune every 16 hours or so; this "scooter," as Voyager scientists called it, could be a cloud plume rising above a deeper cloud deck.

Long, bright clouds, similar to cirrus clouds on Earth, were seen high in Neptune's atmosphere. At low northern latitudes, Voyager captured images of cloud streaks casting their shadows on cloud decks below.

The strongest winds on any planet were measured on Neptune. Most of the winds there blow westward, or opposite to the rotation of the planet. Near the Great Dark Spot, winds blow up to 2,000 kilometers (1,200 miles) an hour.

The magnetic field of Neptune, like that of Uranus, turned out to be highly tilted – 47 degrees from the rotation axis and offset at least 0.55 radii (about 13,500 kilometers or 8,500 miles) from the physical center. Comparing the magnetic fields of the two planets, scientists think the extreme orientation may be characteristic of flows in the interiors of both Uranus and Neptune – and not the result in Uranus's case of that planet's sideways orientation, or of any possible field reversals at either planet. Voyager's studies of radio waves caused by the magnetic field revealed the length of a Neptunian day. The spacecraft also detected auroras, but much weaker than those on Earth and other planets.

Triton, the largest of the moons of Neptune, was shown to be not only the most intriguing satellite of the Neptunian system, but one of the most interesting in all the solar system. It shows evidence of a remarkable geologic history, and Voyager 2 images showed active geyser-like eruptions spewing invisible nitrogen gas and dark dust particles several kilometers into the tenuous atmosphere. Triton's relatively high density and retrograde orbit offer strong evidence that Triton is not an original member of Neptune's family but is a captured object. If that is the case, tidal heating could have melted Triton in its originally eccentric orbit, and the moon might even have been liquid for as long as one billion years after its capture by Neptune.

An extremely thin atmosphere extends about 800 kilometer (500 miles) above Triton's surface. Nitrogen ice particles may form thin clouds a few kilometers above the surface. The atmospheric pressure at the surface is about 14 microbars, 1/70,000th the surface pressure on Earth. The surface temperature is about 38° Kelvin (-391° degrees Fahrenheit), the coldest temperature of any body known in the solar system.

The new moons found at Neptune by *Voyager* are all small and remain close to Neptune's equatorial plane. Names for the new moons were selected from mythology's water deities by the International Astronomical Union; they are Naiad, Thalassa, Despina, Galatea, Larissa, and Proteus.

Voyager 2 solved many of the questions scientists had about Neptune's rings. Searches for "ring arcs," or partial rings showed that Neptune's rings actually are complete, but are so diffuse and the material in them so fine that they could not be fully resolved from Earth. From the outermost in, the rings have been designated Adams, Plateau, Le Verrier and Galle.

Interstellar Mission

The spacecraft are continuing to return data about interplanetary space and some of our stellar neighbors near the edges of the Milky Way.

As the *Voyagers* cruise gracefully in the solar wind, their fields, particles and waves instruments are studying the space around them. The spacecraft are collecting data on the strength and orientation of the Sun's magnetic field; the composition, direction and energy spectra of the solar wind particles and interstellar cosmic rays; the strength of radio emissions that are thought to be originating at the heliopause, beyond which is interstellar space; and the distribution of hydrogen within the outer heliopause. In May 1993, scientists concluded that the plasma wave experiment was picking up radio emissions that originate at the heliopause – the outer edge of our solar system.

The heliopause is the outermost boundary of the solar wind, where the interstellar medium restricts the outward flow of the solar wind and confines it within a magnetic bubble called the heliosphere. The solar wind is made up of electrically charged atomic particles, composed primarily of ionized hydrogen, that stream outward from the Sun.

Exactly where the heliopause is has been one of the great unanswered questions in space physics. By studying the radio emissions, scientists now theorize the heliopause exists some 90 to 120 astronomical units (AU) from the Sun (One AU is equal to 150 million kilometers (93 million miles), or the distance from the Earth to the Sun.

The *Voyagers* have also become space-based ultraviolet observatories and their unique location in the universe gives astronomers the best vantage point they have ever had for looking at celestial objects that emit ultraviolet radiation.

The cameras on the spacecraft have been turned off and the ultraviolet instrument is the only experiment on the scan platform that is still functioning. Voyager scientists expect to continue to receive data from the ultraviolet spectrometers at least until the year 2000. At that time, there may not be enough electrical power for the heaters to keep the ultraviolet instrument warm enough to operate.

Yet there are several other fields and particle instruments that can continue to send back data as long as the spacecraft stay alive. They include: the cosmic ray subsystem, the low-energy charge particle instrument, the magnetometer, the plasma subsystem, the plasma wave subsystem and the planetary radio astronomy instrument. However, the Plasma instrument sensitivity on *Voyager 1* is severely degraded, limiting the science value of the data. Data from all the instruments are transmitted to Earth in real time, at 160 bits per second, and captured by 34-meter DSN stations. After transmission of the data to JPL, it is made available in electronic files to the science teams located around the country for their processing and analysis.

Flight controllers believe both spacecraft will continue to operate and send back valuable data until at least the year 2020. On February 17, 1998, *Voyager 1* passed the *Pioneer 10* spacecraft to become the most distant human-made object in space.

Project Team

At JPL, the position of Voyager project manager has been held successively by H.M. "Bud" Schurmeier (1972-76), John Casani (1976-77), Robert Parks (1978-79), Raymond Heacock (1979-81), Esker Davis (1981-82), Richard Laeser (1982-86), Norman Haynes (1987-89), George Textor (1989-97) and Ed Massey (1998 to present). Dr. Edward C. Stone is the Voyager project scientist. The assistant project scientist for the Jupiter flyby was Dr. Arthur L. Lane, followed by Dr. Ellis D. Miner for the Saturn, Uranus and Neptune encounters.

	Diameter		Distance from Sun	
Jupiter	142,984 km	(88,846 mi)	778,000,000 km	(483,000,000 mi)

Jupiter's Moons			Distance From Planet Center	
Metis	40 km	(25 mi)	128,000 km	(79,500 mi)
Adrastea	24x20x14 km	(14x12x9 mi)	129,000 km	(80,100 mi)
Amalthea	270x166x150 km	(165x103x95 mi)	181,300 km	(112,600 mi)
Thebe	110x90km	(65x55 mi)	222,000 km	(138,000 mi)
Io	3,630 km	(2,225 mi)	422,000 km	(262,000 mi)
Europa	3,138 km	(1,949 mi)	661,000 km	(414,500 mi)
Ganymede	5,262 km	(3,269 mi)	1,070,000 km	(664,900 mi)
Callisto	4,800 km	(3,000 mi)	1,883,000 km	(1,170000 mi)
Leda	16 km	(10 mi)	11,094,000 km	(6,900,000 mi)
Himalia	186 km	(115 mi)	11,480,000 km	(7,133,000 mi)
Lysithia	36 km	(20 mi)	11,720,000 km	(7,282,000 mi)
Elara	76 km	(47 mi)	11,737,000 km	(7,293,000 mi)
Ananke	30 km	(18 mi)	21,200,000 km	(13,173,000 mi)
Carme	40 km	(25 mi)	22,600,000 km	(14,043,000 mi)
Pasiphae	50 km	(31 mi)	23,500,000 km	(14,602,000 mi)
Sinope	36 km	(22 mi)	23,700,000 km	(14,727,000 mi)

	Diameter		Distance from Sun	
Saturn	120,536 km	(74,900 mi)	1.4 billion km	(870 mi)llion mi)

Saturn's Moons			Distance from Planet Center	
Atlas	40x20 km	(24x12 mi)	137,670 km	(85,500 mi)
Prometheus	140x100x80 km	(85x60x50 mi)	139,353 km	(86,600 mi)
Pandora	110x90x80 km	(70x55x50 mi)	141,700 km	(88,500 mi)
Epimetheus	140x120x100 km	(85x70x60 mi)	151,472 km	(94,124 mi)
Janus	220x200x160 km	(135x125x100 mi)	151,422 km	(94,093 mi)
Mimas	392 km	(243 mi)	185,520 km	(115,295 mi)
Enceladus	520 km	(320 mi)	238,020 km	(147,900 mi)
Tethys	1,060 km	(660 mi)	294,660 km	(183,100 mi)
Telesto	34x28x26 km	(20x17x16 mi)	294,660 km	(183,100 mi)
Calypso	34x22x22 km	(20x13x13 mi)	294,660 km	(183,100 mi)
Dione	1,120 km	(695 mi)	377,400 km	(234,500 mi)
Helene	36x32x30 km	(22x20x19 mi)	377,400 km	(234,900 mi)
Rhea	1,530 km	(950 mi)	527,040 km	(327,500 mi)
Titan	5,150 km	(3,200 mi)	1,221,860 km	(759,300 mi)
Hyperion	410x260x220 km	(250x155x135 mi)	1,481,000 km	(920,300 mi)
Iapetus	1,460 km	(910 mi)	3,560,830 km	(2,212,900 mi)
Phoebe	220 km	(135 mi)	12,952,000 km	(8,048,000 mi)

	Diameter		Distance from Sun	
Uranus	51,118 km	(31,764 mi)	3 billion km	(1.8 billion mi)

Uranus's Moons			Distance from Planet Center	
Cordelia	26 km	(16 mi)	49,800 km	(30,950 mi)
Ophelia	30 km	(18 mi)	53,800 km	(33,400 mi)
Bianca	42 km	(26 mi)	59,200 km	(36,800 mi)
Juliet	62 km	(38 mi)	61,800 km	(38,400 mi)
Desdemona	54 km	(33 mi)	62,700 km	(38,960 mi)
Rosalind	84 km	(52 mi)	64,400 km	(40,000 mi)
Portia	108 km	(67 mi)	66,100 km	(41,100 mi)
Cressida	54 km	(32 mi)	69,900 km	(43,400 mi)
Belinda	66 km	(40 mi)	75,300 km	(46,700 mi)
Puck	154 km	(95 mi)	86,000 km	(53,000 mi)
Miranda	472 km	(293 mi)	129,900 km	(80,650 mi)
Ariel	1,158 km	(720 mi)	190,900 km	(118,835 mi)
Umbriel	1,172 km	(728 mi)	265,969 km	(165,300mi
Titania	1,580 km	(981 mi)	436,300 km	(271,100mi
Oberon	1,524 km	(947 mi)	583,400km	(362,500 mi)

	Diameter	Distance from Sun	
Neptune	49,528 km(30,776 mi)	4.5 billion km	(2.7 billion mi)

Neptune's Moons			Distance from Planet Center	
Naiad	54 km	(33 mi)	48,000 km	(29,827 mi)
Thalassa	80 km	(50 mi)	50,000 km	(31,000 mi)
Despina	180 km	(110 mi)	52,500 km	(32,600 mi)
Galatea	150 km	(95 mi)	62,000 km	(38,525 mi)
Larissa	190 km	(120 mi)	73,600 km	(45,700 mi)
Proteus	400 km	(250 mi)	117,600 km	(73,075 mi)
Triton	2,700 km	(1,680 mi)	354,760km	(220,500 mi)
Nereid	340km	(210 mi)	5,509 090 km	(3,423,000 mi)

GALILEO
MISSION

FLIGHT PROJECTS OFFICE

Galileo Fact Sheet

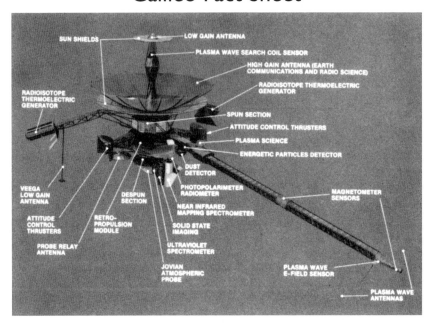

Orbiter

Objectives:
- To Investigate the Chemical Composition and Physical State of Jupiter's Atmosphere
- To Investigate the Chemical Composition and Physical State of the Jovian Satellites
- To Investigate the Structure and Physical Dynamics of the Jovian Magnetosphere

Features:
- Dual Spin Spacecraft to Provide Both Stable Platform for Remote Sensing Instruments and Spinning Platform for Fields and Particles Measurements
- Probe to Penetrate the Atmosphere of Jupiter and Orbiter to Provide 22+ Months of Orbital Science
- Venus-Earth-Earth Gravity Assist Trajectory to Achieve Velocity to Reach Jupiter

Probe

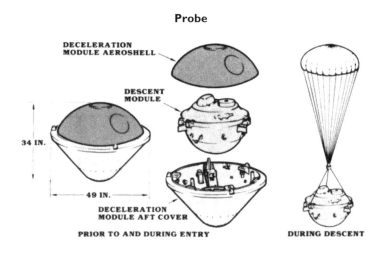

Mission Schedule

MILESTONES	89	90	91	92	93	94	95	96	97
LAUNCH (PERIOD OPENING)	12 OCT ▽								
VENUS		9 FEB ▽							
EARTH 1			8 DEC ▽						
GASPRA				29 OCT ▽					
EARTH 2					8 DEC ▽				
IDA					28 AUG ▽				
PROBE RELEASE							7 JULY ▽		
Io - RELAY - JOI							7 DEC ▽		
JUPITER TOUR							▭▭▭		OCT

1989 VEEGA Trajectory

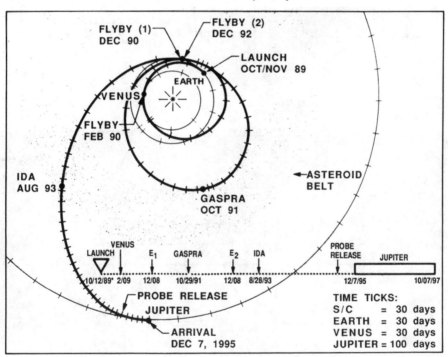

Probe Mission

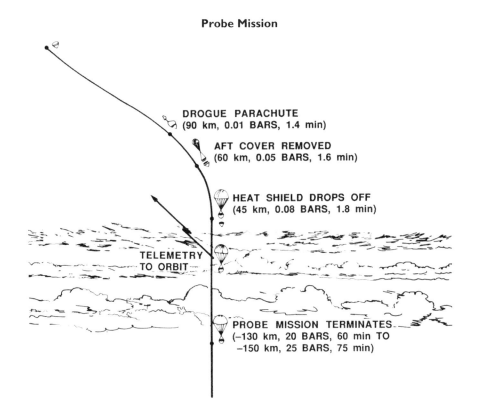

DROGUE PARACHUTE
(90 km, 0.01 BARS, 1.4 min)

AFT COVER REMOVED
(60 km, 0.05 BARS, 1.6 min)

HEAT SHIELD DROPS OFF
(45 km, 0.08 BARS, 1.8 min)

TELEMETRY
TO ORBIT

PROBE MISSION TERMINATES
(–130 km, 20 BARS, 60 min TO
–150 km, 25 BARS, 75 min)

NASA Facts

National Aeronautics and
Space Administration

Jet Propulsion Laboratory
California Institute of Technology
Pasadena, CA 91109

Galileo Mission to Jupiter

NASA's *Galileo* spacecraft continues to make discoveries about the giant planet Jupiter, its moons and its surrounding magnetic environment after more than five years in orbit around Jupiter. The mission was named for the Italian Renaissance scientist Galileo Galilei, who discovered Jupiter's major moons in 1610.

Mission Overview

Galileo's primary mission at Jupiter began when the spacecraft entered into orbit around Jupiter in December 1995, and its descent probe, which had been released five months earlier, dove into the giant planet's atmosphere. The primary mission included a 23-month, 11-orbit tour of the Jovian system, including 10 close encounters of Jupiter's major natural satellites, or moons.

Although the primary mission was completed in December 1997, the mission has been extended three times to take advantage of the spacecraft's durability with 24 more orbits. The extensions have enabled additional encounters with all four of Jupiter's major moons: Io, Europa, Ganymede and Callisto, as well as the small moon Amalthea. *Galileo* will cross Amalthea's orbital plane a second time before making a mission-ending plunge into Jupiter's atmosphere in September 2003.

Galileo was the first spacecraft ever to measure Jupiter's atmosphere directly with a descent probe, and the first to conduct long-term observations of the Jovian system from orbit around Jupiter. It found evidence for subsurface liquid layers of salt water on Europa, Ganymede and Callisto, and it documented extraordinary levels of volcanic activity on Io. During the interplanetary cruise, *Galileo* became the first spacecraft to fly by an asteroid and the first to discover the moon of an asteroid. It was also the only direct observer as fragments from the Shoemaker-Levy 9 comet slammed into Jupiter in July 1994.

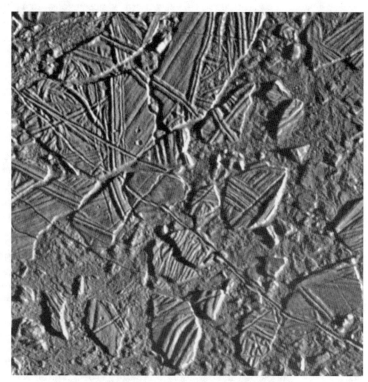

Ice rafts on Jupiter's moon Europa, photographed by the Galileo spacecraft during a flyby February 20, 1997.

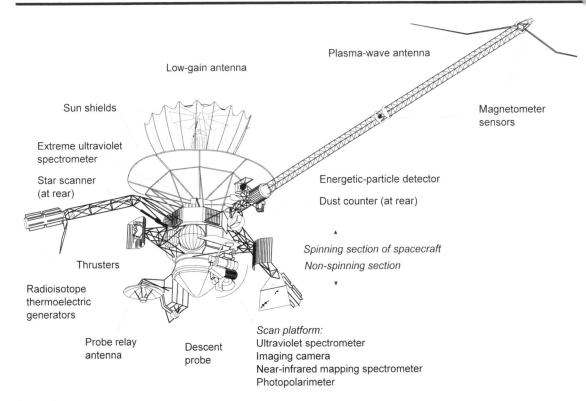

Plasma-wave antenna

Low-gain antenna

Sun shields

Magnetometer
sensors

Extreme ultraviolet
spectrometer

Star scanner
(at rear)

Energetic-particle detector

Dust counter (at rear)

Spinning section of spacecraft
Non-spinning section

Thrusters

Radioisotope
thermoelectric
generators

Probe relay
antenna

Descent
probe

Scan platform:
Ultraviolet spectrometer
Imaging camera
Near-infrared mapping spectrometer
Photopolarimeter

Launch

The *Galileo* spacecraft and its two-stage Inertial Upper Stage (IUS) were carried into Earth orbit on October 18, 1989 by space shuttle *Atlantis* on mission STS-34. The solid-fuel upper stage then accelerated the spacecraft out of Earth orbit toward the planet Venus for the first of three planetary flybys, or "gravity assists," designed to boost *Galileo* toward Jupiter. In a gravity assist, the spacecraft flies close enough to a planet to be propelled by its gravity, creating a "slingshot" effect for the spacecraft. The *Galileo* mission had originally been designed for a direct flight of about 3½ years to Jupiter, using a planetary three-stage IUS. When this vehicle was canceled, plans were changed to use a liquid-fuel Centaur upper stage. Due to safety concerns after the *Challenger* accident, NASA canceled use of the Centaur on the space shuttle, and *Galileo* was moved to the two-stage IUS; this, however, made it impossible for the spacecraft to fly directly to Jupiter. To save the project, Galileo engineers designed a new and remarkable six-year interplanetary flight path using planetary gravity assists.

Venus and Earth Flybys

After flying past Venus at an altitude of 16,000 kilometers (nearly 10,000 miles) on February 10, 1990, the spacecraft swung past Earth at an altitude of 960 kilometers (597 miles) on December 8, 1990. The spacecraft returned for a second Earth swingby on December 8, 1992, at an altitude of 303 kilometers (188 miles). With this, *Galileo* left Earth for the third and final time and headed toward Jupiter.

The flight path provided opportunities for scientific observations. At Venus, scientists obtained the first views of mid-level clouds and gained new information about the atmosphere's dynamics. They also made many Earth observations, mapped the surface of Earth's Moon, and observed its north polar regions.

Because of the modification in *Galileo*'s trajectory, the spacecraft was exposed to a hotter environment than originally planned. To protect it from the Sun, project engineers devised a set of sun shades and pointed the top of the spacecraft toward the Sun, with the umbrella-like high-gain antenna stowed until well after the first Earth flyby in December 1990. Flight controllers stayed in touch with the spacecraft through a pair of low-gain antennas, which send and receive data at a much slower rate.

High-Gain Antenna Problem

The spacecraft was scheduled to deploy its 4.8-meter (16-foot)-diameter high-gain antenna in April 1991 as *Galileo* moved away from the Sun and the risk of overheating ended. The antenna, however, failed to deploy fully.

A special team performed extensive tests and determined that a few (probably three) of the antenna's 18 ribs were held by friction in the closed position. Despite exhaustive efforts to free the ribs, the antenna would not deploy. From

1993 to 1996, extensive new flight and ground software was developed, and ground stations of NASA's Deep Space Network were enhanced in order to perform the mission using the spacecraft's low-gain antennas.

Asteroid Flybys

Galileo became the first spacecraft ever to encounter an asteroid when it passed Gaspra on October 29, 1991. It flew within just 1,601 kilometers (1,000 miles) of the stony asteroid's center at a relative speed of about 8 kilometers per second (18,000 mph). Pictures and other data revealed a cratered, complex, irregular body about 20 by 12 by 11 kilometers (12.4 by 7.4 by 6.8 miles), with a thin covering of dirt-like "regolith" and a possible magnetic field.

On August 28, 1993, *Galileo* flew by a second asteroid, this time a larger, more distant asteroid named Ida. Ida is about 55 kilometers (34 miles) long by 20 and 24 kilometers (12 and 15 miles). Like Gaspra, Ida may have a magnetic field. Scientists made a dramatic discovery when they found that Ida boasts its own moon, making it the first asteroid known to have a natural satellite. The tiny moon, named Dactyl, has a diameter of only about 1.5 kilometers (less than a mile). Scientists studied Dactyl's orbit in order to estimate Ida's density.

Comet Event

The discovery of Comet Shoemaker-Levy 9 in March 1993 provided an exciting opportunity for *Galileo*'s science teams and other astronomers. The comet was breaking up as it orbited Jupiter and was headed to dive into the giant planet's atmosphere in July 1994. The *Galileo* spacecraft, approaching Jupiter, was the only observation platform with a direct view of the impact area on Jupiter's far side. Despite the uncertainty of the predicted impact times, *Galileo* team members pre-programmed the spacecraft's science instruments to collect data and were able to obtain spectacular images of the comet impacts.

Jupiter Arrival

On July 13, 1995, *Galileo*'s descent probe, which had been carried aboard the parent spacecraft, was released and began a five-month free fall toward Jupiter. The probe had no engine or thrusters, so its flight path was established by pointing of the *Galileo* orbiter before the probe was released. Two weeks later, *Galileo* used its 400-newton main rocket engine for the first time as it readjusted its flight path to arrive at the proper point at Jupiter.

Arrival day on December 7, 1995, turned out to be an extremely busy 24-hour period. When *Galileo* first reached Jupiter and while the probe was still approaching the planet, the orbiter flew by two of Jupiter's major moons – Europa and Io. *Galileo* passed Europa at an altitude of about 33,000 kilometers (20,000 miles), while the Io approach was at an altitude of about 900 kilometers (600 miles). About four hours after leaving Io, the orbiter made its closest approach to Jupiter, encountering 25 times more radiation than the level considered deadly for humans.

Descent Probe

Eight minutes later, the orbiter started receiving data from the descent probe, which slammed into the top of the Jovian atmosphere at a comet-like speed of 170,000 km/hr (106,000 mph). In the process, the probe withstood temperatures twice as hot as a layer roughly corresponding to the Sun's surface. The probe slowed by aerodynamic braking for about two minutes, then deployed its parachute and dropped its heat shield.

Like the other gas giant outer planets, Jupiter has no solid surface. The wok-shaped probe kept transmitting data for nearly an hour as it parachuted down through Jupiter's atmosphere. Temperatures exceeding 150 degrees Celsius (302 degrees Fahrenheit) eventually caused electronics to fail at a depth corresponding to about 22 times sea-level air pressure on Earth, a pressure more than double what had been the mission objective. That descent took the probe about 130 kilometers (81 miles) below the level corresponding to Earth sea-level air pressure. As it descended, the probe relayed information to the orbiter about sunlight and heat flux, pressure, temperature, winds, lightning and the composition of the atmosphere.

An hour after receiving the last transmission from the probe, at a point about 200,000 kilometers (130,000 miles) above the planet, the *Galileo* spacecraft fired its main engine to brake into orbit around Jupiter.

This first loop around Jupiter lasted about seven months. *Galileo* fired its thrusters at its farthest point in the orbit to keep it from coming so close to the giant planet on later orbits. This adjustment prevented possible damage to spacecraft sensors and computer chips from Jupiter's intense radiation environment.

During this first orbit, new software was installed to give the spacecraft extensive new onboard data-processing capabilities. It enabled data compression, permitting the spacecraft to transmit up to 10 times the number of pictures and other measurements that would have been possible otherwise.

In addition, hardware changes on the ground and adjustments to the spacecraft-to-Earth communication system increased the average telemetry rate tenfold. Although the problem with the high-gain antenna prevented some of the mission's original objectives from being met, the great majority of them were. So many objectives were achieved that scientists feel *Galileo* has produced considerably more science than ever envisioned at the project's start 20 years ago. Data compression features enabled the science of the mission to be possible and was a significant achievement.

Orbital Tour

During its primary mission orbital tour, *Galileo*'s itinerary included four flybys of Jupiter's moon Ganymede, three of Callisto and three of Europa. These encounters were about 100 to 1,000 times closer than those performed by NASA's *Voyager 1* and 2 spacecraft during their Jupiter flybys in 1979. During each flyby, *Galileo*'s instruments scanned and scrutinized the surface and features of each moon. After about a week of intensive observation, with its tape recorder full of data, the spacecraft spent the next one to two months in orbital "cruise," sending to Earth data stored on the onboard tape recorder.

Mission Extensions

Galileo's prime mission ended in December 1997. Since then, three mission extensions have taken advantage of the spacecraft's ability to continue returning valuable scientific discoveries.

The two-year Galileo Europa Mission used eight close encounters with Europa to examine that moon intensively, then made four passes near Callisto and two near Io. This first extended mission added significantly to the evidence that a liquid ocean has existed and probably still exists beneath Europa's icy surface. The spacecraft came so close to Europa that it could see features the size of a school bus. *Galileo* also returned information about thunderstorms in Jupiter's atmosphere during this extended mission. The spacecraft is exposed to harsh radiation from Jupiter's radiation belts whenever it comes near the planet, so examination of Io, closest in of Jupiter's four major moons, was saved until after the Europa flybys. As *Galileo* approached Io, engineers worked through the night to counteract radiation effects to the onboard computer. *Galileo* provided new insight into Io's intense volcanic activity and captured images of an actively erupting fire fountain.

A second extended mission, the Galileo Millennium Mission, was initially approved through 2000, then extended further to gain more discoveries and send the spacecraft to a controlled impact into Jupiter's atmosphere in 2003. During the Galileo Millennium Mission, *Galileo* has made additional close flybys of all four of Jupiter's major moons. A major benefit to science was obtained by having *Galileo* still functioning well when NASA's Saturn-bound *Cassini* spacecraft passed Jupiter in December 2000. Observations and measurements by *Galileo* and *Cassini* were coordinated to examine Jupiter's huge magnetosphere and other parts of the Jupiter system in ways that neither spacecraft could have done alone. During this time, *Galileo* became the first interplanetary spacecraft to make an interstellar discovery, namely that the star cluster Delta Velorium was binary, with one star periodically eclipsing the other. The discovery was made because the star scanner failed to recognize this standard star. Later, *Galileo* flybys over Io's north and south poles in 2001 provided measurements useful for determining whether that moon generates its own magnetic field.

Close Encounters by the Galileo Orbiter

Orbit	Target	Date (Universal Time)	Altitude (kilometers and miles)	
0	Io	Dec. 7, 1995	897 km	(558 mi)
1	Ganymede	June 27, 1996	835 km	(519 mi)
2	Ganymede	Sept. 6, 1996	261 km	(162 mi)
3	Callisto	Nov. 4, 1996	1136 km	(706 mi)
4	Europa	Dec. 19, 1996	692 km	(430 mi)
5	none			
6	Europa	Feb. 20, 1997	586 km	(364 mi)
7	Ganymede	April 5, 1997	3102 km	(1928 mi)
8	Ganymede	May 7, 1997	1603 km	(996 mi)
9	Callisto	June 25, 1997	418 km	(260 mi)
10	Callisto	Sept. 17, 1997	535 km	(333 mi)
11	Europa	Nov. 6, 1997	2043 km	(1270 mi)
12	Europa	Dec. 16, 1997	201 km	(125 mi)
13	none			
14	Europa	March 29, 1998	1644 km	(1022 mi)
15	Europa	May 31, 1998	2515 km	(1562 mi)
16	Europa	July 21, 1998	1834 km	(1140 mi)
17	Europa	Sept. 26, 1998	3582 km	(2226 mi)
18	Europa	Nov. 22, 1998	2271 km	(1411 mi)
19	Europa	Feb. 1, 1999	1439 km	(894 mi)
20	Callisto	May 5, 1999	1321 km	(821 mi)
21	Callisto	June 30, 1999	1048 km	(651 mi)
22	Callisto	Aug. 14, 1999	2299 km	(1429 mi)
23	Callisto	Sept. 16, 1999	1052 km	(654 mi)
24	Io	Oct. 11, 1999	611 km	(380 mi)
25	Io	Nov. 26, 1999	301 km	(187 mi)
26	Europa	Jan. 3, 2000	351 km	(218 mi)
27	Io	Feb. 22, 2000	198 km	(123 mi)
28	Ganymede	May 20, 2000	809 km	(502 mi)
29	Ganymede	Dec. 28, 2000	2338 km	(1452 mi)
30	Callisto	May 25, 2001	138 km	(86 mi)
31	Io	Aug. 6, 2001	194 km	(120 mi)
32	Io	Oct. 16, 2001	184 km	(114 mi)
33	Io	Jan. 17, 2002	102 km	(63 mi)
34	Amalthea	Nov. 5, 2002	160 km	(99 mi)
Planned Encounter				
35	Jupiter	Sept. 21, 2003	impact	

In November 2002, *Galileo* swung closer to Jupiter than ever before, flying by the moon Amalthea, which is less than one-tenth the size of Io and less than half as far from Jupiter. Scientists used measurements of *Galileo*'s radio signal to estimate the mass and density of Amalthea. They are also studying data collected as *Galileo* flew through Jupiter's gossamer ring to gain new understanding of the magnetic forces and the energetic charged particles close to the planet. *Galileo* sampled for the first time the innermost regions of the magnetosphere and radiation belts, a region equivalent to Earth's Van Allen belts.

Galileo's final orbit is an elongated loop away from Jupiter. In September 2003, the spacecraft will head back for a direct impact and burn up as it plows into Jupiter's dense atmosphere.

Spacecraft

Orbiter

The *Galileo* orbiter weighed 2,223 kilograms (2½ tons) at launch and measured 5.3 meters (17 feet) from the top of the low-gain antenna to the bottom of the descent probe. The orbiter features an innovative "dual-spin" design. Most spacecraft are stabilized in flight either by spinning around a major axis, or by maintaining a fixed orientation in space, referenced to the Sun and another star. As the first dual-spin planetary spacecraft, *Galileo* combines these techniques. A spinning section rotates at about 3 rpm, while a non-spinning section provides a fixed orientation for cameras and other remote sensors. A star scanner on the spinning side determines orientation and spin rate; gyroscopes on the non-spinning side provide the basis for measuring turns and pointing instruments.

The power supply, propulsion module and most of the computers and control electronics are mounted on the spinning section. Two low-gain antennas mounted on the spinning section supported communications during the Earth-Venus-Earth leg of the flight. One was pointed upward toward the Sun, with the other was mounted on a deployable arm to point down. The upward-pointing antenna is currently carrying the communications load, while the lower antenna has been restowed and there are no plans to use it again.

The *Galileo* orbiter spacecraft carries 11 scientific instruments. Another seven were on the descent probe.

The orbiter's spinning section carries instruments to study magnetic fields and charged particles. The instruments include magnetometer sensors mounted on an 11-meter (36-foot)-long boom to minimize interference from the spacecraft's electronics, a plasma instrument to detect low- and medium-energy charged particles, and a plasma wave detector to study electromagnetic waves generated by the particles. There are also a high-energy particle detector, a detector of cosmic and Jovian dust, an extreme ultraviolet detector associated with the ultraviolet spectrometer, and a heavy ion counter to assess potentially hazardous charged-particle environments the spacecraft flies through.

Galileo's non-spinning section carries instruments such as cameras that need to be held steady. These instruments include the camera system; the near-infrared mapping spectrometer to make multispectral images for atmosphere and surface chemical analysis; the ultraviolet spectrometer to study gases; and the photopolarimeter-radiometer to measure radiant and reflected energy. The camera system obtains images of Jupiter's satellites at resolutions from 20 to 1,000 times better than the best possible from NASA's *Voyager* spacecraft; its charge-coupled-device sensor is much more sensitive than previous spacecraft cameras and is able to detect a broader color band. *Galileo's* non-spinning section also carries a dish antenna that picked up the descent probe's signals during its fall into Jupiter's atmosphere.

The spacecraft's propulsion module consists of twelve 10-newton thrusters and a single 400-newton engine, which use monomethyl-hydrazine as fuel and nitrogen tetroxide as the oxidizer. The propulsion system was developed and built by Messerschmitt-Bolkow-Blohm and provided by the Federal Republic of Germany as NASA's major international partner on *Galileo*.

Because radio signals take on the order of 45 minutes (plus or minus a few minutes depending on where the spacecraft is relative to Earth) to travel from Earth to Jupiter and back, the *Galileo* spacecraft was designed to operate from computer instructions sent to it in advance and stored in spacecraft memory. A single master sequence of commands can cover a period ranging from weeks to months of quiet operations between flybys of Jupiter's moons. During busy encounter operations, one sequence of commands covers only about a week. *Galileo's* internal processor was an old rebuilt version of the RCA 1802 chip, similar to those used to run the old PacMan computer games of the 1980's. The flight team designed a special form of the software language called "machine language" in order to change its functionality over time. After subtracting the power needed to run essential spacecraft systems, the power left for the experiments was on the order of 60 watts (enough to power a light bulb).

The flight software includes fault-protection sequences designed to automatically put *Galileo* in a safe state in case of computer glitches or other unforeseen circumstances. Electrical power is provided by two radioisotope thermoelectric generators. Heat produced by the natural radioactive decay of plutonium is converted to electricity (570 watts at launch, 432 in *Galileo's* final days in 2003) to operate the orbiter's equipment. This is the same type of power source was used on other NASA missions including *Viking* to Mars, *Voyager* and *Pioneer* to the outer planets, and *Cassini* to Saturn.

Descent Probe

Galileo's descent probe had a mass of 339 kilograms (750 pounds). It was slowed and protected by a deceleration module consisting of an aeroshell and an aft cover designed to block heat generated by friction during atmospheric entry. Inside the aeroshells were a descent module and its 2.5-meter (8-foot) parachute. The descent module carried a radio transmitter and seven scientific instruments. These were devices to measure temperature, pressure and deceleration, atmospheric composition, clouds, particles, and light and radio emissions from lightning and energetic particles in Jupiter's radiation belts.

Science Results

Among the Galileo mission's key science findings since reaching Jupiter are the following:

- The descent probe measured atmospheric elements and found that their relative abundances were different than on the Sun, indicating Jupiter's evolution after the planet formed out of the solar nebula.
- Jupiter's atmosphere has numerous, large thunderstorms concentrated in specific zones above and below the equator, where winds are highly turbulent. Although individual lightning strokes appear less frequently than on Earth, they are up to 1,000 times more powerful than terrestrial lightning. Clouds of ammonia, the first ever detected in a planet's atmosphere, form only from material brought up from the lower regions of the atmosphere, that is in "fresh" clouds. It is still a mystery why only fresh clouds, covering about 1 percent of Jupiter, show the ammonia signature while the remaining clouds, which probably also contain ammonia, lose their sharp spectral signature.
- Evidence supports a theory that liquid oceans exist under Europa's icy surface. There are places where recognizable features that once were whole have been separated from each other by new, smooth ice. These areas indicate that when the older features were separated, they floated for a time in liquid water, much as icebergs float in Earth's polar regions. Scientists believe the water later froze solid, creating "rafts" of ice, similar to some seen on a smaller scale at Earth's polar regions. There are also indications of volcanic ice flows, with liquid water flowing across the surface of Europa. These discoveries are particularly intriguing, since liquid water is a key ingredient in the process that may lead to the formation of life. Europa is criss-crossed by faults and ridges, with regions where large pieces of crust have separated and shifted. Faults are breaks in the moon's outer crust where the crust on either side has shifted. Ridges indicate that sections of the crust have also moved.
- Galileo magnetic data provide evidence not only that Europa still has a liquid salt water layer under its ice, but that such layers also exist, farther below the surface, on Ganymede and Callisto. Magnetometer readings show that magnetic fields around those moons vary in ways indicating the presence of electrically conducting layers – possibly oceans – under the surface.
- Europa, Io and Ganymede all have metallic cores. Processes on these three inner moons have permitted denser elements to separate out and sink to the moons' respective centers. On the other hand, the composition of the more distant moon Callisto is fairly uniform throughout, indicating it did not follow the same evolutionary path as the other three moons.
- Europa, Ganymede and Callisto each provide evidence of having a surface-bound exosphere, the term for a thin atmospheric layer composed of ions – electrically charged gases, and neutral gases surrounding the moon, and loosely bound to the surface. The presence of these gases proves the efficacy of a process previously well-known only in the laboratory called "sputtering." It refers to collisions of charged particles from the magnetosphere with the icy surface that release water vapor and other molecules.
- Ganymede generates a magnetic field, just as Earth does. In fact, Ganymede is the first moon of any planet known to possess an intrinsic magnetic field, though other of Jupiter's moons have secondary magnetic fields induced by the strong magnetic field of Jupiter itself. Ganymede's magnetosphere – the small magnetic "bubble" created by its field within the surrounding, more powerful, magnetic field of Jupiter – is actually somewhat larger than Mercury's, a planet of similar size. There even appears to be trapped radiation in a miniature radiation belt similar to Earth's Van Allen radiation belts on magnetic field lines close to Ganymede. The discovery of a magnetic field within Ganymede challenges theoretical models of how planetary magnetic fields arose on Earth and other celestial bodies.
- Galileo made observations of satellite surfaces that revealed, among many phenomena, the following: patterns of long, looping cracks that extend for hundreds of miles over the surface of Europa, caused by successive warming and fissuring of the surface; a strikeslip fault on Europa as long as the California segment of the San Andreas fault; a surface on Ganymede formed by high tectonic activity, with faulting and fracturing; evidence of extensive though still mysterious erosion of Callisto's primordial surface that creates a smoother appearance for some of the craters found there.
- Io's extensive volcanic activity may be 100 times greater than that found on Earth. It is continually modifying its surface. Many changes have been recorded since Voyager's visit in 1979, and other major changes have been seen during the course of the Galileo mission. During one four-month interval, an area the size of the state of Arizona was blanketed by volcanic debris thrown out of the volcano Pillan. Imaging and spectral analyses have shown that most of the eruptions on Io must be composed of liquid silicate rock, which contain silicon-oxygen compounds. The temperatures of these lavas are too high for other materials, such as sulfur, and in fact are even significantly

hotter than most eruptions on Earth today. The composition of these hot lavas may be more similar to a type of volcanism that occurred on the Earth more than 3 billion years ago.

- *Galileo* characterized the complex plasma environment of Io that is derived from the volcanic activity. The spacecraft showed: that plasma (a fluid-like gas composed solely of charged particles) flowing in the donut-shaped torus of Io, flows in until it stalls in the polar regions of Io, and that the plasma performs in a similar fashion in the wake created as plasma from Jupiter flows around Io. The mission measured the intense currents generated by the interaction of the plasma with Io and its neutral particle cloud, discovered that energetic electrons bounce along the field lines that connect Io with the charged particle region of Jupiter's atmosphere. *Galileo* also identified the process through which material lost from Io's surroundings is transported outward from Jupiter through a mechanism previously only theorized, namely the interchange of structures known as "magnetic flux tubes" that carry low-density plasma in from large distances.
- Jupiter's ring system is formed by dust kicked up as interplanetary meteoroids smash into the planet's four small inner moons. The outermost ring is actually two rings, one embedded within the other.
- *Galileo* was the first spacecraft to dwell in a giant planet magnetosphere long enough to identify its global structure and to investigate its dynamics. Despite some superficial similarities of behavior, Jupiter exhibits dynamics that differ from those familiar from Earth's magnetosphere. Unlike Earth, where the aurora are generated as Earth and the Sun interact, it has become clear that Jupiter's main auroral oval links to those regions of the magnetosphere (the middle-magnetosphere) dominated by the powerful effects of Jupiter's rotation. Jupiter exhibits strong dawn-to-dusk asymmetries in the flow of plasma that are not found in Earth's magnetosphere. The sub-storm process, a terrestrial phenomena that is a result of the interaction of Earth and the Sun, may occur in Jupiter's magnetosphere, with consequences that differ greatly from those familiar at Earth.

Orbiter Scientific Experiments

The following is a list of the science instruments on each part of the spacecraft, with the name of the principal investigator responsible for the instrument and notes on the experiment's main object of study.

Remote sensing instruments on non-spinning section:

- Camera – Dr. Michael Belton, National Optical Astronomy Observatories. Galilean satellites, high-resolution, atmospheric small-scale dynamics.
- Near-infrared mapping spectrometer – Dr. Robert Carlson, Jet Propulsion Laboratory. Surface, atmospheric composition thermal mapping.
- Photopolarimeter-radiometer – Dr. James Hansen, Goddard Institute for Space Studies. Atmospheric particles, thermal / reflected radiation.
- Ultraviolet spectrometer / extreme ultraviolet explorer – Dr. Ian Stewart, University of Colorado. Atmospheric gases, aerosols.

Instruments studying magnetic fields and charged particles, located on spinning section:

- Magnetometer – Dr. Margaret Kivelson, University of California, Los Angeles. Strength and fluctuations of magnetic fields.
- Energetic particle detector – Dr. Donald Williams, Johns Hopkins Applied Physics Laboratory. Distribution of electrons, protons and ions of high energy.
- Plasma investigation – Dr. Lou Frank, University of Iowa. Composition, energy, distribution of ions and electrons of medium and low energy.
- Plasma wave subsystem – Dr. Donald Gurnett, University of Iowa. Electromagnetic waves and wave-particle interactions.
- Dust-detection subsystem – Dr. Harald Krueger, Max Planck Institut fur Kernphysik. Mass, velocity, charge of particles smaller than a micrometer in size.

Engineering Experiment:

- Heavy ion counter – Dr. Edward Stone, California Institute of Technology. Spacecraft's charged-particle environment. Energy distribution of heavy ions.

Radio Science:

- Celestial mechanics – Dr. John Anderson, Jet Propulsion Laboratory. Masses and internal structures of bodies from spacecraft tracking.
- Propagation – Dr. H. Taylor Howard, Stanford University. Size and atmospheric structure of Jupiter's moons from radio propagation.

Scientific Experiments on Descent Probe

- Atmospheric structure – Dr. Alvin Seiff, San Jose State University Foundation. Temperature, pressure, density, molecular weight profiles.

- Neutral mass spectrometer – Dr. Hasso Niemann, NASA Goddard Space Flight Center. Chemical composition.
- Helium abundance – Dr. Ulf von Zahn, Institut fur Atmospharenphysik, Universitat Rostock. Helium / hydrogen ratio.
- Nephelometer – Dr. Boris Ragent, San Jose State University Foundation. Clouds and particles.
- Net flux radiometer – Dr. Larry Sromovsky, University of Wisconsin. Thermal / solar energy profiles.
- Lightning and radio emissions / energetic particles – Dr. Louis Lanzerotti, Bell Laboratories, and Dr. Klaus Rinnert, Max Planck-Institut fur Aeronomie; Harald Fischer, Institut fur Reine und Angewandte Kernphysik, Universitat Kiel. Lightning detection, energetic particles.
- Doppler wind experiment – Dr. David Atkinson, University of Idaho. Measure winds, learn their energy source.

Ground System

Galileo communicates with Earth via NASA's Deep Space Network, a worldwide system of large antenna complexes with receivers and transmitters located in Australia, Spain and California's Mojave Desert, linked to a network control center at the Jet Propulsion Laboratory in Pasadena, California. Through this network, the spacecraft receives commands, sends science and engineering data, and is tracked by doppler and ranging measurements.

Doppler measurements detect changes in the frequency of the spacecraft's radio signal that reveal how fast the spacecraft is moving toward or away from Earth. In ranging measurements, radio signals transmitted from Earth are marked with a code which is returned by the spacecraft, enabling ground controllers to keep track of the distance to the spacecraft.

Management

The Galileo project is managed for NASA's Office of Space Science, Washington, D.C., by the Jet Propulsion Laboratory, Pasadena, a division of the California Institute of Technology. JPL designed and built the *Galileo* orbiter, and operates the mission.

At NASA Headquarters, Dr. Barry Geldzahler is Galileo program manager and Dr. Denis Bogan is program scientist.

At JPL, the position of project manager has been held successively by John Casani, Richard Spehalski, Bill O'Neil, Bob Mitchell, Jim Erickson, Dr. Eilene Theilig and, currently, Dr. Claudia J. Alexander. Dr. Torrence V. Johnson is project scientist.

NASA's Ames Research Center, Moffett Field, California, managed the descent probe, which was built by Hughes Aircraft Co., El Segundo, California. The position of probe manager was held successively by Joel Sperans, Benny Chinn and Marcie Smith. The probe scientist is Dr. Richard E. Young.

Galileo's experiments are being carried out by more than 100 scientists from the United States, Great Britain, Germany, France, Canada and Sweden.

The Galileo Messenger

Issue 34   June 1994

Ida's Moon Discovered

CONTENTS

About 14 minutes before closest approach to Ida, this image was taken by Galileo's charge coupled device (CCD) camera, illustrating the 37 to 1 size difference between Ida and its moon (far right). Although the satellite appears to be "next" to Ida, it is actually slightly in the foreground. This image, along with the Near Infrared Mapping Spectrometer data, helped scientists triangulate the bodies to determine that the satellite is about 100 kilometers from the center of Ida.

Discovery of Ida's Moon Indicates Possible "Families" of Asteroids

Although *Galileo* flew by Asteroid Ida last August 28, some of the images it took are just now being transmitted and analyzed. That is why scientists were recently surprised to discover that Ida is not alone in space, but has a moon in orbit around it.

The discovery was first made by Ann Harch of the *Galileo* camera team, who noticed a bright object near Ida on some new images that were processed on February 17. The team considered and eliminated the possibility that the object was a planet, star, or something other than a moon. A few days later, the Near Infrared Mapping Spectrometer (NIMS) team also noticed some odd data while analyzing Ida's mineral content. They compared notes with the camera team and realized they had, indeed, found a moon.

Using images taken by the two *Galileo* instruments and comparing sighting angles at different times, the scientists determined the as-yet-unnamed moon is located about 100 kilometers from Ida's center. NIMS data also indicate that the rocks and soil on the surface of the tiny moon (only about 1.5 kilometers long) have roughly equal mixtures of olivine, orthopyroxene, and clinopyroxene, while Ida's surface is predominately olivine with a bit of orthopyroxene.

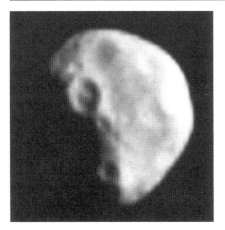

This close up of Ida's 1.5-kilometer-wide moon is the most detailed picture of the recently discovered natural satellite of Asteroid 243 Ida taken by the Galileo Solid State Imaging camera during its encounter with the asteroid on August 28, 1993.

The two are about the same temperature – 200° K. More data are needed to determine the characteristics of the moon's orbit, which in turn will help to calculate Ida's density.

Shortly after the discovery, *Galileo* scientists eliminated the idea that the moon is a passing body caught in Ida's gravity. They also doubt that the moon is a piece of Ida knocked loose by a smaller projectile, especially since their bulk compositions differ slightly. Instead, scientists are theorizing that the two are siblings of a "family" of asteroids formed hundreds of millions of years ago when a larger, 100-kilometer-wide asteroid was shattered in a great collision. Instead of fragments shooting straight out from the impact, the exploding asteroid may have produced jets of material carrying two or more objects out together. Those objects would then be captured, gravitationally, around each other.

Thus, a family of asteroids could have been created as a result of such an impact. Ida belongs to the Koronis family that travels in the main Asteroid Belt between Mars and Jupiter. Gaspra, the asteroid visited by *Galileo* in October 1991, is a member of the Flora family.

The discovery of Ida's moon "probably means they [asteroidal moons] are quite common," said astronomer Michael J.S. Belton, who leads the *Galileo* camera team. Many scientists suspect that a significant fraction of asteroids may have satellites. He noted that scientists believe they are on the verge of answering many questions about the existence and origin of asteroids and their satellites.

Galileo finished transmitting Ida data in June – including additional images of Ida's moon – and scientists expect to be able to determine more about the origin, composition, size, and orbit of Ida's moon, as well as the dynamics of collisions that played a central role in shaping the planets.

How Can an Asteroid Have a Moon?

Sir Isaac Newton first described how any two objects, no matter what size or how far apart they are from each other, exert an attractive force upon one another. Since gravity is, relatively, a very weak force (compared to electricity and magnetism, with which we have familiar experiences), we don't recognize that tables, baseballs, buildings, and even people all gravitationally attract each other. These forces are fantastically small, but it is interesting to note that the gravitational attraction between a parent holding a child is stronger than that of any one of the planets (except Earth!) on that child. In other words, even an object with little mass can exert a greater force than another much more massive object, if the smaller object is much closer than the larger one. In mathematical terms, the force of gravity falls off as the inverse of the distance squared. So, if the distance between two objects is doubled, the attractive force is one-quarter the value.

Similarly, Ida and its moon attract each other. Because they are far from other bodies compared to their mutual separation, they influence each other very strongly. Any two bodies under the right circumstances will orbit each other, even if they also orbit about a third body, such as the Sun. (The Sun, of course, is in orbit about the galaxy, which in turn is in orbit about other galaxies, and so on.) Bodies actually orbit about their mutual center of mass, called the barycenter, which is located along a line connecting their centers, and at a distance from either center inversely proportional to the mass of that object. Calculations have shown that asteroids like Ida can have satellites in stable orbits out to about 100 to 200 times their radius, so Ida could have satellites out to several thousand kilometers. However, only the one moon, 1993 (243) 1, has been seen as yet.

— *Jan Ludwinshi, Mission Design Team Chief*

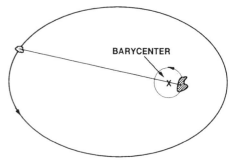

Any two objects, even if small and distant, will orbit one another if they are far from other objects.

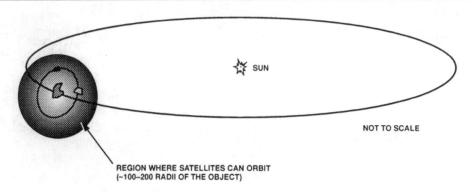

REGION WHERE SATELLITES CAN ORBIT
(~100-200 RADII OF THE OBJECT)

If a planet or asteroid is orbiting the Sun, the planet or asteroid can have a satellite if the satellite is close enough that the planet's or asteroid's gravity influences it more than the Sun's gravity.

Name That Moon!

There is no precedent for naming an asteroid's moon. Responsibility for naming Solar System bodies lies with the International Astronomical Union's (IAU) Working Group on Planetary System Nomenclature.

The Galileo Project is now soliciting names for Ida's moon, which currently carries the temporary designation 1993 (243) 1: 1993 is the year Galileo photographed it; 243 is the numerical designation of Ida; and the 1 notes that it is the first satellite discovered around Ida. The name chosen for Ida's satellite should relate in some way to Ida, either through mythology (Ida was a nymph who cared for the infant Jupiter) or similarity in name (Lupino has already been suggested; remember actress Ida Lupino?).

All suggested names and accompanying rationale for the choice must be received by the Galileo Project by July 31, 1994. The Project (through the Near Infrared Mapping Spectrometer and Solid State Imaging teams) will then make a recommendation to the IAU in August. As the discoverers of the satellite, we trust that the IAU will give our recommendation special consideration.

Please send your suggested names to:

Name Ida's Satellite
Project Galileo
Mail Stop 264-419
Jet Propulsion Laboratory
4800 Oak Grove Drive
Pasadena, CA 91109

We appreciate and will seriously consider all suggestions.

Galileo Fact Sheet – Jupiter 95

CONTENTS

Mission Summary

Galileo is now in the final stretch of its six-year trip to Jupiter. It consists of two parts: a deep space craft to orbit the planet for a two-year study of the entire Jovian system, and a probe to enter and study the atmosphere's structure and properties. Orbiter and probe separated July 13; both will reach the planet December 7, 1995.

Galileo's study of the Jovian atmosphere, its giant magnetosphere, and its four major satellites (all different and three of them larger than Earth's Moon) will help scientists understand the solar system's history and evolution.

On December 7, the *Galileo* orbiter will fly close by Jupiter's volcanic satellite Io, then receive probe science data as the probe descends by parachute through the clouds. Finally, the orbiter will begin a two year series of orbits around Jupiter, flying close by Ganymede (4), Callisto (3), and Europa (3). There will be one flyby per orbit, for a total of ten separate close encounters. During the two year tour, *Galileo* will also study Jupiter's rings, monitor Io's volcanoes, and study four of its inner minor satellites.

Launched October 18, 1989, by the space shuttle *Atlantis* with an Inertial Upper Stage booster, the Galileo mission used the gravity fields of Venus and Earth to accelerate the spacecraft enough to reach Jupiter. It also flew by two asteroids, making the first close observations of such bodies and discovering the first asteroid moon. Then, in July 1994, it observed the comet Shoemaker-Levy 9 fragments impacting Jupiter's night side.

Flight Highlights

The *Galileo* orbiter is a dual-spin spacecraft whose main body rotates at about 3 rpm. This portion of the vehicle carries six fields and particles sensors, thrusters, communications, and many other engineering subsystems. The lower part can be held in fixed orientation or rotated; it carries the camera and three other remote sensing instruments, the probe relay antenna, and other electronics.

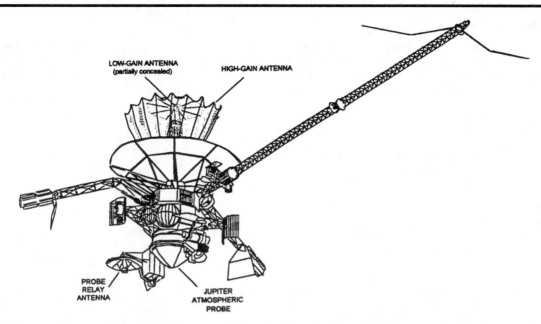

LOW-GAIN ANTENNA
(partially concealed)

HIGH-GAIN ANTENNA

PROBE
RELAY
ANTENNA

JUPITER
ATMOSPHERIC
PROBE

Galileo's atmospheric probe has a deceleration module with a heat shield to slow and protect the descent module and its parachute during entry into the atmosphere.

Galileo's Orbital Tour

I = Io
E = EUROPA
G = GANYMEDE
C = CALLISTO

SUN

Galileo

MILLION KILOMETERS

GALILEO'S JUPITER ORBITS, 1995 - 97

The descent module has six scientific instruments and a radio relay transmitter that will send data up to the orbiter during the 75-minute descent mission. It will measure atmospheric temperature, pressure, composition, and structure.

The orbiter was designed to transmit its scientific data at a very high rate from Jupiter, using a deployable, umbrella-like antenna. The antenna is only partly deployed, however, and the Jupiter data will be transmitted, like those from *Galileo*'s previous observations, over a smaller antenna at a lower data rate. Modifications to the ground receiving systems of NASA's Deep Space Network (DSN) on Earth and to ground and spacecraft software will permit *Galileo* to achieve most of its scientific objectives at Jupiter.

Management

The Galileo mission was developed and is managed by NASA's Jet Propulsion Laboratory (JPL). NASA's Ames Research Center manages the probe mission, and Hughes Aircraft Company built the probe. The orbiter's retropropulsion module and some scientific instruments were supplied by Germany. The JPL-managed DSN supports the Galileo mission with its stations in Australia, California, and Spain.

JPL Public Information Office 818-354-5011
July 1995

Major Mission Characteristics

Launch	10/18/89
Venus Flyby	02/10/90
Earth Flyby 1	12/08/90
Gaspra Asteroid Flyby	10/29/91
Earth Flyby 2	12/08/92
Ida Asteroid Flyby	08/28/93
Comet Shoemaker-Levy Observations	07/94
Probe Release	07/13/95
Jupiter Arrival, Io Flyby, Probe Descent / Relay and Orbit Insertion	12/7 - 8/95
Orbital Tour / Playbacks (includes 10 satellite flybys)	5/96 - 12/7/97

Spacecraft Characteristics

Category	Orbiter	Probe
Mass, kilograms (pounds)	2,223 (4,980)	339 (746)
Rocket propellant (usable)	925 kg (2,035 lb)	—
Height, meters (feet)	6.15 (20.5)	0.86 (3.1)
Science instruments	10*	5*
Instrument mass	118 kg (260lb)	30 kg (66 lb)

* Two radio-science experiments use spacecraft radio and DSN stations as their instrument.

NATIONAL AERONAUTICS AND SPACE ADMINISTRATION

Galileo Jupiter Arrival
Press Kit

December 1995

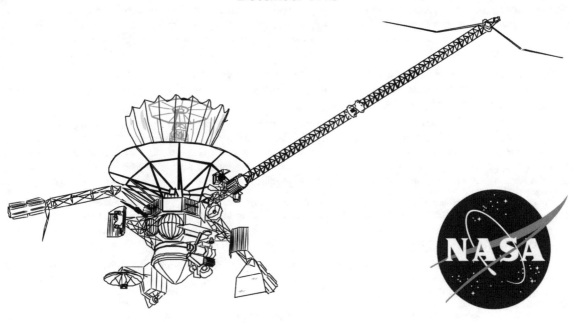

Contacts

Douglas Isbell, Headquarters, Washington, DC	Policy / Program Management	(202) 358-1753
Franklin O'Donnell, Jet Propulsion Laboratory, Pasadena, CA	Galileo Mission	(818) 354-5011
Ann Hutchison, Ames Research Center, Mountain View, CA	Atmospheric Descent Probe	(415) 604-4968

CONTENTS

RELEASE: 95-xxx

Galileo's Mission at Jupiter Poised to Begin

Reaching its ultimate destination after an already eventful space journey of more than six years and 2.3 billion miles, NASA's *Galileo* mission will arrive at the giant planet Jupiter on December 7, 1995.

The jam-packed arrival day calls for the *Galileo* spacecraft and its recently separated atmospheric probe to carry out a detailed choreography of data gathering and critical engineering events almost 20 years in the making. By day's end, *Galileo* should be the first spacecraft to enter orbit around one of the solar system's giant outer planets, and the first to send an instrumented probe directly into one of their atmospheres. The *Galileo* orbiter spacecraft then will begin at least two years of close-up observations of Jupiter, its moons, its faint rings and its powerful radiation, magnetic field and dust environment.

Galileo's scientific instruments represent the most capable payload of experiments ever sent to another planet. The data they will return promises to revolutionize our understanding of the Jovian system and reveal important clues about the formation and evolution of the solar system.

Jupiter is 318 times more massive and 1,400 times more voluminous than Earth, but only one-fourth as dense, since it is composed primarily of hydrogen (89 percent) and helium (10 percent). The fifth planet from the Sun is known primarily for the banded appearance of its upper atmosphere and its centuries-old Great Red Spot, a massive, hurricane-like storm as big as three Earths. Jupiter generates the biggest and most powerful planetary magnetic field, and it radiates more heat from internal sources than it receives from the Sun.

"In many ways, Jupiter is like a miniature solar system in itself," says Dr. Wesley T. Huntress, Associate Administrator for Space Science at NASA Headquarters, Washington, DC. "Within Jupiter's constellation of diverse moons, its intense magnetic field, and its swarms of dust and charged particles, the *Galileo* mission should uncover new clues about how the Sun and the planets formed, and about how they continue to interact and evolve."

The 2,223-kilogram (2½-ton) *Galileo* orbiter spacecraft carries 10 scientific instruments; the 339-kilogram (746-pound) probe carries six more instruments. The spacecraft radio link to Earth and the probe-to-orbiter radio link serve as instruments for additional scientific investigations.

The chain of key mission events for *Galileo* on Jupiter arrival day begins with a close flyby of the moon Io by the *Galileo* orbiter at a distance of just 600 miles (1,000 kilometers). The probe has been traveling a separate path toward atmospheric entry since it was released by the *Galileo* spacecraft on July 13.

Io's gravity will change *Galileo*'s direction to help it go into orbit around Jupiter. Due to the high radiation in this interior region of the Jovian system, this is the closest that *Galileo* is planned to come to this moon, although the spacecraft will observe Io during many subsequent orbits.

Four hours later, the orbiter will link up by radio with the previously released *Galileo* probe as it floats via parachute downward through the top level of Jupiter's atmosphere. By that time, the conical probe will have slammed into the upper fringe of Jupiter's atmosphere at a top speed of 106,000 mph and endured deceleration forces as high as 230 times Earth's gravity. Dropping downward on its eight-foot diameter parachute, the probe will make the first direct measurements of Jupiter's atmosphere and clouds, and it may encounter lightning or even water rain as it descends more than 125 miles (200 kilometers) from the top of Jupiter's clouds.

The *Galileo* orbiter will record the measurements radioed from the probe for up to 75 minutes, before finally turning away to prepare for a crucial 49-minute-long burn of its main rocket engine that will insert the spacecraft into Jovian orbit. The probe below is expected to succumb a few hours later to the increasingly intense heat it will find deep below the clouds.

The orbiter will then begin its tour of at least 11 orbits of the Jovian system, including 10 close encounters with three of the four Galilean satellites (four with Ganymede and three each with Callisto and Europa), and observing Io's erupting volcanoes. In mid-March, *Galileo* will fire its main rocket engine for one last major burn to put itself into an orbit away from the most intense Jovian radiation environment.

The Galileo mission had originally been designed for a direct flight to Jupiter of about two-and-a-half years. Changes in the launch system after the Space Shuttle *Challenger* accident, including replacement of the Centaur upper-stage rocket with the Inertial Upper Stage (IUS), precluded this direct trajectory. *Galileo* engineers designed a new interplanetary flight path using several gravity-assist swingbys (once past Venus and twice around Earth) called the Venus-Earth-Earth-Gravity-Assist or VEEGA trajectory.

Galileo was launched aboard Space Shuttle *Atlantis* and an IUS on October 18, 1989. In addition to its Earth and Venus flybys, *Galileo* became the first spacecraft ever to fly closely by two asteroids, Gaspra and Ida. During the second asteroid encounter, two of *Galileo*'s 10 science instruments discovered a small moon – later named Dactyl – orbiting around Ida, the first time such an object has been confirmed. *Galileo*'s instruments also performed the only direct observations of the impact of the fragments of Comet Shoemaker-Levy 9 with Jupiter in July 1994, providing key insights into the early stages of the impact evolution.

Communications to and from *Galileo* are conducted through NASA's Deep Space Network, using tracking stations in California, Spain and Australia. A combination of new, specially developed software for *Galileo*'s onboard computer and improvements to ground-based signal receiving hardware in the Deep Space Network will enable the mission to accomplish at least 70 percent of its original science goals using only its small, low-gain antenna, despite the failure of its high-gain antenna to unfurl properly in April 1991. The data return will include an average of two to three images per day once the spacecraft begins transmitting imaging data to Earth in July 1996.

The total cost of the Galileo mission, from the start of planning through the end of mission in December 1997, is $1.354 billion, including $892 million in spacecraft development costs. *Galileo*'s arrival in the Jovian system represents the culmination of a project that began formally in 1977, and the realization of a dream of planetary scientists since the earliest days of the field. "The *Pioneer* and *Voyager* spacecraft that flew by Jupiter so quickly in the 1970s stunned us with their pictures of rings, active volcanoes on the moon Io and other unexpected findings," Huntress said. "Right now, we can still only imagine the discoveries that will flow from *Galileo* as it travels for months in this most unusual and unearthly environment." NASA's Jet Propulsion Laboratory, Pasadena, California, built the *Galileo* orbiter spacecraft and manages the overall mission. *Galileo*'s atmospheric probe is managed by NASA's Ames Research Center, Mountain View, California.

(End of General Release)

Media Services Information

NASA Television Transmission

NASA Television is available through the Sapient 2 satellite on transponder 5, channel 9, 69 degrees west longitude, frequency 3880 MHz, audio subcarrier 6.8 MHz, horizontal polarization. The schedule for television transmissions on Jupiter arrival day, December 7, 1995, will be available from the Jet Propulsion Laboratory, Pasadena, California; Ames Research Center, Mountain View, California; Kennedy Space Center, Florida; and NASA Headquarters, Washington, DC.

Status Reports

Status reports on mission activities before, during and after Jupiter arrival will be issued by the Jet Propulsion Laboratory's Public Information Office. They may be accessed on line as noted below.

Briefings

A pre-arrival press conference to outline plans for Jupiter orbit insertion will be held at the Jet Propulsion Laboratory on November 9, 1995. Press conferences are scheduled at 10 a.m. and at 6:45 p.m. PST, December 7, 1995, during Jupiter arrival. A press conference on quick-look science findings from the atmospheric descent probe is planned from the Ames Research Center on December 19, 1995.

Internet Information

Extensive information on the Galileo mission, including an electronic copy of this press kit, press releases, fact sheets, status reports and images, is available from the Jet Propulsion Laboratory's World Wide Web home page at http://www.jpl.nasa.gov. In addition to offering such public affairs materials, the JPL home page links to the Galileo Project's Web home page, http://www.jpl.nasa.gov/galileo, which offers additional information on the mission.

The general JPL site may also be accessed via Internet using anonymous file transfer protocol (FTP) at the address ftp.jpl.nasa.gov. Users should log on with the user name "anonymous" and enter their E-mail address as the password. For users without Internet access, the site may additionally be accessed by modem at (818) 354-1333.

Galileo Quick Look Facts

Times below are all Earth-received times, allowing for the time it takes the transmissions from the spacecraft to reach Earth (approximately 52 minutes).

Launch and deployment: STS-34 Atlantis and IUS	October 18, 1989
Venus flyby (about 16,000 kilometers / 9,500 miles)	February 10, 1990
Earth 1 flyby (about 1,000 kilometers / 620 miles)	December 8, 1990
Asteroid Gaspra flyby (about 1,600 kilometers / 950 miles)	October 29, 1991
Earth 2 flyby (about 300 kilometers / 190 miles)	December 8, 1992
Asteroid Ida flyby (about 2,400 kilometers / 1,400 miles)	August 28, 1993
Probe release	July 12, 1995 11:07 p.m. PDT
Jupiter arrival (probe and orbiter)	December 7, 1995
Io flyby (about 1,000 kilometers / 6,200 miles)	December 7, 1995 10:38 a.m. PST
Probe atmospheric entry and relay	December 7, 1995 2:56 p.m. PST
Probe mission duration:	40-75 minutes
Jupiter Orbit Insertion (JOI)	December 7, 1995 5:19 p.m. PST
Orbital tour of Galilean satellites	December 1995-November 1997

Orbiter satellite encounters:

- Io: December 7, 1995 (no imaging or spectral data will be taken at this opportunity)
- Ganymede: July 4 and September 6, 1996; April 5 and May 7, 1997
- Callisto: November 4, 1996; June 25 and September 17, 1997
- Europa: December 19, 1996; February 20 and November 26, 1997

Orbiter End of Mission	December 7, 1997.6
Mission Cost	$1.35 billion plus international contribution estimated at $110 million.

Galileo Orbiter and Jupiter Atmospheric Probe

	Orbiter	**Probe**
Mass, kilograms (pounds)	2,223 (4,890)	339 (746)
Usable propellant mass	925 (2,035)	—
Height	6.15 meters (20.5 feet)	86 cm (34 inch)
Instrument payload	12 experiments	7 experiments
Payload mass, kilograms (pounds)	118 (260)	30 (66)
Electric power	Radioisotope thermoelectric generators (570-470 watts)	Lithium-sulfur battery 730 watt-hours

Shuttle *Atlantis* (STS-34) Crew (Johnson Space Center)
Donald E. Williams, Commander
Michael J. McCulley, Pilot
Shannon W. Lucid, Mission Specialist
Franklin R. Chang-Diaz, Mission Specialist
Ellen S. Baker, Mission Specialist

Galileo *Management*
Galileo Project (Jet Propulsion Laboratory)
William J. O'Neil, Project Manager
Neal E. Ausman, Jr., Mission Director
Matthew R. Landano, Deputy Mission Director
Dr. Torrence V. Johnson, Project Scientist

Atmospheric Probe (Ames Research Center)
Marcie Smith, Probe Manager
Dr. Richard E. Young, Probe Scientist

Galileo Program (NASA Headquarters)
Donald Ketterer, Program Manager
Dr. Jay Bergstralh, Program Scientist
Dr. Wesley Huntress Jr., Associate Administrator for Space Science

Galileo Mission Overview
(1995 to end of mission)

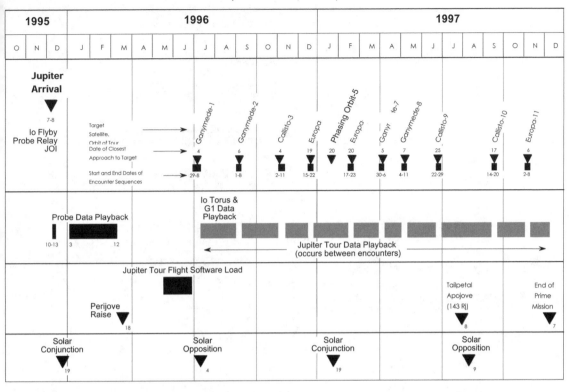

Atmospheric Probe Mission

On July 13, 1995, the *Galileo* spacecraft spun up to 10 rpm and aimed its cone-shaped Jupiter atmospheric entry probe toward its Jupiter entry point 51 million miles (82 million kilometers) away. Guillotine-like cable cutters sliced though umbilicals connecting the two, and the probe was released from the main spacecraft for its solo flight to Jupiter.

On December 7, 1995, the probe's descent into Jupiter will provide the first-ever on-the-spot measurements of Jupiter or any other outer planet. Instruments on board will identify the chemical components of Jupiter's atmosphere and their proportions, and search for clues to Jupiter's history and the origin of the solar system.

Six hours before entry the command unit signals the probe to "wake up," and three hours later instruments begin collecting data on lightning, radio emissions, and charged particles.

The probe will strike the atmosphere at an angle of only 8 degrees to the horizon – steep enough so it won't skip out again into space, yet shallow enough to survive the heat and jolting deceleration of entry. The probe will enter the equatorial zone traveling the same direction as the planet's rotation.

During its entry, the incandescent gas cap ahead of the probe would be as bright as the Sun to an observer and twice as hot (15,555 degrees Celsius or 28,000 degrees Fahrenheit) as the solar surface. With the exception of nuclear radiation, entry will be like flying through a nuclear fireball. The probe will be subjected to wrenching forces as it decelerates from 106,000 mph to 100 mph (about 170,000 to about 160 km/hr) in just two minutes – a force estimated at up to 230 times Earth's gravity.

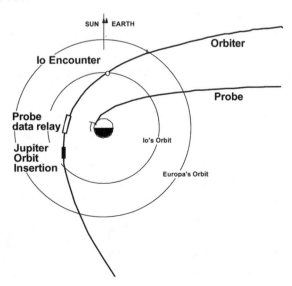

Once it survives the hazards of its blazing entry, the probe will be operating in Jupiter's primarily hydrogen and helium atmosphere. For part of the descent through Jupiter's three main cloud layers, the probe will be immersed in gases at or below room temperature. It may encounter hurricane winds of up to 200 mph (about 320 km/hr) and lightning and heavy rain at the base of the water clouds believed to exist within the planet's atmosphere.

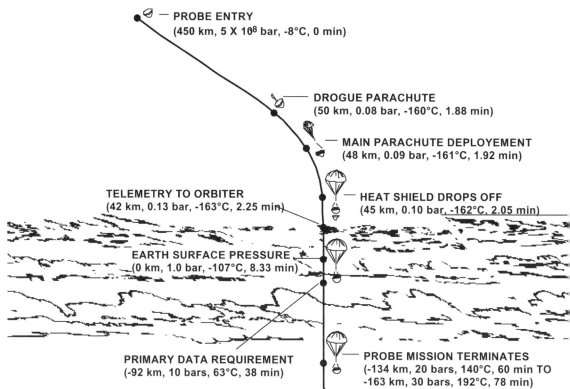

PROBE ENTRY
(450 km, 5 X 10^8 bar, -8°C, 0 min)

DROGUE PARACHUTE
(50 km, 0.08 bar, -160°C, 1.88 min)

MAIN PARACHUTE DEPLOYEMENT
(48 km, 0.09 bar, -161°C, 1.92 min)

TELEMETRY TO ORBITER
(42 km, 0.13 bar, -163°C, 2.25 min)

HEAT SHIELD DROPS OFF
(45 km, 0.10 bar, -162°C, 2.05 min)

EARTH SURFACE PRESSURE
(0 km, 1.0 bar, -107°C, 8.33 min)

PRIMARY DATA REQUIREMENT
(-92 km, 10 bars, 63°C, 38 min)

PROBE MISSION TERMINATES
(-134 km, 20 bars, 140°C, 60 min TO
-163 km, 30 bars, 192°C, 78 min)

The *Galileo* orbiter, about 214,000 kilometers (133,000 miles) above, will acquire the probe signal within 50 seconds. It will receive the probe's science data transmissions and store them both in the spacecraft tape recorder and in computer memory for later transmission to Earth.

The probe will pass through the white cirrus clouds of ammonia crystals on the highest cloud deck. Beneath this ammonia layer probably lie reddish-brown clouds of ammonium hydrosulfides. Once past this layer, the probe is expected to reach heavier water clouds. This lowest cloud layer may act as a buffer between what some scientists believe will be uniformly mixed regions below the clouds and the turbulent whirl of gases above. Eventually, the probe will pass below these clouds, where rising pressure and temperature will melt and vaporize the craft.

The probe's observations from within the clouds of Jupiter are uniquely valuable to the scientists, and every effort is being made to secure them. The entire probe data collection will be tape recorded for later playback. It also will be stored in a compressed and shortened version in onboard computer memory. The spacecraft computer will be able to store almost 70 minutes of the planned 75-minute probe observing time and capture the critical part of the atmospheric data. Playback of the first 40 minutes of probe data, stored in the spacecraft's computer memory, will occur in December. Because the Sun will be between Earth and Jupiter and causing radio interference with *Galileo* in late December, full playback of the probe data will be delayed until early 1996. The December playback is planned on a "best effort" basis due to the very small Sun-Earth-spacecraft angles and will be repeated twice more in January as the communication angles improve.

Orbiter Mission

Galileo's flyby of the Jovian satellite Io at 10:38 a.m. PST (Earth-received time) on December 7 provides the spacecraft a gravity assist to aid getting into its own orbit around Jupiter. After the atmospheric probe mission is completed, the *Galileo* orbiter will turn from receiving probe data to performing its own mission. It will carry out a 49-minute rocket firing that will slow the spacecraft to allow it to be captured into orbit around Jupiter. Three months later, another rocket firing will lift the spacecraft's orbit out of the high-radiation environment of Jupiter's charged particle belts, which could damage *Galileo*'s electronics.

Galileo's two-year orbital tour of the Jovian system is an elaborate square dance requiring the spacecraft to swing around one moon to reach the next. The first Ganymede encounter in July 1996 is the first of these satellite swingbys, and will shorten and change the shape of *Galileo*'s ensuing orbit. Each time the orbiter flies closely past one of the major inner moons, *Galileo*'s course will be changed due to the satellite's gravitational effects. Careful targeting allows each flyby to direct the spacecraft on to its next satellite encounter and the spacecraft's next orbit around Jupiter. *Galileo* will fly by Ganymede four times, Callisto three, and Europa three. Io gets only the one close pass, on arrival day, because *Galileo* cannot linger long in the hazardous radiation environment in which Io resides without damaging the spacecraft's electronics.

Galileo's Eleven Orbit Trek Around Jupiter

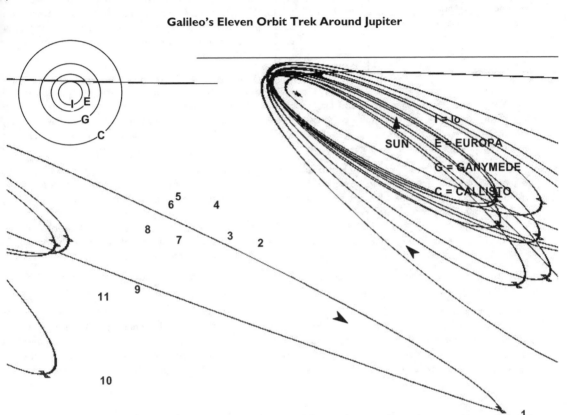

These satellite encounters will be at altitudes as close as 200 kilometers (125 miles) above the surfaces of the moons, and typically 100 to 300 times closer than the *Voyager* satellite flybys, to determine surface chemical composition, geological features and geophysical history. *Galileo*'s scanning instruments will scrutinize the surface and features of each. After a week or so of satellite observation, with its tape recorder full of data, the spacecraft will spend the next months in orbit playing out the information to Earth. Throughout the 23-month orbital tour, *Galileo* will continuously gather data on the Jovian magnetospheric and dust environment.

Early in *Galileo*'s first seven-month orbit of Jupiter, the complete set of data radioed by the atmospheric probe will be played back to Earth. Time will also be taken to radio new data processing software to the spacecraft to give *Galileo* new data compression and storage capabilities, and allow the spacecraft to return data at a low rate using the low-gain antenna.

Galileo's Orbital Tour, 1995-1997

The spacecraft's orbital tour consists of 11 different elliptical orbits around Jupiter. Each orbit (except one) includes a close flyby and gravity assist at Ganymede, Callisto or Europa, near the inner (Jupiter) end of the orbit. The outer end of the orbit will vary from 5 to almost 20 million kilometers (3.2 to more than 12 million miles). No close flyby is planned for Orbit 5, when *Galileo* is out of communication due to solar conjunction, when the Sun will be between Jupiter and Earth. Distant scientific encounters with additional satellites are scheduled for a number of orbits, and the spacecraft will observe Io at medium range on every orbit.

The Jovian System

Jupiter is the largest planet in the solar system. Its radius is 44,400 miles (71,500 kilometers), more than 11 times Earth's, and its mass is 318 times that of our planet. It is made mostly of light elements, principally hydrogen and

Orbit	Satellite Encounter	Date	Altitude, km (miles)	
1	Ganymede	July 4, 1996	500	(300)
2	Ganymede	September 6	259	(161)
3	Callisto	November 4	1,102	(685)
4	Europa	December 19	693	(431)
(5)	(Solar conjunction)		(no close flyby)	
6	Europa	February 20, 1997	587	(365)
7	Ganymede	April 5	3,056	(1,899)
8	Ganymede	May 7	1,580	(982)
9	Callisto	June 25	416	(258)
10	Callisto	September 17	524	(326)
11	Europa	November 6	1,124	(698)

helium. Its atmosphere and clouds are deep and dense, and a significant amount of energy is emitted from its interior. It has no solid surface. Its gases become hotter and denser with increasing depth.

Early Earth-based telescopic observations showed bands and spots in Jupiter's atmosphere; one storm system, the Red Spot, has been seen to persist over three centuries. The light and dark bands and some of the spots have disappeared and reappeared over periods of many years, and as the quality of Jupiter observation has improved so has the amount of variability seen in the clouds.

Atmospheric features were seen in greatly improved detail with the *Pioneer* and *Voyager* missions of the 1970's. The *Voyager* encounters in the spring and summer of 1979 allowed the observation of short-term variations in real time as Jupiter turned beneath the spacecraft's cameras. Astronomers using Earth-based infrared telescopes have recently studied the nature and vertical dynamics of deeper clouds, and the new Earth- and orbit-based telescopes observe Jupiter's atmospheric developments and climate changes, most notably during the Comet Shoemaker-Levy 9 impacts.

Sixteen Jovian satellites are known. The four largest, discovered by Galileo in 1610, are about the size of small planets, and were seen by *Voyager*'s experimenters to have the varied terrain of small worlds. The innermost of these, Io, has active sulfurous volcanoes, discovered by *Voyager 1* and further observed by *Voyager 2*, Earth-based infrared astronomy and the Hubble Space Telescope. Io and Europa are about the size and density of Earth's moon (3-4 times the density of water) and probably mostly rocky inside. Europa may also exhibit surface activity. Ganymede and Callisto, further out from Jupiter, are the size of Mercury but less than twice as dense as water; their interiors are probably about half ice and half rock, with mostly ice or frost surfaces which show distinct and interesting features.

Of the others, eight are in inclined, highly eccentric orbits far from the planet, and four (three discovered by the *Voyager* missions in 1979) are close to the planet. *Voyager* also discovered a thin ring system at Jupiter in 1979.

Jupiter has the strongest planetary magnetic field known; the resulting region of its influence, called the magnetosphere, is a huge teardrop-shaped bubble in the solar wind pointing away from the Sun. The inner part of the magnetically-constrained charged particle belt is doughnut-shaped, but farther out it flattens into a disk. The magnetic poles are offset and tilted relative to Jupiter's axis of rotation, so the field appears to wobble around with Jupiter's rotation (about every 10 hours), sweeping up and down across the inner satellites and making waves throughout the magnetosphere.

Jupiter's Satellites

Name	Discovery	Mean dist. to Jupiter, km	Period, days	Radius, km	Notes
Metis	Voyager, 1979	127,960	0.3	(20)*	
Adrastea	Voyager, 1979	128,980	0.3	12 x 8	
Amalthea	Barnard, 1892	181,300	0.5	135 x 75	
Thebe	Voyager, 1979	221,900	0.7	(50)	
Io	Galileo, 1610	421,660	1.8	1,815	density 3.57**; volcanic
Europa	Galileo, 1610	670,900	3.5	1,569	density 2.97**; icy crust
Ganymede	Galileo, 1610	1,070,000	7.2	2,631	density 1.94**; deep ice crust
Callisto	Galileo, 1610	1,883,000	16.7	2,400	density 1.86**; deep ice crust
Leda	Kowal, 1974	11,094,000	239	(8)	long, tilted elliptical orbit
Himalia	Perrine, 1904	11,480,000	250	(90)	in "family" with Leda
Lysithea	Nicholson, 1938	11,720,000	259	(20)	in Leda "family"
Elara	Perrine, 1905	11,737,000	260	(40)	in Leda "family"
Ananke	Nicholson, 1951	21,200,000	631	(15)	retrograde in long, highly tilted, elliptical orbit
Carme	Nicholson, 1938	22,600,000	692	(22)	in "family" with Ananke
Pasiphae	Melotte, 1908	23,500,000	735	(35)	in Ananke "family"
Sinope	Nicholson, 1914	23,700,000	758	(20)	in Ananke "family"

* Radius numbers in parentheses are uncertain by more than 10%
** Density is in grams per cubic centimeter (water's density is 1)

Jupiter's Rings

- Inner "Halo" ring, about 100,000 to 122,800 kilometers from Jupiter's center.
- "Main" ring, 122,800-129,200 kilometers from center.
- Outer "Gossamer" ring, 129,200 to about 214,200 kilometers.
- Note that satellites Metis, Adrastea and Amalthea orbit in the outer part of the ring region.
- The Jovian satellites are named for Greek and Roman gods. Names of new moons are conferred by the International Astronomical Union.

Interplanetary Cruise Science

Galileo has already returned a wealth of surprising new information from the "targets of opportunity" it has observed on the way to Jupiter. Two first-ever asteroid encounters yielded close-up images of the asteroids Gaspra and Ida, and the extraordinary discovery of a moon (later named Dactyl) orbiting Ida.

Lunar Science

In 1992, Galileo revisited the north pole of the Moon explored by early spacecraft, imaging the region for the first time in infrared color and providing new information about the distribution of minerals on the lunar surface. The spacecraft flew within 68,000 miles (110,000 kilometers) of the Moon on December 7, 1992, obtaining multispectral lunar images, calibrating Galileo's instruments by comparing their data to those of previous lunar missions, and getting additional baselines for comparing our Moon with the Jovian satellites Galileo will be exploring.

A major result of Galileo's first lunar flyby was the confirmation of the existence of a huge ancient impact basin in the southern part of the Moon's far side. The presence of this basin was inferred from Apollo data in the 1970s, but its extent had never been mapped before. Galileo imaged the Moon's north pole at several different wavelengths (including infrared), a feat never before accomplished. Scientists found evidence that the Moon has been more volcanically active than researchers previously thought.

The Near-Infrared Mapping Spectrometer (NIMS) imaged the polar region in 204 wavelengths, another first in lunar mapping. The spacecraft also collected spectral data for dark mantle deposits, areas of local explosive volcanic eruptions. These maria deposits are more compositionally diverse toward the near side of the Moon. Specifically, scientists discovered that titanium is present in low to intermediate amounts toward the far side, suggesting that the far side has a thicker crust. This type of spectral data also allows scientists to determine the sequence of meteoric impacts and the thickness of ancient lava flows.

In observing the features of the Imbrium impact basin on the near side of the Moon, the imaging team found the Moon to have been more volcanically active earlier than previously thought. They found hidden maria, or "cryptomaria," are overlain by other features visible only through special spectral bands. Nearly 4 billion years ago, the impact in the Imbrium basin threw out a tremendous amount of rock and debris that blanketed the Moon and caused erosion of the highland terrain. This blanketing and sculpture can be seen in Galileo's images of the north pole.

Gaspra

Nine months into its two-year Earth-to-Earth orbit, Galileo entered the asteroid belt, and on October 29, 1991, it accomplished the first-ever asteroid encounter. It passed about 1,600 kilometers (1,000 miles) from the stony asteroid Gaspra at a relative speed of about 8 kilometers per second (18,000 mph); scientists collected pictures of Gaspra and other data on its composition and physical properties. These revealed a cratered, complex, and irregular body about 19 by 12 by 11 kilometers (12 by 7.5 by 7 miles), with a thin covering of dirt-like regolith and a possible magnetic field.

Ida

On August 28, 1993, Galileo had a second asteroid encounter, this time with Ida, a larger, more distant body than Gaspra. There, Galileo made the discovery of the first moon of an asteroid. Ida is about 55 kilometers or 34 miles long; like Gaspra, it is very irregular in shape; it rotates every 4.6 hours around an offset axis. Apparently, like Gaspra, it may have a magnetic field.

The closest-approach distance was about 2,400 kilometers (1,500 miles), with a relative speed of nearly 12.6 kilometers per second or 28,000 mph.

Ida's satellite, later named Dactyl, was found in a camera frame and an infrared scan. The 1.5-kilometer satellite was estimated to be about 100 kilometers (60 miles) from the center of the asteroid.

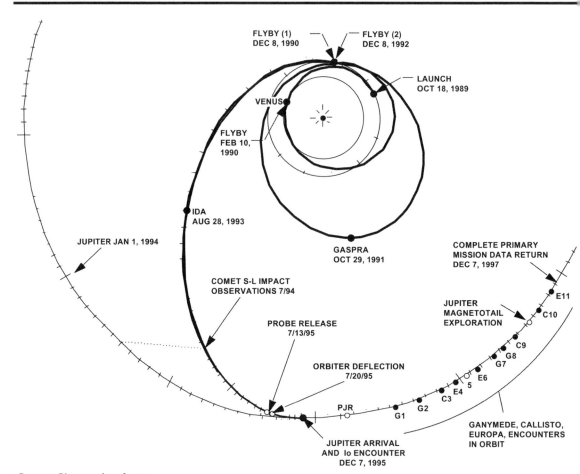

FLYBY (1)
DEC 8, 1990

FLYBY (2)
DEC 8, 1992

LAUNCH
OCT 18, 1989

VENUS

FLYBY
FEB 10,
1990

IDA
AUG 28, 1993

JUPITER JAN 1, 1994

GASPRA
OCT 29, 1991

COMPLETE PRIMARY
MISSION DATA RETURN
DEC 7, 1997

E11

COMET S-L IMPACT
OBSERVATIONS 7/94

JUPITER
MAGNETOTAIL
EXPLORATION

C10

C9

G8

PROBE RELEASE
7/13/95

G7

ORBITER DEFLECTION
7/20/95

E6

5

E4

C3

PJR

G2

G1

GANYMEDE, CALLISTO,
EUROPA, ENCOUNTERS
IN ORBIT

JUPITER ARRIVAL
AND Io ENCOUNTER
DEC 7, 1995

Comet Shoemaker-Levy

The discovery of Comet Shoemaker-Levy 9 in May 1993 provided an exciting new opportunity for *Galileo*'s science team.

The *Galileo* spacecraft, approaching Jupiter, was the only observation platform capable of making measurements in line of sight to the comet's impact area on Jupiter's far side. Although there was no additional funding available for this new target of opportunity, an observation program was planned for *Galileo*'s remote sensing instruments. All of *Galileo*'s observations had to be programmed in advance into the spacecraft computer, notwithstanding uncertainties in the predicted impact times. The data were stored on the spacecraft (one tape load plus some computer memory space) for playback at the 10 bits-per-second rate. Playback continued, with necessary interruptions for other activities, until late January 1995.

Galileo's imaging system used different methods to cover the time uncertainties (amounting to hours) of the impacts for different events. Repeated imaging, rather like a very slow motion picture, captured the very last impact (fragment W) which appeared to last 26 minutes. A smeared image, producing a streak representing the night-side impact fireball among smears showing Jupiter and some satellites, provided brightness histories of two events, the impact of fragments K and N. The photopolarimeter-radiometer detected three events. The infrared spectrometer detected two events providing critical information on the size, temperature and altitude of the impact fireball and the heating of the atmosphere by the impact "fallback." Galileo scientists have combined their data to produce interpretive histories of the 90-second impact events.

Interplanetary Dust

In the summer of 1995, *Galileo* found itself flying through the most intense interplanetary dust storm ever measured. It was the largest of several dust storms encountered by *Galileo* since December 1994, when the spacecraft was still almost 110 million miles (about 175 million kilometers) from Jupiter. The dust particles are apparently emanating from somewhere in the Jovian system and may be the product of volcanoes on Jupiter's moon, Io, or could be coming from Jupiter's faint ring system. Probably no larger than those found in cigarette smoke, the particles may also be leftover material from Comet Shoemaker-Levy's collision with Jupiter.

Scientists believe the particles are electrically charged and then accelerated by Jupiter's powerful magnetic field. Calculations indicate the dust is speeding through interplanetary space at between nearly 90,000 and 450,000 mph (40 and 200 kilometers per second), depending on particle size. Even at such high speeds, these tiny particles pose no danger to the *Galileo* spacecraft because they are so tiny.

Chances of understanding the nature of these dust storms are improving since, after the onset of the current storm, *Galileo* flight engineers commanded the spacecraft to collect and transmit dust data as often as three times a day. The normal collection rate had been twice per week. When *Galileo* enters Jupiter orbit, its first measurements should provide key information about the origins of this strange phenomena.

Galileo's New Telecommunications Strategy

Engineers and scientists involved with the Galileo mission to Jupiter have successfully devised several creative techniques to enable the spacecraft to achieve the majority of its scientific objectives despite the failure of its main communications antenna to open as commanded.

Upgrades to *Galileo*'s onboard computer software and its ground-based communications hardware have been developed and tested by JPL in response to what would have been a profound loss for the orbiter portion of the mission. (The spacecraft's Jupiter atmospheric probe mission can be executed without the new techniques, but the upgrades will enhance the orbiter's ability to reliably record and re-transmit the unique data to be collected by the probe as it descends on a parachute.) The new telecommunications strategy hinges on more effective use of the spacecraft's low-gain antenna, which is limited to a very low data rate compared to the main, high-gain antenna.

The switch to the low-gain antenna and its lower data rate means that far fewer data bits will be returned from Jupiter. However, new software on the spacecraft will increase *Galileo*'s ability to edit and compress the large quantity of data collected by the spacecraft and then transmit it to Earth in a shorthand form. On Earth, new technology is being used to greatly sharpen the hearing of the telecommunications equipment that will be receiving *Galileo*'s whisper of a signal from Jupiter. Together, these efforts should enable *Galileo* to fulfill at least 70 percent of its original scientific objectives.

The High-Gain Antenna

The 4.8-meter (16-foot)-wide, umbrella-like high-gain antenna is mounted at the top of the spacecraft. When unfurled, the antenna's hosiery-like wire mesh stretches over 18 umbrella ribs to form a large parabolic dish. *Galileo* was to have used this dish to radio its scientific data from Jupiter. This high-performance, X-band antenna was designed to transmit data back to Earth at rates of up to 134,400 bits of digital information per second (the equivalent of about one imaging frame each minute).

Galileo's original mission plan called for the high-gain antenna to open shortly after launch. For the Venus-Earth-Earth Gravity-Assist (VEEGA) trajectory mission, however, the heat-sensitive high-gain antenna had to be left closed and stowed behind a large sun shade to protect it during the spacecraft's passage through the inner solar system. During this portion of *Galileo*'s journey, two small, heat-tolerant low-gain antennas provided the spacecraft's link to Earth. One of these S-band antennas, mounted on a boom, was added to the spacecraft expressly to bolster *Galileo*'s telecommunications during the flight to Venus. The other primary low-gain antenna mounted to the top of the high-gain was destined to become the only means through which *Galileo* will be able to accomplish its mission.

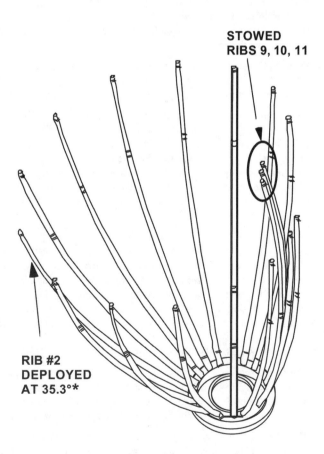

**STOWED
RIBS 9, 10, 11**

**RIB #2
DEPLOYED
AT 35.3°***

* AFTER HAMMERING RIB #2 NOW AT 43°

The Antenna Problem

On April 11, 1991, after *Galileo* had traveled far enough from the heat of the Sun, the spacecraft executed stored computer commands designed to unfurl the large high-gain antenna. But telemetry received minutes later showed that something went wrong. The motors had stalled and the antenna had only partially opened.

In a crash effort over the next several weeks, a team of more than 100 technical experts from JPL and industry analyzed *Galileo's* telemetry and conducted ground testing with an identical spare antenna. They deduced that the problem was most likely due to the sticking of a few antenna ribs, caused by friction between their standoff pins and sockets.

The excessive friction between the pins and sockets has been attributed to etching of the surfaces that occurred after the loss of a dry lubricant that had been bonded to the standoff pins during the antenna's manufacture in Florida. The antenna was originally shipped to JPL by truck in its own special shipping container. In December 1985, the antenna, again in its own shipping container, was sent by truck to NASA's Kennedy Space Center (KSC) in Florida to await launch. After the *Challenger* accident, *Galileo* and its antenna had to be shipped back to JPL in late 1986. Finally, they were reshipped to KSC for integration and launch in 1989. The loss of lubricant is believed to have occurred due to vibration the antenna experienced during those cross-country truck trips.

Extensive analysis has shown that, in any case, the problem existed at launch and went undetected; it is not related to sending the spacecraft on the VEEGA trajectory or the resulting delay in antenna deployment.

Attempts to Free the Antenna

While diagnosis of the problem continued, the Galileo team sent a variety of commands intended to free the antenna. Most involved turning the spacecraft toward and away from the Sun, in the hope that warming and cooling the apparatus would free the stuck hardware through thermal expansion and contraction. None of these attempts succeeded in releasing the ribs.

Further engineering analysis and testing suggested that "hammering" the antenna deployment motors – turning them on and off repeatedly – might deliver the force needed to free the stuck pins and open the antenna. After more than 13,000 hammerings between December 1992 and January 1993, engineering telemetry from the spacecraft showed that additional deployment force had been generated, but it had not freed the ribs. Other approaches were tried, such as spinning the spacecraft up to its fastest rotation rate of 10 rpm and hammering the motors again, but these efforts also failed to free the antenna.

Project engineers believe the state of the antenna has been as well-defined as long-distance telemetry and laboratory tests will allow. After the years-long campaign to try to free the stuck hardware, the project has determined there is no longer any significant prospect of the antenna being deployed.

Nevertheless, one last attempt will be made in March 1996, after the orbiter's main engine is fired to raise *Galileo's* orbit around Jupiter. This "perijove raise maneuver" will deliver the largest acceleration the spacecraft will have experienced since launch, and it follows three other mildly jarring events: the release of the atmospheric probe, the orbiter deflection maneuver that follows probe release, and the Jupiter orbit insertion engine firing. It is possible, but extremely unlikely, that these shocks could jar the stuck ribs enough to free the antenna. This will be the last attempt to open the antenna before radioing the new software to the spacecraft to inaugurate the advanced data compression techniques designed specifically for use with the low-gain antenna.

The Low-Gain Antenna

The difference between *Galileo* sending its data to Earth using the high-gain antenna and the low-gain is like the difference between the concentrated light from a spotlight versus the light emitted diffusely from a bare bulb. If unfurled, the high-gain would transmit data back to ground-based Deep Space Network (DSN) collecting antennas in a narrowly focused beam. The low-gain antenna transmits in a comparatively unfocused broadcast, and only a tiny fraction of the signal actually reaches DSN receivers. Because the received signal is 10,000 times fainter, data must be sent at a lower rate to ensure that the contents are clearly understood.

New Software on the Spacecraft

Key to the success of the mission are two sets of new flight software. The first set, called Phase 1, began operating in March 1995 and was designed expressly to partially back-up and ensure receipt of the most important data collected from the atmospheric probe. Once the critical scientific data from the probe is safely returned to Earth, a second set of new software will be radioed and loaded onto the spacecraft in March 1996.

This Phase 2 software will provide programs to shrink the voluminous science data the *Galileo* orbiter will collect and store on its tape recorder during its two-year mission, while retaining the scientifically important information, and return that data at the lower data rate.

Without any new enhancements, the low-gain antenna's data transmission rate at Jupiter would be limited to only 8-16 bits per second (bps), compared to the high-gain's 134,400 bps. However, the innovative Phase 2 software changes, when coupled with hardware and software adaptations at Earth-based receiving stations, will increase the data rate from Jupiter by as much as 10 times, to 160 bps. The data compression methods will allow retention of the most interesting and scientifically valuable information, while minimizing or eliminating less valuable data (such as the dark background of space) before transmission. Two different methods of data compression will be used. In both methods, the data are compressed on board the spacecraft before being transmitted to Earth.

The first method, called "lossless" compression, allows the data to be reformatted back to their original state once on the ground. This technique is routinely used in personal computer modems to increase their effective transmission rates. The second compression method is called "lossy," a term used to describe the dissipation of electrical energy, but which in this case refers to the loss of some original data through mathematical approximations used to abbreviate the total amount of data to be sent to the ground. Lossy compression will be used to shrink imaging and plasma wave data down to as little as $1/80^{th}$ of its original volume.

Customizing Receivers on Earth

S-Band telecommunication was once the standard for space missions, and several S-band performance-enhancing capabilities were implemented at DSN tracking stations in the 1980s. For Galileo and its S-band low-gain antenna, these capabilities are being restored at the Canberra, Australia, 70-meter antenna. Because Australia is in the southern hemisphere and Jupiter is in the southern sky during Galileo's tour, the Canberra complex will receive most of Galileo's data.

Another critical, ongoing DSN upgrade will be the addition of so-called Block V receivers at the tracking stations. These receivers, which are being installed for multi-mission use, will allow all of Galileo's signal power to be dedicated to the data stream by suppressing the traditional carrier signal, thus allowing use of higher data rates.

Finally, starting early in the orbital tour, the 70-meter and two 34-meter DSN antennas at Canberra will be arrayed to receive Galileo's signal concurrently, with the received signals electronically combined. The arraying technique allows more of the spacecraft's weak signal to be captured, thereby enabling a higher data rate, which translates into the receipt of more data. In addition, other arraying is planned: the 64-meter Parkes Radio Telescope in Australia will be arrayed with the Canberra antennas, as will the 70-meter DSN antenna in Goldstone, California, when its view of Galileo overlaps with Canberra's.

The Tape Recorder Problem

Galileo's tape recorder is a key link in techniques developed to compensate for the loss of use of Galileo's high-gain antenna. The tape recorder is to be used to store information, particularly imaging data, until it can be compressed and edited by spacecraft computers and radioed via Galileo's low-gain antenna back to Earth.

On October 11, 1995, with just weeks to go before Jupiter arrival, the tape recorder malfunctioned. Data from the spacecraft showed the recorder failed to cease rewinding after recording an image of Jupiter.

A week later, following extensive analysis, the spacecraft tape recorder was tested and proved still operational, but detailed study of engineering data indicates that the tape recorder can be unreliable under some operating conditions. The problem appears to be manageable, however, and should not jeopardize return of the full complement of images of Jupiter and its moons that are to be stored on the recorder for playback over the course of the mission.

On October 24, the spacecraft executed commands for the tape recorder to wind an extra 25 times around a section of tape possibly weakened when the recorder had been stuck in rewind mode with the tape immobilized for about 15 hours. Due to uncertainty about its condition, spacecraft engineers have declared that portion near the end of the tape reel is "off-limits" for future data recording. The extra tape wound over it secures that area of tape, eliminating any stresses that could tear the tape at this potential weak spot. Unfortunately, the approach image of Jupiter that Galileo took October 11 was stored on the portion of tape that is now off-limits, and will not be played back. More significantly, project officials also decided not to take pictures of Io and Europa on December 7, including what would have been the closest encounter of Io (from a distance of 600 miles or 1,000 kilometers). Instead, the tape recorder will be completely devoted that day to gathering data from Galileo's Jupiter atmospheric probe.

Analysis of the tape recorder's condition continues so spacecraft engineers can fully understand its capabilities and potential weaknesses. Currently, the prospects look very good for finding ways to reliably operate the recorder with little loss to the orbital mission objectives.

Science Saved and Science Lost

Very few of Galileo's original measurement objectives have had to be completely abandoned as a result of the high-gain antenna problem. For the most part, science investigations on the spacecraft have adapted to the lower data rates using a variety of techniques, depending on the nature of the experiment. The new software and DSN receiver hardware will increase the information content of the data that will be returned by at least 100 times more than what would have been possible otherwise.

The onboard data processing made possible by the Phase 2 software will allow the spacecraft to store and transmit nearly continuous observations of the Jovian magnetosphere and extensive spectral measurements of the planet and its satellites in the infrared, visible, and ultraviolet, including more than 1,500 high-resolution images.

While tens of thousands of images would be required for large-scale movies of Jupiter's atmospheric dynamics, the hundreds of images allocated to atmospheric imaging will allow in-depth study of several individual features in the clouds of Jupiter. Cooperative observations with Hubble Space Telescope investigators and ground-based observers has long been planned as part of the Galileo mission to provide information on the global state of Jupiter's atmosphere.

Like a tourist allotted one roll of film per city, the Galileo team will select its observations carefully at each encounter to ensure the maximum amount of new and interesting scientific information is returned. The imaging campaign will focus on the planet and the four large Galilean moons, but it will also cover the four inner minor satellites and Jupiter's rings. Ten close satellite encounters will be conducted: one Io flyby (on approach), three of Europa, three of Callisto, and four of Ganymede. Five additional mid-range encounters (from closer than 80,000 kilometers, or about 50,000 miles) will be conducted with these moons.

For the orbiter portion of the mission, it is useful to realize that *Galileo*, with its sophisticated instruments, closer satellite flybys, and long duration in Jovian orbit, was specifically designed to answer many of the questions that the *Pioneer* and *Voyager* spacecraft were unable to answer. None of those characteristics have been affected by the loss of the high-gain antenna; only the total volume of data has been reduced.

As a result, when *Galileo* examines a class of phenomena, fewer samples of that class can be studied, and often, the spectral or temporal resolution will be reduced to lessen the total volume of data. The resulting information, however, will nevertheless provide unique insight into the Jovian system.

Some specific impacts from the loss of the high-gain antenna include: elimination of color global imaging of Jupiter once per orbit; elimination of global studies of Jupiter's atmospheric dynamics such as storms, clouds, and latitudinal bands (efforts to image atmospheric features, including the Great Red Spot, are still planned, however); a reduction in the spectral and spatial coverage of the moons, which provided context for study of high-resolution observations of their key features; and reduction of much of the so-called fields and particles microphysics (requiring high temporal- and spectral-frequency sampling of the environment by all instruments) during the cruise portion of each orbit. Most of the fields and particles microphysics, however, will be retained during the satellite encounters.

Imaging Playback

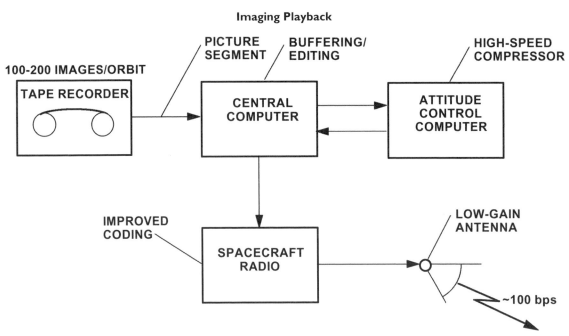

Highlights of Jupiter Science to be Returned Via *Galileo*'s Low-Gain Antenna:

- 100 percent of probe data
- Nearly continuous, real-time survey of Jovian magnetosphere for two years
- Approximately 1,500 images of the four Galilean satellites, four inner minor satellites, Jupiter and its rings

- Ten very close encounters: Europa (3), Callisto (3) and Ganymede (4)
- Five Voyager-class (less than 80,000 kilometers) encounters with Galilean satellites

Ground Systems and Spacecraft Operations

Galileo communicates with Earth via NASA's Deep Space Network (DSN), which has a complex of large antennas with receivers and transmitters located in the California desert, in Australia and in Spain, linked to a network control center at JPL in Pasadena, California. The spacecraft receives commands, sends science and engineering data, and is tracked by doppler and ranging measurements through this network. Mission control responsibilities include commanding the spacecraft, interpreting the engineering and scientific data it sends in order to understand how it is performing and responding, and analyzing navigation data obtained by the Deep Space Network. The controllers use a set of complex computer programs to help them control the spacecraft and interpret the data.

The *Galileo* spacecraft carries out its complex operations, including maneuvers, scientific observations and communications, in response to stored sequences which are sent up to the orbiter periodically through the Deep Space Network in the form of command loads.

The spacecraft status and health are monitored through data from 1,418 onboard measurements. The Galileo flight team interprets these data into trends to avert or work around equipment failure. Their conclusions become an important input, along with scientific plans, to the sequence design process. The telemetry monitoring is supported by computer programs written and used in the mission support area.

Navigation is the process of estimating, from radio range and doppler measurements, the position and velocity of the spacecraft to predict its flight path and to design course-correcting maneuvers. These calculations must be done with computer support. The Galileo mission, with its complex gravity assist flight to Jupiter and 10 gravity assist satellite encounters in the Jovian system, is extremely dependent on consistently accurate navigation.

In addition to the programs which directly operate the spacecraft and are periodically transmitted to it, the mission operations team uses software amounting to 650,000 lines of programming code in the sequence design process; 1,615,000 lines in the telemetry interpretation; and 550,000 lines of code in navigation. These all had to be written, checked, tested, used in mission simulations and, in many instrument cases, revised before the mission could begin.

Science investigators are located variously at JPL or at their home laboratories, linked by computer communications. From either location, they are involved in developing the sequences affecting their experiments and, in some cases, helping to change preplanned sequences to follow up on unexpected discoveries with second looks.

The Spacecraft

The Galileo mission and systems were designed to investigate three broad aspects of the Jovian system: the planet's atmosphere, the satellites and the magnetosphere. The spacecraft was constructed in three segments, which help focus on these areas: 1) the atmospheric probe; 2) a non-spinning section of the orbiter carrying cameras and other remote sensors; 3) the spinning main section of the orbiter spacecraft which includes the fields and particles instruments, designed to sense and measure the environment directly as the spacecraft flies through it. The spinning section also carries the communications antennas, the propulsion module, flight computers and most support systems.

This innovative "dual spin" design allows part of the orbiter to rotate constantly at three rpm, and part of the spacecraft to remain fixed. This means that the orbiter can easily accommodate magnetospheric experiments (which need to take measurements while rapidly sweeping about) while also providing stability and a fixed orientation for cameras and other sensors. The spin rate can be increased to 10 revolutions per minute for additional stability during major propulsive maneuvers.

Galileo's atmospheric probe weighs 339 kilograms (746 pounds), and includes a deceleration module to slow and protect the descent module, which carries out the scientific mission.

The deceleration module consists of an aeroshell and an aft cover, designed to block the heat generated by friction during the sharp deceleration of atmospheric entry. Inside the shells are the descent module and its 2.5-meter (8-foot) parachute. The descent module carries a radio-relay transmitter and six scientific instruments. Operating at 128 bits per second, each of the dual L-band transmitters send nearly identical streams of scientific data to the orbiter. Probe electronics are powered by batteries with an estimated capacity of about 18 amp-hours on arrival at Jupiter.

Probe instruments include an atmospheric structure group of sensors measuring temperature, pressure and deceleration; a neutral mass spectrometer and a helium-abundance detector supporting atmospheric composition studies; a nephelometer for cloud location and cloud-particle observations; a net-flux radiometer measuring the difference, upward versus downward, in radiant energy flux at each altitude; and a lightning / radio-emission instrument with an energetic particle detector, measuring light and radio emissions associated with lightning and energetic particles in Jupiter's radiation belts.

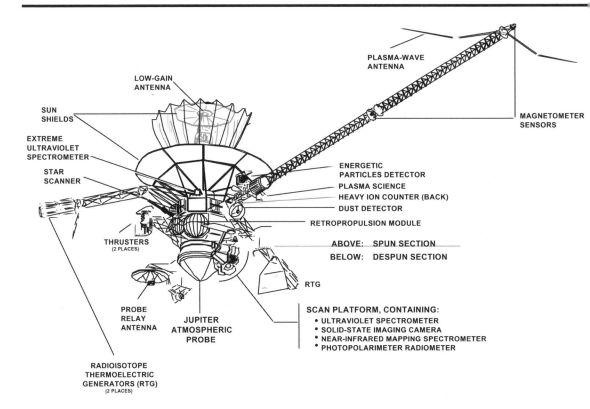

The *Galileo* orbiter spacecraft, in addition to supporting the probe activities, will support all the scientific investigations of Jupiter's satellites and magnetosphere, and remote observation of the giant planet itself.

At launch, the orbiter weighed about 2,223 kilograms (4,900 pounds), not counting the upper stage-rocket adapter, but including about 925 kilograms of usable rocket propellant. This propellant is used in almost 30 relatively small maneuvers during the long gravity-assisted flight to Jupiter, three large thrust maneuvers including the one that puts the craft into its Jupiter orbit, and the 30 or so trim maneuvers planned for the satellite tour phase. It is also consumed in the small pulses that turn and orient the spacecraft.

The propulsion module consists of twelve 10-newton thrusters, a single 400-newton engine, the monomethyl-hydrazine fuel, nitrogen-tetroxide oxidizer, and pressurizing gas tanks, tubing, valves and control equipment. (A thrust of 10 newtons would support a weight of about one kilogram or 2.2 pounds at Earth's surface.) The propulsion system was developed and built by Messerschmitt-Bolkow-Blohm (MBB) and provided by Germany as a partner in Project Galileo.

In addition to the scientific data acquired by its 10 instruments, the *Galileo* orbiter acquires and can transmit a total of 1,418 engineering measurements of internal operating conditions including temperatures, voltages, computer states and counts. The spacecraft transmitters will operate at S-band frequency (2,295 megahertz).

Two low-gain antennas (one pointed upward or toward the Sun, and one on a deployable arm to point down, both mounted on the spinning section) supported communications during the Earth-Venus-Earth leg of the flight. The top-mounted antenna is currently carrying the communications load, including science data and playbacks, in place of the high-gain antenna, and is the basis of the redesigned Jupiter sequences. The other low-gain antenna has been restowed after supporting operations during the early VEEGA phase, and is not expected to be used again.

Because radio signals take more than one hour to travel from Earth to Jupiter and back, the *Galileo* spacecraft was designed to operate from programs sent to it in advance and stored in spacecraft memory. A single master sequence program can cover from weeks to months of quiet operations between planetary and satellite encounters. During busy encounter operations, one program covers only about a week.

These sequences operate through flight software installed in the principal spacecraft computers. In the command and data subsystem software, there are about 35,000 lines of code, including 7,000 lines of automatic fault protection software, which operates to put the spacecraft in a safe state if an untoward event such as an onboard computer glitch were to occur. The articulation and attitude control software has about 37,000 lines of code, including 5,500 lines devoted to fault protection.

Electrical power is provided to *Galileo*'s equipment by two radioisotope thermoelectric generators. Heat produced by natural radioactive decay of plutonium-238 dioxide is converted to electricity (570 watts at launch, 485 at the end of the mission) to operate the orbiter equipment for its eight-year baseline mission. This is the same type of power source used by the two *Voyager* spacecraft missions to the outer planets, the *Pioneer* Jupiter spacecraft, and the twin *Viking* Mars landers.

Scientific instruments to measure fields and particles, together with the main antenna, the power supply, the propulsion module, most of the computers and control electronics, are mounted on the spinning section. The instruments include magnetometer sensors, mounted on an 11-meter (36-foot) boom to minimize interference from the spacecraft; a plasma instrument detecting low-energy charged particles and a plasma wave detector to study waves generated by the particles; a high-energy particle detector; and a detector of cosmic and Jovian dust. It also carries the Heavy Ion Counter, an engineering experiment added to assess the potentially hazardous charged-particle environments the spacecraft flies through, and an added Extreme Ultraviolet detector associated with the UV spectrometer on the scan platform.

The despun section carries instruments and other equipment whose operation depends on a steady pointing capability. The instruments include the camera system; the near-infrared mapping spectrometer to make multispectral images for atmospheric and moon surface chemical analysis; the ultraviolet spectrometer to study gases; and the photopolarimeter-radiometer to measure radiant and reflected energy. The camera system will obtain images of Jupiter's satellites at resolutions from 20 to 1,000 times better than *Voyager's* best, largely because it will be closer. The CCD sensor in *Galileo's* camera is more sensitive and has a broader color detection band than the vidicons of *Voyager*. This section also carries an articulated dish antenna to track the atmospheric probe and pick up its signals for recording and relay to Earth.

Technology Benefits Derived from Galileo

The research and development necessary to build and fly *Galileo* has produced several technological innovations.

Charge coupled devices like those in *Galileo's* television systems are used in some home video cameras, yielding sharper images than ever conceived of in the days before the project began. In addition, radiation-resistant components developed for *Galileo* are now used in research, businesses, and military applications where radiation environment is a concern. Another advance, integrated circuits resistant to cosmic rays, has helped to handle disturbances to computer memory that are caused by high-energy particles; these disturbances plague extremely high-speed computers on Earth and all spacecraft.

Program / Project Management

Galileo's scientific experiments are being carried out by more than 100 scientists from six nations. These experimenters use dedicated instruments and the radio subsystems on the *Galileo* orbiter and probe. In addition, NASA has appointed 17 interdisciplinary scientists whose studies reach across more than one *Galileo* instrument data set.

The Galileo Project is managed for NASA's Office of Space Science by the Jet Propulsion Laboratory, a division of the California Institute of Technology. This responsibility includes designing, building, testing, operating and tracking *Galileo*. Germany has furnished the orbiter's retro-propulsion module and some of the instruments and is participating in the scientific investigations. The radioisotope thermoelectric generators were designed and built by the General Electric Company for the U.S. Department of Energy.

NASA's Ames Research Center, Mountain View, California, is responsible for the atmosphere probe, which was built by Hughes Aircraft Company, El Segundo, California. At Ames, the probe manager is Marcie Smith and the probe scientist is Dr. Richard E. Young.

At JPL, William J. O'Neil is project manager, Dr. Torrence V. Johnson is project scientist, Neal E. Ausman Jr. is mission director, and Matthew R. Landano is deputy mission director.

At NASA Headquarters, the program manager is Donald Ketterer. The program scientist is Dr. Jay Bergstralh. Dr. Wesley T. Huntress Jr. is Associate Administrator for the Office of Space Science.

Galileo Scientific Experiments

Probe:

Experiment / Instrument	Principal Investigator	Objectives
Atmospheric Structure	Alvin Seiff, NASA Ames Research Center	Temperature, pressure density, molecular weight profiles
Neutral Mass Spectrometer	Hasso Niemann, NASA Goddard SFC	Chemical composition
Helium Abundance	Ulf von Zahn, Bonn University, FRG	Helium / hydrogen ratio

Nephelometer	Boris Ragent, NASA Ames	Clouds, solid / liquid particles
Net Flux Radiometer	Larry Sromovsky, Univ. of Wisconsin	Thermal / solar energy profiles
Lightning / Energetic Particles	Louis Lanzerotti, Bell Laboratories	Detect lightning, measure energetic particles

Orbiter (Despun):

Experiment / Instrument	Principal Investigator	Objectives
Solid State Imaging Camera	Michael Belton, NOAO (Team Leader)	Galilean satellites at 1-kilometer resolution or better, other bodies correspondingly
Near-Infrared Mapping Spectrometer	Robert Carlson, JPL	Surface / atmospheric composition, thermal mapping
Ultraviolet Spectrometer (includes extreme UV sensor on spun section)	Charles Hord, University of Colorado	Atmospheric gases, aerosols, etc.
Photopolarimeter Radiometer	James Hansen, Goddard Institute for Space Studies	Atmospheric particles, thermal / reflected radiation

Orbiter (Spinning):

Experiment / Instrument	Principal Investigator	Objectives
Magnetometer	Margaret Kivelson, UCLA	Strength and fluctuations of magnetic fields
Energetic Particles	Donald Williams, Johns Hopkins APL	Electrons, protons, heavy ions in atmosphere
Plasma	Lou Frank, University of Iowa	Composition, energy, distribution of ions
Plasma Wave	Donald Gurnett, University of Iowa	Electromagnetic waves and wave-particle interactions
Dust	Eberhard Grun, Max Planck Inst. fur Kernphysik	Mass, velocity, charge of submicron particles
Radio Science: Celestial Mechanics	John Anderson, JPL (Team Leader)	Masses and motions of bodies from spacecraft tracking
Radio Science: Propagation	H. Taylor Howard, Stanford University (Team Leader)	Satellite radii, atmospheric structure, from radio propagation

Engineering Experiment:		
Heavy Ion Counter	Edward Stone, Caltech	Spacecraft charged particle environment

Interdisciplinary Investigators

Frances Bagenal, University of Colorado
Andrew F. Cheng, The Johns Hopkins University
Fraser P. Fanale, University of Hawaii
Peter Gierasch, Cornell University
Donald M. Hunten, University of Arizona
Andrew P. Ingersoll, California Institute of Technology
Wing-Huen Ip, NSPO / RDD, Taipei
Michael McElroy, Harvard University
David Morrison, NASA Ames Research Center
Glenn S. Orton, Jet Propulsion Laboratory
Tobias Owen, State University of New York
Alain Roux, Centre de Recherches en Physique de l'Environment
Christopher T. Russell, University of California at Los Angeles
Carl Sagan, Cornell University
Gerald Schubert, University of California at Los Angeles
William H. Smyth, Atmospheric & Environmental Research, Inc.
James Van Allen, University of Iowa

NATIONAL AERONAUTICS AND SPACE ADMINISTRATION

Galileo End of Mission
Press Kit

September 2003

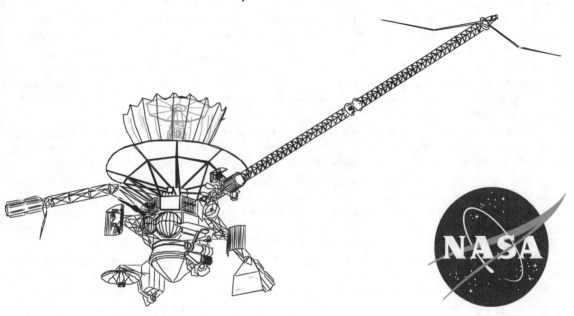

Media Contacts

Donald Savage Headquarters, Washington, D.C.	Policy / Program Management	(202) 358-1727 donald.savage@hq.nasa.gov
Carolina Martinez Jet Propulsion Laboratory, Pasadena, California	Galileo Mission	(818) 354-9382 carolina.martinez@jpl.nasa.gov
D.C. Agle Jet Propulsion Laboratory, Pasadena, California	Galileo Mission	(818) 393-9011 david.c.agle@jpl.nasa.gov

CONTENTS

GENERAL RELEASE:

Galileo to Taste Jupiter Before Taking Final Plunge

In the end, the *Galileo* spacecraft will get a taste of Jupiter before taking a final plunge into the planet's crushing atmosphere, ending the mission on Sunday, September 21. The team expects the spacecraft to transmit a few hours of science data in real time leading up to impact.

The spacecraft has been purposely put on a collision course with Jupiter to eliminate any chance of an unwanted impact between the spacecraft and Jupiter's moon Europa, which *Galileo* discovered is likely to have a subsurface ocean. The long-planned impact is necessary now that the onboard propellant is nearly depleted. Without propellant, the spacecraft would not be able to point its antenna toward Earth nor adjust its trajectory, so controlling the spacecraft would no longer be possible.

"It has been a fabulous mission for planetary science, and it is hard to see it come to an end," said Dr. Claudia Alexander, Galileo project manager at NASA's Jet Propulsion Laboratory, Pasadena, California. "After traversing almost 3 billion miles and being our watchful eyes and ears around Jupiter, we're keeping our fingers crossed that even in its final hour, *Galileo* will still give us new information about Jupiter's environment."

Although scientists are hopeful to get every bit of data back for analysis, the likelihood of getting anything is unknown because the spacecraft has already endured more than four times the cumulative dose of harmful Jovian radiation it was designed to withstand. The spacecraft will enter an especially high-radiation region again as it approaches Jupiter.

Launched in the cargo bay of Space Shuttle *Atlantis* in 1989, the mission has produced a string of discoveries while circling the solar system's largest planet, Jupiter, 34 times. *Galileo* was the first mission to measure Jupiter's atmosphere directly with a descent probe and the first to conduct long-term observations of the Jovian system from orbit. It found evidence of subsurface liquid layers of saltwater on Europa, Ganymede and Callisto and it examined a diversity of volcanic activity on Io. *Galileo* is the first spacecraft to fly by an asteroid and the first to discover a moon of an asteroid.

The prime mission ended six years ago, after two years of orbiting Jupiter. NASA extended the mission three times to continue taking advantage of *Galileo*'s unique capabilities for accomplishing valuable science. The mission was possible because it drew its power from two long-lasting radioisotope thermoelectric generators provided by the Department of Energy.

From launch to impact, the spacecraft has traveled 4,631,778,000 kilometers (about 2.8 billion miles).

Its entry point into the giant planet's atmosphere is about ¼ degree south of Jupiter's equator. If there were observers floating along at the cloud tops, they would see *Galileo* streaming in from a point about 22 degrees above the local horizon. Streaming in could also be described as screaming in, as the speed of the craft relative to those observers would be 48.2 kilometers per second (nearly 108,000 mph). That is the equivalent of traveling from Los Angeles to New York City in 82 seconds. In comparison, the *Galileo* atmospheric probe, aerodynamically designed to slow down when entering, and parachute gently through the clouds, first reached the atmosphere at a slightly more modest 47.6 kilometers per second (106,500 mph).

"This is a very exciting time for us as we draw to a close on this historic mission and look back at its science discoveries. *Galileo* taught us so much about Jupiter but there is still much to be learned and for that we look with promise to future missions," said Dr. Charles Elachi, director of JPL.

JPL, a division of the California Institute of Technology in Pasadena, manages the Galileo mission for NASA's Office of Space Science, Washington, D.C.

Additional information about the mission and its discoveries is available on line at http://galileo.jpl.nasa.gov.

(End of General Release)

Media Services Information

NASA Television Transmission

NASA Television is broadcast on AMC-9, transponder 9C, C Band, 85 degrees west longitude, frequency 3880.0 MHz, vertical polarization, audio monaural at 6.8 MHz. The schedule for television transmission of video animations, B-roll and live-interview opportunities will be available from the Jet Propulsion Laboratory, Pasadena, California, and NASA Headquarters, Washington, D.C.

News Releases

The Jet Propulsion Laboratory's Media Relations Office will issue news releases on the *Galileo* end of mission. They may be accessed on line as noted below.

Briefings

A space science update about the *Galileo* end of mission will be held at NASA Headquarters in Washington, D.C., on September 17, 2003. Additional details about the update will be available from NASA Headquarters and the Jet Propulsion Laboratory.

Internet Information

Extensive information about the Galileo mission, including an electronic copy of this Press Kit, press releases, fact sheets, status reports and images, is available from the Jet Propulsion Laboratory's World Wide Web home page at http://www.jpl.nasa.gov. Sign up for e-mail subscriptions to news releases is available at the home page.

Additional information on Galileo is available at http://galileo.jpl.nasa.gov.

Quick Facts

Spacecraft

- Dimensions: 5.3 meters (17 feet) high; magnetometer boom extends 11 meters (36 feet) to one side
- Weight: 2,223 kilograms (2½ tons, or 4,902 pounds), including 118 kilograms (260 pounds) of science instruments and 925 kilograms (2,040 pounds) of propellant
- Power: 570 watts (at launch) from radioisotope thermoelectric generators
- Science instruments: Solid state imaging camera, near-infrared mapping spectrometer, ultraviolet spectrometer, photopolarimeter radiometer, magnetometer, energetic particles detector, plasma investigation, plasma wave subsystem, dust detector, heavy ion counter

Atmospheric Probe

- Size: 127 centimeters (50 inches) diameter, 91 centimeters (36 inches) high
- Weight: 339 kilograms (750 pounds)
- Science instruments: Atmospheric structure, neutral mass spectrometer, helium abundance, nephelometer, net flux radiometer, lightning / energetic particles, doppler wind experiment

Mission

- Launch: October 18, 1989 from Kennedy Space Center, Florida, on space shuttle *Atlantis* on mission STS-34
- Primary mission: October 1989 to December 1997
- Extended missions: Three, from 1997 to 2003
- Venus flyby: February 10, 1990, at altitude of 16,000 kilometers (10,000 miles)
- Earth flybys: December 8, 1990, at altitude of 960 kilometers (597 miles); December 8, 1992 at altitude of 303 kilometers (188 miles)
- Asteroid Gaspra flyby: October 29, 1991, at 1,601 kilometers (1,000 miles)
- Comet Shoemaker-Levy 9: Impacts of comet fragments into Jupiter observed while en route in July 1994
- Asteroid Ida flyby: August 28, 1993, at 2,400 kilometers (1,400 miles)
- Atmospheric probe release: July 12, 1995
- Probe speed into Jupiter's atmosphere: 47.6 km/sec (106,000 mph)
- Jupiter arrival and orbit insertion: December 7, 1995
- Probe atmospheric entry and relay: December 7, 1995
- Number of Jupiter orbits during entire mission: 34
- Number of flybys of Jupiter moons: Io 7, Callisto 8, Ganymede 8, Europa 11, Amalthea 1
- Total distance traveled from launch to final impact: 4,631,778,000 kilometers (approx. 2.8 billion miles)
- Speed of spacecraft at time of impact: 48.2 kilometers per second (nearly 108,000 mph)

Program

- Cost: Total from start of planning through end of mission is $1.39 billion. International contribution estimated at an additional $110 million
- Partners: More than 100 scientists from United States, Great Britain, Germany, France, Canada and Sweden carried out *Galileo*'s experiments. NASA's Ames Research Center, Mountain View, California, responsible for atmosphere probe, built by Hughes Aircraft Company, El Segundo, California. Radioisotope thermoelectric generators designed and built by General Electric Co. for the U.S. Department of Energy
- Approximate number of people who worked on some portion of the Galileo mission: 800

Top 10 Science Results

1. The descent probe measured atmospheric elements and found that their relative abundances were different than on the Sun, indicating Jupiter's evolution since the planet formed out of the solar nebula.
2. *Galileo* made a first observation of ammonia clouds in another planet's atmosphere. The atmosphere seems to create ammonia ice particles of material from lower depths, but only in "fresh" clouds.
3. Io's extensive volcanic activity may be 100 times greater than that found on Earth. The heat and frequency of eruption are reminiscent of early Earth.
4. Io's complex plasma interactions in Io's atmosphere include support for currents and coupling to Jupiter's atmosphere.
5. Evidence supports a theory that liquid oceans exist under Europa's icy surface.
6. Ganymede is the first satellite known to possess a magnetic field.
7. *Galileo* magnetic data provide evidence that Europa, Ganymede and Callisto have a liquid saltwater layer.
8. Europa, Ganymede, and Callisto all provide evidence of a thin atmospheric layer known as a "surface-bound exosphere."
9. Jupiter's ring system is formed by dust kicked up as interplanetary meteoroids smash into the planet's four small inner moons. The outermost ring is actually two rings, one embedded with the other.
10. *Galileo* was the first spacecraft to dwell in a giant planet magnetosphere long enough to identify its global structure and to investigate its dynamics.

Technical Firsts

- First flyby and imaging of an asteroid (Gaspra and later, Ida).
- Discovered first moon around an asteroid (the moon Dactyl orbits asteroid Ida).
- Only direct observation of Comet Shoemaker-Levy 9's impact into Jupiter's atmosphere.
- First spacecraft to deploy an entry probe into an outer planet's atmosphere. The descent probe measured the atmospheric composition and structure of Jupiter's atmosphere and provided clues to the origin of Jupiter and giant planets in other star systems.
- First and so far the only spacecraft to orbit an outer planet. (*Cassini* at Saturn in 2004 will become the second spacecraft to enter orbit around one of the outer planets.)

Mission Overview

NASA's *Galileo* spacecraft was designed to study the large, gaseous planet Jupiter, its moons and its surrounding magnetosphere, which is a magnetic bubble surrounding the planet. The craft was named for the Italian Renaissance scientist who discovered Jupiter's major moons in 1610.

The primary mission at Jupiter began when the spacecraft entered orbit in December 1995 and its descent probe, which had been released five months earlier, dove into the giant planet's atmosphere. Its primary mission included a 23-month, 11-orbit tour of the Jovian system, including 10 close encounters of Jupiter's major moons.

Although the primary mission was completed in December 1997, the mission has been extended three times since then. *Galileo* had 35 encounters of Jupiter's major moons – 11 with Europa, 8 with Callisto, 8 with Ganymede, 7 with Io and 1 with Amalthea. The mission will end when the spacecraft impacts Jupiter on Sunday, September 21, 2003.

Launch

The *Galileo* spacecraft and its two-stage Inertial Upper Stage were carried into Earth orbit on October 18, 1989 by space shuttle *Atlantis* on mission STS-34. The two-stage Inertial Upper Stage solid rocket then accelerated the spacecraft out of Earth orbit toward the planet Venus for the first of three planetary "gravity assists" designed to boost *Galileo* toward Jupiter. In a gravity assist, the spacecraft flies close enough to a planet to be propelled by its gravity, creating a "slingshot" effect for the spacecraft. The Galileo mission had originally been designed for a direct flight of about three and a half years to Jupiter, using a three-stage Inertial Upper Stage. When that booster was canceled, plans were changed to a Centaur upper stage, and ultimately to the two-stage Inertial Upper Stage, which precluded a direct trajectory. To save the project, *Galileo* engineers designed a flight path using planetary gravity assists.

Venus and Earth Flybys

After flying past Venus at an altitude of 16,000 kilometers (nearly 10,000 miles) on February 10, 1990, the spacecraft swung past Earth at an altitude of 960 kilometers (597 miles) on December 8, 1990. That flyby increased *Galileo's* speed enough to send it on a two-year elliptical orbit around the Sun. The spacecraft returned for a second Earth swingby on December 8, 1992, at an altitude of 303 kilometers (188 miles). With this, *Galileo* left Earth for the third and final time and headed toward Jupiter.

The flight path provided opportunities for scientific observations. Scientists obtained the first views of mid-level clouds on Venus and confirmed the presence of lightning on that planet. They also made many Earth observations, mapped the surface of Earth's Moon, and observed its north polar regions.

Because of the modification in *Galileo*'s trajectory, the spacecraft was exposed to a hotter environment than originally planned. To protect it from the Sun, project engineers devised a set of sunshades and pointed the top of the spacecraft toward the Sun, with the umbrella-like high-gain antenna furled until well after the first Earth flyby in December 1990. Flight controllers stayed in touch with the spacecraft through a pair of low-gain antennas, which send and receive data at a much slower rate.

High-Gain Antenna

The spacecraft was scheduled to deploy its 4.8-meter (16-foot)-diameter high-gain antenna in April 1991 as *Galileo* moved away from the Sun and the risk of overheating ended. The antenna, however, failed to deploy fully.

A special team performed extensive tests and determined that a few (probably three) of the antenna's 18 ribs were held by friction in the closed position. Despite exhaustive efforts to free the ribs, the antenna would not deploy. From 1993 to 1996, extensive new flight and ground software was developed, and ground stations of NASA's Deep Space Network were enhanced in order to perform the mission using the spacecraft's low-gain antennas.

Asteroid Flybys

Galileo became the first spacecraft ever to encounter an asteroid when it passed Gaspra on October 29, 1991. It flew within just 1,601 kilometers (1,000 miles) of the stony asteroid's center at a relative speed of about 8 kilometers per second (18,000 mph). Pictures and other data revealed a cratered, complex, irregular body about 20 by 12 by 11 kilometers (12.4 by 7.4 by 6.8 miles), with a thin covering of dust and rubble.

On August 28, 1993, *Galileo* carried out a second asteroid encounter, this time with a larger, more distant asteroid named Ida. Ida is about 55 kilometers (34 miles) long and 24 kilometers (15 miles) wide. Observations indicated that both Ida and Gaspra have magnetic fields, although Ida is older and its surface is covered with craters. Scientists discovered that Ida boasts its own moon, making it the first asteroid known to have a natural satellite. The tiny moon, named Dactyl, has a diameter of only about 1.5 kilometers (less than a mile). By determining Dactyl's orbit, scientists estimated Ida's density.

Comet Shoemaker-Levy

The discovery of Comet Shoemaker-Levy 9 in March 1993 provided an exciting opportunity for *Galileo*'s science teams and other astronomers. The comet was breaking up as it orbited Jupiter, and was headed to dive into the giant planet's atmosphere in July 1994.

Galileo Encounters at Jupiter

Orbit	Target	Date	Altitude	
0	Io	December 7, 1995	897 km	(558 mi)
1	Ganymede	June 27, 1996	835 km	(519 mi)
2	Ganymede	September 06, 1996	261 km	(162 mi)
3	Callisto	November 04, 1996	1136 km	(706 mi)
4	Europa	December 19, 1996	692 km	(430 mi)
5	[none]			
6	Europa	February 20, 1997	586 km	(364 mi)
7	Ganymede	April 05, 1997	3102 km	(1928 mi)
8	Ganymede	May 07, 1997	1603 km	(996 mi)
9	Callisto	June 25, 1997	418 km	(260 mi)
10	Callisto	September 17, 1997	535 km	(333 mi)
11	Europa	November 06, 1997	2043 km	(1270 mi)
12	Europa	December 16, 1997	201 km	(125 mi)
13	[none]			
14	Europa	March 29, 1998	1644 km	(1022 mi)
15	Europa	May 31, 1998	2515 km	(1562 mi)
16	Europa	July 21, 1998	1834 km	(1140 mi)
17	Europa	September 26, 1998	3582 km	(2226 mi)

Orbit	Target	Date	Altitude	
18	Europa	November 22, 1998	2271 km	(1411 mi)
19	Europa	February 01, 1999	1439 km	(894 mi)
20	Callisto	May 05, 1999	1321 km	(821 mi)
21	Callisto	June 30, 1999	1048 km	(651 mi)
22	Callisto	August 14, 1999	2299 km	(1429 mi)
23	Callisto	September 16, 1999	1052 km	(654 mi)
24	Io	October 11, 1999	611 km	(380 mi)
25	Io	November 26, 1999	301 km	(187 mi)
26	Europa	January 03, 2000	351 km	(218 mi)
27	Io	February 22, 2000	198 km	(123 mi)
28	Ganymede	May 20, 2000	809 km	(502 mi)
29	Ganymede	December 28, 2000	2338 km	(1452 mi)
30	Callisto	May 25, 2001	138 km	(86 mi)
31	Io	August 06, 2001	194 km	(120 mi)
32	Io	October 16, 2001	184 km	(114 mi)
33	Io	January 17, 2002	102 km	(63 mi)
34	Amalthea	November 05, 2002	160 km	(99 mi)
35	Jupiter	September 21, 2003	[impact]	

The *Galileo* spacecraft, approaching Jupiter, was the only observation platform with a direct view of the comet's impact area on Jupiter's far side. Despite the uncertainty of the predicted impact times, Galileo team members preprogrammed the spacecraft's science instruments to collect data and were able to obtain spectacular images of the comet impacts.

Trajectory and Key Mission Events

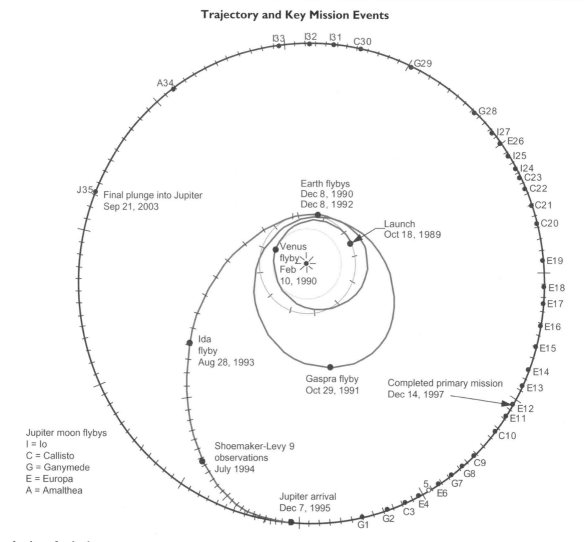

Jupiter Arrival

On July 13, 1995, *Galileo*'s descent probe, which had been carried aboard the parent spacecraft, was released and began a five-month free fall toward Jupiter. The probe had no engine or thrusters, so its flight path was established by pointing of the *Galileo* orbiter before the probe was released. Two weeks later, *Galileo* used its main rocket engine for the first time as it readjusted its flight path to arrive at the proper point at Jupiter.

Arrival day on December 7, 1995, turned out to be an extremely busy 24-hour period. When *Galileo* first reached Jupiter and while the probe was still approaching the planet, the orbiter flew by two of Jupiter's major moons – Europa and Io. *Galileo* passed Europa at an altitude of about 33,000 kilometers (20,000 miles), while the Io approach was at an altitude of about 900 kilometers (600 miles).

About four hours after leaving Io, the orbiter made its closest approach to Jupiter, encountering 25 times more radiation than the level considered deadly for humans.

Descent Probe

Eight minutes later, the orbiter started receiving data from the descent probe, which slammed into the top of the Jovian atmosphere at a comet-like speed of 170,000 km/hr (106,000 mph). In the process, the probe withstood temperatures twice as hot as the Sun's surface. The probe slowed by aerodynamic braking for about two minutes before deploying its parachute and dropping a heat shield.

The wok-shaped probe floated down about 200 kilometers (125 miles) through the clouds, transmitting data to the orbiter on sunlight and heat flux, pressure, temperature, winds, lightning and atmospheric composition. Fifty-eight minutes into its descent, high temperatures silenced the probe's transmitters. The probe sent data from a depth with a pressure 23 times that of the average on Earth's surface, more than twice the mission requirement.

An hour after receiving the last transmission from the probe, at a point about 200,000 kilometers (130,000 miles) above the planet, the *Galileo* spacecraft fired its main engine to brake into orbit around Jupiter.

This first orbit lasted about seven months. *Galileo* fired its thrusters at its farthest point in the orbit to keep it from coming so close to the giant planet on later orbits. This adjustment helped mitigate possible damage to spacecraft sensors and computer chips from Jupiter's intense radiation environment.

During this first orbit, new software was installed which gave the orbiter extensive new onboard data processing capabilities. It permitted data compression, enabling the spacecraft to transmit up to 10 times the number of pictures and other measurements that would have been possible otherwise.

In addition, hardware changes on the ground and adjustments to the spacecraft-to-Earth communication system increased the average telemetry rate tenfold. Although the problem with the high-gain antenna prevented some of the mission's original objectives from being met, the great majority were. So many new objectives were achieved that scientists feel *Galileo* has produced considerably more science than ever envisioned at the project's start 20 years ago.

Orbital Tour

During its primary mission orbital tour, *Galileo*'s itinerary included four flybys of Jupiter's moon Ganymede, three of Callisto and three of Europa. These encounters were about 100 to 1,000 times closer than those performed by NASA's *Voyager 1* and 2 spacecraft during their Jupiter flybys in 1979. *Galileo*'s instruments scanned and scrutinized the surface and features of each moon. After about a week of intensive observation, with its tape recorder full of data, the spacecraft spent the next one to two months – until the next encounter in orbital "cruise" – playing back the information in transmissions to Earth.

Extended Missions

A two-year extension, the Galileo Europa Mission, began in December 1997 and included intensive study of Europa through eight consecutive close encounters. The flybys added to knowledge about Europa's frozen surface and the intriguing prospect that liquid oceans may lie underneath. The Galileo Europa Mission provided a valuable opportunity to make additional flybys of the volcanic moon Io. The prime mission provided only one opportunity for close-up study of Io. *Galileo* had two encounters of Io during the Europa Mission, gathering new information on Io's volcanic activity. In addition, *Galileo* studied Callisto in four flybys. The Galileo Millennium Mission added another year of operations, including more flybys of Io and Ganymede, plus joint studies with the *Cassini* spacecraft as it passed Jupiter in December 2000 for a gravity assist toward Saturn.

Spacecraft

The *Galileo* orbiter weighed 2,223 kilograms at launch (2½ tons) and measured 5.3 meters (17 feet) from the top of the low-gain antenna to the bottom of the probe. The orbiter features an innovative "dual-spin" design. Most spacecraft are stabilized in flight either by spinning around a major axis, or by maintaining a fixed orientation in space, referenced to the Sun and another star. As the first dual-spin planetary spacecraft, *Galileo* combines these techniques. A spinning section rotates at about 3 rpm, and a "despun" section is counter-rotated to provide a fixed orientation for cameras and other remote sensors. A star scanner on the spinning side determines orientation and spin rate; gyroscopes on the despun side provide the basis for measuring turns and pointing instruments.

The power supply, propulsion module and most of the computers and control electronics are mounted on the spinning section. The spinning section also carries instruments to study magnetic fields and charged particles. These instruments include magnetometer sensors mounted on an 11-meter (36-foot) boom to minimize interference from the spacecraft's electronics; a plasma instrument to detect low-energy charged particles; and a plasma wave detector to study electromagnetic waves generated by the particles. There are also a high-energy particle detector and a detector of cosmic and Jovian dust, an extreme ultraviolet detector associated with the ultraviolet spectrometer, and a heavy ion counter to assess potentially hazardous charged particle environments the spacecraft flies through.

Galileo's de-spun section carries instruments that need to be held steady. These instruments include the camera system; the near-infrared mapping spectrometer to make multispectral images for atmosphere and surface chemical analysis; the ultraviolet spectrometer to study gases; and the photopolarimeter-radiometer to measure radiant and reflected energy. The camera system obtains images of Jupiter's satellites at resolutions from 20 to 1,000 times better than the best possible from NASA's *Voyager* spacecraft; its charge-coupled-device (CCD) sensor is much more sensitive than previous spacecraft cameras and is able to detect a broader color band. *Galileo*'s de-spun section also carries a dish antenna that picked up the descent probe's signals during its fall into Jupiter's atmosphere.

The spacecraft's propulsion module consists of twelve 10-newton (2.25-pound-force) thrusters and a single 400-newton (90-pound-force) engine which use monomethyl-hydrazine fuel and nitrogen-tetroxide oxidizer. The propulsion system was developed and built by Messerschmitt-Bolkow-Blohm (MBB) and provided by the Federal Republic of Germany as NASA's major international partner on *Galileo*.

Because radio signals take more than one hour to travel from Earth to Jupiter and back, the *Galileo* spacecraft was designed to operate from computer instructions sent to it in advance and stored in spacecraft memory. A single master sequence of commands can cover a period ranging from weeks to months of quiet operations between flybys of Jupiter's moons. During busy encounter operations, one sequence of commands covers only about a week.

Galileo Spacecraft

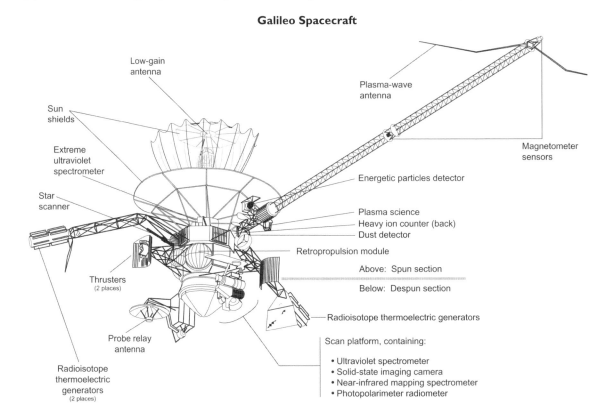

These sequences operate through flight software installed in the spacecraft computers, with built-in automatic fault protection software designed to put *Galileo* in a safe state in case of computer glitches or other unforeseen circumstances. Electrical power is provided by two radioisotope thermoelectric generators. Heat produced by natural radioactive decay of plutonium is converted to electricity (570 watts at launch, 485 at the end of the mission) to operate the orbiter spacecraft's equipment. This is the same type of power source used on other NASA missions including *Viking* to Mars, *Voyager* and *Pioneer* to the outer planets, *Ulysses* to study the Sun, and *Cassini* to Saturn.

Descent Probe

Galileo's descent probe had a mass of 339 kilograms (750 pounds), and included a deceleration module to slow and protect the descent module. The deceleration module consisted of an aeroshell and an aft cover designed to block heat generated by friction during atmospheric entry. Inside the aeroshells were the descent module and its 2.5-meter (8-foot) parachute. The descent module carried a radio transmitter and seven scientific instruments. These were devices to measure temperature, pressure and deceleration, atmospheric composition, clouds, particles, and light and radio emissions from lightning and energetic particles in Jupiter's radiation belts.

Orbiter Instruments

The *Galileo* orbiter spacecraft carries 11 scientific instruments. Another seven were on the descent probe. One engineering instrument on the orbiter, originally for measurements to aid design of future spacecraft, also collects scientific information.

End of Mission Sequence of Events

On the day of impact, it will take *Galileo*'s radio signals approximately 52 minutes to travel between the spacecraft and Earth. The time at which radio signals reach Earth indicating that an event has occurred is known as "Earth-received time." All times quoted below are in Earth-received time at the Jet Propulsion Laboratory in Pasadena, California. All times are Pacific Daylight Time, which is seven hours behind Universal Time.

At 12:52 a.m. PDT on Sunday, September 21, 2003, the 70-meter-diameter (230-foot) Deep Space Network tracking station antenna near Madrid, Spain, will listen to the spacecraft. The science instruments will be configured and begin to send their data in real time to Earth. *Galileo* is at a distance of 965,000 kilometers (600,000 miles).

About eight hours later, *Galileo* will cross the volcanic satellite Io's orbit at a distance of 422,000 kilometers (262,000 miles). The spacecraft has ventured inside this distance only twice: as it arrived at Jupiter in 1995 and again in 2002, when it made a flyby of the small inner moon Amalthea.

By 9:42 a.m., the intensity of radiation interference will reach a point where even a bright star like Vega can no longer reliably be seen by the attitude control star scanner. The software will now be told to expect to see no more stars.

At 11:31 a.m., *Galileo* will be 143,000 kilometers (89,000 miles) above the clouds and the magnetometer instrument will take its final data for the mission. At this distance from Jupiter, the magnetic field is so strong that the instrument, even in its most robust configuration, would produce a signal that would be completely saturated, and of no further scientific value.

Seventeen minutes later (11:48 a.m.), the spacecraft will pass the orbit of the tiny satellite Amalthea. As the spacecraft passes Amalthea, a special measurement will be taken using the star scanner. During a previous flyby of this small body on November 5, 2002, flashes of light were seen by the star scanner that might indicate the presence of rocky debris circling Jupiter in the vicinity of the satellite. Though on this final pass *Galileo* will not be near Amalthea, the measurement may help confirm or constrain the extent of this hypothesized orbital debris.

At about 12:17 p.m., *Galileo* will pass the orbits of the innermost moons, Adrastea and Metis. *Galileo* will be just 57,500 kilometers (35,700 miles) above the clouds, closing fast, and picking up speed. A few minutes later the *Galileo* Orbiter will join the *Galileo* probe in going closer to Jupiter than any other man-made object. At 43,000 kilometers altitude (26,725 miles), the spacecraft is now at a distance that is $1/9$th of the span between Earth and its own Moon.

At 12:42 p.m. with roughly seven minutes to go, *Galileo* will move from day to night as it passes into Jupiter's shadow, and, one minute later, it will pass behind the limb of the giant planet as seen from Earth. Only 9,283 kilometers (5,768 miles) above the clouds, the path of the spacecraft will take it out of sight of ground controllers. The last data ever to be received from the *Galileo* spacecraft will now been sent. The remaining few minutes of the craft will be spent in darkness.

At approximately 12:49 p.m., *Galileo* will reach the end of its nearly 14-year odyssey.

Program / Project Management

The Galileo project is managed for NASA's Office of Space Science, Washington, D.C., by the Jet Propulsion Laboratory, Pasadena, a division of the California Institute of Technology. JPL designed and built the *Galileo* orbiter, and operates the mission.

At NASA Headquarters, Dr. Barry Geldzahler is Galileo program manager and Dr. Denis Bogan is program scientist.

At JPL, the position of project manager has been held successively by John Casani, Richard Spehalski, Bill O'Neil, Bob Mitchell, Jim Erickson, Eilene Theilig and, currently, Dr. Claudia J. Alexander. Dr. Torrence V. Johnson is project scientist.

NASA's Ames Research Center, Moffett Field, California, managed the descent probe, which was built by Hughes Aircraft Co., El Segundo, California. The position of probe manager was held successively by Joel Sperans, Benny Chinn and Marcie Smith. The probe scientist is Dr. Richard E. Young.

9-15-03

NATIONAL AERONAUTICS AND SPACE ADMINISTRATION

The Jupiter Millennium Mission
Press Kit

The Galileo and Cassini Encounter at the Fifth Planet

October 2000

Contacts

Donald Savage	Policy / Program Management	(202) 358-1753
Headquarters, Washington, D.C.		donald.savage@hq.nasa.gov
Guy Webster	Cassini and Galileo Missions	(818) 354-6278
Jet Propulsion Laboratory, Pasadena, CA		guy.webster@jpl.nasa.gov

CONTENTS

RELEASE:

Spacecraft Double-Team the King of Planets

Two NASA spacecraft are teaming up to scrutinize Jupiter during the next few months to gain a better understanding of the planet's stormy atmosphere, diverse moons, faint rings and vast bubble of electrically charged gas.

The joint studies of the solar system's largest planet by the *Galileo* and *Cassini* spacecraft will also resemble the passing of a baton from the durable veteran to the promising rookie, say mission controllers at NASA's Jet Propulsion Laboratory in Pasadena, California.

Galileo has been running laps around Jupiter since December 1995, continuing to produce scientific discoveries after surviving more than double the orbital time and triple the radiation exposure originally intended for it. It will pass close to Jupiter's largest moon, Ganymede, on December 28.

Cassini left Earth on October 15, 1997, bound for Saturn with a dozen scientific instruments to carry into orbit there and a European-made probe, *Huygens*, to drop onto Saturn's biggest moon in 2004. *Cassini* will make its closest approach to Jupiter on December 30. It will still be nearly 10 million kilometers (6 million miles) away, well outside the orbits of Jupiter's four large moons – Io, Europa, Callisto and Ganymede – but within the orbits of nine small ones.

Cassini began transmitting Jupiter pictures and data this month.

"We have a chance to make observations with a well-instrumented spacecraft that has more capabilities than any spacecraft that has previously visited Jupiter," said Robert Mitchell, JPL's Cassini program manager. "Fortunately, *Galileo* is still operating there, so we can get a synergistic effect in studies of Jupiter by having spacecraft at two different locations in the vicinity of Jupiter at the same time. That's not something we could have counted on in 1995."

One joint study will examine how the "solar wind" of charged particles speeding away from the Sun buffets Jupiter's magnetosphere, the bubble of charged gas rotating around Jupiter under the control of the planet's magnetic field. In November, *Cassini* will be in the solar wind upstream of where the wind hits the magnetosphere, while *Galileo* will be inside the magnetosphere. *Cassini* will monitor fluctuations in the solar wind while *Galileo* watches the response of Jupiter's magnetosphere to those fluctuations.

During the past five years, *Galileo* has measured frequent changes in the density of particles in the magnetosphere, but researchers have not had the opportunity to connect the effects to specific changes in the solar wind, said Dr. Torrence Johnson, Galileo project scientist at JPL.

JPL physicist Dr. Scott Bolton, on science teams for both *Cassini* and *Galileo*, said, "Having two spacecraft there at once is possibly the only chance in our lifetime to simultaneously connect changes in the solar wind to conditions inside Jupiter's giant magnetosphere."

Getting a better grasp on how Jupiter's magnetosphere acts and reacts will advance understanding of the smaller magnetosphere surrounding Earth and larger ones affecting areas of the galaxy where stars are being born, Bolton said. Disturbances in Earth's magnetosphere can disrupt electrical and communications systems.

Another study taking advantage of dual vantage points will focus on a stream of dust, finer than particles in cigarette smoke, originating from volcanoes on Jupiter's moon Io. Patterns in the stream as it passes first one satellite, then another, could give information about the dust's movement. Researchers also hope to identify its composition, which would be a sampling of material from Io.

Both spacecraft will study eclipses of Jupiter's large moons. While the moons are in the shadow of Jupiter, glows can be seen that are overwhelmed by reflected sunlight at other times. Excitation of the moons' thin atmospheres by energetic particles in Jupiter's magnetosphere causes the glows. Researchers hope to learn more about gases on the moons by studying these glows.

Cassini will study Jupiter's atmosphere from October through March as the craft approaches from the sunny side, then recedes from the dark side of the planet. "If we're lucky, we may even see a storm arise, and see how it starts and how it evolves," said Dr. Dennis Matson, Cassini project scientist at JPL. The Jupiter studies will also provide a dress rehearsal, checking out equipment and procedures for *Cassini*'s main mission at Saturn, Matson said.

JPL manages the Cassini and Galileo missions for NASA's Office of Space Science, Washington DC. JPL is a division of the California Institute of Technology, Pasadena, California. Cassini is a cooperative endeavor of NASA, the European Space Agency and the Italian Space Agency.

More information on the joint spacecraft study of Jupiter is available at http://www.jpl.nasa.aov/jupiterflyby. An expanded press kit on the study is available at http://www.jpl.aov/presskits/jupiterflyby.

(End of General Release)

Media Services Information

NASA Television Transmission

NASA Television is broadcast on the satellite GE-2, transponder 9C, C Band, 85 degrees west longitude, frequency 3880.0 MHz, vertical polarization, audio monaural at 6.8 MHz. The schedule for television transmission of video animations, B-roll and live-interview opportunities will be available from the Jet Propulsion Laboratory, Pasadena, California, and NASA Headquarters, Washington, DC.

News Releases

News releases, image advisories and status reports about the *Cassini* and *Galileo* studies of Jupiter will be issued by the Jet Propulsion Laboratory's Media Relations Office. They may be accessed on line as noted below.

Briefings

A science briefing about the *Galileo* and *Cassini* joint studies of Jupiter will be held at the Jet Propulsion Laboratory, Pasadena, California, on December 30, 2000, the date of *Cassini*'s closest approach to Jupiter.

Additional details about the briefing will be available closer to those dates from NASA Headquarters and the Jet Propulsion Laboratory.

Internet Information

Extensive information about the Cassini and Galileo missions and their joint studies of Jupiter, including an electronic copy of this Press Kit, press releases, fact sheets, status reports and images, is available from the Jet Propulsion Laboratory's World Wide Web home page at http://www.jpl.nasa.gov/. Sign-up for e-mail subscriptions to news releases is available at the home page.

A special Internet site for the Jupiter Millennium Flyby offers graphics and educational material, and will provide updates of pictures and data gathered by the spacecraft and by related Jupiter research. The site is at http://www.jpl.nasa.gov/jupiterflyby.

The Cassini Program also maintains a home page at http://www.jpl.nasa.gov/cassini, and the Galileo Project maintains one at http://galileo.jpl.nasa.gov.

Cassini Quick Facts

Spacecraft

- Spacecraft dimensions: 6.7 meters (22 feet) high; 4 meters (13.1 feet) wide
- Weight: 5,712 kilograms (12,593 pounds) with fuel, *Huygens* probe and adapter. Unfueled orbiter alone weighs 2,125 kilograms (4,685 pounds)
- Science instruments: camera; magnetic field studies; dust and ice grain analysis; infrared energy measurement; chemical composition of Saturn, its moons and rings; neutral and charged particle measurement; radar mapping; and gravitational wave searches
- Power: 885 watts (at launch) from radioisotope thermoelectric generators

Huygens Probe

- Probe dimensions: 2.7 meters (8.9 feet) in diameter
- Weight: 320 kilograms (705 pounds)
- Science instruments: spectrometer to identify atmospheric makeup; aerosol collector for chemical analysis; imager; sensors to measure atmospheric structure; wind speed measurements; sensors to measure conditions at impact site

Launch

- Launch vehicle: Titan IVB / Centaur Upper Stage

Mission

- Launch: October 15, 1997, from Cape Canaveral Air Force Station, Florida.
- Venus flybys: April 26, 1998, at 337 kilometers (209 miles); June 24, 1999, at 623 kilometers (387 miles)
- Earth flyby: August 18, 1999, at 1,175 kilometers (730 miles)
- Jupiter flyby: December 30, 2000, at 9.7 million kilometers (6 million miles)
- Saturn arrival date: July 1, 2004
- Huygens probe Titan entry date: November 27, 2004
- Distance traveled to reach Saturn: 3.5 billion kilometers (2.2 billion miles)

Program

- Partners: NASA, the European Space Agency (ESA), Italian Space Agency (ASI, for Agenzia Spaziale Italiana); total of 17 countries involved
- U.S. states in which *Cassini* work has been carried out: 33
- Cost of mission: Total about $3.26 billion, including $1.422 billion pre-launch development, $704 million mission operations, $54 million tracking and $422 million launch vehicle, for U.S. cost of $2.6 billion, plus $500 million ESA, $160 million ASI.

Galileo Quick Facts

Spacecraft

- Spacecraft dimensions: 5.3 meters (17 feet) high; magnetometer boom extends 11 meters (36 feet) to one side
- Orbiter weight: 2,223 kilograms (4,902 pounds), including 118 kilograms (260 pounds) of science instruments and 925 kilograms (2040 pounds) of usable rocket propellant
- Orbiter science instruments: camera; spectrometers to identify atmospheric makeup; atmospheric particles and reflectance studies; magnetic fields measurements; investigations of charged particles and dust; plasma wave detection
- Power: 570 watts (at launch) from radioisotope thermal generators

Descent Probe

- Probe size: 127 centimeters (50 inches) diameter, 91 centimeters (36 inches) high
- Probe weight: 339 kilograms (750 pounds)
- Probe instruments: Devices to measure temperature, pressure, atmospheric composition, clouds, particles, and energy from lightning

Launch

- Launch vehicle: Space Shuttle *Atlantis*, on mission STS-34, carried *Galileo* and its two-stage Inertial Upper Stage into Earth orbit

Mission

- Launch: October 18, 1989, from Kennedy Space Center, Florida
- Venus flyby: February 10, 1990, at 16,000 kilometers (10,000 miles)
- Earth flybys: December 8, 1990, at 960 kilometers (597 miles); December 8, 1992, at 303 kilometers (188 miles)
- Probe descent into Jupiter's atmosphere: December 7, 1995
- *Galileo* orbiter in Jupiter orbit since: December 7, 1995
- Current orbit number during fall 2000: 29

Program

- Cost of mission: $1.39 billion plus foreign contribution of $110 million

Millennium Flyby Science Objectives

The *Cassini* and *Galileo* spacecraft will study several aspects of Jupiter and its surrounding environment from October 2000 through March 2001, before, during and after *Cassini*'s closest approach to Jupiter on December 30, 2000. Some of the scientific observations will take advantage of having two different vantage points in the vicinity of the planet at the same time. Some will make the most of using *Cassini*'s measurement and transmission capabilities during its flyby to answer questions raised by longer-term *Galileo* observations. Here are some of the science objectives for Jupiter studies during the next few months:

- Examine interactions between the solar wind and Jupiter's magnetosphere by having one spacecraft inside the magnetosphere while the other is upstream in the solar wind.
- Detail Jupiter's atmospheric dynamics with *Cassini* imaging instruments in visible, infrared and ultraviolet wavelengths. Enough images are planned to produce a movie of Jovian atmospheric dynamics.
- Observe Io and Ganymede during eclipses with both spacecraft to characterize surface gases and learn about surface properties, such as how fluffy or porous the surface is.

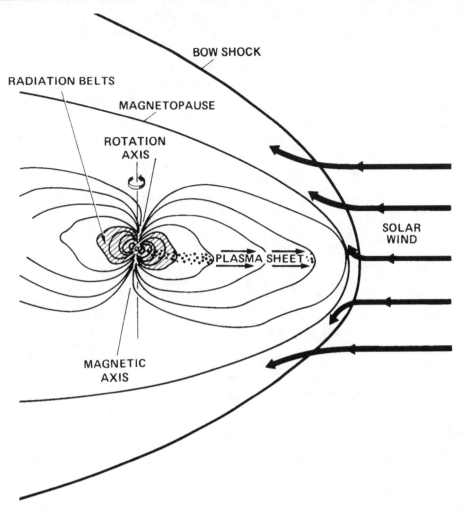

- Observe microscopic dust particles from Io with both spacecraft for information about their movement and composition.
- Study any changes in Io torus for connections with Io's volcanoes.
- Use *Cassini* spectrometers to study the surface composition of Jupiter's moons.
- Map changes in the highly directional "synchrotron" radio emissions from Jupiter's magnetosphere. Synchrotron radiation is the emission of very directional radio waves by electrons that are spiraling rapidly under the influence of the magnetic field.
- Determine rotation period of the small moon Himalia.
- Study particle size in rings.

Telescopes on the ground will join *Cassini* and *Galileo* in studying Jupiter. Jupiter's synchrotron radiation will be studied in the radio wavelengths using the Very Large Array radio telescope in Socorro, New Mexico. A bit farther afield, in Earth orbit, the Hubble Space Telescope will be watching Jupiter as well. Observations coordinated between *Cassini*, *Galileo* and Hubble will study how Jupiter's aurora changes in response to changes in the solar wind near Jupiter and changes in Jupiter's magnetosphere.

Middle-school and high-school students participating in the Goldstone Apple Valley Radio Telescope project will also be watching Jupiter in radio wavelengths. Students will collect ground-based observations of Jupiter throughout the fall and winter using two radio telescopes located in Goldstone, California. The students' observations will be combined with *Cassini* and *Galileo* observations to model the radiation environment of Jupiter's magnetosphere.

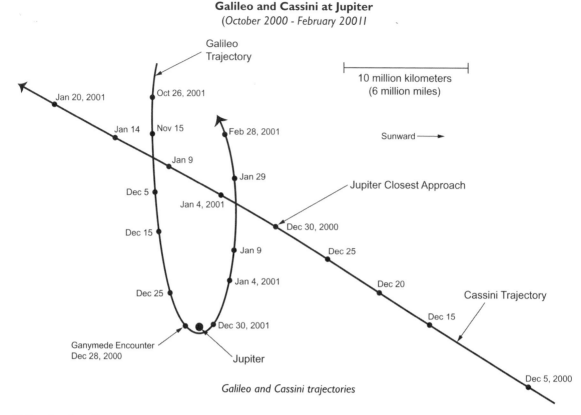

Galileo and Cassini trajectories

Why Jupiter?

Jupiter predominates among our solar system's planets. It is more than twice as massive as all the other planets put together. It resembles the Sun in its hydrogen-rich composition and in having its own coterie of 17 diverse worlds in orbit around it.

The Planet

Jupiter is the fourth brightest natural object we see in the sky from Earth after the Sun, Moon and Venus so it has captured the attention of sky watchers since prehistoric times. It was named for the ruler and most powerful god of Roman mythology. Jupiter's size helps make it bright. Its volume is as great as 1,400 Earths. However, Jupiter's overall density is only a little more than the density of water. It is a gas planet, not a rocky one like Earth.

Through a telescope, Jupiter appears as a yellowish disc crossed with parallel orange-red bands. Since the mid-1600s, astronomers have noted spots moving across Jupiter's face as the planet rotates. Some of the spots and other cloud features last for years, much longer than clouds and storm systems do on Earth. Best known is the "Great Red Spot," an oval about three times as wide as Earth. It has likely persisted for more than three centuries.

Astronomers have used these moving spots to roughly gauge the planet's rotation period. A Jupiter "day" is less than 10 hours, making the biggest planet of the solar system also the quickest-spinning one. The fast twirling causes Jupiter to bulge out in the middle. Its diameter is about 10 percent greater at the equator than it is from pole to pole.

Four NASA spacecraft flew past Jupiter in the 1970s: *Pioneer 10*, *Pioneer 11*, *Voyager 1* and *Voyager 2*. Their instruments returned data about Jupiter's atmosphere. The atmosphere is predominantly hydrogen and helium, with less than one percent from a combination of other ingredients, including methane, ammonia, phosphorus, water vapor, and various hydrocarbons.

How We Compare		
Characteristic	**Earth**	**Jupiter**
Equatorial diameter	12,756 kilometers (7,926 miles)	142,984 kilometers (88,846 miles)
Mean density (grams per cubic centimeter)	5.52	1.33
Orbital revolution (Earth years)	1	11.86
Rotation period (hours)	23.93	9.92
Atmosphere	77% nitrogen 21 % oxygen	86% hydrogen 14% helium

Jupiter's atmosphere displays alternating patterns of dark belts and light zones. The locations and sizes of the belts and zones change gradually with time. Within these belts and zones are clouds and storm systems that have raged for years. The Great Red Spot giant storm rotates once counter-clockwise every six days. Since it is in the southern hemisphere of the planet, this rotational direction indicates it is a high-pressure zone, unlike Earth's cyclones, which are low-pressure zones.

Jupiter radiates more energy into space than it receives from the Sun. Its interior has a sustained temperature of approximately 20,000 degrees Kelvin (35,500 degrees Fahrenheit). The interior heat is generated by the slow gravitational compression of the planet. The heat flows outward from the planet's core, driving the circulation present in the Jovian atmosphere. This circulation process, called convection, involves the rise of hot gas from the Jovian interior to higher atmospheric layers where it cools and flows back down.

Jupiter's fast rotation combines with convection to drive the planet's global wind patterns. The light and dark bands visible through telescopes provide evidence of these processes. The white zones are colored by ammonia ice crystals while the dark belts are colored by unknown materials, collectively called chromophores. Theoretical studies indicate there are three cloud layers in the upper reaches of Jupiter's atmosphere. From the top down, the clouds are composed of ammonia ice, ammonium hydrosulfide, and water ice.

At some relatively unclouded and dry areas, heat from the planet's lower levels glows through brightly in infrared wavelengths. *Galileo*'s atmospheric probe descended into one of these "hot spots" on December 7, 1995.

The Jovian System

Just as the Sun has its solar system, Jupiter has its Jovian system. Studies of Jupiter's system have been advancing our understanding of the broader solar system for nearly four centuries.

In 1610, Galileo Galilee discovered Jupiter's four largest natural satellites, or moons: Io, Europa, Ganymede and Callisto. These four moons are now collectively called the Galilean satellites. This discovery was the first observational evidence of a center of motion not apparently centered on the Earth. It was a major argument in favor of Copernicus' heliocentric theory of the motions of the planets, which was an unpopular theory in early 17th century theology. Galileo's support of the Copernican theory got him arrested for heresy by the Inquisition and led to his life imprisonment under house arrest.

Researchers have now found 17 moons circling Jupiter. The latest discovery was announced in June 2000. They range from Ganymede, larger than the planets Mercury or Pluto, to rocks less than 50 kilometers (30 miles) across. NASA's *Galileo* spacecraft, orbiting Jupiter since 1995, has returned a steady stream of information about features of the Galilean satellites, such as the volcanoes on Io and ice on Europa. The spacecraft has provided several lines of evidence that Europa may cloak a deep ocean of liquid water under its surface ice. That has fueled speculation about possible life there, since life as we know it requires liquid water.

Besides moons, the Jovian system includes dark rings and a huge magnetosphere of energetic particles moving under the influence of the planet's strong magnetic field.

Jupiter's Moons

Jupiter's natural satellites take their names from women, nymphs and boys linked to the god Jupiter, or Zeus, in Roman and Greek mythology.

Four small moons orbit closer to Jupiter than any of the Galilean moons do.

Metis, the innermost, and Adrastea, next closest in, were both discovered through data from the *Voyager 1* flyby in 1979. Metis and Adrastea lie within Jupiter's main ring and may be the sources of material for the ring.

Amalthea, next out, was discovered in 1892 by Edward Emerson Barnard. It has an irregular shape, much longer in one dimension than others. Some of its many craters are extremely large relative to the size of the moon. The surface shows a dark red color, possibly from a dusting of sulfur from Io's volcanoes. Bright patches of unidentified green material appear on major slopes of Amalthea.

Thebe, the fourth innermost of Jupiter's satellites, also was discovered through *Voyager 1* data. Thebe takes about 16 hours to complete each orbit around Jupiter.

Active lava flows are resurfacing portions of Io, the most volcanically active world in the solar system. Plumes from the volcanoes extend to more than 300 kilometers (190 miles) above the surface, with material being ejected at speeds up to a kilometer (0.6 miles) per second. Io's volcanoes apparently result from heating of the moon by tidal pumping. Io is perturbed in its orbit by Europa and Ganymede, two other large moons nearby. As a consequence of the forced eccentric orbit, the distance between Io and Jupiter changes as Io orbits, and the large gravitational pull from Jupiter results in tidal bulging as great as 100 meters (330 feet) on Io's surface.

Jupiter's Moons

Satellite	Discovery	Orbit (from Jupiter)		Diameter	Orbital Period
Metis	1979	128,000 km	(80,000 miles)	40 km (25 miles)	7 hours
Andrastea	1979	129,000 km	(80,000 miles)	20 km (12 miles)	7 hours
Amalthea	1892	181,300 km	(113,000 miles)	270 ×166 km (167 ×103 miles)	12 hours
Thebe	1979	222,000 km	(138,000 miles)	100 × 90 km (62 × 56 miles)	16 hours
Io	1610	422,000 km	(262,000 miles)	3,630 km (2,256 miles)	1.77 days
Europa	1610	671,000 km	(416,000 miles)	3,138 km (2,256 miles)	3.55 days
Ganymede	1610	1,070,000 km	(665,000 miles)	5,262 km (3,270 miles)	7.15 days
Callisto	1610	1,883,000 km	(1,170,000 miles)	4,800 km (2,983 miles)	16.7 days
Leda	1974	11,094,000 km	(6,895,000 miles)	16 km (10 miles)	239 days
Himalia	1904	11,480,000 km	(7,135,000 miles)	186 km (116 miles)	251 days
Lysithea	1938	11,720,000 km	(7,284,000 miles)	36 km (22 miles)	260 days
Elara	1905	11,737,000 km	(7,295,000 miles)	76 km (47 miles)	260 days
Ananke	1951	21,200,000 km	(13,176,000 miles)	30 km (19 miles)	631 days
Carme	1938	22,600,000 km	(14,046,000 miles)	40 km (25 miles)	692 days
Pasiphae	1908	23,500,000 km	(14,605,000 miles)	50 km (31 miles)	735 days
Sinope	1914	23,700,000 km	(14,730,000 miles)	36 km (22 miles)	758 days
S/1999 11	2000	24,200,000 km	(15,040,000 miles)	12 km (7 miles)	774 days

Io is located within an intense radiation belt of electrons and ions trapped in Jupiter's magnetic field. As the magnetosphere rotates with Jupiter, it sweeps past Io and strips away about 1,000 kilograms (1 ton) of material per second. The material forms a torus, a doughnut-shaped cloud of ions that glow in the ultraviolet. The torus' heavy ions migrate outward, and their pressure inflates the Jovian magnetosphere to more than twice its expected size. Some of the more energetic sulfur and oxygen ions fall along the magnetic field into the planet's atmosphere, resulting in auroras.

Europa shows a bright surface of ice, with few craters but many intersecting linear features, apparently left from fissuring of the ice. Gravity measurements by *Galileo* show the uppermost layers of Europa to a depth of about 100 kilometers (60 miles) to consist of water, either frozen or liquid. Internal heating from tidal forces similar to Io's may keep some of the water ice melted.

Pictures from the *Galileo* spacecraft show large rafts of ice that have apparently broken away from each other and rearranged themselves. That is one line of evidence for a fluid layer underneath. Another comes from magnetic field features that could be explained by a layer of electrically conductive saltwater, but not by solid ice.

Ganymede is the largest moon in our solar system. Like Callisto, the next farthest out, Ganymede likely has a rocky core with a water-ice mantle and a crust of rock and ice.

Ganymede has had a complex geological history. It has mountains, valleys, craters and lava flows. Ganymede is mottled by both light and dark regions. It is heavily cratered in the dark regions, implying ancient origin. The bright regions show a different kind of terrain, grooved with ridges and troughs. The grooved features were apparently formed more recently than the dark, cratered area, perhaps by tension from global tectonic processes.

Callisto, second largest of Jupiter's moons, orbits just beyond the planet's main radiation belt. Callisto is the most heavily cratered satellite in the solar system. Its crust dates back 4 billion years, shortly after the solar system was formed.

Callisto has the lowest density of the Galilean satellites, though it is still denser than Jupiter itself. From recent observations made by the *Galileo* spacecraft, Callisto appears to have an icy crust about 200 kilometers (124 miles) thick. Beneath the crust is a possible salty ocean more than 10 kilometers (6 miles) deep. The *Galileo* data suggest the interior is composed of compressed rock and ice with the percentage of rock increasing as depth increases. Meteorites have punched holes in Callisto's crust, causing water to spread over the surface and form bright rays and rings around the craters.

Jupiter's nine outer moons fall into two distinct orbital groups. Leda, Himalia, Lysithea, and Elara all orbit at approximately 11 million kilometers (7 million miles) from Jupiter. Ananke, Carme, Pasiphae, Sinope, and the most recently discovered, still unnamed moon orbit at approximately 23 million kilometers (14 million miles). None of these 8 outer satellites has been explored in any detail or with enough resolution to gain a full understanding of the nature of the terrain.

Jupiter's Rings

The three known rings around Jupiter are thin and very dark. Only about five percent of the sunlight falling on the rings is reflected back. Because of this combination, the rings remained hidden from human discovery until the *Voyager I* spacecraft saw them backlit by the sun in 1979.

Jupiter's Rings

Ring	Distance from Jupiter*	Ring Width
Halo	92,000 km (57,178 miles)	30,500 km (18,956 miles)
Main	122,500 km (76,134 miles)	6,440 km (4,002 miles)
Gossamer	128,940 km (80,136 miles)	100,000 km (62,150 miles)

* Distance is measured from Jupiter's center to the ring's inner edge.

The rings' dark appearance suggests they are composed of grains of rocky material and contain little or no ice. This differs from Saturn's vast ring system, which is primarily water ice in composition.

Atmospheric and magnetic drag doesn't allow particles in Jupiter's rings to remain in orbit for long. Data from the *Galileo* spacecraft shows that ring material is being constantly resupplied by dust formed through micrometeorite impacts on Jupiter's four innermost satellites.

Jupiter's Magnetosphere

The layer of metallic hydrogen within fast-spinning Jupiter generates an enormous magnetic field around the planet. Earth also has a magnetic field, generated by turning of Earth's iron core, but Jupiter's is much stronger, the strongest of any planet's.

Jupiter's magnetosphere is a tear-shaped bubble of charged particles (electrons and ions) constrained under the influence of the magnetic field. It is the largest thing in the solar system except the solar wind, which reaches across the entire solar system. Jupiter's magnetosphere tails outward past the orbit of Saturn and, if it shined in the visible wavelengths of the spectrum instead of at radio wavelengths, it would appear two to three times wider than the disc of the Sun or Moon to viewers on Earth.

The inner part of Jupiter's magnetosphere is doughnut-shaped, but farther out it flattens into a disc. The magnetic poles are tilted relative to Jupiter's axis of rotation, so the field appears to wobble around with Jupiter's fast rotation, sweeping up and down across the inner moons and making waves throughout the magnetosphere.

The solar wind, a variable flow of electrically charged particles blowing outward from the Sun, interacts with the magnetosphere. The field deflects most of the solar wind particles, and the wind shapes the magnetosphere. The boundary between the solar wind and the magnetosphere is called the "magnetopause." A "bow shock," named for the wave that builds up before a ship's bow as it plows through water, is formed in the solar wind upstream from the magnetopause. Downstream, on the side of the planet away from the Sun, the drag of the solar wind draws the magnetosphere out into an elongated "magnetotail."

Cassini Spacecraft and Mission

The Cassini mission to Saturn is the one of the most ambitious efforts in planetary space exploration ever mounted. A joint endeavor of NASA and the European Space Agency (ESA), with additional involvement of the Italian Space Agency, Agenzia Spaziale Italiana, *Cassini* will study the Saturnian system in detail over a four-year period beginning in 2004.

The sophisticated robotic spacecraft will go into orbit around the ringed planet, and will deliver a scientific probe called *Huygens* to be released from the main spacecraft to parachute through the atmosphere to the surface of Saturn's largest and most interesting moon, Titan.

Cassini's journey began on October 15, 1997, with liftoff of a Titan IVB / Centaur launch vehicle from Pad 40 at the Cape Canaveral Air Force Station in Florida.

In maneuvers called gravity assist flybys, *Cassini* has flown twice past Venus then once past Earth. These flybys increased the spacecraft's speed as it approached and flew past each planet.

Jupiter Flyby

Those three gravity assists could not give *Cassini* quite as much cumulative boost as it needs to reach Saturn. The main reason *Cassini* will fly near Jupiter in December 2000 is for a final gravity assist. The opportunity to do some science at Jupiter is a side benefit. The Jupiter studies will also put *Cassini's* instruments, systems and personnel through a thorough checkout in preparation for Saturn.

Cassini's distance from Jupiter at the time of its closest approach, at 5:12 a.m. Eastern Standard Time on December 30, 2000, will be about 9.7 million kilometers (6 million miles). That is well beyond the orbits of Jupiter's four largest moons, but within the orbits of nine small ones. It is the distance calculated to give the spacecraft just the right gravitational assist needed for the final leg of its trip to Saturn.

After a nearly seven-year journey, *Cassini* will arrive at Saturn July 1, 2004, where it will enter orbit and begin its detailed scientific observations of the Saturnian system.

Saturn is the second largest planet in the solar system. Like the other gaseous outer planets — Jupiter, Uranus and Neptune — its atmosphere is made up mostly of hydrogen and helium. Saturn's distinctive, bright rings are made up of ice and rock particles ranging in size from grains of sand to boxcars. More moons of greater variety orbit Saturn than any other planet. So far, observations from Earth and by spacecraft have found 18 Saturnian satellites ranging from small asteroid-size bodies to the aptly named Titan, which is larger than the planet Mercury. The 12 scientific instruments on the *Cassini* orbiter will conduct in-depth studies of the planet, its moons, rings and magnetic environment. The six instruments on the *Huygens* probe, which will be dispatched from *Cassini* during its first orbit of Saturn, will provide our first direct sampling of Titan's atmospheric chemistry and the first photographs of its hidden surface.

Saturn's face appears placid, but masks a windswept atmosphere where an equatorial jet stream blows at 1,800 km/hr (1,100 mph) and swirling storms roil beneath the cloud tops. Early explorations by NASA's *Pioneer* and *Voyager* spacecraft between 1979 and 1981 also found Saturn to possess a huge and complex magnetic environment where trapped protons and electrons interact with each other, the planet, the rings and the surfaces of many of the satellites. Saturn's famous rings were found to consist of not just a few monolithic bands but thousands of rings and ringlets with particles sometimes herded into complicated orbits by the gravitational interaction of small moons previously unseen from Earth.

Haze-covered Titan offers a tantalizing mix of a nitrogen-methane atmosphere and a surface that may feature chilled lakes of ethane or coatings of sticky brown organic condensate that has rained down from the atmosphere. Standing on Titan's surface beneath an orange sky, a visitor from Earth likely would find a cold, exotic world with a pungent odor reminiscent of a petroleum processing facility. Because Titan and Earth share so much in atmospheric composition, Titan is thought to hold clues to how the primitive Earth evolved into a life-bearing planet.

Mission at Saturn

Upon reaching Saturn on July 1, 2004, *Cassini* will fire its main engine for 96 minutes to brake the spacecraft's speed and allow it to be captured as a satellite of Saturn. Passing through the dusty, outermost E-ring, *Cassini* will swing in close to the planet to an altitude only one-sixth the diameter of Saturn itself to begin the first of some six dozen orbits during the rest of its four-year mission.

On November 6, 2004, *Cassini* will release the European-built *Huygens* probe toward Titan. On November 27, *Huygens* will enter Titan's atmosphere, deploy its parachutes and begin its scientific observations during a descent of up to two and a half hours through that moon's dense atmosphere. Instruments onboard will measure the temperature, pressure, density and energy balance in the atmosphere. As the probe breaks through the cloud deck, a camera will capture pictures of the Titan panorama. Titan's surface properties will be observed, and about 1,000 images of the clouds and surface will be returned. In the final moments of descent, a spotlight will illuminate the surface for the camera.

If the probe survives landing at about 25 km/hr (15 mph), it can possibly return data from Titan's surface, where the atmospheric pressure is 1.6 times that of Earth's. The probe could touch down on solid ground, ice or even splash down in a lake of ethane or methane. One instrument on board will discern whether *Huygens* is bobbing in liquid, and other instruments on board would tell the chemical composition of that liquid. Throughout its mission, *Huygens* will radio data collected by its instruments to the *Cassini* orbiter to be stored and then relayed to Earth.

If the probe continues to send data to *Cassini* from Titan's surface, it will be able to do so for only about 30 minutes, when the probe's battery power is expected to run out.

After the *Huygens* probe mission is complete, *Cassini*'s focus will be on taking measurements with the orbiter's 12 instruments and returning the information to Earth. During the course of *Cassini*'s mission, it will execute dozens of close flybys of Titan and several close flybys of selected icy moons. In addition, the orbiter will make other, more-distant flybys of Saturn's moons. *Cassini*'s orbits will also allow it to study Saturn's polar regions in addition to the planet's equatorial zone.

Cassini Spacecraft

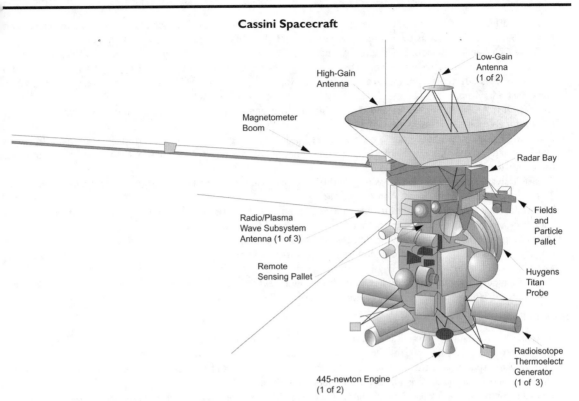

The Cassini Spacecraft

The *Cassini* spacecraft, including the orbiter and the *Huygens* probe, is one of the largest, heaviest and most complex interplanetary spacecraft ever built. The orbiter alone weighs 2,125 kilograms (4,685 pounds). With the 320-kilogram (705-pound) *Huygens* probe and a launch vehicle adapter attached, and 3,132 kilograms (6,905 pounds) of propellants loaded, the spacecraft weighed 5,712 kilograms (12,593 pounds) at launch. Only the two *Phobos* spacecraft sent to Mars by the former Soviet Union were heavier.

The *Cassini* spacecraft stands more than 6.7 meters (22 feet) high and is more than 4 meters (13 feet) wide. The magnetometer instrument is mounted on an 11-meter (36-foot) long boom that extends outward from the spacecraft. Three other rod-like booms that each reach 10 meters (about 32 feet) outward from the spacecraft are antennas for the radio plasma wave subsystem.

The complexity of the spacecraft is necessitated both by its flight path to Saturn and by the ambitious program of scientific observations to be undertaken once the spacecraft reaches its destination. It has 22,000 wire connections and more than 14 kilometers (8.7 miles) of cabling. Because of the very dim sunlight at Saturn's orbit, solar arrays are not feasible, so electrical power is supplied by a set of radioisotope thermoelectric generators, which use heat from the natural decay of plutonium-238 to generate electricity to run *Cassini*'s systems. These power generators are of the same design as those used on the *Galileo* and *Ulysses* missions.

Science Experiments

Equipment for 12 science experiments is carried onboard the *Cassini* orbiter. Another six fly on the *Huygens* probe. Descriptions of the instruments and the names of the principal investigators or team leaders are:

Orbiter:

- Imaging Science Subsystem: Takes pictures in visible, near-ultraviolet and near-infrared light. Dr. Carolyn Porco, University of Arizona, Tucson, Arizona.
- *Cassini* Radar: Maps surface of Titan using radar imager to pierce the veil of haze. Also used to measure heights of surface features. Dr. Charles Elachi, NASA's Jet Propulsion Laboratory, Pasadena, California.
- Radio Science Subsystem: Searches for gravitational waves in the universe; studies the atmosphere, rings and gravity fields of Saturn and its moons by measuring telltale changes in radio waves sent from the spacecraft. Dr. Arvydas J. Kliore, NASA's Jet Propulsion Laboratory, Pasadena, California.
- Ion and Neutral Mass Spectrometer: Examines neutral and charged particles near Titan, Saturn and the icy satellites to learn more about their extended atmospheres and ionospheres. Dr. J. Hunter Waite, Southwest Research Institute, San Antonio, Texas.

- Visible and Infrared Mapping Spectrometer: Identifies the chemical composition of the surfaces, atmospheres and rings of Saturn and its moons by measuring colors of visible light and infrared energy given off by them. Dr. Robert H. Brown, University of Arizona, Tucson, Arizona.
- Composite Infrared Spectrometer: Measures infrared energy from the surfaces, atmospheres and rings of Saturn and its moons to study their temperature and composition. Virgil Kunde, NASA's Goddard Space Flight Center, Greenbelt, Maryland.
- Cosmic Dust Analyzer: Studies ice and dust grains in and near the Saturn system. Dr. Eberhard Grim, Max Planck Institut fur Kernphysik, Heidelberg, Germany.
- Radio and Plasma Wave Spectrometer: Investigates plasma waves (generated by ionized gases flowing out from the Sun or orbiting Saturn), natural emissions of radio energy and dust. Dr. Donald A. Gurnett, University of Iowa, Iowa City, Iowa.
- *Cassini* Plasma Spectrometer: Explores plasma (highly ionized gas) within and near Saturn's magnetic field. Dr. David T. Young, University of Michigan, Ann Arbor, Michigan.
- Ultraviolet Imaging Spectrograph: Measures ultraviolet energy from atmospheres and rings to study their structure, chemistry and composition. Dr. Larry Esposito, University of Colorado, Boulder, Colorado.
- Magnetospheric Imaging Instrument: Images Saturn's magnetosphere and measures interactions between the magnetosphere and the solar wind, a flow of ionized gases streaming out from the Sun. Dr. Stamatios M. Krimigis, Johns Hopkins University Applied Physics Laboratory, Baltimore Maryland.
- Dual Technique Magnetometer: Studies Saturn's magnetic field and its interactions with the solar wind, the rings and the moons of Saturn. Dr. David J. Southwood, Imperial College of Science & Technology, London, England.

Huygens Probe:

- Descent Imager and Spectral Radiometer: Makes images and measures temperatures of particles in Titan's atmosphere and on Titan's surface. Dr. Martin Tomasko, University of Arizona, Tucson, Arizona.
- *Huygens* Atmospheric Structure Instrument: Explores the structure and physical properties of Titan's atmosphere. Dr. Marchello Fulchignoni, Paris Observatory, Meudon, France.
- Gas Chromatograph and Mass Spectrometer: Measures the chemical composition of gases and suspended particles in Titan's atmosphere. Dr. Hasso B. Neimann, NASA's Goddard Space Flight Center, Greenbelt, Maryland.
- Aerosol Collector Pyrolyzer: Examines clouds and suspended particles in Titan's atmosphere. Dr. Guy M. Israel, Service d'Aeronomie du Centre National de la Recherche Scientifique, Verrieres-le-Buisson, France.
- Surface Science Package: Investigates the physical properties of Titan's surface. Dr. John C. Zarnecki, University of Kent, England.
- Doppler Wind Experiment: Studies Titan's winds from their effect on the probe during its descent. Dr. Michael K. Bird, University of Bonn, Germany.

Cassini Signature Disk

In August 1997, a small digital versatile disk (DVD) was installed aboard the *Cassini* spacecraft during processing at NASA's Kennedy Space Center. The disk contains a record of 616,400 handwritten signatures from 81 countries around the globe. Signatures were received from people of all ages and backgrounds.

Mail came from individuals, families and, often, from entire schools of students. Signatures came from the very young, just learning to write, and from the very old, whose hands were no longer steady. Signatures were sent in behalf of loved ones who had died in the recent past. Even the signatures of Jean-Dominique Cassini and Christiaan Huygens were obtained from letters they wrote during the 17th century.

Sorting, counting and scanning the signatures was performed by volunteers from The Planetary Society, Pasadena, California.

Galileo Spacecraft and Mission

NASA's *Galileo* spacecraft was designed to study the large, gaseous planet Jupiter, its moons, and its surrounding magnetosphere, which is a magnetic bubble surrounding the planet. The craft was named for the Italian Renaissance scientist who discovered Jupiter's major moons in 1610 with the first astronomical telescope.

Mission Overview

Galileo's primary mission at Jupiter began when the spacecraft entered into orbit around Jupiter in December 1995, and its descent probe, which had been released five months earlier, dove into the giant planet's atmosphere. Its primary mission included a 23-month, 11-orbit tour of the Jovian system, including 10 close encounters of Jupiter's major moons.

Although the primary mission was completed in December 1997, the mission has been extended twice since then. The first extended mission, known as the Galileo Europa mission, included 14 additional encounters of Jupiter's major moons − eight with Europa, four with Callisto and two with Io. The Galileo Millennium mission, in progress through the end of 2000, has included additional flybys of Io and Ganymede.

Launch

The *Galileo* spacecraft and its two-stage Inertial Upper Stage (IUS) were carried into Earth orbit on October 18, 1989 by space shuttle *Atlantis* on mission STS-34. The two-stage IUS solid rocket then accelerated the spacecraft out of Earth orbit toward the planet Venus for the first of three planetary "gravity assists" designed to boost *Galileo* toward Jupiter. In a gravity assist, the spacecraft flies close enough to a planet to be propelled by its gravity, creating a "slingshot" effect for the spacecraft. The Galileo mission had originally been designed for a direct flight of about three and a half years to Jupiter, using a three-stage IUS. When this vehicle was canceled, plans were changed to a Centaur upper stage, and ultimately to the two-stage IUS, which precluded a direct trajectory. To save the project, Galileo engineers designed a flight path using planetary gravity assists.

Venus and Earth Flybys

After flying past Venus at an altitude of 16,000 kilometers (nearly 10,000 miles) on February 10, 1990, the spacecraft swung past Earth at an altitude of 960 kilometers (597 miles) on December 8, 1990. That flyby increased *Galileo*'s speed enough to send it on a two-year elliptical orbit around the Sun. The spacecraft returned for a second Earth swingby on December 8, 1992, at an altitude of 303 kilometers (188 miles). With this, *Galileo* left Earth for the third and final time and headed toward Jupiter.

The flight path provided opportunities for scientific observations. Scientists obtained the first views of mid-level clouds on Venus and confirmed the presence of lightning on that planet. They also made many Earth observations, mapped the surface of Earth's Moon, and observed its north polar regions.

Because of the modification in *Galileo*'s trajectory, the spacecraft was exposed to a hotter environment than originally planned. To protect it from the Sun, project engineers devised a set of sunshades and pointed the top of the spacecraft toward the Sun, with the umbrella-like high-gain antenna furled until well after the first Earth flyby in December 1990. Flight controllers stayed in touch with the spacecraft through a pair of low-gain antennas, which send and receive data at a much slower rate.

High-Gain Antenna Problem

The spacecraft was scheduled to deploy its 4.8-meter-diameter (16-foot) high-gain antenna in April 1991 as *Galileo* moved away from the Sun and the risk of overheating ended. The antenna, however, failed to deploy fully.

A special team performed extensive tests and determined that a few (probably three) of the antenna's 18 ribs were held by friction in the closed position. Despite exhaustive efforts to free the ribs, the antenna would not deploy. From 1993 to 1996, extensive new flight and ground software was developed, and ground stations of NASA's Deep Space Network were enhanced in order to perform the mission using the spacecraft's low-gain antennas.

Asteroid Flybys

Galileo became the first spacecraft ever to encounter an asteroid when it passed Gaspra on October 29, 1991. It flew within just 1,601 kilometers (1,000 miles) of the stony asteroid's center at a relative speed of about 8 kilometers per second (18,000 mph). Pictures and other data revealed a cratered, complex, irregular body about 20 by 12 by 11 kilometers (12.4 by 7.4 by 6.8 miles), with a thin covering of dirt-like "regolith."

On August 28, 1993, *Galileo* carried out a second asteroid encounter, this time with a larger, more distant asteroid named Ida. Ida is about 55 kilometers (34 miles) long and 24 kilometers (15 miles) wide. Observations indicated that both Ida and Gaspra have magnetic fields, although Ida is older and its surface is covered with craters. Scientists discovered that Ida boasts its own moon, making it the first asteroid known to have a natural satellite. The tiny moon, named Dactyl, has a diameter of only about 1.5 kilometers (less than a mile). By determining Dactyl's orbit, scientists estimated Ida's density.

Comet Shoemaker-Levy

The discovery of Comet Shoemaker-Levy 9 in March 1993 provided an exciting opportunity for *Galileo*'s science teams and other astronomers. The comet was breaking up as it orbited Jupiter, and was headed to dive into the giant planet's atmosphere in July 1994.

The *Galileo* spacecraft, approaching Jupiter, was the only observation platform with a direct view of the impact area on Jupiter's far side. Despite the uncertainty of the predicted impact times, Galileo team members pre-programmed the spacecraft's science instruments to collect data and were able to obtain spectacular images of the comet impacts.

Jupiter Arrival

On July 13, 1995, *Galileo*'s descent probe, which had been carried aboard the parent spacecraft, was released and began a five-month free fall toward Jupiter. The probe had no engine or thrusters, so its flight path was established by pointing of the *Galileo* orbiter before the probe was released. Two weeks later, *Galileo* used its main rocket engine for the first time as it readjusted its flight path to arrive at the proper point at Jupiter.

Arrival day on December 7, 1995, turned out to be an extremely busy 24-hour period. When *Galileo* first reached Jupiter and while the probe was still approaching the planet, the orbiter flew by two of Jupiter's major moons Europa and Io. *Galileo* passed Europa at an altitude of about 33,000 kilometers (20,000 miles), while the Io approach was at an altitude of about 900 kilometers (600 miles). About four hours after leaving Io, the orbiter made its closest approach to Jupiter, encountering 25 times more radiation than the level considered deadly for humans.

Descent Probe

Eight minutes later, the orbiter started receiving data from the descent probe, which slammed into the top of the Jovian atmosphere at a comet-like speed of 170,000 km/hr (106,000 mph). In the process, the probe withstood temperatures twice as hot as the Sun's surface. The probe slowed by aerodynamic braking for about two minutes before deploying its parachute and dropping a heat shield.

The wok-shaped probe floated down about 200 kilometers (125 miles) through the clouds, transmitting data to the orbiter on sunlight and heat flux, pressure, temperature, winds, lightning and atmospheric composition. Fifty-eight minutes into its descent, high temperatures silenced the probe's transmitters. The probe sent data from a depth with a pressure 23 times that of the average on Earth's surface, more than twice the mission requirement.

An hour after receiving the last transmission from the probe, at a point about 200,000 kilometers (130,000 miles) above the planet, the *Galileo* spacecraft fired its main engine to brake into orbit around Jupiter.

This first orbit lasted about seven months. *Galileo* fired its thrusters at its farthest point in the orbit to keep it from coming so close to the giant planet on later orbits. This adjustment helped mitigate possible damage to spacecraft sensors and computer chips from Jupiter's intense radiation environment.

During this first orbit, new software was installed which gave the orbiter extensive new onboard data processing capabilities. It permitted data compression, enabling the spacecraft to transmit up to 10 times the number of pictures and other measurements than would have been possible otherwise.

In addition, hardware changes on the ground and adjustments to the spacecraft-to-Earth communication system increased the average telemetry rate tenfold. Although the problem with the high-gain antenna prevented some of the mission's original objectives from being met, the great majority were. So many new objectives were achieved that scientists feel *Galileo* has produced considerably more science than ever envisioned at the project's start 20 years ago.

Orbital Tour

During its primary mission orbital tour, *Galileo*'s itinerary included four flybys of Jupiter's moon Ganymede, three of Callisto and three of Europa. These encounters were about 100 to 1,000 times closer than those performed by NASA's *Voyager 1* and 2 spacecraft during their Jupiter flybys in 1979. *Galileo*'s instruments scanned and scrutinized the surface and features of each moon. After about a week of intensive observation, with its tape recorder full of data, the spacecraft spent the next one to two months until the next encounter in orbital "cruise," playing back the information in transmissions to Earth.

Extended Mission

A two-year extension, the Galileo Europa Mission, began in December 1997 and included intensive study of Europa through eight consecutive close encounters. The flybys added to knowledge about Europa's frozen surface and the intriguing prospect that liquid oceans may lie underneath. In addition, *Galileo* studied Callisto in four flybys, and approached Io twice, gathering new information on Io's volcanic activity. The Galileo Millennium Mission added another year of operations, including more flybys of Io and Ganymede, plus the joint studies with the *Cassini* spacecraft as it passes Jupiter in December 2000 for a gravity assist toward Saturn.

Spacecraft

The *Galileo* orbiter weighed 2,223 kilograms at launch (2½ tons) and measured 5.3 meters (17 feet) from the top of the low-gain antenna to the bottom of the probe. The orbiter features an innovative "dual-spin" design. Most spacecraft are stabilized in flight either by spinning around a major axis, or by maintaining a fixed orientation in space, referenced to the Sun and another star. As the first dual-spin planetary spacecraft, *Galileo* combines these techniques. A spinning section rotates at about 3 rpm, and a "despun" section is counter-rotated to provide a fixed orientation for cameras and other remote sensors. A star scanner on the spinning side determines orientation and spin rate; gyroscopes on the despun side provide the basis for measuring turns and pointing instruments.

The power supply, propulsion module and most of the computers and control electronics are mounted on the spinning section. The spinning section also carries instruments to study magnetic fields and charged particles. These instruments include magnetometer sensors mounted on an 11-meter (36-foot) boom to minimize interference from the spacecraft's electronics; a plasma instrument to detect low-energy charged particles; and a plasma wave detector to study electromagnetic waves generated by the particles. There is also a high-energy particle detector and a detector of cosmic and Jovian dust, an extreme ultraviolet detector associated with the ultraviolet spectrometer, and a heavy ion counter to assess potentially hazardous charged particle environments the spacecraft flies through.

Galileo Spacecraft

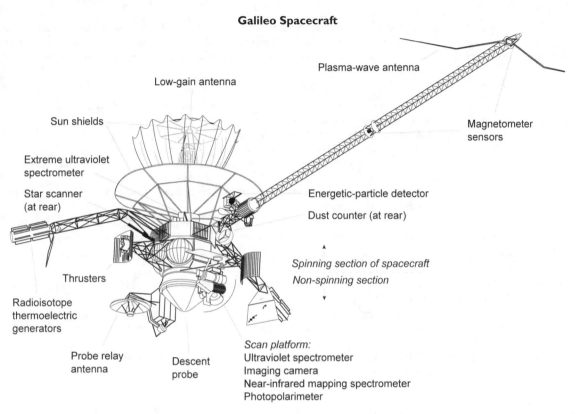

Plasma-wave antenna

Low-gain antenna

Sun shields

Extreme ultraviolet
spectrometer

Magnetometer
sensors

Star scanner
(at rear)

Energetic-particle detector

Dust counter (at rear)

Spinning section of spacecraft
Non-spinning section

Thrusters

Radioisotope
thermoelectric
generators

Probe relay
antenna

Descent
probe

Scan platform:
Ultraviolet spectrometer
Imaging camera
Near-infrared mapping spectrometer
Photopolarimeter

Galileo's de-spun section carries instruments that need to be held steady. These instruments include the camera system; the near-infrared mapping spectrometer to make multispectral images for atmosphere and surface chemical analysis; the ultraviolet spectrometer to study gases; and the photopolarimeter-radiometer to measure radiant and reflected energy. The camera system obtains images of Jupiter's satellites at resolutions from 20 to 1,000 times better than the best possible from NASA's *Voyager* spacecraft; its charge-coupled-device (CCD) sensor is much more sensitive than previous spacecraft cameras and is able to detect a broader color band. *Galileo*'s de-spun section also carries a dish antenna that picked up the descent probe's signals during its fall into Jupiter's atmosphere.

The spacecraft's propulsion module consists of twelve 10-newton (22.5-pound-force) thrusters and a single 400-newton (90-pound-force) engine which use monomethyl-hydrazine fuel and nitrogen-tetroxide oxidizer. The propulsion system was developed and built by Messerschmitt-Bolkow-Blohm (MBB) and provided by the Federal Republic of Germany as NASA's major international partner on *Galileo*.

Because radio signals take more than one hour to travel from Earth to Jupiter and back, the *Galileo* spacecraft was designed to operate from computer instructions sent to it in advance and stored in spacecraft memory. A single master sequence of commands can cover a period ranging from weeks to months of quiet operations between flybys of Jupiter's moons. During busy encounter operations, one sequence of commands covers only about a week.

These sequences operate through flight software installed in the spacecraft computers, with built-in automatic fault protection software designed to put *Galileo* in a safe state in case of computer glitches or other unforeseen circumstance. Electrical power is provided by two radioisotope thermoelectric generators. Heat produced by natural radioactive decay of plutonium is converted to electricity (570 watts at launch, 485 at the end of the mission) to operate the orbiter spacecraft's equipment. This is the same type of power source used on other NASA missions including *Viking* to Mars, *Voyager* and *Pioneer* to the outer planets, *Ulysses* to study the Sun, and *Cassini* to Saturn.

Descent Probe

Galileo's descent probe had a mass of 339 kilograms (750 pounds), and included a deceleration module to slow and protect the descent module. The deceleration module consisted of an aeroshell and an aft cover designed to block heat generated by friction during atmospheric entry. Inside the aeroshells were the descent module and its 2.5-meter (8-foot) parachute. The descent module carried a radio transmitter and seven scientific instruments. These were devices to measure temperature, pressure and deceleration, atmospheric composition, clouds, particles, and light and radio emissions from lightning and energetic particles in Jupiter's radiation belts.

Orbiter Instruments

The *Galileo* orbiter spacecraft carries 11 scientific instruments. Another seven were on the descent probe. One engineering instrument on the orbiter, originally for measurements to aid design of future spacecraft, also collects scientific information. The orbiter instruments, what they study, and their principal investigators or team leaders are:

Remote Sensing Instruments on the Non-Spinning Section:

- Solid State Imaging Camera: Galilean satellites, high-resolution, atmospheric small-scale dynamics. Dr. Michael Belton, National Optical Astronomy Observatories, Tucson, Arizona.
- Near-Infrared Mapping Spectrometer: Surface and atmospheric composition thermal mapping. Dr. Robert Carlson, NASA's Jet Propulsion Laboratory, Pasadena, California.
- Photopolarimeter-Radiometer: Atmospheric particles, thermal / reflected radiation. Dr. James Hansen, Goddard Institute for Space Studies, New York, New York
- Ultraviolet Spectrometer / Extreme Ultraviolet Spectrometer Experiment: Atmospheric gases, aerosols. Dr. Ian Stewart, University of Colorado, Boulder, Colorado.

Instruments on the Spinning Section, Studying Magnetic Fields and Charged Particles:

- Magnetometer: Strength and fluctuation of magnetic fields. Dr. Margaret Kivelson, University of California, Los Angeles, California.
- Energetic Particle Detector: Electrons, protons, heavy ions. Dr. Donald Williams, Johns Hopkins University Applied Physics Laboratory, Laurel, Maryland.
- Plasma Investigation: Composition, energy, distribution of ions. Dr. Lou Frank, University of Iowa, Iowa City, Iowa.
- Plasma Wave Subsystem: Electromagnetic waves and wave-particle interactions. Dr. Donald Gurnett, University of Iowa, Iowa City, Iowa.
- Dust-Detection Subsystem: Mass, velocity, charge of submicrometer particles. Dr. Eberhard Grun, Max Planck Institut fur Kernphysik, Heidelberg, Germany.

Radio Science:

- Celestial Mechanics: Masses and internal structures of bodies from spacecraft tracking. John Anderson, NASA's Jet Propulsion Laboratory, Pasadena, California.
- Propagation: Satellite radii and atmospheric structure from radio propagation. H. Taylor Howard, Stanford University, Stanford, California.

Engineering Experiment:

- Heavy Ion Counter: Spacecraft charged-particle environment, ion flow from solar flares. Dr. Edward Stone, California Institute of Technology, Pasadena, California.

Galileo Discoveries

The Galileo mission has returned a wealth of scientific findings since it arrived at Jupiter December 7, 1995. It began accomplishing a string of technical firsts even earlier.

- Jupiter has many large thunderstorms, with lightning strokes up to 1,000 times more powerful than those on Earth.
- Jupiter's atmosphere has a wide range of cloudiness and water vapor, just as Earth has humidity zones ranging from the Sahara to the tropics.
- Evidence of liquid water oceans beneath Europa's surface, which boosts the odds of life existing there. Iceberg-like "rafting," or floating ice chunks are apparent.
- Europa is criss-crossed by faults and ridges.
- Europa, Ganymede and Io have metallic cores with separated layers; Callisto has no core and less separation between layers.
- Europa is surrounded by an ionosphere, a cloud of electrically-charged gases.
- Ganymede has a very thin hydrogen atmosphere, a magnetic field, and magnetosphere. It is the first satellite known to possess a magnetic field.
- Ganymede's surface has faults, fractures, and some evidence of volcanic flows.
- Callisto's surface is blanketed by powder-like debris, perhaps created by some unknown erosional process.

• Io is very active volcanically (perhaps 10 times more than Earth), and its surface is continually changing. During one four-month period observed by *Galileo*, volcanic debris spread over an area the size of Arizona.

Technical Firsts

• First flyby and imaging of an asteroid (Gaspra and later, Ida).
• Discovered first moon around an asteroid (the moon Dactyl orbits asteroid Ida).
• Only direct observation of Comet Shoemaker-Levy 9's impact into Jupiter's atmosphere.
• First spacecraft to deploy an entry probe into an outer planet's atmosphere. The *Galileo* probe measured the atmospheric composition and structure of Jupiter's atmosphere and provided clues to the origin of Jupiter and giant planets in other star systems.
• First and so far the only spacecraft to orbit an outer planet. (*Cassini* at Saturn in 2004 will become the second spacecraft to enter orbit around one of the outer planets.)

Deep Space Network

NASA's scientific investigations of the solar system are accomplished mainly through the use of robotic spacecraft, such as *Galileo* and *Cassini*. The Deep Space Network provides the two-way communications link that guides and controls spacecraft and brings back images and other scientific data they collect. The Deep Space Network encompasses complexes strategically placed on three continents. The largest and most sensitive scientific telecommunications system in the world, it also performs radio and radar astronomy observations for the exploration of the solar system and the universe. It is managed and operated for NASA by the Jet Propulsion Laboratory, a division of the California Institute of Technology.

The network features three deep-space communications complexes placed approximately 120 degrees apart around the world: at Goldstone in California's Mojave Desert; near Madrid, Spain; and near Canberra, Australia. This configuration ensures that an antenna is always within sight of a given spacecraft, day and night, as the Earth rotates. Each complex contains up to 10 deep space stations equipped with large parabolic reflector antennas.

Antennas and Facilities

Each of the network's complexes has one 70-meter (230-foot)-diameter antenna. These are the largest and most sensitive antennas, capable of tracking spacecraft traveling more than 16 billion kilometers (10 billion miles) from Earth. The surface of the 70-meter reflector must remain accurate within a fraction of the signal wavelength, meaning that the precision across the 3,850-square-meter (4,600-square-yard) surface is maintained within one centimeter (0.4 inch). The dish reflector and its mount weigh nearly 2.7 million kilograms (2,970 U.S. tons). Each complex also has a 34-meter (112-foot)-diameter high-efficiency antenna, incorporating more recent advances in antenna design and mechanics. The reflector surface is precision shaped for maximum signal gathering capability.

The most recent additions to the Deep Space Network are several 34-meter beam waveguide antennas. On earlier DSN antennas, sensitive electronics were centrally mounted on the hard-to-reach reflector structure, making upgrades and repairs difficult. On beam waveguide antennas, however, such electronics are located in a below-ground pedestal room, with the radio signal brought from the reflector to this room through a series of precision-machined radio frequency reflective mirrors. Not only does this architecture provide the advantage of easier access for enhancements and maintenance, but it also allows for better thermal control for critical electronic components and for more electronics to be placed in the antenna to support operation at multiple frequencies.

There is also one 26-meter (85-foot)-diameter antenna at each complex for tracking Earth-orbiting satellites, which are in orbits primarily 160 to 1,000 kilometers (100 to 620 miles) above Earth. The two-axis astronomical mount allows these antennas to point low on the horizon to pick up fast-moving Earth-orbiting satellites as soon as they come into view.

All of the antennas communicate directly with the control center at JPL in Pasadena, California, the operations hub for the network. The control center staff directs and monitors operations, transmits commands and oversees the quality of spacecraft telemetry and navigation data delivered to network users. In addition to the complexes and the control center, a ground communications facility provides communications linking the three complexes to the control center at JPL, to flight control centers in the United States and overseas, and to scientists around the world. Voice and data traffic between various locations is sent via land lines, submarine cable, microwave links and communications satellites.

The Radio Link

The Deep Space Network's radio link to spacecraft is basically the same as other point-to-point microwave communications systems, except for the very long distances involved and the very low spacecraft signal strength. "Very low" might be an understatement: The total signal power arriving at a network antenna from a spacecraft encounter among the outer planets can be 20 billion times weaker than the power level in a modern digital wristwatch battery. The extreme weakness of the signal results from restrictions placed on the size, weight and power supply of the spacecraft by the cargo area and weight-lifting limitations of the launch vehicle. Consequently, the design of the radio link is the result of engineering tradeoffs between spacecraft transmitter power and antenna diameter, and the sensitivity that can be built into the ground receiving system.

Typically, a spacecraft signal is limited to 20 watts, or about the same power required to light a refrigerator bulb. When the signal arrives at Earth from outer space, say from the neighborhood of Saturn, it is spread over an area with a diameter equal to about 1,000 Earth diameters. As a result, the ground antenna is able to receive only a very small part of the signal power, which is degraded by background radio noise, or static. Noise is radiated naturally from nearly all objects in the universe, including Earth and the sun. Noise is also inherently generated in all electronic systems, including the Deep Space Network's own detectors. Since there will always be noise amplified with the signal, the ability of the ground receiving system to separate the noise from the signal is critical. The network uses state-of-the-art, low-noise receivers and telemetry coding techniques to create unequaled sensitivity and efficiency.

International Team

The Cassini program is an international cooperative effort involving NASA, the European Space Agency (ESA), the Italian space agency, Agenzia Spaziale Italiana (ASI), and several separate European academic and industrial contributors. The Cassini partnership represents an undertaking whose scope and cost would not likely be borne by any single nation, but it made possible through shared investment and participation. Hundreds of scientists and engineers from 16 European countries and 33 U.S. states make up team that developed and will fly and receive data from *Cassini* and *Huygens*.

At NASA Headquarters, Mark Dahl is Cassini program executive and Dr. Jay Bergstralh is Cassini program scientist.

In the United States, the mission is managed for NASA's Office of Space Science by the Jet Propulsion Laboratory, Pasadena, California. JPL is a division of the California Institute of Technology. At JPL, Robert Mitchell is the Cassini program manager, and Dr. Dennis L. Matson is the Cassini project scientist.

The major U.S. contractor is Lockheed-Martin, whose contributions include the launch vehicle and upper stage, spacecraft propulsion module and the radioisotope thermoelectric generators. NASA's Lewis Research Center managed development of the Centaur upper stage.

Development of the *Huygens* Titan probe is managed by the European Space Technology and Research Center (ESTEC). ESTEC's prime contractor, Aerospatiale in Toulouse, France, assembled the probe with equipment supplied by many European countries. At ESA, Michel Verdant is Huygens project manager, and Dr. Jean-Pierre Lebreton is project scientist.

At ASI, Enrico Flameni is the project manager for *Cassini*'s radio antenna and other contributions to the spacecraft.

Galileo

Galileo's scientific experiments are being carried out by more than 100 scientists from six nations.

The Galileo Project is managed for NASA's Office of Space Science by the Jet Propulsion Laboratory, Pasadena, California. JPL is a division of the California Institute of Technology. This responsibility includes designing, building, testing, operating and tracking *Galileo*. Germany furnished the orbiter's retro-propulsion module and some of the instruments and is participating in the scientific investigations.

NASA's Ames Research Center, Moffett Field, California, was responsible for the atmosphere probe, which was built by Hughes Aircraft Company, El Segundo, California.

At NASA Headquarters, the Galileo program manager is Dr. Paul Hertz, and the program scientist is Dr. Denis Bogart.

At JPL, Galileo project manager is Jim Erickson, and project scientist is Dr. Torrence Johnson.

Deep Space Network

Gael Squibb is director of JPL's Telecommunications and Mission Operations Directorate. Richard Coffin is deputy director. Joe Wackley is manager of operations.

CASSINI-HUYGENS MISSION

Public Information Office   California Institute of Technology
Jet Propulsion Laboratory   National Aeronautics And Space Administration
Telephone (818) 354-5011   Pasadena, California 91109

Contact: Mary Beth Murrill
         (818) 354-6478

FOR IMMEDIATE RELEASE July 24, 1995

European Cassini Hardware Delivered to NASA

A saucer-shaped, gold-colored space probe nearly 3 meters (9 feet) in diameter has been delivered to NASA by the European Space Agency and is now being readied for testing on the graphite and aluminum framework of the Saturn-bound *Cassini* spacecraft.

The engineering model of Europe's *Huygens* probe, delivered to NASA's Jet Propulsion Laboratory, is twin to the actual flight model that is destined to parachute to the exotic surface of Saturn's largest moon, Titan. The reflective fabric that blankets the probe will provide thermal and micrometeorite protection to the device in space. The flight model of the *Huygens* probe will be installed on the *Cassini* spacecraft at Cape Canaveral, Florida, prior to launch on October 6, 1997.

A second important delivery of European hardware for *Cassini* was received on July 21 with the arrival of Italy's specially tailored engineering model of the 10-meter (13-foot)-diameter high-gain telecommunications antenna. The antenna, provided by the Italian Space Agency, Agenzia Spaziale Italiana, will serve as the *Cassini* spacecraft's "voice box" and "ears," sending data back and receiving commands from Earth during the 11-year-long mission. The multi-channel antenna is also a crucial part of several of *Cassini*'s scientific investigations, including imaging radar and gravity experiments.

"With these major deliveries from our European partners, we remain on schedule to meet our Saturn launch date," said Cassini project manager Richard J. Spehalski at JPL. "These critical hardware contributions enable us to proceed into technical qualification of the *Cassini* orbiter and *Huygens* probe hardware." Mated together, the probe and orbiter hardware will undergo months of structural and space environmental testing.

The *Cassini* spacecraft will fly a looping, seven-year-long course, swinging twice past Venus and once past Earth and Jupiter to reach Saturn, about 1.4 billion kilometers (nearly 1 billion miles) away. Once there, *Cassini* will release the *Huygens* probe, with six instruments onboard, to explore Titan's atmosphere and surface. Over the course of its two-and-a-half hour parachute descent to Titan's surface, the *Huygens* probe is expected to return a wealth of data to the *Cassini* orbiter for relay back to Earth.

Titan is a moon the size of a small planet. Its chemically complex atmosphere is primarily nitrogen and is rich in hydrocarbons, resembling the early atmosphere of Earth. Lakes or small oceans of liquid ethane may surround a continent-sized surface feature on Titan recently discovered by scientists using the orbiting Hubble Space Telescope. Studies of Titan by the *Huygens* probe will not only provide insight into the history and status of this unique body, but will also provide clues to the early history of Earth.

With the *Huygens* probe portion of the mission completed, the *Cassini* spacecraft itself will remain in orbit around Saturn to return nearly four years of scientific information about the planet, its rings, moons and magnetic environment. *Cassini* will also conduct numerous flybys of Titan to gather more information about its atmosphere and global surface characteristics. An imaging radar instrument that can see through Titan's opaque atmospheric haze will gather data to produce photograph-like images of the surface.

Many of Saturn's moons will be targets of intensive study, as will the dynamics, structure and make-up of the planet's rings. Saturn itself, made up mostly of hydrogen and helium, will be characterized in detail, and *Cassini* will study how charged particles from the Sun interact with the large, complex magnetic environment that envelops Saturn.

The *Huygens* Probe is named for 17th century Dutch scientist Christiaan Huygens, who discovered Titan in 1659. The *Cassini* spacecraft is named for Italian-French astronomer Jean-Dominque Cassini, who discovered four more of Saturn's moons and in 1675 found the gap – now called the Cassini Division – that separates two of Saturn's more prominent rings.

The Cassini mission is an international project jointly managed by NASA, the European Space Agency and Agenzia Spaziale Italiana. Contractors, academic institutions and space and science agencies from 17 countries are participating in the mission.

The overall Cassini mission is managed by the Jet Propulsion Laboratory for NASA's Office of Space Science, Washington, D.C.

Public Information Office
Jet Propulsion Laboratory
California Institute of Technology
National Aeronautics And Space Administration
Pasadena, California 91109

Telephone (818) 354-5011

http://www.jpl.nasa.gov

Contact: Mary Beth Murrill

FOR IMMEDIATE RELEASE   April 21, 1997

NASA's *Cassini* spacecraft, due for launch toward the planet Saturn in early October, arrived today at NASA's Kennedy Space Center (KSC) in Florida.

The spacecraft was shipped from NASA's Jet Propulsion Laboratory in Pasadena, California, by a U.S. Air Force C-17 air cargo plane. The spacecraft will now undergo final integration and testing prior to being taken to Launch Complex 40 for mating to an Air Force Titan IV launch vehicle.

Saturn is best known for its complex ring system and a complex atmosphere with very high winds. The *Cassini* spacecraft will deploy an instrumented probe called *Huygens* to explore Saturn's moon Titan, itself the size of a small planet. *Huygens* will ride a parachute through Titan's dense atmosphere, which may have important similarities to the early atmosphere of Earth. Studies of Saturn's atmosphere along with its rings, moons and magnetic environment will help produce a better understanding of planetary evolution.

At KSC, after post-arrival inspections of the spacecraft have been completed, integration of the 12 science instruments not already installed will be finished. Next, the large parabolic high-gain antennaand then the propulsion module will be mated to the spacecraft. At that point, an integrated functional test will be run to verify that all of these systems are operating properly together. Finally, the *Huygens* Probe, which up to now has been undergoing its component integration and associated testing separately, will be mated with the *Cassini* spacecraft, fully completing spacecraft integration.

Cassini was built and is managed for NASA by JPL. The European Space Agency (ESA) is contributing the *Huygens* Probe. The high-gain antenna and elements of several of *Cassini*'s science instruments are being provided by the Italian Space Agency (ASI).

Cassini is scheduled for launch on October 6, 1997, at 5:38 a.m. Eastern Standard Time to begin its nearly seven-year journey to the outer solar system. Once it reaches Saturn, the spacecraft is expected to complete 60 orbits of the planet and its moons during a four-year primary mission.

######

NATIONAL AERONAUTICS AND SPACE ADMINISTRATION

Cassini Launch
Press Kit

October 1997

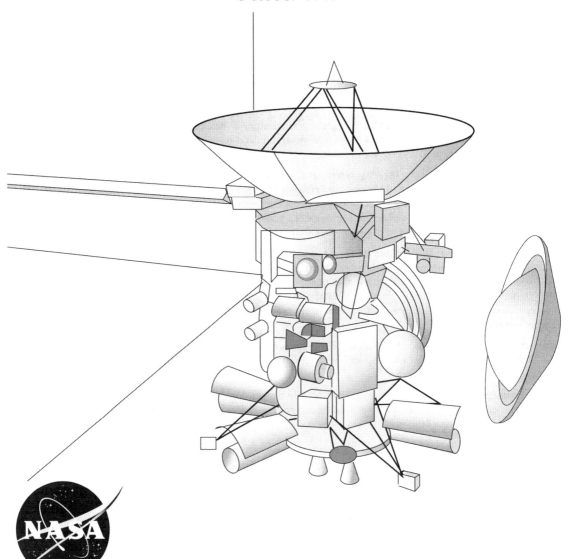

Contacts

Douglas Isbell, Headquarters, Washington, DC	Policy / Program Management	(202) 358-1753
Donald Savage, Headquarters, Washington, DC	Cassini Nuclear Safety	(202) 358-1727
Franklin O'Donnell, Jet Propulsion Laboratory, Pasadena, California	Cassini Mission	(818) 354-5011
Mary Beth Murrill, Jet Propulsion Laboratory, and Nuclear Safety, Pasadena, California	Cassini Mission	(818) 354-6478
George Diller, Kennedy Space Center, Florida	Launch Operations	(407) 867-2468

CONTENTS

RELEASE:

Cassini to Survey Realm of Saturn and Titan

The planet Saturn, its icy rings, the enigmatic moon Titan, other moons and the huge magnetic bubble that surrounds most of them are the prime scientific targets of the international Cassini mission, the most ambitious planetary exploration ever mounted. Final preparation of *Cassini* is now under way for a pre-dawn launch from Cape Canaveral, Florida, on October 6, 1997.

The mission also entails the first descent of a probe to a moon of another planet, sending the *Huygens* probe to the surface of Saturn's moon Titan – by far the most distant landing ever attempted on another object in the solar system.

Cassini, in development since October 1989, is a cooperative endeavor of NASA, the European Space Agency (ESA) and the Italian Space Agency, or Agenzia Spaziale Italiana (ASI). The mission will send a sophisticated robotic spacecraft equipped with 12 scientific experiments to orbit Saturn for a four-year period and to study the Saturnian system in detail. The ESA-built *Huygens* probe that will parachute into Titan's thick atmosphere carries another six scientific instrument packages.

"Saturn, with its rings, 18 known moons and its magnetic environment, is a lot like a solar system in miniature form," said Dr. Wesley T. Huntress, NASA's associate administrator for space science. "It represents an unsurpassed laboratory where we can look for answers to many fundamental questions about the physics, chemistry and evolution of the planets and the conditions that give rise to life. *Cassini* and the *Huygens* probe represent our best efforts yet in our ongoing exploration of the solar system."

The launch period for *Cassini*'s nearly seven-year journey to Saturn opens on October 6 and closes November 15, 1997. The launch is scheduled for October 13 at 4:55 a.m. Eastern Daylight Time (08:55 Universal Time). The launch window runs for 140 minutes each day and moves earlier by about six minutes daily. A U.S. Air Force Titan IVB / Centaur launch system, the most powerful launch vehicle in the U.S. fleet, will loft the *Cassini* spacecraft onto the interplanetary trajectory that will deliver the spacecraft to Saturn almost seven years later, on July 1, 2004. *Cassini*'s primary mission concludes in July 2008.

In the early 1980s, the international science community selected a Saturn orbiter and Titan probe as the next logical step in the exploration of the solar system, similar to the Galileo mission to Jupiter. The *Cassini* orbiter and *Huygens* probe are the result of years of collaborative planning by space scientists and engineers in the United States and Europe.

Cassini is a follow-on mission to the brief reconnaissance of Saturn performed by the *Pioneer 11* spacecraft in 1979 and the *Voyager 1* and 2 encounters of 1980 and 1981. Those highly successful flybys produced volumes of new information, discoveries and questions about Saturn, its environs and its family of rings and moons. Those encounters, along with recent key findings from the Hubble Space Telescope and ground-based observatories, continue to entice scientists who view the Saturnian system as a one-stop treasure trove with countless clues to the history of planetary and solar system evolution.

The mission is named for two 17th century astronomers. Italian-French astronomer Jean-Dominique Cassini (born Gian Domenico Cassini in his native Italy) made several key discoveries about Saturn between 1671 and 1684; he established that Saturn's rings are split largely into two parts by a narrow gap, known since as the Cassini Division. Dutch scientist Christiaan (cq) Huygens discovered Titan in 1655 and was responsible for many important Saturn findings.

Saturn is the second largest planet in the solar system. Like the other gaseous outer planets – Jupiter, Uranus and Neptune – its atmosphere is made up mostly of hydrogen and helium. Saturn's placid-looking, butterscotch-colored face masks a windswept atmosphere where jet streams blow at 1,800 km/hr (1,100 mph) and swirling storms roil just beneath the cloud tops. Spacecraft passing by Saturn found a huge and complex magnetic environment, called a magnetosphere, where trapped protons and electrons interact with each other, the planet, the rings and the surfaces of many of the moons.

The bright rings for which Saturn is best known have been found to consist of not just a few but hundreds of rings and ringlets, broad and thin, made up of ice and rock particles ranging in size from grains of sand to boxcars. The *Voyager* spacecraft observations in 1980 and 1981 found that the particles are herded into complicated orbits by the gravitational interaction of small moons unseen from Earth. Continuously jostled by their different orbital speeds and competing gravitational tugs from nearby moons and Saturn itself, the particles bump and grind away at each other, spreading out into a broad, thin sheet that is less than 100 meters (about 330 feet) thick, but almost as broad as the distance between the Earth and its Moon. So-called "shepherd moons" were found orbiting near the edges of some rings. Like sheep dogs working the edges of a flock, these moons gravitationally herd in and contain ring particles that would otherwise spread out into deep space.

More moons of greater variety orbit Saturn than any other planet. These natural satellites range from Titan, which is larger than either Mercury or Pluto and nearly the size of Mars, to bodies so puny that astronomers call them "moonlets." There are likely many more moons than the 18 that have been confirmed.

Although it is believed to be too cold to support life, haze-covered Titan is thought to hold clues to how the primitive Earth evolved into a life-bearing planet. It has an Earth-like, nitrogen-based atmosphere and a surface that many scientists believe probably features chilled lakes of ethane and methane (which may also pool in subsurface reservoirs). Scientists believe that the moon's surface is probably coated with the residue of a sticky brown organic rain. Titan's orange haze chemically resembles smog, and it is thought to be composed of naturally occurring, smoke-like hydrocarbon particles. Standing on Titan's surface beneath this orange sky, a visitor from Earth likely would find a cold, exotic world with a pungent odor reminiscent of a petroleum processing facility. *Huygens* will provide our first direct sampling of Titan's atmospheric chemistry and the first detailed photographs of its hidden surface.

In maneuvers called gravity assist swingbys, *Cassini* will fly twice past Venus then once each past Earth and Jupiter on its way to Saturn. The spacecraft's speed relative to the Sun increases as it approaches and swings past each planet, giving *Cassini* the cumulative boost it needs to reach its ultimate destination.

On November 6, 2004, *Cassini* will release the disc-shaped *Huygens* probe toward Titan. After a three-week ballistic free fall toward Titan, the 2.7-meter-diameter (8.9-foot), battery powered *Huygens* will enter Titan's atmosphere, deploy its parachutes and begin its scientific observations. Data gathered during *Huygens'* 2½-hour descent through Titan's dense atmosphere will be radioed to the *Cassini* spacecraft and relayed to Earth. Instruments on the descent probe will measure the chemistry, temperature, pressure, density and energy balance in the atmosphere. As the probe breaks through the cloud deck, a camera will capture panoramic pictures of Titan. Titan's surface properties will be measured, and more than 500 images of the clouds and surface will be returned. In the final moments of descent, a spotlight will illuminate the surface for spectroscopic measurements of its composition.

If the probe survives landing – which should occur at a fairly low speed of about 25 km/hr (15 mph) – it may return data from Titan's surface, where the atmospheric pressure is 1.6 times that of Earth's and the temperature is -179° C (-290° F). The exact conditions it will encounter are unknown; the probe could touch down on solid ground, rock-hard ice, or even splash down in a lake of ethane and methane. One instrument on board will discern whether *Huygens* is bobbing in liquid, and other instruments onboard could measure the chemical composition of that liquid.

During the course of the *Cassini* orbiter's mission, it will execute more than 40 targeted close flybys of Titan, many as close as 950 kilometers (about 590 miles) above the surface. This will permit high-resolution mapping of Titan's surface with the Titan imaging radar instrument, which can see through the opaque haze covering that moon to produce vivid photograph-like images.

The *Cassini* spacecraft, including the orbiter and *Huygens* probe, is the most complex interplanetary spacecraft ever built. Its sophisticated instruments are state-of-the-art and represent the best technical efforts of the United States and 16 European nations involved in the mission. Because of *Cassini*'s challenging mission, long voyage and the value of its promised scientific return, each component and the system as a whole has undergone an unprecedented program of rigorous testing for quality and performance to assure the highest possible probability of mission success.

Because of the very dim sunlight at Saturn's orbit, solar arrays are not feasible. Electrical power is supplied to the orbiter by a set of radioisotope thermoelectric generators (RTGs), which convert the heat from the natural decay of plutonium-238, in the form of plutonium dioxide, to electricity for *Cassini*'s systems. The same material is used in 117 radioisotope heater units (RHUs) placed on *Cassini* and *Huygens* to keep electronics systems at their operating temperatures. RHUs were most recently used on the Mars Pathfinder mission's *Sojourner* rover to keep the system from failing during cold Martian nights. *Huygens* will draw its electrical power from a set of five batteries during its entry and descent into Titan's atmosphere.

Telecommunications with *Cassini* during the mission will be carried out through the giant dish antennas of NASA's Deep Space Network, with complexes located in California, Spain and Australia. *Cassini*'s high-gain antenna is provided by ASI. Data from the *Huygens* probe will be relayed to an ESA operations complex in Darmstadt, Germany. NASA's Lewis Research Center managed development of the Centaur upper stage.

Development of the *Huygens* probe was managed by an ESA team located at the European Space Technology and Research Center (ESTEC) in Noordwijk, the Netherlands. The *Cassini* orbiter was designed, developed and assembled at NASA's Jet Propulsion Laboratory (JPL), located in Pasadena, California. JPL is a division of the California Institute of Technology. The overall mission is managed by JPL for NASA's Office of Space Science, Washington, DC.

(End of General Release)

Media Services Information

NASA Television Transmission

NASA Television is broadcast on the satellite GE-2, transponder 9C, C Band, 85 degrees west longitude, frequency 3880.0 MHz, vertical polarization, audio monaural at 6.8 MHz. The schedule for television transmissions for the *Cassini* launch will be available from the Jet Propulsion Laboratory, Pasadena, California; Johnson Space Center, Houston, Texas; Kennedy Space Center, Florida, and NASA Headquarters, Washington, DC.

Status Reports

Status reports on mission activities for *Cassini* will be issued by the Jet Propulsion Laboratory's Public Information Office. They may be accessed on line as noted below. Daily audio status reports are available by calling (800) 391-6654 or (818) 354-4210.

Launch Media Credentialing

Requests to cover the *Cassini* launch must be faxed in advance to the NASA Kennedy Space Center news room at (407) 867-2692. Requests must be on the letterhead of the news organization and must state the editor's assignment to cover the *Cassini* launch.

Briefings

A pre-launch briefing on the mission and science objectives of *Cassini* will be held at Kennedy Space Center, and is planned for 10 a.m. EDT on October 4, 1997.

Internet Information

Extensive information on *Cassini*, including an electronic copy of this Press Kit, press releases, fact sheets, status reports and images, is available from the Jet Propulsion Laboratory's World Wide Web home page at http://www.jpl.nasa.gov/. The Cassini Program also maintains a home page at http://www.jpl.nasa.gov/cassini.

Quick Facts

Spacecraft

- Spacecraft dimensions: 6.7 meters (22 feet) high; 4 meters (13.1 feet) wide
- Weight: 5,712 kilograms (12,593 pounds) with fuel, Huygens probe, adapter, etc.; 2,125 kilograms (4,685 pounds) unfueled orbiter alone
- Science instruments: camera; magnetic field studies; dust and ice grain analysis; infrared energy measurement; chemical composition of Saturn, its moons and rings; neutral and charged particle measurement; radar mapping; and radio wave searches
- Power: 885 watts (633 watts at end of mission) from radioisotope thermoelectric generators

Huygens Probe

- Probe dimensions: 2.7 meters (8.9 feet) in diameter
- Weight: 320 kilograms (705 pounds)
- Science instruments: spectrometer to identify atmospheric makeup; aerosol collector for chemical analysis; imager; sensors to measure atmospheric structure; wind speed measurements; sensors to measure conditions at impact site.

Launch Vehicle

- Type: Titan IVB / Centaur
- Weight: 1 million kilograms (2.2 million pounds)

Mission

- Launch: October 13, 1997 from Cape Canaveral Air Station, Florida. Primary launch period runs through November 15, 1997. On October 13, 140-minute window opens at 4:55 a.m. EDT. Opening of window moves about 6 minutes earlier daily thereafter
- Venus flybys: April 25, 1998 at 300 kilometers (190 miles); June 23, 1999 at 1,530 kilometers (950 miles)
- Earth flyby: August 17, 1999 at 800 kilometers (500 miles) or higher
- Jupiter flyby: December 30, 2000 at 10 million kilometers (6 million miles)
- Saturn arrival date: July 1, 2004
- *Huygens* probe Titan entry date: November 27, 2004
- Distance traveled to reach Saturn: 3.5 billion kilometers (2.2 billion miles)
- Saturn's average distance from Earth: 1.43 billion kilometers (890 million miles)
- One-way light time to Saturn during orbital tour: 68 to 85 minutes
- Huygens' entry speed into Titan's atmosphere: 20,000 km/hr (about 12,400 mph)

Program

- Partners: NASA, the European Space Agency (ESA), Italian Space Agency (Agenzia Spaziale Italiana (ASI)); total of 17 countries involved
- U.S. states in which *Cassini* work was carried out: 33
- Number of people who have worked on some portion of *Cassini / Huygens*: More than 5,000
- Cost of mission: $1.422B pre-launch development; $755M mission operations; $54M tracking; $422M launch vehicle; $500M ESA; $160M ASI; total about $3.3 billion

Saturn at a Glance

General

- One of five planets known to the ancients; Roman god of agriculture, also linked to Kronos, Greek god of time; father of Jupiter, king of the gods
- Yellowish color; at times the 5th brightest planet in night sky

Physical Characteristics

- Second largest planet in the solar system, after Jupiter
- Equatorial diameter 120,536 kilometers (74,898 miles) at cloud tops; polar diameter 108,728 kilometers (67,560 miles), making it the most oblate planet
- Density 0.69 (water = 1)
- Volume 764 times that of Earth, but only 95 times more massive

Orbit

* Sixth planet from the Sun, between Jupiter and Uranus
* Mean distance from Sun 1.43 billion kilometers (890 million miles)
* Brightness of sunlight at Saturn 1 percent of that at Earth
* Revolves around Sun once every 29.42 Earth years
* Saturn's interior rotates once every 10 hours, 39.4 minutes
* Poles tilted 29 degrees

Environment

* Saturn's atmosphere above the clouds is approximately 94 percent hydrogen and 6 percent helium
* Winds near Saturn's equator blow toward east at 500 meters per second (1,100 mph)
* Temperature at Saturn's cloud tops -139° C (-218° F)

Previous Exploration

* *Pioneer 11* flyby September 1, 1979
* *Voyager 1* flyby November 12, 1980
* *Voyager 2* flyby August 25, 1981

Rings

* Saturn's main ring system would barely fit in the space between Earth and its Moon
* B-ring often contains radial spokes of dust-sized material which are regenerated frequently
* Cassini Division between the B-ring and A-ring is sparsely populated with ring material
* E-ring is densest at the orbit of Enceladus and may be fed by Enceladus eruptions.

The Rings of Saturn

Ring	Distance		Width	
D	66,970 km	(41,610 mi)	7,540 km	(4,690 mi)
C	74,510 km	(46,300 mi)	17,490 km	(10,870 mi)
B	92,000 km	(57,170 mi)	25,580 km	(15,890 mi)
A	122,170 km	(75,910 mi)	14,610 km	(9,080 mi)
F	140,180 km	(87,100 mi)	50 km	(30 mi)
G	170,180 km	(105,740 mi)	500 to several 1,000s km (300 to several 1,000s mi)	
E	181,000 km	(112,000 mi)	302,000 km	(188,000 mi)

Distance is from Saturn's center to closest edge of ring

The Known Moons of Saturn (18)

Moon	Diameter		Distance		Discoverer
Pan	20 km	(10 mi)	133,583 km	(83,004 mi)	1990: Showalter
Atlas	32 km	(20 mi)	137,640 km	(85,530 mi)	1980: Terrile (Voyager)
Prometheus	100 km	(62 mi)	139,350 km	(86,590 mi)	1980: Collins (Voyager)
Pandora	84 km	(52 mi)	141,700 km	(88,050 mi)	1980: Collins (Voyager)
Epimetheus	119 km	(74 mi)	151,422 km	(94,089 mi)	1966: Walker
Janus	178 km	(111 mi)	151,472 km	(94,120 mi)	1966: Dolfus
Mimas	392 km	(244 mi)	185,520 km	(115,280 mi)	1789: Herschel
Enceladus	499 km	(310 mi)	238,020 km	(147,900 mi)	1789: Herschel
Tethys	1,060 km	(658 mi)	294,660 km	(183,090 mi)	1684: Cassini
Telesto	22 km	(14 mi)	294,660 km	(183,090 mi)	1980: Smith
Calypso	20 km	(12 mi)	294,660 km	(183,090 mi)	1980: Smith
Dione	1,120 km	(696 mi)	377,400 km	(234,500 mi)	1684: Cassini
Helene	35 km	(22 mi)	378,400 km	(235,100 mi)	1980: Laques and Lecacheux
Rhea	1,528 km	(949 mi)	527,040 km	(327,490 mi)	1672: Cassini
Titan	5,150 km	(3,200 mi)	1,221,850 km	(759,220 mi)	1655: Huygens
Hyperion	283 km	(176 mi)	1,481,100 km	(920,300 mi)	1848: Bond
Iapetus	1,436 km	(892 mi)	3,561,300 km	(2,212,900 mi)	1671: Cassini
Phoebe	220 km	(140 mi)	12,952,000 km	(8,048,000 mi)	1898: Pickering

Pan "PAN"   Telesto "tel-LESS-toe"
Atlas "AT-luss"   Calypso "kuh-LIP-soh"
Prometheus "pro-MEE-thee-uss"   Dione "DIE-oh-nee"
Pandora "pan-DOR-uh"   Helene "huh-LEE-nee"
Epimetheus "epp-ee-MEE-thee-uss"   Rhea "REE-uh"
Janus "JANE-uss"   Titan "TIE-tun"
Mimas "MY-muss"   Hyperion "high-PEER-ee-on"
Enceladus "en-SELL-uh-duss"   Iapetus "eye-APP-eh-tuss"
Tethys "TEE-thiss"   Phoebe "FEE-bee"

Why Saturn?

Saturn offers a rich scientific environment to explore. The planet and the ring system serve as a physical model for the disc of gas and dust that surrounded the early Sun and from which the planets formed. The success of searches for other planetary systems elsewhere in our galaxy partly depends upon how well we understand the early stages of formation of planets.

Detailed knowledge of the history and processes now occurring on Saturn's elaborately different moons will provide valuable data for studying how each of the solar system's planets evolved to their present states. Represented among Saturn's collection of moons is a huge variety of chemical, geologic and atmospheric processes. Physics and chemistry are the same everywhere, and the knowledge gained about Saturn's magnetosphere or Titan's atmosphere will have applications here on Earth.

Chief among *Cassini*'s goals within Saturn's system, however, is the unmasking of Titan.

After decades of speculation and experiment in the modern age, scientists still seek fundamental clues to the question of how life began on Earth. Most experts suspect that life arose by chance combinations of complex carbon compounds in a primeval soup. But all studies of life's origin are hampered by ignorance about the chemical circumstances on the young Earth.

In our solar system, only Earth and Titan have atmospheres rich in nitrogen. Earth's siblings in the inner solar system, Venus and Mars, possess carbon dioxide atmospheres, while Jupiter and Saturn resemble the Sun in their high content of hydrogen and helium. Hydrocarbons like the methane present on Titan may have been abundant on the young Earth.

The importance of Titan in this connection is that it may preserve, in deep-freeze, many of the chemical compounds that preceded life on Earth. Some scientists believe we will find that Titan shares more in common with the early Earth than Earth itself does today.

The results from *Cassini*'s instruments and the *Huygens* probe, along with the results of our continuing explorations of Mars, Europa and the variety of life-bearing environments on Earth, will significantly enhance scientific efforts to solve the mystery of our origins.

Cassini and Planetary Exploration

The *Cassini-Huygens* mission is an enterprise that, from the initial vision to the completion of the mission, will span nearly 30 years. The formal beginning was in 1982, when a joint working group was formed by the Space Science Committee of the European Science Foundation and the Space Science Board of the National Academy of Science in the United States. The charter of the group was to study possible modes of cooperation between the U.S. and Europe in the field of planetary science. Their precept was that the mission would be beneficial for the scientific, technological and industrial sectors of their countries. As a result of their involvement in the studies, European scientists proposed a Saturn orbiter and Titan probe mission to the European Space Agency (ESA), suggesting a collaboration with NASA.

In 1983, the U.S. Solar System Exploration Committee recommended that NASA include a Titan probe and radar mapper in its core program and also consider a Saturn orbiter. In 1984-85, a joint NASA / ESA assessment of a Saturn orbiter-Titan probe mission was completed. In 1986, ESA's Science Program Committee approved *Cassini* for initial Phase A study, with a conditional start in 1987.

In 1987-88, NASA carried out further work on designing and developing the standardized *Mariner Mark II* spacecraft and a set of outer planets missions that would be accomplished with the new spacecraft line. The program was an early effort to reduce the costs of planetary exploration by producing multiple spacecraft for different missions but made with the basic spacecraft components off the same assembly line. *Cassini* and the Comet Rendezvous / Asteroid Flyby (CRAF) were the first two missions chosen for further study. At the same time in Europe, a Titan probe Phase A study was carried out by ESA in collaboration with a European industrial consortium led by Marconi Space Systems. The Titan probe was renamed *Huygens* by ESA as its first medium-sized mission of its Horizon 2000 space science program.

In 1989, funding for CRAF and Cassini was approved by the U.S. Congress. NASA and ESA simultaneously released announcements of opportunity for scientists to propose scientific investigations for the missions. In 1992, a funding cap was placed on the Mariner Mark II program, effectively ending the new spacecraft line and, at the same time, canceling the CRAF mission. *Cassini* was restructured to cut the cost of the mission and to simplify the spacecraft and its operation.

The design of *Cassini* is the result of extensive tradeoff studies which considered cost, mass, reliability, durability, suitability and availability of hardware. To forestall the possibility of mechanical failures, moving parts were eliminated from the spacecraft wherever the functions could be performed satisfactorily without them. Thus, early designs that included moving science instrument platforms or turntables were discarded in favor of instruments fixed to the spacecraft body, whose pointing requires rotation of the entire spacecraft. Tape recorders were replaced with solid state recorders. Mechanical gyroscopes were replaced with hemispherical resonator gyroscopes. An articulated probe relay antenna was discarded in favor of using the high-gain antenna to capture the radio signal of the *Huygens* probe. A deployable high-gain antenna of the type used for *Galileo* was considered and abandoned.

Project engineers, both those who designed and built the hardware and those who operate the spacecraft, relied heavily on extensive past experience to provide a spacecraft design more sophisticated, reliable and capable than any other spacecraft ever built for exploration of the planets. Because of that care in design, the *Cassini* spacecraft is easier to operate and will return more scientific data about its targets than has been possible in any previous planetary mission.

The research and development that Cassini has funded has provided key technologies that have enabled many of NASA's new "faster, better, cheaper" missions. All of NASA's new Discovery class missions, such as Mars Pathfinder, so far have used innovative technology derived from Cassini, and spacecraft being developed for NASA's New Millennium rely heavily upon fundamental new technologies brought forth by Cassini.

It is the ability to perform synergistic science that sets Cassini apart from other missions. The very complex interactions that are in play in systems such as those found at Jupiter and Saturn can best be addressed by instrument platforms such as *Galileo* and *Cassini*. Many phenomena to be studied are often sensitive to a large number of parameters; a measurement might have to take into account simultaneous dependencies on location, time, directions to the Sun and planet, the orbital configurations of certain moons, magnetic longitude and latitude and solar wind conditions. To deal with such complexity, the right types of instruments must be on the spacecraft to make all the necessary and relevant measurements, and all the measurements must be made essentially at the same time. Identical conditions very seldom recur. A succession or even a fleet of less well equipped spacecraft could not obtain the same results. The need for a broadly based, diverse collection of instruments is the reason why the *Cassini-Huygens* spacecraft is so large.

The Cassini and Huygens missions, featuring the intertwined work of NASA, ESA and ASI, have become models for future international space science cooperation.

The Saturn System

Saturn is easily visible to the naked eye, and was known to ancient peoples around the world. It was not until the invention of the telescope, however, that Saturn's characteristic rings began to come into focus.

Historical Observations

The Italian astronomer Galileo was the first to look at Saturn through a telescope in 1609-10. Viewed through Galileo's crude instrument, Saturn was a puzzling sight. Unable to make out the rings, Galileo thought he saw two sizable companions close to the planet. Having recently discovered the major moons of Jupiter, he supposed that Saturn could have large moons, too. ". . . [T]o my very great amazement, Saturn was seen to me to be not a single star, but three together, which almost touch each other," he wrote at the time.

Galileo was even more astonished when, two years later, he again looked at Saturn through his telescope only to find that the companion bodies had apparently disappeared. "I do not know what to say in a case so surprising, so unlooked-for and so novel," he wrote in 1612. The rings were simply "invisible" because he was now viewing them edge-on. Two years later, they again reappeared, larger than ever. He concluded that what he saw were some sort of "arms" that grew and disappeared for unknown reasons. He died never knowing that he had been the first to observe Saturn's rings.

Nearly half a century later, the Dutch scientist Christiaan Huygens solved the puzzle that vexed Galileo. Thanks to better optics, Huygens was able to pronounce in 1659 that the companions or arms decorating Saturn were not appendages, but rather the planet "is surrounded by a thin, flat ring, which nowhere touches the body." His theory was received with some opposition, but was confirmed by the observations of Robert Hooke and Italian-French astronomer Jean Dominique Cassini.

While observing Saturn, Huygens also discovered the moon Titan. A few years later, Cassini added several other key Saturn discoveries. Using new telescopes, Cassini discovered Saturn's four other major moons – Iapetus, Rhea, Tethys and Dione.

In 1675, Cassini discovered that Saturn's rings are split largely into two parts by a narrow gap – known since as the "Cassini Division." In the 19th century, J.E. Keeler, pursuing theoretical studies developed by James Clerk Maxwell, showed that the ring system was not a uniform sheet but made up of small particles that orbited Saturn.

The first detection of Saturn's magnetic field came with the flyby of Saturn by NASA's *Pioneer 11* spacecraft in 1979. Then, in 1980 and 1981, the NASA *Voyager 1* and *Voyager 2* spacecraft flew through Saturn's system to reveal storms and subtle latitudinal banding in the planet's atmosphere, several more small moons and a breathtaking collection of thousands of ringlets. The *Voyager*s found ring particles ranging in size from nearly invisible dust to icebergs the size of a house. The spacing and width of the ringlets were discovered to be orchestrated at least in part by gravitational tugs from a retinue of orbiting moons and moonlets, some near ring edges but most far beyond the outermost main rings. *Voyager* instruments confirmed a finding from ground-based instruments that the rings contain water ice, which may cover rocky particles.

Saturn has been a frequent target of the Hubble Space Telescope, which has produced stunning views of long-lived hurricane-like storms in Saturn's atmosphere. The world's major telescopes, including Hubble, were recently trained on Saturn to observe the phenomenon known to astronomers as a Saturn ring plane crossing. The rings were seen edge-on from Earth's perspective on May 22 and August 10, 1995, and on February 11, 1996. Ring plane crossings provide astronomers with unique views of Saturn's system.

These observations showed that the ring plane was not absolutely flat; the tilt of the F-ring distorts the appearance of the rings, causing one side to appear brighter than the other during ring plane crossings. Searches for new moons turned up several suspects, but most are now believed to be bright "knots" in the F-ring. Of particular interest were these ring arcs, natural "satellites" in the F-ring that appeared cloud-like and spread over a small area, instead of sharp pinpoints. The origin of these clumps of material in the F-ring is not well understood.

The faint, outermost E-ring is also easier to detect when viewed edge-on due to the greater amount of material in the line-of-sight. Thus, observations made over the course of the ring plane crossing provided new information on the thickness of the rings. New information gathered on the location and density of material in the rings was used by designers to plan the most advantageous and safest course for *Cassini*'s flight through the E-ring upon arrival at Saturn in 2004.

The Hubble Space Telescope is an important tool for studying Saturn, its rings, moons and magnetosphere in support of the Cassini mission. Hubble observations of Saturn's atmosphere were made after storms were discovered by ground-based observers. First in 1990 and again in 1994, apparent upwellings of ammonia clouds appeared and then were spread around the planet by prevailing winds.

Hubble observations of Titan indicate that color differences in Titan's hemispheres seen during the *Voyager* flybys in 1980 and 1981 have since reversed themselves. Some Hubble observations have studied chemical processes in Titan's atmosphere. Images made in the infrared have looked through Titan's clouds and allowed some mapping of its surface. Hubble has also contributed new information about the processes in Saturn's magnetosphere through ultraviolet measurements of Saturn's auroras.

Scientists using Hubble expect to study the planet and the ring system as it opens up to our view during the course of *Cassini*'s cruise to the planet. Hubble investigations will place *Cassini* results in the context of the decades-long studies of the planet and help direct *Cassini* instruments for studies of Saturn's system during its orbital tour.

Saturn's Place in the Solar System

Studies of star formation indicate that our solar system formed within a giant collection of gases and dust, drawn together by gravitational attraction and condensed over many millions of years into many stars. The giant gas cloud condensed into rotating pools of higher density in a process called gravitational collapse. These rotating pools of material condensed more rapidly until their temperatures and densities were great enough to form stars. Surrounding each new star, the leftover material flattened into a disc rotating approximately in the plane of the star's equator. This material can eventually form planets – and this is apparently what occurred in our solar system.

The composition of the planets is largely controlled by their temperatures, which in turn is determined by their distances from the Sun. Compounds with high melting points were the first to condense, followed by silicates. These formed the rocky cores of all the planets.

While hydrogen is the predominant element in the universe and in our solar system, other gases are present, including water, carbon dioxide and methane. As ices, these predominate in the cooler outer solar system. Gases also collected as envelopes around the planets and moons. (Where conditions are right, large quantities of water in its liquid state can form, as exhibited by Earth's oceans, Mars' flood plains and a possible ocean beneath the surface of Europa, a moon of Jupiter.)

The large outer planets contain much of the primordial cloud's gases not trapped by the Sun. Hydrogen is the most abundant material in the Sun and in all the large gaseous planets – Jupiter, Saturn, Uranus and Neptune. Each of these giant planets has many moons. These moons form systems of natural satellites, creating the equivalent of miniature solar systems around the gas giants formed by processes similar to those responsible for the system around the Sun.

Saturn is similar to Jupiter in size, shape, rotational characteristics and moons, but Saturn is less than one-third the mass of Jupiter and is almost twice as far from the Sun. Saturn radiates more heat than it receives from the Sun. This is true of Jupiter as well, but Jupiter's size and cooling rate suggest that it is still warm from the primordial heat generated from condensation during its formation. Slightly smaller Saturn, however, has had time to cool – so some mechanism, such as helium migrating to the planet's core, is needed to explain its continuing radiation of heat.

Voyager measurements found that the ratio of helium to molecular hydrogen in Saturn is 0.06, compared to Jupiter's value of 0.13 (which is closer to the solar abundance and that of the primordial solar nebula). Helium depletion in Saturn's upper atmosphere is believed to result from helium raining down to the lower altitudes; this supports the concept of helium migration as the heat source in Saturn. *Cassini*'s measurements of Saturn's energy, radiation and helium abundance will help explain the residual warmth.

Saturn's visible features are dominated by atmospheric clouds. They are not as distinct as Jupiter's clouds, primarily because of a haze layer covering the planet that is a result of cooler temperatures due to the weaker incoming sunlight. This reduced solar radiation, and correspondingly greater influence from escaping internal heat, lead to greater wind velocities on Saturn. Both Saturn's and Jupiter's weather are thus driven by heat from below.

The Planet Saturn

Saturn is the sixth planet from the Sun. Compared with Earth, Saturn is 9.5 times farther away from the Sun. From Saturn, the Sun is about $1/10^{th}$ the size of the Sun we see from Earth. Sunlight spreads as it travels through space; an area on Earth receives 90 times more sunlight than an equivalent area on Saturn. Because of this fact, the same light-driven chemical processes in Saturn's atmosphere take 90 times longer than they would at Earth. The farther away from the Sun, the slower a planet travels in its orbit, and the longer it takes to complete its orbit about the Sun. Saturn's year is equal to 29.46 Earth years.

Saturn's orbit is not circular but slightly elliptical in shape; as a result, Saturn's distance from the Sun changes as it orbits the Sun. This elliptical orbit causes a small change in the amount of sunlight that reaches the surface of this gaseous planet at different times in the Saturn year, and may affect the planet's upper atmospheric composition over that period.

Saturn's period of rotation around its axis depends on how it is measured. The cloud tops show a rotation period of 10 hours, 15 minutes at the equator, but the period is 23 minutes longer at higher latitudes. A radio signal that has been associated with Saturn's magnetic field shows a period of 10 hours, 39.4 minutes.

This high rotation rate creates a strong centrifugal force that causes an equatorial bulge and a flattening of Saturn's poles. As a result, Saturn's cloud tops at the equator are about 60,330 kilometers (37,490 miles) from the center, while the cloud tops at the poles are only about 54,000 kilometers (33,550 miles) from the center. Saturn's volume is 764 times the volume of Earth.

Saturn has the lowest density of all the planets because of its vast, distended, hydrogen-rich outer layer. Like the other giant planets, Saturn contains a liquid core of heavy elements including iron and rock of about the same volume as Earth, but having three or more times the mass of Earth. This increased density is due to compression resulting from the pressure of the liquid and atmospheric layers above the core, and is caused by gravitational compression of the planet.

The core of molten rocky material is believed to be covered with a thick layer of metallic liquid hydrogen and, beyond that, a layer of molecular liquid hydrogen. The great overall mass of Saturn produces a very strong gravitational field, and at levels just above the core the hydrogen is compressed to a state that is liquid metallic, which conducts electricity. (On Earth, liquid hydrogen is usually made by cooling the hydrogen gas to very cold temperatures, but on Saturn, liquid hydrogen is very hot and is formed under several million times the atmospheric pressure found at Earth's surface.) This conductive liquid metallic hydrogen layer, which is also spinning with the rest of the planet, is believed to be the source of Saturn's magnetic field. Turbulence or convective motion in this layer of Saturn's interior may create Saturn's magnetic field.

One unusual characteristic of Saturn's magnetic field is that its axis is the same as that of the planet's rotation. This is different from that of five other notable magnetic fields: those of Mercury, Earth, Jupiter, Uranus and Neptune. Current theory suggests that when the axes of rotation and magnetic field are aligned, the magnetic field cannot be maintained. Scientists do not understand the alignment of Saturn's strong magnetic field with its rotation axis, which does not fit with theories of how planets' magnetic fields are generated.

On Earth, there is a definite separation between the land, the oceans and the atmosphere. Saturn, on the other hand, has only layers of hydrogen that transform gradually from a liquid state deep inside into a gaseous state in the atmosphere, without a well-defined boundary. This is an unusual condition that results from the very high pressures and temperatures found on Saturn. Because the pressure of the atmosphere is so great, the atmosphere is compressed so much where the separation would be expected to occur that it actually has a density equal to that of the liquid. This condition is referred to as "supercritical." It can happen to any liquid and gas if compressed to a point above critical pressure. Saturn thus lacks a distinct surface; when making measurements, scientists use as a reference point the altitude where the pressure is 1 bar (or one Earth atmosphere). This pressure level is near Saturn's cloud tops.

The major component of Saturn's atmosphere is hydrogen gas. If the planet were composed solely of hydrogen, there would not be much of interest to study. However, the composition of Saturn's atmosphere includes 6 percent helium gas by volume; in addition, $1/10,000^{th}$ of 1 percent is composed of other trace elements. Using spectroscopic analysis, scientists find that these atmospheric elements can interact to form ammonia, phosphine, methane, ethane, acetylene, methylacetylene and propane. Even a small amount is enough to freeze or liquefy and make clouds of ice or rain possessing a variety of colors and forms.

With the first pictures of Saturn taken by the *Voyager* spacecraft in 1980, the clouds and the winds were seen to be almost as complex as those that *Voyager* found on Jupiter just the year before. Scientists have made an effort to label the belts and zones seen in Saturn's cloud patterns. The banding results from convective flows in the atmosphere driven by temperature – very much the same process that occurs in Earth's atmosphere, but on a grander scale and with a different heat source.

Saturn has different rotation rates in its atmosphere at different latitudes. Differences of 1,500 km/hr (more than 900 mph) were seen between the equator and nearer the poles, with higher speeds at the equator. This is five times greater than the wind velocities found on Jupiter.

Saturn's cloud tops reveal the effects of temperature, winds and weather many kilometers below. Hot gases rise; as they rise, they cool and can form clouds. As these gases cool, they begin to sink; this convective motion is the source of the billowy clouds seen in the cloud layer. Cyclonic storms observed in the cloud tops of Saturn are much like the smaller versions we see in weather satellite images of Earth's atmosphere.

Temperature variations in Saturn's atmosphere are the driving force for the winds and thus cloud motion. The lower atmosphere is hotter than the upper atmosphere, causing gases to move vertically, and the equator is warmer than the poles because it receives more direct sunlight. Temperature variations, combined with the planet's rapid rotation rate, are responsible for the fast horizontal motion of winds in the atmosphere.

Titan

Titan presents an environment which appears to be unique in the solar system, with a thick organic hazy atmosphere containing organic (or carbon-based) compounds, an organic ocean or lakes and a rich soil filled with frozen molecules similar to what scientists believe led to the origin of life on Earth. In the three centuries since the discovery of Titan we have come to see it as a world strangely similar to our own, yet located almost 1½ billion kilometers (900 million miles) from the Sun. With a thick, nitrogen-rich atmosphere, possible seas and a tar-like permafrost, Titan is thought to harbor organic compounds that may be important in the chain of chemistry that led to life on Earth.

Titan has been described as having an environment similar to that of Earth before biological activity forever altered the composition of Earth's atmosphere. The major difference on Titan, however, is the absence of liquid water and Titan's very low temperature. Thus there is no opportunity for aqueous chemistry at Earth-like temperatures – considered crucial for the origin of life as we know it. Scientists say the surface temperatures on Titan are thought to be cold enough to preclude any biological activity whatsoever at Titan.

As on Earth, the dominant atmospheric constituent in Titan's atmosphere is nitrogen. Methane represents about 6 percent of the atmospheric composition. Titan's surface pressure is 1.6 bars – more than 50 percent greater than that on Earth, despite Titan's smaller size. The surface temperature was found by Voyager to be -179° C (-290° F), indicating that there is little greenhouse warming.

The opacity of Titan's atmosphere is caused by naturally produced photochemical smog. *Voyager's* infrared spectrometer detected many minor constituents produced primarily by photochemistry of methane, which produces hydrocarbons such as ethane, acetylene and propane. Methane also interacts with nitrogen atoms, forming "nitriles" such as hydrogen cyanide. With Titan's smoggy sky and distance from the Sun, a person standing on Titan's surface in the daytime would experience a level of daylight equivalent to about $1/1,000^{th}$ the daylight at Earth's surface.

What is the source of molecular nitrogen, the primary constituent of Titan's atmosphere as we see it today? Was the molecular nitrogen accumulated as Titan formed, or was is it the byproduct of ammonia that formed with Titan? Did it come from comets? This important question can be investigated by looking for argon in Titan's atmosphere. Both argon and nitrogen condense at similar temperatures. If molecular nitrogen from the solar nebula – the cloud of gas that formed the Sun – was the source of nitrogen on Titan, then the ratio of argon to nitrogen in the solar nebula should be preserved on Titan. Such a finding would mean that we have truly found a sample of the "original" planetary atmosphere.

Some of the hydrocarbons found at Titan spend time as the aerosol haze in the atmosphere that obscures the surface. Many small molecules of compounds such as hydrogen cyanide and acetylene combine to form larger chains in a process called polymerization, resulting in additional aerosols. Eventually they drift to Titan's surface. Theoretically the aerosols should accumulate on the surface and, over the life of the solar system, produce a global ocean of ethane, acetylene, propane and other constituents with an average depth of up to 1 kilometer (about two-thirds of a mile). A large amount of liquid methane mixed with ethane could theoretically provide an ongoing source of methane in the atmosphere, similar to the way oceans on Earth supply water to the atmosphere. Radar and near-infrared data from ground-based studies show, however, that there is no global liquid ocean on Titan, although there could be lakes and seas. Titan appears to have winds.

The surface of Titan was not visible to *Voyager* at the wavelengths available to *Voyager's* cameras. What knowledge exists about the appearance of the surface of Titan comes from Earth-based radar measurements and more recent images acquired with the Hubble Space Telescope at wavelengths longer than those of *Voyager's* cameras. Hubble images from 1994 and later reveal brightness variations suggesting that Titan has a large continent-sized region on its surface that is distinctly brighter than the rest of the surface at both visible and near-infrared wavelengths. Preliminary studies suggest that a simple plateau or elevation change of Titan's surface cannot explain the image features; the brightness differences must be partly due to a different composition and/or roughness of material. Like other moons in the outer solar system, Titan is expected to have a predominantly water ice crust. Water at the temperatures found in the outer solar system is as solid and strong as rock. There are weak spectral features that suggest ice on Titan's surface, but some dark substance is also present. Scientists conclude that something on the surface is masking the water ice.

Titan's size alone suggests that it may have a surface similar to Jupiter's moon Ganymede – somewhat modified by ice tectonics, but substantially cratered and old. If Titan's tectonic activity is no more extensive than that of Ganymede, circular crater basins may provide storage for lakes of liquid hydrocarbons. Impacting meteorites would create a layer of broken, porous surface materials, called regolith, which may extend to a depth of 1 to 3 kilometers (about 1 to 2 miles). This regolith could provide subsurface storage for liquid hydrocarbons as well. In contrast to Ganymede, Titan may have incorporated as much as 15 percent ammonia as it formed in the colder region of Saturn's orbit. As Titan's water ice surface froze, ammonia-water liquid would have been forced below the surface. This liquid will be buoyant relative to the surface water ice crust, however; ammonia-water magma thus may have forced its way along cracks to the surface, forming exotic surface features. Density measurements suggest that Titan is made up of roughly half rocky silicate material and half water ice. Methane and ammonia could have been mixed with the water ice during Titan's formation. The formation of Titan by accretion was at temperatures warm enough for Titan to differentiate; rocky material sank to form a dense core, covered by a mantle of water, ammonia and methane ices.

The mixture of ammonia with water could ensure that Titan's interior is still partially unsolidified, as the ammonia would effectively act like antifreeze. Radioactive decay in the rocky material in the core could heat the core and mantle, making it possible that a liquid layer could exist today in Titan's mantle.

Methane could be trapped in Titan's water ice crust, which could provide a possible long term source for the methane in Titan's atmosphere if it were freed by ongoing volcanic processes. Due to Titan's thick natural smog, *Voyager* was prevented from viewing the surface; the images showed a featureless orange face. Spectroscopic observations by *Voyager's* infrared spectrometer revealed traces of ethane, propane, acetylene and other organic molecules in addition to methane. These hydrocarbons are produced by the combination of solar ultraviolet light and electrons from Saturn's rapidly rotating magnetosphere striking Titan's atmosphere. Hydrocarbons produced in the atmosphere eventually condense out and rain down on the surface, thus there may exist lakes of ethane and methane, perhaps enclosed in the round bowls of impact craters. Titan's hidden surface may have exotic features such as mountains sculpted by hydrocarbon rain, and perhaps rivers, lakes and "waterfalls" of exotic liquids. Water and ammonia magma from Titan's interior may occasionally erupt, spreading across the surface.

Titan's orbit takes it both inside and outside the magnetosphere of Saturn. When Titan is outside the magnetosphere and exposed to the solar wind, its interaction may be similar to that of other bodies in the solar system such as Mars, Venus or comets (these bodies have substantial interaction with the solar wind, and, like Titan, have atmospheres but no strong internal magnetic fields).

The interaction of Titan with the magnetosphere provides a way for both the magnetospheric plasma to enter Titan's atmosphere and for atmospheric particles to escape Titan. *Voyager* results suggested that this interaction produces a torus of neutral particles encircling Saturn, making Titan a potentially important source of plasma to Saturn's magnetosphere. The characteristics of this torus are yet to be explored, and will be studied by the *Cassini* orbiter. The interaction of ice particles and dust from Saturn's rings will play a special role as the dust moves out towards Titan's torus and becomes charged by collisions. When the dust is charged it behaves partially like a neutral particle orbiting Titan according to Kepler's laws (gravity driven), and partially like a charged particle moving with Saturn's magnetosphere. The interaction of dust with Saturn's magnetosphere will provide scientists with a detailed look at how dust and plasma interact.

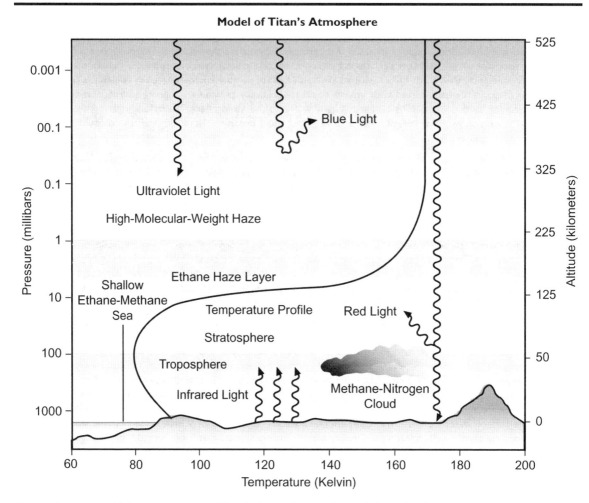

Model of Titan's Atmosphere

Electrical storms and lightning may exist in Titan's skies. *Cassini* will search for visible lightning and listen for "whistler" emissions that can be detected when lightning discharges a broad band of electromagnetic emission, part of which can propagate along Saturn's magnetic field lines. These emissions have a decreasing tone with time (because the high frequencies arrive before the low frequencies). Lightning whistlers have been detected in both Earth's and Jupiter's magnetosphere. They can be detected by radio and plasma wave instruments from large distances and also can be used to estimate the frequency of lightning.

Titan may have its own internally generated magnetic field. Recent results from the *Galileo* spacecraft at Jupiter indicate the possibility of an internally generated magnetic field associated with the moon Ganymede. For Titan there are two possibilities: a magnetic field could be induced from the interaction of Titan's substantial atmosphere with the flow of Saturn's magnetosphere (like Venus's interaction with the solar wind); or a magnetic field could be generated internally from dynamo action in a metallic molten core (like Earth's). (Under the dynamo theory, a magnetic field is created by the circling motion of electrically charged material in the core.) In addition to being important to understanding the Titan interaction with Saturn's magnetosphere, a Titan magnetic field, if generated internally, would help scientists define the natural satellite's interior structure.

The Rings

From a distance, the rings of Saturn look like majestically symmetrical hoops surrounding the planet. Up close, however, from the views provided by the *Voyager* spacecraft, the rings turn out to be a still splendid but somewhat unruly population of ice and rock particles jostling against each other or being pushed and pulled into uneven orbits by bigger particles and by Saturn's many moons.

The mass of all the ring particles measured together would comprise a moon about the size of Mimas, one of Saturn's medium-small moons. The rings may, in fact, be at least partly composed of the remnants of such a moon or moons, torn up by gravitational forces.

Their precise origin is a mystery. It is not known if rings formed around Saturn out of the initial solar system nebula, or after one or more moons were torn apart by Saturn's gravity. If the rings were the result of the numerous comets

Saturn's Rings

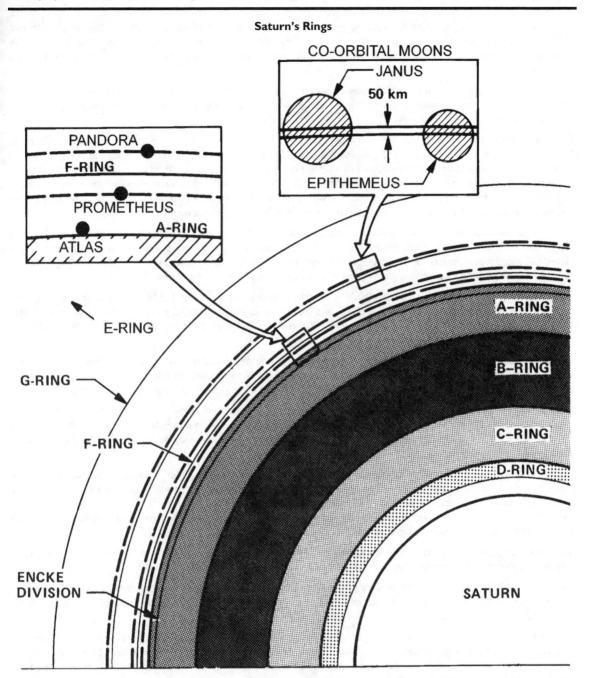

captured and destroyed by Saturn's gravity, why are Saturn's bright rings so different in nature from the dark rings of neighboring planets? Over the lifetime of the rings, they must have been bombarded continually by comets and meteors – and therefore they should have accumulated a great amount of carbonaceous and silicate debris – yet water ice is the only material positively identified in spectra of the rings.

The effects of torque and gravitational drag – along with the loss of momentum through collisions – should have produced a system only one-tenth to one-hundredth the age of the solar system itself. If this hypothesis is correct, then we cannot now be observing a ring system around Saturn that formed when the solar system formed.

In fact, Saturn's rings – as well as the rings of all the other large planets – may have formed and dissipated many times since the beginning of the solar system. An extraordinarily complex structure is seen across the entire span of Saturn's ring system. The broad B-ring, for instance, often contains numerous "spokes" – radial, rotating features that may be caused by a combination of magnetic and electrostatic forces. The individual rings themselves defy definition; the count in high-resolution images suggests anywhere from 500 to 1,000 separate rings. Named in order of discovery, the labels that scientists have assigned to the major rings do not indicate their relative positions. From the planet outward, they are known as the D-, C-, B-, A-, F-, G- and E-rings.

The possibility of numerous, small natural satellites within Saturn's ring system was a puzzle the *Voyager* mission had hoped to solve. *Voyager*'s best-resolution studies of the ring system were aimed at revealing any bodies larger than about 10 kilometers (about 6 miles) in diameter; but only three were found and none were located within the main ring complex.

The *Voyager* high-resolution studies did, however, detect signs of small moonlets not actually resolved in the images. When a small, dense body passes near a section of low-density ring material, its gravitational pull distorts the ring and creates what are known as "edge waves."

Cassini will be able to perform a number of the experiments that *Voyager* used to detect other gravitational effects on Saturn's ring material. One experiment involves "watching" as a beam of light (or, in one case, radio waves) passes through the ring, then observing the effects of the ring material on the beam. As the beam passes through the ring material, it may be attenuated or even extinguished. This "occultation" experiment provides an extremely high-resolution study of a single path across the rings – with resolutions up to about 100 meters (about 330 feet), *Cassini* will obtain far more detailed information on ring structures than the *Voyager* instruments.

Voyager's instruments did detect many minute ring structures and found that the F-ring was far more complex than images had suggested. The data set, furthermore, showed that the B-ring was quite opaque in regions, and the Cassini "Division" was not at all empty. It also provided a direct measurement of the maximum thickness of the ring system in several locations, finding it to be much less than 100 meters (about 330 feet) thick there.

Many narrow ringlets were found with slightly eccentric, non-circular shapes and orbits. These eccentric ringlets generally lie in gaps in the mass of nearly circular rings that make up the majority of Saturn's ring system. *Voyager* also found very few truly empty "gaps" in the ring system. Moonlets inside the rings do appear to clear lanes within the ring plane, giving the rings their grooved appearance.

Voyager cameras found shepherd moons that tend to contain ring particles that would otherwise spread. Also seen were density waves that move though portions of the ring plane like a crowd starting a "wave" in a stadium. This phenomenon is due to the effects of one or more moons or moonlets gravitationally tugging on ring particles.

Gravitational interactions with moons seem to create most of the structure visible in the rings, but some structural detail exists even where there is no gravitational interaction with a moon. Some poorly understood fluid physics may be responsible for some unexplained structure in the rings. Other ring features may be explained by moonlets or large particles in the rings that have gone undetected.

Voyager images showed dark, radial structures on the rings. These so-called "spokes" were seen as they formed and rotated about the planet. Spokes seemed to appear rapidly – as a section of ring rotated out of the darkness near the dawn terminator – and then dissipate gradually, rotating around toward the dusk terminator of the ring. A spoke's formation time seemed to be very short; in some imaging studies they were seen to grow more than 6,000 kilometers (about 3,700 miles) in distance in just 5 minutes.

The spokes in Saturn's rings have an unexplained link with the planet's magnetic field, and are likely just one visible manifestation of many interactions the rings have with Saturn's electromagnetic fields.

Ground-based infrared studies of the spectra of the A- and B-rings show that they are composed largely, perhaps even exclusively, of water ice. The spectral characteristics of the rings are also very similar to those of several of Saturn's inner moons.

Studies of the main rings show that the ring system is not completely uniform in its makeup, and that some sorting of materials within the A- and B-rings exists. Why such a non-uniform composition exists is unknown. The E-ring is somewhat bluish in color – and thus different in makeup from the main rings. It is believed that the moon Enceladus is the source of E-ring material.

Since ring particles larger than about 1 millimeter represent a considerable hazard to the *Cassini* spacecraft, the mission plan will include efforts to avoid dense particle areas of Saturn's ring plane. The spacecraft will be oriented to provide maximum protection for itself and its sensitive instrumentation packages. Even with such protective measures, passage through the ring plane out beyond the main rings will allow *Cassini*'s instruments to make important measurements of the particles making up the less dense regions of the Cassini ring plane. These studies could provide considerable insight into the composition and environment of the ring system and Saturn's icy natural satellites.

The Icy Moons

The moons of Saturn are diverse – ranging from the planet-like Titan to tiny, irregular objects only tens of kilometers (or miles) in diameter. These bodies are all (except perhaps Phoebe) believed to hold not only water ice, but also other components such as methane, ammonia and carbon dioxide. Saturn has at least 18 moons; there are likely other small, undiscovered natural satellites in the planet's system.

Many of the smallest moons were discovered during the *Voyager* spacecraft flybys. The 18th moon, Pan, was found nearly 10 years after the flybys during close analysis of *Voyager* images; it is embedded in the Encke Gap within Saturn's A-ring. Saturn's ring plane crossings — when the obscuring light from Saturn's bright rings dims as the rings move to an edge-on orientation from Earth's perspective — represent the ideal time for discovering new moons. Images obtained by the Hubble Space Telescope during the ring plane crossings in 1995 and 1996, however, did not reveal any unambiguous discoveries of new moons.

Before the advent of spacecraft exploration, scientists expected the moons of the outer planets to be geologically dead. They assumed that heat sources were not sufficient to have melted the moons' mantles enough to provide a source of liquid, or even semi-liquid, ice or ice silicate slurries. The *Voyager* and *Galileo* spacecraft have radically altered this view by revealing a wide range of geological processes on the moons of the outer planets. For example, Saturn's moon Enceladus may be currently active. Several of Saturn's medium-sized moons are large enough to have undergone internal melting with subsequent differentiation and resurfacing.

Many moons in the outer solar system show the effects of tidal interactions with their parent planets, and sometimes with other moons. This gravitational pushing and pulling can heat a moon's interior, causing tectonic, volcanic or geyser-like phenomena. Another factor is the presence of non-ice components, such as ammonia hydrate or methanol, which lower the melting point of near-surface materials. Partial melts of water ice and various other components — each with their own melting point and viscosity — provide material for a wide range of geological activity.

Because the surfaces of so many moons of the outer planets exhibit evidence of geological activity, planetary scientists have begun to think in terms of unified geological processes occurring on the planets and their moons. For example, partial melts of water ice with other materials could produce flows of liquid or partially molten slurries similar to lava flows on Earth that result from the partial melting of silicate rock mixtures. The ridged and grooved terrains on moons such as Saturn's Enceladus and Tethys may have resulted from tectonic activities found to occur throughout the solar system. The explosive volcanic eruptions possibly occurring on Enceladus may be similar to those occurring on Earth.

Saturn's Major Moons

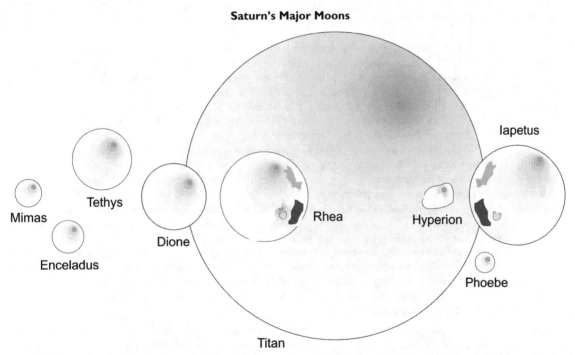

Several of Saturn's moons underwent periods of melting and active geology within a billion years of their formation and then became quiescent. For nearly a billion years after their formation, the moons all underwent intense bombardment and cratering. The bombardment tapered off to a slower rate, but still continues. By counting the number of craters on a moon's surface, and making certain assumptions about the amount and frequency of impacting material, geologists are able to estimate when a specific portion of a moon's surface was formed.

Meteorites bombarding icy bodies change the surfaces by excavating and exposing fresh material. Impacts can also cause some surface or subsurface materials to turn to vapor, and the subsequent escape of those vaporized materials can create a slag deposit enriched in opaque, dark materials. Both the moons of Jupiter and the medium-sized moons of Saturn tend to be brighter on the hemispheres leading in the direction of orbital motion (each moon's so-called "leading" side, as opposed to its "trailing" side); this effect is thought to be due to the roughing-up and coating that the leading side receives from meteor bombardment.

Saturn's six largest moons after Titan are smaller than Titan and Jupiter's giant Galilean moons, but they are still sizable and represent a unique class of icy satellite. Tethys, Rhea, Iapetus, Mimas and Enceladus are thought to be largely water ice, possibly mixed with ammonia or other volatiles. They have smaller amounts of rocky silicates than the Galilean moons.

* Saturn's innermost, medium-sized moon, Mimas, is covered with craters, including one named Herschel that covers a third of the moon's diameter. There is a suggestion of grooves in its surface that may be features caused by the impact that created Herschel. The craters on Mimas tend to be high-rimmed, bowl-shaped pits. Crater count of Mimas suggest that it has undergone several episodes of resurfacing.

* The next moon outward from Saturn is Enceladus, an object that was known from telescope measurements to reflect nearly 100 percent of the visible light it receives. This leads scientists to believe Enceladus' surface is probably pure water ice. *Voyager 2* images reveal a surface that had been subjected in the recent geological past to extensive resurfacing. Grooved formations similar to those on the Galilean moon Ganymede are prominent.

The lack of impact craters on part of Enceladus leads scientists to conclude the age of its surface is less than a billion years. Some form of ice volcanism may be currently occurring on Enceladus. One possible source of heating is tidal interactions, perhaps with the moon Dione.

Enceladus is possibly responsible for the formation of Saturn's E-ring. This ring is a tenuous collection of icy particles that extends from inside the orbit of Enceladus to beyond the orbit of Dione, but is much thicker than Saturn's other rings. The maximum thickness position of the ring coincides with the orbital position of Enceladus. If some form of volcanism is occurring on the surface, it could provide a source of particles for the ring.

* Tethys is covered with impact craters, including Odysseus, an enormous impact structure more than 400 kilometers (250 miles) in diameter, or more than one-third of the moon's diameter of 1,060 kilometers (658 miles). The craters tend to be flatter than those on Mimas or Earth's Moon, probably because they have been smoothed out over the eons under Tethys' stronger gravitational field. Evidence for episodes of resurfacing is seen in regions that have fewer craters and higher reflectivity. In addition, there is a huge trench formation, the Ithaca Chasma, which may be a degraded form of the grooves found on Enceladus.

* Dione is about the same size as Tethys but more dense, and shows a wide variety of surface features. Next to Enceladus among Saturn's moons, it has the most extensive evidence for internal activity. It has enough rocky material in its makeup to produce heat from natural radioactivity, which could be the cause of its internal activity. Most of the surface is heavily cratered, but gradations in crater density indicate that several periods of resurfacing occurred during the first billion years of Dione's existence. The leading side of the moon is about 25 percent brighter than the other, due possibly to more intensive bombardment by micrometeorites on the darker hemisphere. Wispy streaks, which are about 50 percent brighter than the surrounding areas, are believed to be the result of internal activity and subsequent flows of erupting material.

* Rhea appears to be superficially very similar to Dione. Bright, wispy streaks cover one hemisphere. No resurfacing events seem to have occurred in its early history, however. Some regions lack large craters, while other regions have a preponderance of such impacts. The larger craters may be due to a population of larger debris that was more prevalent during an earlier era of collisions.

* When the astronomer Cassini discovered Iapetus in 1672, he noticed almost immediately that at one point in its orbit around Saturn it was very bright, but on the opposite side of the orbit, the moon nearly disappeared. He correctly deduced that the trailing hemisphere is composed of highly reflective material, while the leading hemisphere is strikingly darker. (This sets Iapetus apart from Saturn's other moons and the moons of Jupiter, which tend to be brighter on their leading hemispheres.) *Voyager* images show that the bright side, which reflects nearly 50 percent of the light it receives, is fairly typical of a heavily cratered icy satellite. The other side, which is the face that Iapetus puts forth as it moves forward in its orbit, is coated with a much darker, redder material that has a reflectivity of only about 3 to 4 percent.

Scientists still do not agree on whether the dark material originated from an outside source or was created from Iapetus' own interior. One scenario for the outside deposit of material would involve dark particles being ejected from the little moon Phoebe and drifting inward to coat Iapetus. The major problem with this model is that the dark material on Iapetus is redder than Phoebe, although the material could have undergone chemical changes after its expulsion from Phoebe that made it redder. One observation lending credence to an internal origin is the concentration of material on crater floors, which implies that something is filling in the craters. In one model proposed by scientists, methane could erupt from the interior and then become darkened by ultraviolet radiation.

Iapetus is odd in other respects. It is the only large Saturn moon in a highly inclined orbit, one that takes it far above and below the plane in which the rings and most of the moons orbit. It is less dense than objects of similar brightness, which implies that it has a higher fraction of ice or possibly methane or ammonia in its interior.

Small Moons

The Saturn system has a number of unique, small moons. Two types of objects have been found only in Saturn's system: the co-orbital moons and the Lagrangians. A third type of object, shepherding moons, have been found only at Saturn and Uranus. All three groups of moons are irregularly shaped and probably consist primarily of ice.

Three shepherds – Atlas, Pandora and Prometheus – are believed to play a key role in defining the edges of Saturn's A- and F-rings. The orbit of Atlas lies several hundred kilometers (or miles) from the outer edge of the A-ring. Pandora and Prometheus, which orbit on either side of the F-ring, constrain the width of this narrow ring and may cause its kinky appearance.

The co-orbital moons Janus and Epimetheus, discovered in 1966 and 1978, respectively, exist in an unusual dynamic situation. They move in almost identical orbits. Every four years, the inner moon (which orbits slightly faster than the outer one) overtakes its companion. Instead of colliding, the moons exchange orbits. The four-year cycle then begins again. Perhaps these two moons were once part of a larger body that disintegrated after a major collision.

Saturn's Magnetosphere

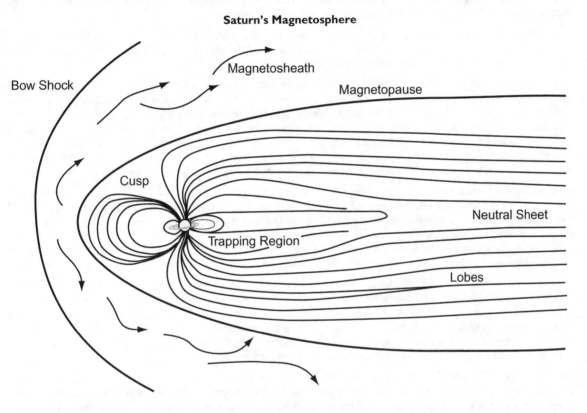

The three other small moons of Saturn – Calypso, Helene and Telesto – orbit in the Lagrangian points of larger moons, one associated with Dione and two with Tethys. The Lagrangian points are locations within an object's orbit in which a less massive body can move in an identical, stable orbit. The points lie about 60 degrees in front of and behind the larger body. (Although no other known moons in the solar system are Lagrangians, the Trojan asteroids orbit in two of the Lagrangian points of Jupiter. Mars also has an asteroid at one Lagrangian point.)

Telescope observations showed that the surface of Hyperion, which lies between the orbits of Iapetus and Titan, is covered with ice. Because Hyperion is not very reflective, its ice must be mixed with a significant amount of darker, rocky material. The color of Hyperion is similar to that of the dark side of Iapetus and D-type asteroids; all three bodies may be rich in primitive material rich in organic compounds. Although Hyperion is only slightly smaller than Mimas, it has a highly irregular shape which, along with a battered appearance, suggests that it has been subjected to intense bombardment and fragmentation. There is also evidence that Hyperion is in a chaotic rotation – probably the result of tidal forces during close passages of nearby Titan.

Saturn's outermost moon, Phoebe, a dark object with a surface composition probably similar to that of C-type asteroids, is in a wrong-way orbit compared to the other moons (this is called a "retrograde" orbit). The orbit is also highly inclined, taking Phoebe high above and below the plane on which the rings and most of the moons orbit. These peculiar orbital characteristics alone suggest that Phoebe may be a captured asteroid. *Voyager* images show a mottled appearance. Although it is smaller than Hyperion, Phoebe has a nearly spherical shape.

Pan, the 18th known moon of Saturn, was discovered in 1990 in *Voyager 2* images that were taken in 1981. This small object is embedded within the A-ring and helps to clear the particles out of the Encke division.

The Magnetosphere

Saturn, its moons and rings sit inside an enormous bubble in the solar wind created by the planet's strong magnetic field. This "sphere of influence" of Saturn's magnetic field – called a magnetosphere – resembles a similar, but smaller, magnetic bubble surrounding Earth. The same kind of structure envelopes Jupiter and several other bodies in the solar system. A supersonic "solar wind" of electrically charged particles blows outward from the Sun within our star's own magnetosphere. The pressure of this solar wind against a planet's magnetosphere compresses the sunward side of the sphere and blows past the planet, giving the magnetosphere a drawn-out shape like that of a comet.

Inside Saturn's vast magnetospheric bubble is a mixture of particles including electrons, various species of ions and neutral atoms and molecules, several populations of very energetic charged particles (like those in Earth's Van Allen Belts) and charged dust grains. The charged particles and dust grains all interact with both the steady and the fluctuating electric and magnetic fields present throughout the magnetosphere.

The primary sources of particles in Saturn's magnetosphere are thought to be the moons Dione and Tethys, which are bombarded by the more energetic particles. But the solar wind, ionosphere, rings, Saturn's atmosphere, Titan's atmosphere and the other icy moons are sources as well.

The mysterious "spokes" in the rings of Saturn are probably caused by electrodynamic interactions between the tiny charged dust particles in the rings and the magnetosphere. Aurora, which exist on Saturn as well as Earth, are produced when trapped charged particles spiral down along magnetic field lines leading to the planet and collide with gases in the atmosphere. Scientists generally agree that the internal magnetic fields of the giant planets arise from an electrical field generated somewhere inside the planets' liquid interiors. Saturn's interior is probably quite exotic because of the great pressures caused by its large size. There may be a molten rocky core, but wrapped around this core scientists would expect to find layers of uncommon materials such as liquid metallic hydrogen and perhaps liquid helium.

Making measurements close to the planet over a wide range of latitudes and longitudes, *Cassini* will measure the details of the gravitational and magnetic field and tell us more about Saturn's interior. The spacecraft's magnetometer will measure the strength and direction of the magnetic field throughout the magnetosphere – close to the planet where the field is nearly dipolar, and further from the planet where electrical currents create a non-dipolar field. The magnetometer will measure the planet's magnetic field with sufficient accuracy to determine if it is indeed symmetrical. If so, the basic tenets of the dynamo theory invented to explain planetary magnetic fields may need to be reexamined.

Among all the planets with magnetic fields, there are two main sources of energy driving magnetospheric processes: the planet's rotation and the solar wind. In turn there are two types of large-scale plasma flow within the magnetosphere – co-rotation and convection. The nature of the large-scale circulation of particles in the magnetosphere depends on which source is dominant. At Earth, the energy is derived primarily from the solar wind; at Jupiter it is derived from the planet's rapid rotation rate. Saturn's magnetosphere is especially interesting because it is somewhere in between; both energy sources should play an important role.

The rotation of Saturn's magnetic field with the planet creates a large electric field that extends into the magnetosphere. The combination of this electric field and Saturn's magnetic field create electromagnetic forces that cause charged plasma particles to "co-rotate" (rotate together with Saturn and its internal magnetic field) as far out as Rhea's orbit.

The other large-scale flow of charged particles, convection, is caused by the solar wind pulling the lines of the magnetic field toward the tail of the magnetosphere. This leads to a plasma flow from day side to night side on open field lines, as well as a return flow from night side to day side on closed field lines (particularly near the equatorial plane).

Saturn has an ionosphere, a thin layer of partially ionized gas at the top of the sunlit atmosphere. Collisions between particles in the atmosphere and the ionosphere create a frictional drag that causes the ionosphere to rotate together with Saturn and its atmosphere.

Mission Overview

The Cassini mission will span 11 years, including 6½ years of interplanetary cruise and four years in orbit at Saturn. Along the way the spacecraft will fly by several planets, making use of their gravitational energy to speed its flight toward Saturn.

Launch

Launch Periods. Given the relative positions of the planets and the trajectory *Cassini* must fly, the primary launch period opens on October 6, 1997, and continues through November 15. Launch is scheduled for October 13, 1997.

The best conditions exist from October 6 through November 4; launching between those dates would result in *Cassini* arriving at Saturn in July 2004. The period from November 5 to 15 is considered a contingency period; *Cassini's* launch is still feasible but less desirable, as more spacecraft propellant would be needed to refine its flight path and the arrival at Saturn would be delayed by many months. A launch from November 5 through 9 would place the spacecraft at Saturn in December 2004; launching from November 10 through 13, *Cassini* would arrive at Saturn in July 2005. Launching on November 14 or 15 would deliver the spacecraft to Saturn in December 2005.

Secondary and backup trajectories exist in the unplanned event that *Cassini* is not launched during the primary launch period. Both require the spacecraft to fly a Venus-Earth-Earth gravity assist (VEEGA) trajectory instead of the Venus-Venus-Earth-Jupiter (VVEJGA) trajectory required for the primary launch period.

The secondary launch period is a 45-day period between November 28, 1997, and January 11, 1998. The spacecraft's arrival at Saturn would be October 13, 2006, after a flight of 8.8 years. The significant difference between this mission and the primary one is a longer interplanetary cruise and poorer illumination of Saturn's rings due to their changing angle relative to the Sun and Earth.

There is also a backup launch period from March 19 through April 5, 1999, requiring a flight of 9.8 years to Saturn with an arrival date of December 22, 2008. This mission entails a much longer flight time to Saturn. Studies of Saturn's rings would be significantly degraded due to the angle of the rings. Due to the long flight time, electrical power output from the radioisotope thermoelectric generators would be degraded. This would result in fewer instruments being allowed to operate at a given time, or in lower power states, or with less engineering support for a suite of instruments.

Daily Window. The launch window on October 13 opens at 4:55 a.m. Eastern Daylight Time and extends for 140 minutes. After that date, the opening time of the launch window moves earlier by about six minutes daily. The length of the window on subsequent days remains at 140 minutes per day.

Launch Vehicle

The Titan IVB launch vehicle consists of two upgraded solid rocket motors (known as SRMUs); a two-stage liquid-propellant core; a liquid-fuel upper stage; and a payload fairing.

The three-segment SRMUs are 3.2 meters (10.5 feet) in diameter and 34.2 meters (112.4 feet) long. Each motor contains 312,458 kilograms (688,853 pounds) of hydroxyl terminated polybutadiene (HTPB) propellant and provides a maximum thrust of 7.56 million newtons (1.7 million pounds) at sea level. Flight control is achieved by directing the thrust through a gimbaled nozzle controlled by hydraulic actuators. Six staging rockets on each SRMU ensure positive separation from the core following SRMU burnout.

The first stage of the 3.04-meter-diameter (10-foot) core is powered by a twin assembly LR87-AJ-11 engine, while the second stage is powered by a single assembly LR91-AJ-11 engine. Both stages use storable hypergolic propellants (propellants that ignite when they

Cassini's Titan IVB Launch Vehicle

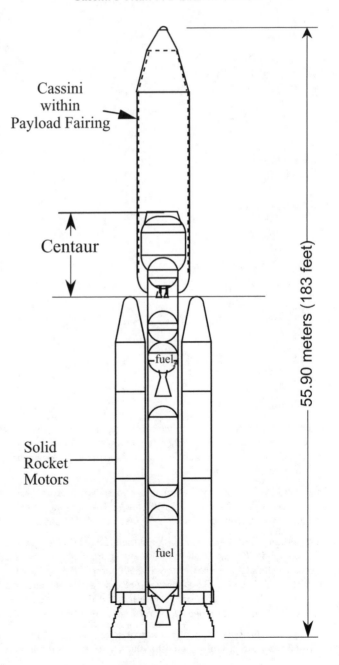

Cassini within Payload Fairing

Centaur

Solid Rocket Motors

fuel

fuel

55.90 meters (183 feet)

come in contact with each other). The fuel is aerozine 50, a 50-50 mixture of hydrazine and unsymmetrical dimethyl hydrazine. During combustion, the fuel is mixed with nitrogen tetroxide as an oxidizer.

Vehicle flight control and navigation for the Titan are provided by the guidance control unit, which includes a flight control subsystem and an inertial measurement subsystem. The first stage separates from the second stage when the second-stage engine ignites. Separation of the second stage and the Centaur upper stage is achieved when four retrorockets fire and a structural adapter is severed by pyrotechnics called "SuperZip."

The payload fairing is 5.9 meters (16.7 feet) in diameter, 20 meters (66 feet) long and has an aluminum structure.

Centaur Upper Stage. The Centaur upper stage, managed by NASA's Lewis Research Center and produced by Lockheed-Martin Astronautics, is 8.9 meters (29.45 feet) long and 4 meters (14 feet) in diameter. It provides 147,000 newtons (33,000 pounds) of engine thrust using cryogenic liquid oxygen and liquid hydrogen.

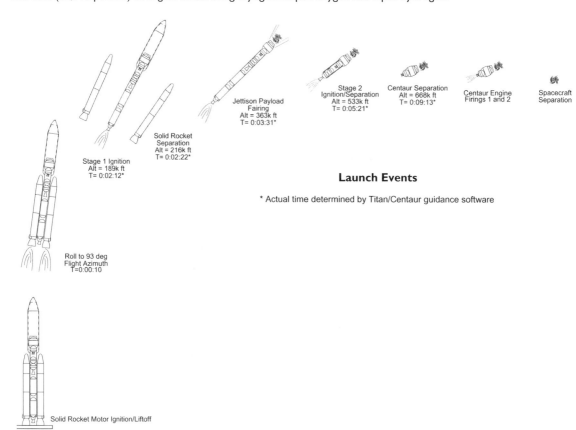

Jettison Payload
Fairing
Alt = 363k ft
T= 0:03:31*

Solid Rocket
Separation
Alt = 216k ft
T= 0:02:22*

Stage 1 Ignition
Alt = 189k ft
T= 0:02:12*

Stage 2
Ignition/Separation
Alt = 533k ft
T= 0:05:21*

Centaur Separation
Alt = 668k ft
T= 0:09:13*

Centaur Engine
Firings 1 and 2

Spacecraft
Separation

Launch Events

* Actual time determined by Titan/Centaur guidance software

Roll to 93 deg
Flight Azimuth
T=0:00:10

Solid Rocket Motor Ignition/Liftoff

Launch Events

Variations in launch date have minimal impact on the launch sequence of events through and including the first firing of the Centaur upper stage. A launch slip would shift the entire sequence equally. But the sequence of events following the first Centaur firing – including the coast period in parking orbit and the Centaur's second firing – would depend on the launch slip.

Launch begins with the ignition of the solid rocket motors, which burn for 2 minutes, 22 seconds to an altitude of approximately 66,000 meters (216,500 feet). The first stage of the liquid-fueled Titan ignites at 2 minutes, 12 seconds into flight at an altitude of about 58,000 meters (189,000 feet).

At an altitude of approximately 110,600 meters (363,000 feet) and 3 minutes, 31 seconds into flight, the payload fairing that surrounds the spacecraft is jettisoned. Ignition of the Titan's second stage occurs at 5 minutes, 21 seconds into flight at an altitude of approximately 162,500 meters (533,000 feet). At approximately 9 minutes, 13 seconds into flight at an altitude of about 203,600 meters (668,000 feet), the Centaur upper stage and the *Cassini* spacecraft separate from the Titan. (The exact time of separation – and the actual start times, burn durations and separation of the Centaur from *Cassini* – are determined by Titan / Centaur guidance software.)

At Titan-Centaur separation, flight control is transferred to the Centaur. Following separation, the Centaur completes a pre-burn sequence and is ignited for the first of two burns in what is referred to as main engine start #1. The burn ends after 131 seconds at main engine cut-off #1. After some settling time, the spacecraft and Centaur are injected into a parking orbit with a perigee (closest point to Earth) of 170 kilometers (105 miles) perigee and an apogee (farthest point from Earth) of 445 kilometers (276 miles), with an orbital inclination between 28.6 and 31.6 degrees, depending on the time of launch. This parking orbit is designed to provide an orbital lifetime of about 20 days in case the Centaur fails to restart its main engine successfully. The Centaur / spacecraft combination will coast in parking orbit until trajectory conditions are right for interplanetary injection. The coast time in parking orbit between the first main engine cutoff and the engine restart will be 8 to 32 minutes if launch takes place in the primary period, depending on the time and day liftoff occurs. The coast duration is about 14 minutes for an October 6 launch date at the opening of the daily window.

Toward the end of this coast period, the Centaur is oriented for its second burn using thrusters. After the Centaur engine is restarted, it will burn for approximately 7 to 8 minutes. Some 30 seconds after the conclusion of this second burn, the Centaur issues a command instructing the *Cassini* spacecraft to prepare for separation from the Centaur. By five minutes after the second Centaur engine cutoff, the Centaur orients *Cassini*'s high-gain antenna to point toward the Sun and begins to roll to the separation attitude. This maneuver is completed by 6 minutes after the second engine cutoff. Six minutes after the second engine cutoff, the *Cassini* spacecraft receives a command instructing it to fire explosive devices separating it from the Centaur. Final separation is completed by 6½ minutes after the second Centaur engine cutoff.

Interplanetary Trajectory

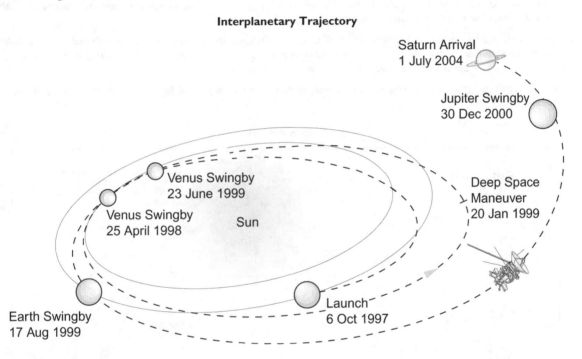

Post-Separation Events

After separating from the Centaur, *Cassini*'s attitude control and articulation subsystem gains control over the spacecraft and ensures that the high-gain antenna is pointed toward the Sun to keep the rest of the spacecraft shaded. The onboard computer then places the spacecraft in a "safe" state so that it could operate autonomously for up to 10 days in the event of telecommunications problems. The Canberra, Australia, complex of NASA's Deep Space Network will be the first to acquire *Cassini*'s signal.

The Centaur executes a collision and contamination avoidance maneuver that prevents it from impacting either Venus or the spacecraft. This burn occurs approximately 20 minutes after separation to allow the *Cassini* spacecraft and the Centaur to drift apart, insuring that the plume from the Centaur's reaction control thrusters does not impact the spacecraft. About eight minutes later, the Centaur executes a "blowdown" maneuver to expel unused propellant and to further reduce the probability of impact with *Cassini* or Venus. About an hour after separation, the Centaur depletes the hydrazine from its reaction control subsystem. The Centaur mission ends approximately 87 minutes after separation. By this time, *Cassini* has its high-gain antenna pointed toward the Sun, is transmitting real-time telemetry via one of its two low-gain antennas, and is awaiting instructions from the ground.

Approximately two days after launch, ground controllers will send to *Cassini* the sequence of computer commands to control the spacecraft for the following week. This sequence will include a trajectory correction maneuver in which

Cassini fires its onboard thrusters to fine tune its flight path. The next four or five command sequences sent to the spacecraft also will control the spacecraft for about one week each.

Cruise Phase

In maneuvers called gravity assist swingbys, *Cassini* will fly twice around Venus and once each past Earth and Jupiter. The spacecraft's speed relative to the Sun increases as it approaches and swings past each planet, giving *Cassini* the cumulative boost it needs to reach Saturn. The gravity assist technique has been used extensively in the planetary exploration program to send spacecraft from one planet to another, and to increase a spacecraft's speed. In orbit around their respective planets, *Galileo* at Jupiter and *Cassini* at Saturn often use gravity assists to navigate from one large moon to the next.

Cassini will make two gravity assist swingbys of Venus – one on April 25, 1998, at an altitude of about 300 kilometers (about 185 miles), and another on June 23, 1999, at an altitude of 1,530 kilometers (950 miles). These will be followed by a swingby of Earth occurring August 17, 1999, at an altitude of 800 kilometers (500 miles) or higher. As the spacecraft heads outward into the solar system it will make a swingby of Jupiter on December 30, 2000, passing within about 9,655,000 kilometers (about 6 million miles) of the planet.

Plans call for *Cassini* to carry out a low-activity flight plan during which only essential engineering and navigation activities such as trajectory correction maneuvers will be performed. The science instruments are planned to be turned off except for a few maintenance activities.

These include a single post-launch checkout of all of *Cassini*'s science instruments, as well as calibration of *Cassini*'s magnetometer during its subsequent Earth flyby. *Huygens* probe health checks are scheduled to occur every six months. No science observations are planned at Venus, Earth or Jupiter.

Two years before Saturn arrival, science instruments on the *Cassini* orbiter will be turned on, calibrated and begin collecting data.

Saturn Arrival

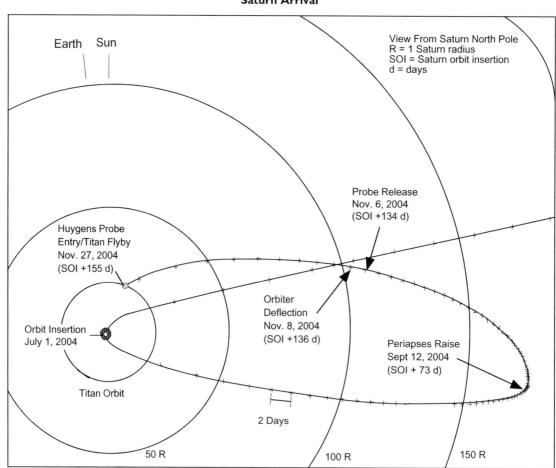

Saturn Arrival

The most critical phase of the mission after launch is Saturn orbit insertion. When *Cassini* reaches Saturn on July 1, 2004, the spacecraft will fire its main engine for 94 minutes to brake its speed and allow it to be captured as a satellite of Saturn. Passing through the dusty outermost ring of Saturn – called the E-ring – *Cassini* will swing close to the planet to begin the first of some six dozen orbits to be completed during the rest of its four-year primary mission.

The arrival period also provides a unique opportunity to observe Saturn's rings and the planet itself, as this is the closest approach the spacecraft will make to Saturn during the entire mission. A maneuver to lower the inclination of *Cassini*'s orbit, called a periapsis-raise maneuver, will be performed in September 2004 to establish the geometry for the entry of the *Huygens* probe when it is released during *Cassini*'s first Titan flyby.

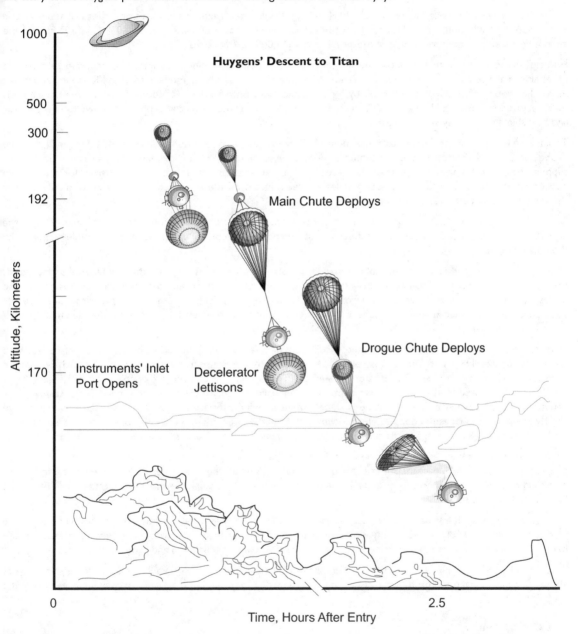

Huygens' Descent to Titan

Main Chute Deploys

Drogue Chute Deploys

Instruments' Inlet Port Opens

Decelerator Jettisons

Altitude, Kilometers

1000
500
300
192
170
0

Time, Hours After Entry

2.5

The Huygens Probe Mission

The *Huygens* probe will be carried to Saturn's system by *Cassini*. Bolted to *Cassini* and fed electrical power through an umbilical cable, *Huygens* will ride along during the seven-year journey largely in a "sleep" mode, awakened every six months for three-hour instrument and engineering checkups.

Some 22 days before it hits the top of Titan's atmosphere, *Huygens* will be released from *Cassini* on November 6, 2004. With its umbilical cut and bolts released, *Huygens* will spring loose from the mother ship and fly on a ballistic trajectory to Titan. The probe will spin at 7 rpm for stability. Onboard timers will switch on the probe systems before the probe reaches Titan's upper atmosphere.

Two days after the probe's release, the orbiter will perform a deflection maneuver; this will keep *Cassini* from following *Huygens* into Titan's atmosphere. This maneuver will also establish the required geometry between the probe and the orbiter for radio communications during the probe descent; in addition, it will also set the initial conditions for *Cassini*'s tour of Saturn's moons, which starts right after the completion of the *Huygens* probe mission.

The *Huygens* probe carries two microwave S-band transmitters and two antennas, both of which will transmit to the *Cassini* orbiter during the probe's descent. One stream of telemetry is delayed by about six seconds with respect to the other to avoid data loss if there are brief transmission outages.

Probe descent will take place November 27, 2004. *Huygens* will enter Titan's atmosphere at a speed of some 20,000 km/hr (about 12,400 mph). It is designed to withstand the extreme cold of space (about -200° C (-330° F)) and the intense heat it will encounter during its atmospheric entry (12,000° C (more than 21,600° F)).

Huygens' parachutes will further slow the descent so the probe can conduct an intensive program of scientific observations all the way down to Titan's surface. When the probe's speed has moderated to 1,400 km/hr (about 870 mph), the probe's after cover is pulled off by a pilot parachute. An 8.3-meter (27-foot)-diameter main parachute is then deployed to ensure a slow and stable descent. The main parachute slows the probe and allows the decelerator and heat shield to fall away when the parachute is released.

To limit the duration of the descent to a maximum of 2½ hours, the main parachute is jettisoned 900 seconds after the probe has entered the top of the atmosphere. A smaller, 3-meter (9.8-foot)-diameter drogue chute deploys to support the probe for the remainder of the descent. The batteries and other resources are sized for a maximum mission duration of 153 minutes – including at least three minutes on the surface, but possibly up to a half an hour there if the descent takes less time than expected.

During the first part of the probe's descent, instruments onboard *Huygens* probe are controlled by a timer. During the final 10 to 20 kilometers (6 to 12 miles) of descent, instruments will be controlled on the basis of altitude measured by the radar altimeter.

Throughout the descent, *Huygens*' atmospheric structure instrument will measure more than six physical properties of the atmosphere. The gas chromatograph and mass spectrometer will determine the chemical composition of the atmosphere as a function of altitude. The aerosol collector and pyrolyzer will capture aerosol particles – fine liquid or solid particles suspended in the atmosphere – heat them and send the resulting vapor to the chromatograph / spectrometer for analysis.

Huygens' descent imager and spectral radiometer will take pictures of cloud formations and Titan's surface, and also determine the visibility within Titan's atmosphere. As the surface looms closer, the instrument will switch on a bright lamp and measure the spectral reflectance of the surface. Throughout the descent, the doppler shift of *Huygens*' radio signal will be measured by the doppler wind experiment onboard the *Cassini* orbiter to determine Titan's atmospheric winds, gusts and turbulence. As the probe is shifted about by winds, the frequency of its radio signal would change slightly in what is known as the doppler effect – similar to how the pitch of a train whistle appears to rise and then fall as the train passes. Such changes in frequency can be used to deduce the wind speed experienced by the probe.

As *Huygens* nears impact, its surface science package will activate a number of instruments to measure surface properties. *Huygens* will impact the surface at about 25 km/hr (15 mph); the chief uncertainty is whether its landing will be a thud or a splash. If *Huygens* lands in liquid, these instruments will measure the liquid's properties while the probe floats for a few minutes.

If *Huygens* lands in liquid ethane it will not be able to return data for very long, because the extremely low temperature of this liquid (about -180° C (-290° F)) would prevent the batteries from operating. In addition, if liquid ethane permeates the probe's science instrument packages, the radio would be badly tuned and probably not operate.

Assuming *Huygens* continues to send data to *Cassini* from Titan's surface, it will be able to do so for a maximum of about 30 minutes, when the probe's battery power is expected to run out and the *Cassini* orbiter disappears over the probe's horizon.

The Orbital Tour

After the end of the *Huygens* mission, *Cassini* will continue its four-year orbital tour, consisting of more than 70 orbits around Saturn that will be shaped by gravity assist flybys of Titan, or by firings of *Cassini*'s thrusters or main engine. The size of these orbits, their orientation to the Saturn-Sun plane and their inclination to Saturn's equator are dictated by various scientific requirements. These include: imaging radar coverage of Titan's surface; flybys of selected icy moons, Saturn or Titan; occultations by Saturn's rings; and crossings of the ring plane.

Cassini will make at least six close targeted flybys of selected icy moons of greatest interest – Iapetus, Enceladus, Dione and Rhea. Images taken with *Cassini*'s high-resolution telescopic cameras during these flybys will show surface features equivalent in size to a baseball diamond. At least two dozen more distant flybys (at altitudes of about 100,000 kilometers (about 60,000 miles)) will be made of the major moons of Saturn other than Titan. The varying inclination angle of *Cassini*'s orbits also will allow studies of Saturn's polar regions in addition to the planet's equatorial zone.

Titan is the only Saturn moon large enough to enable significant gravity assist changes in *Cassini*'s orbits, though some of the other moons can help to accomplish more modest adjustments with their small gravitational effect on *Cassini*'s flight path. Precise navigation and targeting of the point at which *Cassini* flies by Titan will be used to shape the orbital tour in the same way the *Galileo* mission has used its encounters of Jupiter's large moons to shape its Jovian tour.

End of Prime Mission

The prime mission tour concludes on July 1, 2008, four years after Saturn arrival and 33 days after the last Titan flyby, which occurs on May 28, 2008. The aim point of the final flyby is chosen to position *Cassini* for a Titan flyby on July 31, 2008 – providing the opportunity to proceed with more flybys during an extended mission, if resources allow. Nothing in the design of the tour precludes an extended mission.

Mission Operations

Two-way communication with *Cassini* will be through the large dish antennas of NASA's worldwide Deep Space Network. The spacecraft will transmit and receive in the microwave X-band, using either its parabolic high-gain antenna or one of its two low-gain antennas. The high-gain antenna is also used for radio and radar experiments and for receiving signals from the *Huygens* probe.

Because *Cassini*'s science instruments are fixed and the entire spacecraft must be turned to point them, the spacecraft will be frequently reoriented by using either gyroscope-like reaction wheels or the spacecraft's set of small onboard thrusters. Consequently, most science observations will be made without a real-time communications link to Earth. Data will be stored on *Cassini*'s two solid state data recorders, each with a capacity of about 2 gigabits (equivalent to about 200 megabytes each).

Each of *Cassini*'s science instruments is run by a microprocessor capable of controlling the instrument and formatting / packetizing data. Ground controllers will run the spacecraft with a combination of some centralized commands to control system-level resources, and some commands issued by the individual science instruments' microprocessors. Packets of data will be collected from each instrument on a schedule that may vary at different times. Data packets are either stored on *Cassini*'s onboard solid state recorders or transmitted to Earth.

Mission controllers, engineering teams and science teams will monitor telemetry from the spacecraft and look for anomalies in real time. The flight systems operations team retrieves engineering data to determine the health, safety and performance of the spacecraft, and processes the tracking data to determine and predict the spacecraft's trajectory. Data will normally be received by the Deep Space Network with one tracking pass by one antenna per day, with occasional extra coverage for special radio science experiments.

Spacecraft operations during *Cassini*'s interplanetary cruise will be centralized at JPL. During the Saturn tour, a system of distributed science operations will be implemented with centralized spacecraft control continuing at JPL. The concept is to allow scientists to operate their instrument from their home institution with as much ease as possible and with the minimum interaction necessary to collect their data.

Cost-Saving Approaches

Because of the need to reduce the cost of activities before launch and during *Cassini*'s nearly seven-year-long cruise, the development of many operations capabilities has been deferred until after launch.

The following are among choices made to reduce costs and simplify mission operations:

- Limited science. To reduce costs, there is no plan to acquire science data during *Cassini*'s interplanetary cruise, including its flybys of Venus, Earth and Jupiter. The only exception is a gravitational wave experiment; this will attempt to detect gravitational waves emitted by supermassive objects such as quasars, active galactic nuclei or binary black holes.
- Strict modes of operation. *Cassini*'s power system cannot supply enough electricity to run all science instruments simultaneously. In addition, science instruments cannot all be pointed optimally at the same time because of the fact that they are fixed to the body of the spacecraft. Mission planners therefore developed a concept of operational modes designed to reduce operational complexity. Under this approach, instruments will be operated in a series of standard, well-characterized configurations, identifying the on / off state of each instrument, minimum and maximum power and peak data rates allocated for each instrument, and states of the engineering subsystems (radio, recorder, attitude and articulation control, and propulsion).

- Standardized, reusable sequence templates. Ninety-eight percent of *Cassini*'s Saturn tour will be conducted with a small number of reusable software modules and templates that dictate how science instruments are controlled. The remaining two percent (seven days per year) may consist of unique software command sequences developed for special occasions.
- No sequence team. Mission planning and the development of command sequences will be led by "virtual teams" made up of representatives of the various science experiment or subsystem teams involved in a given procedure. This approach replaces the traditional custom of a "sequence team" responsible for developing the sequences of commands used to control the spacecraft and its instruments.

The Cassini program has remained on schedule and within budget since its inception.

The Spacecraft

The *Cassini* spacecraft is a two-part structure, composed of the orbiter and the *Huygens* Titan probe. The orbiter is designed to enter orbit around Saturn; deliver *Huygens* to its destination and relay the probe's data; and conduct at least four years of detailed studies of Saturn's system. *Huygens* is designed to remain primarily dormant throughout *Cassini*'s journey, then spring into action when it reaches the top of Titan's atmosphere. There, *Huygens* will deploy its parachutes and conduct 2½ hours of intensive measurements as it descends through Titan's atmosphere, all the while transmitting its findings to the *Cassini* orbiter for relay back to Earth.

The Cassini Orbiter

The *Cassini* orbiter alone weighs 2,125 kilograms (4,685 pounds). When the 320-kilogram (705-pound) *Huygens* probe and a launch vehicle adapter are attached, and 3,132 kilograms (6,905 pounds) of propellants are loaded, the spacecraft at launch will weight 5,712 kilograms (12,593 pounds). More than half the spacecraft's mass is propellant – much of which is needed for *Cassini*'s 94-minute main engine firing that brakes it into orbit around Saturn.

The spacecraft stands 6.8 meters (22.3 feet) high and is 4 meters (13 feet) wide. The magnetometer instrument is mounted on a 13-meter-long (42-foot) boom that extends outward from the spacecraft; three other rod-like antenna booms, each measuring 10 meters (about 32 feet), extend outward from the spacecraft in a Y-shape. Most of the spacecraft and its instrument housings are covered with multiple-layered, shiny amber-colored or matte-black blanketing material. The blankets protect *Cassini* against the extreme heat and cold of space, and maintain the room temperature operating environment needed for computers and other electronic systems onboard. The blanketing includes layers of material like that used in bullet-proof vests to afford protection against dust-size particles called micrometeoroids that zip through interplanetary space.

Cassini Spacecraft

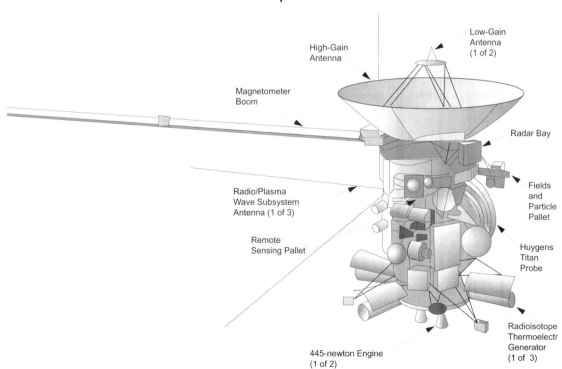

The spacecraft's complexity is necessitated both by its flight path to Saturn and by the ambitious program of scientific observations to be accomplished at Saturn. *Cassini* has some 22,000 wire connections and more than 12 kilometers (7.5 miles) of cabling linking its instruments, computers and mechanical devices.

Cassini's cargo of science instruments, the *Huygens* probe and the enormous quantity of fuel the spacecraft needs to brake into orbit around Saturn make it the largest interplanetary spacecraft ever launched by the United States. (The former Soviet Union's *Phobos 1* and 2 Mars craft each weighed 6,220 kilograms (13,713 pounds).) *Cassini* and *Huygens*, attached and fueled but without the launch vehicle adapter, weigh about 5,577 kilograms (12,295 pounds). More than half of that mass is liquid fuel. *Cassini*'s fuel mass alone is more than the mass of the *Galileo* and *Voyager* spacecraft combined.

The main body of the orbiter is a nearly cylindrical stack consisting of a lower equipment module, a propulsion module and an upper equipment module, and is topped by the fixed 4-meter (13-foot)-diameter high-gain antenna. Attached about halfway up the stack are a remote sensing pallet, which contains cameras and other remote sensing instruments, and a fields and particles pallet, which contains instruments that study magnetic fields and charged particles. The two pallets carry most of the *Cassini* orbiter's science instruments. In general the whole spacecraft must be turned to point the instruments in the proper direction, though three of the instruments provide their own articulation about one axis.

Several booms will be deployed early in *Cassini*'s flight. These include three rod-like plasma wave antennas and an 11-meter (40-foot) spring-loaded magnetometer boom that extends from a canister mounted on the upper equipment module.

Software sequences – detailed instructions stored in the spacecraft's computer – direct the activities of the spacecraft. A typical sequence may operate *Cassini* for a month without the need for intervention from ground controllers. Onboard computers are designed to withstand the radiation environment of deep space, particularly when the Sun is at peak activity. Solar flares, which can last up to several days, can deliver radiation 1,000 times above the usual radiation levels in interplanetary space. *Cassini*'s electronics have undergone customized radiation hardening to ensure that they won't be disrupted or destroyed by such events.

Sophisticated fault protection software resides in the spacecraft's computers to continuously sample and sense the health of the onboard systems. The fault protection system automatically takes corrective action when it determines the spacecraft is at risk due to any onboard failure.

The orbiter receives electrical power from three radioisotope thermoelectric generators (RTGs). These generators produce power by converting heat into electrical energy. RTGs are not reactors, and the radioactive material is neither fissionable nor fusionable. Heat is provided by the natural radioactive decay of plutonium in a ceramic form of mostly plutonium-238; devices called thermocouples turn the heat into electricity to run the spacecraft. Upon arrival at Saturn, the three generators will provide about 675 watts of power. Plutonium dioxide is also used as the heat source in 82 small radioisotope heater units (RHUs) on the *Cassini* orbiter and 35 on the *Huygens* probe; each produces about 1 watt of heat to keep nearby electronics at their operating temperatures. RHUs were most recently used on Mars Pathfinder's *Sojourner* rover to keep its electronics warm during Martian nights. Both RTGs and RHUs have a long and safe heritage of use and high reliability in NASA's planetary exploration program, including the *Voyager* and *Galileo* missions.

Propulsion for major changes to *Cassini*'s trajectory is provided by one of two main engines. These powerful engines use monomethyl-hydrazine as the fuel and nitrogen tetroxide as the oxidizer. Sixteen smaller engines called thrusters use hydrazine to control *Cassini*'s orientation and to make small adjustments to the spacecraft's flight path.

Guidance and control is governed by sensors that recognize reference stars and the Sun, and by onboard computers that determine the spacecraft's position. Using a new type of gyroscope that vibrates rather than spins, the spacecraft can perform turns, twists and propulsion firings while retaining continuous knowledge of its own position. Unlike spacecraft such as *Galileo*, the *Cassini* orbiter is stabilized along all three axes and thus does not normally rotate during its long cruise to Saturn.

The mission's trajectory poses a challenge for controlling the spacecraft's temperature because in the first several years of the mission, the orbiter will be relatively close to the Sun. During this time, the high-gain antenna will be pointed at the Sun and used as a sunshade to shield the rest of the orbiter and probe. Special paints have been used on the antenna to reflect and radiate much of the sunlight received.

Communications with the spacecraft during its passage through the inner solar system will be through one of the orbiter's two low-gain antennas. In late January 2000, as *Cassini* enters the cooler climes of the asteroid belt and beyond, it will turn its high-gain antenna toward Earth and conduct telecommunications through it for the remainder of the mission.

As *Cassini* moves farther from the Sun, extreme cold becomes a concern. At Saturn's distance, the intensity of sunlight is approximately 1 percent that at Earth.

Heat within the spacecraft is retained by using lightweight, multiple-layered insulating blankets that have been tailor-made for the instrument housings and other areas of the orbiter. The blanket's outer layer is a three-ply membrane composed of a Kapton core with an aluminized inner surface and a metallic outer surface. The translucent Kapton has a yellow color, and when backed by a shiny aluminum layer, results in an amber appearance. Up to 26 layers of material are used in the blankets, which also afford protection against dust-sized micrometeoroids which, traveling at speed of 5 to 40 kilometers per second (about 10,000 to 90,000 mph) could otherwise potentially penetrate portions of the spacecraft.

Orbiter Subsystems

The *Cassini* orbiter contains 12 engineering subsystems that govern spacecraft features and functions including wiring, electrical power, computing, telecommunications, guidance and propulsion.

- The structure subsystem is the skeleton that provides mechanical support and alignment for all flight equipment, including the *Huygens* probe. In addition, it provides an equipotential container – an electrical grounding reference – which provides a shield from radio frequency interference, and protection from space radiation and micrometeoroids. The structure subsystem consists of the upper equipment module, which contains the 12-bay electronics bus assembly, instrument pallets and the magnetometer boom; the lower equipment module; plus all the brackets and structure used to attach the *Huygens* probe, the low-gain and high-gain antennas, electrical generators, main rocket engines, reaction control thrusters and other equipment. The structure subsystem also includes the adapter which supports the spacecraft on the Centaur upper stage during launch.

- The radio frequency subsystem provides the telecommunications facilities for the spacecraft, and is also used as part of the radio science instrument. For communications, it produces an X-band carrier signal at a frequency of 8.4 GHz; modulates it with data received from the command and data subsystem; amplifies the X-band carrier power to produce 20 watts from the traveling wave tube amplifiers; and delivers the signal to the antenna subsystem. (The 20 watts is expected to degrade to about 19 watts by the time *Cassini* reaches Saturn.) From the antenna subsystem, the radio frequency subsystem takes signals transmitted from Earth at a frequency of 7.2 GHz in the microwave X-band, demodulates them, and delivers the commands and data to the command and data subsystem for storage and/or execution.

- Regulated electrical power and various small pyrotechnic devices on the spacecraft are controlled by the power and pyrotechnics subsystem. Operating on command from the central computer system, this subsystem distributes electrical power to instruments and other subsystems on the spacecraft at 30 volts DC. The subsystem also regulates a shunt radiator that can be used to dispose of excess heat. Pyrotechnics include squib devices that will be fired to cut cables and other links that hold the *Huygens* probe onto the orbiter.

- The command and data subsystem is *Cassini*'s nervous system – the central processing and delivery clearinghouse of the spacecraft for commands received from the ground and data sent back to Earth. All elements of the subsystem are duplicated with redundant components that can be used in the event of a component failure. The subsystem receives ground commands and data from the radio frequency subsystem, processes the information and distributes it to other subsystems. The command and data subsystem uses one each of the two redundant solid state recorders and flight computers, which are programmed in the Ada programming language. Memory capacity for each solid state recorder is 2 gigabits.

  Scientific and engineering data from science instruments destined for transmission to Earth are first forwarded to the command and data subsystem for processing and formatting for telemetry, and delivery to the radio frequency subsystem. The command and data subsystem contains software routines that protect the spacecraft in the event of a fault. The software also allows the spacecraft to autonomously respond to faults needing immediate action. Memory for the command and data subsystem is 512 kilowords of random access memory (RAM) and 8 kilowords of programmable read-only memory (PROM).

- The attitude and articulation control subsystem is the spacecraft's inner ear, continuously sensing and measuring the spacecraft's orientation on its three axes and the spacecraft's position in space relative to Earth, Sun, Saturn and other targets. It provides measurements and controls pointing for spacecraft instruments, including scans that require the spacecraft to roll while an instrument performs an observation. The attitude and articulation control subsystem encompasses a number of sensors including redundant Sun sensor assemblies; stellar reference units, or star trackers; a Z-axis accelerometer; and two 3-axis gyro inertial reference units. Each unit consists of four gyros, three orthogonal to each other and the fourth skewed equidistant to the other three. The subsystem also contains actuators for the main rocket engine gimbals and for the redundant reaction wheel.

  With two redundant computers programmed in Ada, the subsystem processes commands from the command and data subsystem and produces commands to be delivered to attitude control actuators and/or spacecraft thrusters or main engines to control *Cassini*'s attitude and to make trajectory changes. The attitude and articulation control subsystem memory has 512 kilobytes of RAM and 8 kilobytes of PROM.

- All power and data cabling, except for coaxial cabling and waveguides, make up the cabling subsystem. This network of cabling conducts power from the three electrical generators to the power and pyrotechnics subsystem and to the power shunt radiator. It also conducts data between the command and data subsystem and the other subsystems and assemblies on the spacecraft. In addition, cabling allows engineers to access *Cassini*'s electronics during spacecraft integration and testing.

- The propulsion module subsystem controls the spacecraft's thrust and changes in its attitude. It works under the command of the attitude and articulation control system. Attitude control is provided by the reaction control subsystem, which consists of four clusters of four hydrazine thrusters each. These move the spacecraft to or maintain it in its desired orientation and are used to point the instruments at their targets. The thrusters are also used for executing small spacecraft maneuvers. For larger changes in the spacecraft velocity, the main rocket engine is used. *Cassini* has a primary and redundant pressure-regulated main engine. Each engine is capable of a thrust of approximately 445 newtons (100 pounds of force). The bipropellant main engines burn nitrogen tetroxide and monomethyl-hydrazine. The engines are gimbaled so that the thrust vector can be maintained through the shifting center of mass of the spacecraft. Mounted below the main engines is a retractable cover that is used during cruise to protect the main engines from micrometeoroids. The main engine cover can be extended and retracted multiple times (at least 25 times), and has a pyro-ejection mechanism to jettison the cover should there be a mechanical problem with the cover that interferes with main engine operation. During cruise the cover will remain closed when the main engines are not in use.
- The temperature control subsystem keeps temperatures of the various parts of the spacecraft within allowable limits through a variety of thermal control techniques, many of which are passive. Automatically positioned reflective louvers are located on *Cassini*'s 12-bay electronics bus. Radioisotope heater units are used where constant heat is required. Multilayer insulation blankets cover much of the spacecraft and its equipment. Electric heaters are used in several locations under control of the spacecraft's main computer. Temperature sensors are located at many sites on the spacecraft, and their measurements are used by the command data subsystem to adjust the heaters. The entire spacecraft body and *Huygens* probe are shaded when necessary by the high-gain antenna.
- The mechanical devices subsystem provides a pyrotechnic separation device that releases the spacecraft from the launch vehicle adapter. Springs then push the spacecraft away from the adapter. The subsystem includes a self-deploying 10.5-meter (40-foot) coiled mast stored in a canister which supports the magnetometers. It also includes an articulation system for a backup reaction wheel assembly, a "pin puller" for the rod-like antennas of the radio plasma wave spectrometer's Langmuir probe, and louvers for venting or holding heat from the radioisotope heater units.
- The electronic packaging subsystem consists of the electronics packaging for most of the spacecraft in the form of the 12-bay electronics bus.
- The solid state recorder is the primary memory storage and retrieval device for the orbiter. The spacecraft is equipped with two recorders, each with a capacity of 2 gigabits (1.8 gigabits at end of mission). Before completion of the *Huygens* descent probe's mission, only one recorder can be used at any time to store science data from the mission. After the *Huygens* mission, however, both recorders may be used to record and play back science data. Data such as spacecraft telemetry and memory loads for various subsystems may be stored in separate files, or partitions, on the recorder. All data recorded to and played back from the solid state recorder are handled by the command data subsystem.
- The antenna subsystem provides a directional high-gain antenna that can transmit and receive on four different bands in the microwave spectrum – X, Ka, S and Ku. The high-gain antenna and low-gain antenna #1 are provided by the Italian space agency. Low-gain antenna #1 is located on the dish structure of the high-gain antenna. Low-gain antenna #2 is located on the *Cassini* orbiter body below the attach point for the *Huygens* probe. During the inner solar system cruise, the high-gain antenna is pointed toward the Sun to provide shade for the spacecraft. The two low-gain antennas allow for one or the other to transmit and receive signals in the microwave X-band to and from Earth when the spacecraft is Sun-pointed. The low-gain antennas also provide an emergency telecommunications capability while *Cassini* is at Saturn.

The Huygens Probe

The *Huygens* probe system includes the probe itself, which enters the Titan atmosphere, and support equipment that remains attached to the orbiter. The probe weighs 320 kilograms (705 pounds) and consists of three main elements:

- A spin-eject device, which uses springs to propels the probe away from the orbiter with a relative velocity of 0.3 to 0.4 meter per second (about 1 foot per second) and simultaneously causes the probe to spin about its axis at 7 rpm. This device is part of the support equipment.
- A front shield, 2.7 meters (8.8 feet) in diameter, covered with a special thermal protection material called AQ60, a low-density mat of silica fibers, to protect the probe from the enormous heat generated during entry into Titan's atmosphere.
- An aft cover that uses thermal protection materials to ensure a slow and stable descent. The main parachute slows the probe and allows the decelerator to fall away when it is released. To limit the duration of the descent to a maximum of 2.5 hours, the main parachute is jettisoned at 900 seconds after atmospheric entry, and is replaced by a 3-meter (about 10-foot)-diameter drogue chute for the remainder of the descent.

Huygens Probe

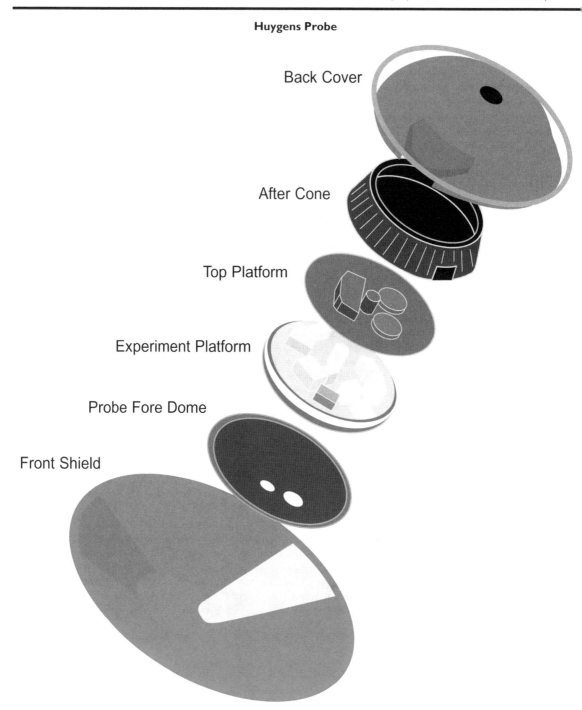

Back Cover

After Cone

Top Platform

Experiment Platform

Probe Fore Dome

Front Shield

Inner structure. The probe's interior consists of two aluminum honeycomb platforms and an aluminum shell. It is linked by fiberglass struts and pyrotechnically operated release mechanisms to the front shield and aft cover. The central equipment platform carries, on both its upper and lower surfaces, the boxes containing the electrical subsystems and the experiments. The upper platform carries the stowed parachute and the transmitter used to radio data in the microwave S-band to the *Cassini* orbiter.

Thermal control. At different phases of the mission, *Huygens* will be subjected to extremes of heat and cold requiring a variety of passive controls to maintain the required temperature conditions. When the spacecraft is nearer the Sun on the Venus and Earth legs of its trajectory, the probe will receive high solar heat input, but will get some protection by being shaded by the orbiter's high-gain antenna. Multi-layered thermal blanketing, which burns off later during Titan atmospheric entry, will also protect the probe from solar heating.

When *Cassini* leaves the inner solar system, the temperature of the probe will be greatly reduced. After separation from the *Cassini* orbiter, *Huygens* will be at its coldest. To ensure that the equipment stays operational, 25 radioisotope heater units (RHUs) are placed in the system. Each RHU, which contains radioactive plutonium dioxide, produces about 1 watt of heat.

During entry into Titan's atmosphere, the front shield may reach temperatures above 1,500° C (2,700° F). Layers of insulation will ensure that the equipment inside stays below 50° C (122° F). Once the chutes are deployed, the probe instruments will be exposed to the cold Titan atmosphere at a temperature of -200° C (-330° F). The internal temperature will be kept within operating limits by thick foam insulation filling the probe and by power dissipation in the experiments and subsystems.

Electrical power. While it is still attached to the *Cassini* orbiter, the *Huygens* probe will obtain power from the orbiter via an umbilical cable. After separation, electrical power is provided by five lithium sulfur dioxide batteries, each with 23 cells. Much of the battery power is used to power the *Huygens* probe's timer for the 22 days of coasting to Titan.

Command and data management. *Huygens*' command and data management subsystem controls the timing and execution of a number of critical events. It times the coast phase, and switches on the probe just before entry. It controls deployment of various components during descent. It distributes commands to other subsystems and to the experiments. It distributes to the experiments information that provides a timeline of conditions that instruments can use to schedule operations. And it collects scientific and engineering data and forwards them to the orbiter during the cruise to Saturn and during the Titan mission.

Probe data relay. The probe data relay subsystem provides the one-way communications link between the *Huygens* probe and the *Cassini* orbiter, and includes equipment installed in each spacecraft. Elements that are part of the probe support equipment on the *Cassini* orbiter include radio frequency electronics (including an ultra-stable oscillator) and a low-noise amplifier. For backup, the *Huygens* probe carries two S-band transmitters, both of which transmit during probe descent, each with its own antenna. The telemetry in one link is delayed by about six seconds with respect to the other to avoid data loss if there are brief transmission outages. Reacquisition of the probe signal would normally occur within this interval.

New Technology

A wealth of new technology was developed and qualified for space flight by or for the Cassini program. Much of this new technology has already been adopted by other space science programs, in some cases at a discounted cost directly attributable to Cassini. This has enabled the development of new classes of low-cost, high-efficiency spacecraft, such as the Discovery and New Millennium spacecraft.

The *Cassini* orbiter advances and extends the technology base of the United States and its partners with several innovations in engineering and information systems. Whereas previous planetary spacecraft used onboard tape recorders to store data, *Cassini* has pioneered a new solid state data recorder with no moving parts. The new recorder eventually will replace tape recorders on all space missions. NASA's Advanced X-ray Astrophysics Facility (AXAF), for example, will use a solid state recorder from the production line established for the Cassini mission. In addition, the recorder has great potential for use in a variety of fields, from aerospace to the entertainment industry, and is expected to find wide applicability in consumer electronics.

Similarly, the main onboard computer that directs operations of the orbiter uses a novel design that draws on new families of electronic chips. Among them are very high-speed integrated circuit (VHSIC) chips developed under a U.S. government-industry research and development initiative. The *Cassini* application GVSC 1750A computer is the first civilian spacecraft application of this technology. The computer system also uses new application-specific integrated circuit (ASIC) parts; each component replaces one hundred or more traditional chips. The ASIC chips allow the development of a data system for *Cassini* 10 times more efficient than earlier spacecraft designs, but at less than one-third the mass and volume. Two missions under NASA's Discovery program, Mars Pathfinder and the Near Earth Asteroid Rendezvous (NEAR), used these chips directly off the *Cassini* production line.

Elsewhere on the *Cassini* orbiter, the power system benefits from an innovative solid state power switch developed for the mission. This switch eliminates rapid fluctuations called transients that usually occur with conventional power switches, with a significantly improved component lifetime. The power switch holds great promise for use in numerous Earth-based applications. A low-mass, low-power, radiation-hardened X-band radio transponder (a combined receiver and transmitter) was developed by the Cassini program. Both Mars Pathfinder and NEAR missions used radio transponders built on the Cassini mission's production line.

The inertial reference units to be used on *Cassini* represent the first space version of a revolutionary new gyro called the hemispherical resonator gyroscope. Gyros commonly used in spacecraft, aircraft and ships are large, very delicate mechanical devices whose many moving parts make them susceptible to failure. This new gyro, which eventually may be used on other spacecraft, promises greater reliability and less vulnerability to failure because it uses no moving parts. A slightly modified *Cassini* gyro was incorporated into the NEAR spacecraft.

Cassini Signature Disk

In August 1997, a small digital versatile disk (DVD) was installed aboard the *Cassini* spacecraft during processing at the Kennedy Space Center. The disk contains a record of 616,400 handwritten signatures from 81 countries around the globe. Signatures were received from people of all ages and backgrounds.

Mail came from individuals, families and, often, from entire schools of students. Signatures came from the very young, just learning to write, and from the very old, whose hands were no longer steady. Signatures were sent in behalf of loved ones who had died in the recent past. Even the signatures of Jean-Dominique Cassini and Christiaan Huygens were obtained from letters they wrote during the 17th century.

Sorting, counting and scanning the signatures was performed over the course of a year by volunteers from the Planetary Society, Pasadena, California. The disk's cover, designed by Charles Kohlhase of the Cassini project, depicts a golden eagle wing feather and various Cassini mission elements to symbolize the signature experience. The feather was chosen to represent both the beauty and power of flight, as well as the quill pen that was used for nearly 14 centuries in writing and signing.

Science Objectives

Cassini's payload represents a carefully chosen set of interrelating instruments that will address many major scientific questions about the Saturn system. The data they return will be analyzed by a team of nearly 300 scientists from the United States and Europe. The Cassini and Huygens mission science objectives are as follows:

Saturn

* Determine the temperature field, cloud properties and composition of Saturn's atmosphere.
* Measure the planet's global wind field, including waves and eddies; make long-term observations of cloud features to see how they grow, evolve and dissipate.
* Determine the internal structure and rotation of the deep atmosphere.
* Study daily variations and relationship between the ionosphere and the planet's magnetic field.
* Determine the composition, heat flux and radiation environment present during Saturn's formation and evolution.
* Investigate sources and nature of Saturn's lightning.

Titan

* Determine the relative amounts of different components of the atmosphere; determine the mostly likely scenarios for the formation and evolution of Titan and its atmosphere.
* Observe vertical and horizontal distributions of trace gases; search for complex organic molecules; investigate energy sources for atmospheric chemistry; determine the effects of sunlight on chemicals in the stratosphere; study formation and composition of aerosols (particles suspended in the atmosphere).
* Measure winds and global temperatures; investigate cloud physics, general circulation and seasonal effects in Titan's atmosphere; search for lightning.
* Determine the physical state, topography and composition of Titan's surface; characterize its internal structure.
* Investigate Titan's upper atmosphere, its ionization and its role as a source of neutral and ionized material for the magnetosphere of Saturn.

Magnetosphere

* Determine the configuration of Saturn's magnetic field, which is nearly symmetrical with Saturn's rotational axis. Also study its relation to the modulation of Saturn kilometric radiation – a radio emission from Saturn that is believed to be linked to the way electrons in the solar wind interact with the magnetic field at Saturn's poles.
* Determine the current systems, composition, sources and concentrations of electrons and protons in the magnetosphere.
* Characterize the structure of the magnetosphere and its interactions with the solar wind, Saturn's moons and rings.
* Study how Titan interacts with the solar wind and with the ionized gases within Saturn's magnetosphere.

The Rings

* Study configuration of the rings and dynamic processes responsible for ring structure.
* Map the composition and size distribution of ring material.
* Investigate the interrelation of Saturn's rings and moons, including imbedded moons.
* Determine the distribution of dust and meteoroid distribution in the vicinity of the rings.
* Study the interactions between the rings and Saturn's magnetosphere, ionosphere and atmosphere.

Icy Moons

- Determine general characteristics and geological histories of Saturn's moons.
- Define the different physical processes that have created the surfaces, crusts or subsurfaces of the moons.
- Investigate compositions and distributions of surface materials, particularly dark, organic-rich materials and condensed ices with low melting points.
- Determine the bulk compositions and internal structures of the moons.
- Investigate interactions of the moons with Saturn's magnetosphere and ring system and possible gas injections into the magnetosphere.

In addition to the science objectives at Saturn, the *Cassini* spacecraft will also conduct a gravity wave investigation through the ASI-provided high-gain antenna during its interplanetary cruise.

Orbiter Science Instruments

The *Cassini* orbiter carries a total of 12 science instruments. Two pallets carry most of the instruments; four instruments are located on the remote sensing experiments pallet, and three are located on the fields and particles experiments pallet. Others are fixed at independent locations on the spacecraft. The experiments include:

- Imaging Science Subsystem, or *Cassini*'s cameras, will photograph a wide variety of targets – Saturn, the rings, Titan and the icy moons – from a broad range of observing distances for various scientific purposes. General science objectives include studying the atmospheres of Saturn and Titan, the rings of Saturn and their interactions with the planet's moons and the surface characteristics of the moons, including Titan. The instrument includes both a narrow-angle and a wide-angle camera. The narrow-angle camera provides high-resolution images of targets of interest, while the wide-angle camera provides more extended spatial coverage at lower resolution. The cameras can also obtain optical navigation frames – images of Saturn's moons against a star background – which are used to keep the spacecraft on the correct trajectory. Team leader is Dr. Carolyn C. Porco of the University of Arizona, Tucson, Arizona.
- Visible and Infrared Mapping Spectrometer will map the surface spatial distribution of the mineral and chemical features of a number of targets, including Saturn's rings, surfaces of the moons, and the atmospheres of Saturn and Titan. The instrument includes a pair of imaging grating spectrometers that are designed to measure reflected and emitted radiation from atmospheres, rings and surfaces to determine their compositions, temperatures and structures. A spectrometer is an optical instrument that splits the light received from objects into its component wavelengths; each chemical has a unique spectral signature and thus can be identified. The instrument obtains information over 352 contiguous wavelengths from 0.35 to 5.1 micrometers; it measures intensities of individual wavelengths and uses the data to infer the composition and other properties of the object that emitted the light. The mapping spectrometer provides images in which every pixel contains high-resolution spectra of the corresponding spot on the target body. Principal investigator is Dr. Robert H. Brown of the University of Arizona, Tucson, Arizona.
- Composite Infrared Spectrometer will measure infrared emissions from atmospheres, rings and surfaces. This spectrometer will create vertical profiles of temperature and gas composition for the atmospheres of Titan and Saturn, from deep in their tropospheres (lower atmospheres), to high in their stratospheres (upper atmospheres). The instrument will also gather information on the thermal properties and composition of Saturn's rings and icy moons. The instrument is a coordinated set of three interferometers designed to measure infrared emissions over wavelengths from 7 to 1,000 micrometers in the mid- and far-infrared range of the electromagnetic spectrum. Each interferometer uses a beam splitter to divide incoming infrared light into two paths. The beam splitter reflects half of the energy toward a moving mirror and transmits half to a fixed mirror. The light is recombined at the detector. As the mirror moves, different wavelengths of light alternately cancel and reinforce each other at a rate that depends on their wavelengths. This information can be used to construct an infrared spectrum. Principal investigator is Virgil G. Kunde of NASA's Goddard Space Flight Center, Greenbelt, Maryland.
- Ultraviolet Imaging Spectrograph is a set of detectors designed to measure ultraviolet light reflected by or emitted from atmospheres, rings and surfaces to determine their compositions, distributions, aerosol content and temperatures. The instrument will also measure fluctuations of sunlight and starlight as the Sun and stars move behind the rings of Saturn and the atmospheres of Saturn and Titan, and will determine the atmospheric concentrations of hydrogen and deuterium. The instrument includes a two-channel, far- and extreme-ultraviolet imaging spectrograph that studies light over wavelengths from 55.8 to 190 nanometers. It also has a hydrogen-deuterium absorption cell and a high-speed photometer. An imaging spectrograph records spectral intensity information in one or more wavelengths of light and then outputs digital data that can be displayed in a visual form, such as a false-color image. The hydrogen-deuterium absorption cell measures the quantity of deuterium, a heavier form of hydrogen. The high-speed photometer determines the radial structure of Saturn's rings by watching starlight through the rings. Principal investigator is Dr. Larry L. Esposito of the University of Colorado, Boulder, Colorado.

- *Cassini* Radar will investigate the surface of Saturn's largest moon, Titan. Titan's surface is covered by a thick, cloudy atmosphere that is hidden to normal optical view, but can be penetrated by radar. The instrument is based on the same imaging radar technology used in missions such as *Magellan* to Venus and the Earth-orbiting Spaceborne Imaging Radar. Scientists hope to determine if oceans exist on Titan and, if so, determine their distribution; investigate the geological features and topography of Titan's solid surface; and acquire data on other targets, such as Saturn's rings and icy moons, as conditions permit.

  The radar will take four types of observations: imaging, altimetry, backscatter and radiometry. In imaging mode, the instrument will bounce pulses of microwave energy off the surface of Titan from different incidence angles and record the time it takes the pulses to return to the spacecraft. These measurements, converted to distances by dividing by the speed of light, will allow construction of visual images of the target surface with a resolution ranging from one third of a kilometer to 1.7 kilometers (about one-fifth mile to one mile).

  Radar altimetry similarly involves bouncing microwave pulses off the surface of the target body and measuring the time it takes the "echo" to return to the spacecraft. In this case, however, the goal will not be to create visual images but rather to obtain numerical data on the precise altitude of surface features. The altimeter resolution is 24 to 27 kilometers horizontally, 90 to 150 meters vertically (about 14 to 16 miles horizontally, 297 to 495 feet vertically).

  In backscatter mode, the radar bounces pulses off Titan's surface and measures the intensity of the energy returning. This returning energy, or backscatter, is always less than the original pulse, because surface features inevitably reflect the pulse in more than one direction. From the backscatter measurements, scientists can draw conclusions about the composition and roughness of the surface.

  In radiometry mode, the radar will operate as a passive instrument, simply recording the heat energy emanating from the surface of Titan. This information can be used to determine the amount of moisture (such as vapors of methane) in Titan's atmosphere, a factor that has an impact on the precision of the other measurements taken by the instrument.

  At altitudes between 22,500 and 9,000 kilometers (about 14,000 to 5,600 miles), the radar will switch between scatterometry and radiometry to obtain low-resolution global maps of Titan's surface roughness, backscatter intensity and thermal emissions. At altitudes between 9,000 and 4,000 kilometers (about 5,600 to 2,500 miles), the instrument will switch between altimetry and radiometry, collecting surface altitude and thermal emission measurements. Below 4,000 kilometers (about 2,500 miles), the radar will switch between imaging and radiometry. Team leader is Dr. Charles Elachi of NASA's Jet Propulsion Laboratory, California Institute of Technology, Pasadena, California.

- Radio Science will use the spacecraft's radio and the ground antennas of NASA's Deep Space Network to study the composition, pressures and temperatures of the atmospheres and ionospheres of Saturn and Titan; the radial structure of and particle size distribution in Saturn's rings; and the masses of objects in the Saturn system and the mass of Saturn's ring system as a whole. Radio science will also be used to search for gravitational waves coming from beyond our solar system. Some of these experiments measure doppler shifts (frequency shifts) and other changes to radio signals that occur when the spacecraft passes behind planets, moons, atmospheres or physical features such as planetary rings. From these measurements, scientists can derive information about the structures and compositions of the occulting bodies, atmospheres and the rings. Team leader is Dr. Arvydas J. Kliore of NASA's Jet Propulsion Laboratory, California Institute of Technology, Pasadena, California.

- *Cassini* Plasma Spectrometer will measure the composition, density, flow velocity and temperature of ions and electrons in Saturn's magnetosphere. The instrument consists of three sensors: an electron spectrometer, an ion beam spectrometer and an ion mass spectrometer. A motor-driven actuator rotates the sensor package to provide 208-degree scanning in the azimuth of the *Cassini* orbiter. The electron spectrometer makes measurements of the energy of incoming electrons; its energy range is 0.7 to 30,000 electron-volts. The ion beam spectrometer determines the energy to charge ratio of an ion; its energy range is 1 electron volt to 50 kilo-electron-volts. The ion mass spectrometer's energy range is 1 electron volt to 50 kilo-electron-volts. Principal investigator is Dr. David T. Young of the Southwest Research Institute, San Antonio, Texas.

- Ion and Neutral Mass Spectrometer will determine the composition and structure of positive ion and neutral particles in the upper atmosphere of Titan and the magnetosphere of Saturn, and will measure the positive ion and neutral environments of Saturn's icy moons and rings. The instrument will determine the chemical, elemental and isotopic composition of the gaseous and volatile components of the neutral particles and the low-energy ions in Titan's atmosphere and ionosphere, Saturn's magnetosphere and the ring environment. Team leader is Dr. J. Hunter Waite of the Southwest Research Institute, San Antonio, Texas.

- Cosmic Dust Analyzer will provide direct observations of small ice or dust particles in the Saturn system in order to investigate their physical, chemical and dynamic properties and study their interactions with the rings, icy moons and magnetosphere of Saturn. The instrument measures the amount, velocity, charge, mass and composition of tiny dust and ice particles. It has two types of sensors – high-rate detectors and a dust analyzer. The two high-rate detectors, intended primarily for measurements in Saturn's rings, count impacts up to 10,000 per second. The dust analyzer determines the electric charge carried by dust particles, the flight direction and impact speed, mass and chemical composition, at rates up to 1 particle per second, and for speeds of 1-100 kilometers per second (up to about 60 miles per second). An articulation mechanism allows the entire instrument to be rotated or repositioned relative to the body of the *Cassini* orbiter. Principal investigator is Dr. Eberhard Grün of the Max Planck Institute für Kernphysik, Heidelberg, Germany.

- Dual Technique Magnetometer will determine the magnetic fields of the planet and moons and study dynamic interactions between different magnetic fields in the planetary environment. The instrument consists of direct-sensing instruments that detect and measure the strength of magnetic fields in the vicinity of the spacecraft. The experiment includes both a flux gate magnetometer and a vector / scalar helium magnetometer. They are used to measure the magnitude and direction of magnetic fields. Since magnetometers are sensitive to electric currents and ferrous components, they are generally placed on an extended boom, as far from the spacecraft as possible. On *Cassini*, the flux gate magnetometer is located midway out on the 11-meter (36-foot) magnetometer boom, and the vector / scalar helium magnetometer is located at the end of the boom. The boom itself, composed of thin, nonmetallic rods, is folded during launch and deployed about two years after launch. The magnetometer electronics are in a bay in the *Cassini* orbiter's spacecraft body. Principal investigator is Dr. David J. Southwood of the Imperial College of Science & Technology, London, England.

- Magnetospheric Imaging Mass Spectrometer is designed to measure the composition, charge state and energy distribution of energetic ions and electrons; detect fast neutral particles; and conduct remote imaging of Saturn's magnetosphere. It is the first instrument ever designed to produce an image of a planetary magnetosphere. This information will be used to study the overall configuration and dynamics of the magnetosphere and its interactions with the solar wind, Saturn's atmosphere, Titan, rings and the icy moons. The instrument will provide images of the ionized gases, called plasma, surrounding Saturn and determine the charge and composition of ions. Like the *Cassini* plasma spectrometer, this instrument has three sensors that perform various measurements: the low-energy magnetospheric measurement system, the charge-energy-mass spectrometer and the ion and neutral camera. The low-energy magnetospheric measurement system will measure low- and high-energy proton, ion and electron angular distributions (the number of particles coming from each direction). The charge-energy-mass spectrometer uses an electrostatic analyzer, a time-of-flight mass spectrometer and microchannel plate detectors to measure the charge and composition of ions. The third sensor, the ion and neutral camera, makes two different types of measurements. It will obtain three-dimensional distributions, velocities and the rough composition of magnetospheric and interplanetary ions. Principal investigator is Dr. Stamatios M. Krimigis of Johns Hopkins University, Baltimore, Maryland.

- Radio and Plasma Wave Science instrument will measure electrical and magnetic fields in the plasma of the interplanetary medium and Saturn's magnetosphere, as well as electron density and temperature. Plasma is essentially a soup of free electrons and positively charged ions, the latter being atoms that have lost one or more electrons. Plasma makes up most of the universe and is created by the heating of gases by stars and other bodies in space. Plasma is distributed by the solar wind; it is also "contained" by magnetic fields (that is, the magnetospheres) of bodies such as Saturn and Titan. The major components of the instrument are an electric field sensor, a magnetic search coil assembly and a Langmuir probe. The electric field sensor is made up of three deployable antenna elements mounted on the upper equipment module of the *Cassini* orbiter. Each element is a collapsible beryllium-copper tube that is rolled up during launch and subsequently unrolled to its 10-meter (about 33-foot) length by a motor drive. The magnetic search coils are mounted on a small platform attached to a support for *Cassini*'s high-gain antenna. The Langmuir probe, which measures electron density and temperature, is a metallic sphere 50 millimeters (about 2 inches) in diameter. The probe is attached to the same platform by a 1-meter (about 3-foot) deployable boom. Principal investigator is Dr. Donald A. Gurnett of the University of Iowa, Iowa City, Iowa.

Huygens Probe Instruments

The *Huygens* descent probe contains a total of six science instruments. They are:

- Descent Imager / Spectral Radiometer uses 13 fields of view, operating at wavelengths of 350 to 1,700 nanometers, to obtain a variety of imaging and spectral observations. Infrared and visible imagers will observe Titan's surface during the latter stages of the descent. Using the *Huygens* probe's rotation, the imagers will build a mosaic of pictures around the landing site. A side-looking visible imager will view the horizon and the underside of any cloud deck. The spectral radiometer will measure concentrations of argon and methane in the atmosphere. It also will determine if the local surface is solid or liquid, and, if solid, its topography. Solar aureole sensors will measure the light intensity around the Sun resulting from scattering by particles suspended in the atmosphere, permitting calculations of their size, number and density. Principal investigator is Dr. Martin G. Tomasko of the University of Arizona, Tucson, Arizona.

- *Huygens* Atmospheric Structure Instrument investigates the physical properties of Titan's atmosphere, including temperature, pressure and atmospheric density as a function of altitude, wind gusts and, in the event of a landing on a liquid surface, wave motion. Comprising a variety of sensors, the instrument will also measure the ion and electron conductivity of the atmosphere and search for electromagnetic wave activity. On Titan's surface, the instrument will be able to measure the conductivity of surface material. The instrument also processes the signal from the *Huygens* probe's radar altimeter to obtain information on surface topography, roughness and electrical properties. Principal investigator is Dr. Marcello Fulchignoni of the Paris Observatory, Meudon, France.

* Aerosol Collector and Pyrolyzer traps particles suspended in Titan's atmosphere using a deployable sampling device. Samples are heated in ovens to vaporize the ice particles and decompose the complex organic materials into their component chemicals. The products are then passed to the gas chromatograph / mass spectrometer for analysis. The instrument will obtain samples at two altitude ranges. The first sample will be taken at altitudes down to 30 kilometers (about 19 miles) above the surface. The second sample will be obtained at an altitude of about 20 kilometers (about 12 miles). Principal investigator is Dr. Guy M. Israel of the Service d'Aeronomie du Centre National de la Recherche Scientifique, Verrieresle-Buisson, France.

* Gas Chromatograph / Mass Spectrometer provides a quantitative analysis of Titan's atmosphere. Atmospheric samples are transferred into the instrument by dynamic pressure as the *Huygens* probe descends through the atmosphere. The mass spectrometer constructs a spectrum of the molecular masses of the gas driven into the instrument. Just before landing, the instrument's inlet port is heated to vaporize material on contact with the surface. Following a safe landing, the instrument can determine Titan's surface composition. The mass spectrometer serves as the detector for the gas chromatograph, for unseparated atmospheric samples and for samples provided by the aerosol collector and pyrolyzer. Portions of the instrument are identical in design to the *Cassini* orbiter's ion and neutral mass spectrometer. Principal investigator is Dr. Hasso B. Neimann of NASA's Goddard Space Flight Center, Greenbelt, Maryland.

* Doppler Wind Experiment uses two ultrastable oscillators, one on the *Huygens* probe and one on the *Cassini* orbiter, to give *Huygens'* radio relay link a stable carrier frequency. Orbiter measurements of changes in probe frequency caused by doppler shift will provide information on the probe's motion. In turn, scientists will be able to derive a height profile of the zonal wind (the component of wind along the line of sight) and its turbulence. Principal investigator is Dr. Michael K. Bird of the University of Bonn, Germany.

* Surface Science Package contains a number of sensors to determine the physical properties and composition of Titan's surface. An acoustic sounder measures the rate of descent, surface roughness and the speed of sound in any liquid. During descent, measurements of the speed of sound will give information on atmospheric composition and temperature. An accelerometer records the deceleration profile at impact, indicating the hardness of the surface. Tilt sensors (liquid-filled tubes with electrodes) measure any pendulum motion of the *Huygens* probe during descent, indicate the *Huygens* probe orientation after landing and measure any wave motion. If the surface is liquid, other sensors measure its density, temperature, refractive index, thermal conductivity, heat capacity and electrical properties. A group of platinum resistance wires, through two of which a heating current can be passed, will measure temperature and thermal conductivity of the surface and lower atmosphere and the heat capacity of the surface material. If the probe lands in liquid, a transducer, pointed downward and operating at 15 kilohertz, will conduct an acoustic sounding of the liquid's depth. The instrument will also provide some crude topographic mapping of the surface as the probe descends the last few meters (or yards) through the atmosphere. Principal investigator is Dr. John C. Zarnecki of the University of Kent, England.

Cassini's Nuclear Safety

The *Cassini* spacecraft derives its electrical power from radioisotope thermoelectric generators (RTGs), lightweight, compact spacecraft power systems that are extraordinarily reliable. RTGs are not nuclear reactors and have no moving parts. They use neither fission nor fusion processes to produce energy. Instead, they provide power through the natural radioactive decay of plutonium (mostly Pu-238, a non-weapons-grade isotope). The heat generated by this natural process is changed into electricity by solid state thermoelectric converters.

RTGs enable spacecraft to operate at significant distances from the Sun or in other areas where solar power systems would not be feasible. They remain unmatched for power output, reliability and durability by any other power source for missions to the outer solar system.

The United States has an outstanding record of safety in using RTGs on 23 missions over the past 30 years. While RTGs have never caused a spacecraft failure on any of these missions, they have been onboard three missions which experienced malfunctions for other reasons. In all cases, the RTGs performed as designed.

More than three decades have been invested in the engineering, safety analysis and testing of RTGs. Safety features are incorporated into the RTG design, and extensive testing has demonstrated that they can withstand physical conditions more severe than those expected from most accidents.

First, the fuel is in the heat-resistant, ceramic form of plutonium dioxide, which reduces its chance of vaporizing in fire or reentry environments. This ceramic-form fuel is also highly insoluble, has a low chemical reactivity, and primarily fractures into large, non-respirable particles and chunks. These characteristics help to mitigate the potential health effects from accidents involving the release of this fuel.

Radioisotope Thermoelectric Generator (RTG)

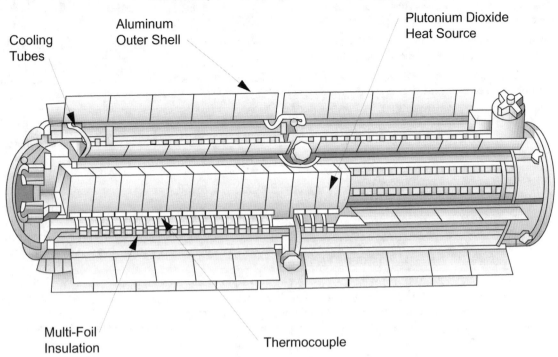

Cooling Tubes

Aluminum Outer Shell

Plutonium Dioxide Heat Source

Multi-Foil Insulation

Thermocouple

Second, the fuel in each RTG is divided among 18 small, independent modular units, each with its own heat shield and impact shell. This design reduces the chances of fuel release in an accident because all modules would not be equally impacted in an accident.

Third, multiple layers of protective materials, including iridium capsules and high-strength graphite blocks, are used to protect the fuel and prevent its accidental release. Iridium metal has a very high melting point and is strong, corrosion-resistant and chemically compatible with plutonium dioxide. These characteristics make iridium useful for protecting and containing each fuel pellet. Graphite is used because it is lightweight and highly heat-resistant.

Potential RTG accidents are sometimes mistakenly equated with accidents at nuclear power plants. It is completely inaccurate to associate an RTG accident with Chernobyl or any other past radiation accident involving nuclear fission. RTGs do not use either a fusion or fission process, and could never explode like a nuclear bomb under any accident scenario. Neither could an accident involving an RTG create the kind of radiation sickness associated with nuclear explosions.

NASA and the Department of Energy, the producer of the RTGs, place the highest priority on assuring the safe use of plutonium in space. Thorough and detailed safety analyses are conducted before launching spacecraft with RTGs, and many prudent steps are taken to reduce the risks involved in missions using RTGs. In addition to NASA's internal safety requirements and reviews, NASA missions that carry nuclear material also undergo an extensive external safety review involving detailed testing and analysis. Further, an independent safety evaluation of the *Cassini* mission has been performed as part of the nuclear launch safety approval process by an Interagency Nuclear Safety Review Panel (INSRP), which is supported by experts from government, industry and academia.

Alternatives

Studies conducted by NASA's Jet Propulsion Laboratory (JPL) have concluded that neither fuel cells nor spacecraft batteries demonstrate the operational life needed for planetary missions, whose duration can exceed 10 years from launch. In addition, the large mass of batteries that would be needed to power a mission such as *Cassini* exceeds current launch vehicle lift capabilities.

JPL's rigorous analysis has also taken into account the advances in solar power technologies that have occurred over the last decade. The conclusion reached by JPL researchers is that solar technology is still not capable of providing sufficient and reliable electrical power for the *Cassini* mission. The mass of solar arrays required would make the spacecraft too heavy for available launch vehicles. Even if a sufficiently powerful launch vehicle were available for an all-solar *Cassini*, other limitations exist with current and near-term solar technologies.

The behavior of solar cells at vast distances from the Sun is not well understood and would add significant risk to the success of a solar-powered mission to Saturn. Saturn is located approximately 1.42 billion kilometers (882 million miles) from the Sun, nearly twice as far from the Sun as Jupiter, the next closest planet.

The size of solar arrays that would be needed, about the size of two tennis courts, would not only be difficult to deploy reliably, but would make turns and other critical maneuvers extraordinarily difficult to perform. This would severely inhibit *Cassini*'s ability to achieve its science objectives.

The large arrays would seriously interfere with the fields of view of many of the science experiments and navigation sensors, further limiting the Cassini mission's ability to achieve the science objectives.

Large arrays could generate serious electromagnetic and electrostatic interference, which would adversely impact the operation of the science experiments and the *Cassini* spacecraft's communications equipment and computers.

Cassini's Earth Swingby

By aiming a spacecraft so that it passes close to a planet or moon, it is possible to boost the spacecraft on to still more distant destinations with greater velocity. This gravity assist maneuver has become an established method of launching massive, instrument-laden spacecraft to the outer planets. *Cassini* will make use of this technique when it swings by Venus twice, then the Earth and Jupiter to reach its ultimate destination of Saturn.

The Earth swingby does not represent a substantial risk to Earth's population because the probability of a reentry during the maneuver is extremely low, less than one in one million. NASA's robotic planetary spacecraft have performed numerous similar maneuvers with extraordinary precision. The redundant design of *Cassini*'s systems and navigational capability allows control of the swingby altitude at Earth to within an accuracy of 3 to 5 kilometers (2 to 3 miles) at an altitude of 800 kilometers (500 miles) or higher.

In addition, NASA has taken specific actions to design the spacecraft and mission in such a way as to ensure the probability of Earth impact is less than one in one million. For example, until seven days before the Earth swingby, the spacecraft is on a trajectory that, without any further maneuvers, would miss the Earth by thousands of kilometers. This trajectory strictly limits the possibility that random external events, such as a micrometeoroid puncture of a spacecraft propellant tank, might lead to Earth impact.

Radiation Hazards of Plutonium-238

Plutonium-238 gives off short-range alpha particles, helium nuclei that usually travel no more than about three inches in air. While the fuel is contained within its iridium capsule, the alpha radiation does not present a hazard, and the external dose resulting from the low levels of gamma and neutron radiation associated with the plutonium dioxide RTG fuel generally do not represent a significant health hazard. External alpha radiation would be stopped by clothing, an outer layer of unbroken skin, or even a sheet of paper. The point at which Pu-238 can become a health hazard is when it is deposited into the body in tiny particle form and becomes lodged there.

If an individual were to inhale plutonium dioxide particles of a sufficiently small size to be deposited and retained in proximity to lung tissue, the alpha radiation could lead to forms of cancer. The ceramic form of plutonium used in RTGs, however, is made to inhibit the fuel from shattering into fine particles that could be readily inhaled.

The ceramic form of plutonium dioxide fuel also has low solubility in water, so it has little potential to migrate in ground water or be taken up by plants. Plutonium dioxide also is highly insoluble in the human digestive system.

A common misconception is that a small amount of plutonium, such as one pound, if evenly distributed over the entire world, could induce lung cancer in every person on Earth. While plutonium can alter or kill living cells if deposited directly onto sensitive human tissue, the important point is that it must be in a form that enables environmental transport and intake by humans. Research has demonstrated that the mechanisms of plutonium dispersion into and transport through the environment, and hence into humans, are extremely difficult and inefficient.

Even in the highly unlikely release of plutonium dioxide from *Cassini*'s RTGs in the event of an accident, independently reviewed analysis shows that the radiation hazard to the average exposed individual would be minuscule, about $1/15,000$ of the lifetime exposure a person receives from natural radiation sources.

The International Team

The Cassini program is an international cooperative effort involving NASA, the European Space Agency (ESA) and the Italian Space Agency, Agenzia Spaziale Italiana (ASI), as well as several separate European academic and industrial contributors. The Cassini partnership represents an undertaking whose scope and cost would not likely be borne by any single nation, but it made possible through shared investment and participation. Hundreds of scientists and engineers from 16 European countries and 33 U.S. states make up team that developed and will fly and receive data from *Cassini* and *Huygens*.

In the United States, the mission is managed for NASA's Office of Space Science by the Jet Propulsion Laboratory (JPL), Pasadena, California. JPL is a division of the California Institute of Technology. At JPL, Richard J. Spehalski is the Cassini program manager, and Ronald F. Draper is the deputy program manager. Dr. Dennis L. Matson is the Cassini project scientist and Dr. Linda J. Spilker is the deputy project scientist. Thomas R. Gavin is spacecraft system manager, William G. Fawcett is science instruments manager, Charles E. Kohlhase is science and mission design manager, and Peter E. Doms is mission and science operations manager.

At NASA Headquarters, Mark Dahl is Cassini program executive and Henry C. Brinton is Cassini program scientist.

The major U.S. contractor is Lockheed Martin, whose contributions include the launch vehicle and upper stage, spacecraft propulsion module and the radioisotope thermoelectric generators. NASA's Lewis Research Center managed development of the Centaur upper stage.

Development of the *Huygens* Titan probe is managed by the European Space Technology and Research Center (ESTEC). ESTEC's prime contractor, Aerospatiale in Toulouse, France, assembled the probe with equipment supplied by many European countries. *Huygens*' batteries and two scientific instruments came from the United States. At ESA, Hamid Hassan is Huygens project manager, and Dr. Jean-Pierre Lebreton is project scientist.

At ASI, Enrico Flameni is the project manager for *Cassini*'s radio antenna and other contributions to the spacecraft.

The U.S. Department of Energy provided *Cassini*'s radioisotope thermoelectric generators. Beverly Cook is program manager for radioisotope power systems at DOE's Office of Space Power Systems, Germantown, Maryland.

The U.S. Air Force provided the Titan IV / Centaur launch vehicle. Launch operations are managed by the 45th Space Wing, Cape Canaveral, Florida, under the command of Brigadier General Randy Starbuck.

10-6-97

NASA Facts

National Aeronautics and
Space Administration

Jet Propulsion Laboratory
California Institute of Technology
Pasadena, CA 91109

August 2001

Cassini Mission to Saturn

CONTENTS

The Cassini mission to Saturn is the most ambitious effort in planetary space exploration ever mounted. A joint endeavor of NASA, the European Space Agency (ESA) and the Italian space agency, Agenzia Spaziale Italiana (ASI), Cassini is sending a sophisticated robotic spacecraft to orbit the ringed planet and study the Saturnian system in detail over a four-year period. Onboard *Cassini* is a scientific probe called *Huygens* that will be released from the main spacecraft to parachute through the atmosphere to the surface of Saturn's largest and most interesting moon, Titan.

Launched in 1997, *Cassini* will reach Saturn in 2004 after an interplanetary cruise spanning nearly seven years. Along the way, it has flown past Venus, Earth and Jupiter in "gravity assist" maneuvers to increase the speed of the spacecraft.

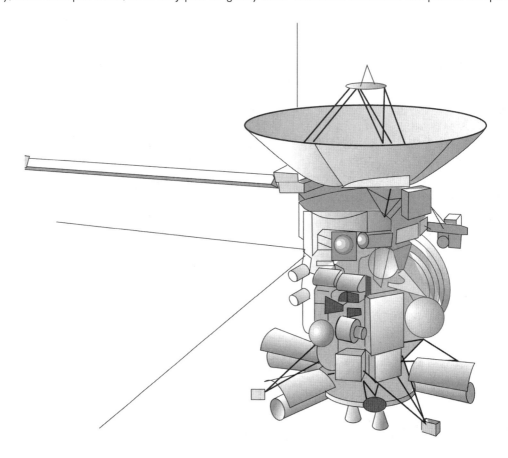

Saturn is the second largest planet in the solar system. Like the other gaseous outer planets – Jupiter, Uranus and Neptune – it has an atmosphere made up mostly of hydrogen and helium. Saturn's distinctive, bright rings are made up of ice and rock particles ranging in size from grains of sand to boxcars. More moons of greater variety orbit Saturn than any other planet. So far, observations from Earth and by spacecraft have found Saturnian satellites ranging from small asteroid-sized bodies to the aptly named Titan, which is larger than the planet Mercury.

The 12 scientific instruments on the *Cassini* orbiter will conduct in-depth studies of the planet, its moons, rings and magnetic environment. The six instruments on the *Huygens* probe, which will be dispatched from *Cassini* during its first orbit of Saturn, will provide our first direct sampling of Titan's atmospheric chemistry and the first photographs of its hidden surface.

Saturn's butterscotch-colored face appears placid at the surface, but it masks a windswept atmosphere where an equatorial jet stream blows at 1,800 km/hr (1,100 mph) and swirling storms roil beneath the cloud tops. Early explorations by NASA's *Pioneer 11* spacecraft in 1979, and the *Voyager 1* and 2 spacecraft in 1980 and 1981, found Saturn to possess a huge and complex magnetic environment where trapped protons and electrons interact with each other, the planet, the rings and the surfaces of many of the satellites. The bright rings for which Saturn is best known were found to consist of not just a few monolithic bands but thousands of rings and ringlets broad and thin, with particles sometimes herded into complicated orbits by the gravitational interaction of small moons previously unseen from Earth.

Haze-covered Titan offers a tantalizing mix of an Earth-like, nitrogen-based atmosphere and a surface that many scientists believe probably features chilled lakes of ethane and a surface coated with sticky brown organic condensate that has rained down from the atmosphere. Standing on Titan's surface beneath an orange sky, a visitor from Earth likely would find a cold, exotic world with a pungent odor reminiscent of a petroleum processing facility. Because Titan and Earth share so much in atmospheric composition, Titan is thought to hold clues to how the primitive Earth evolved into a life-bearing planet.

The Saturnian system offers more variety in scientific targets for study than any other planet in our solar system. Over the course of *Cassini*'s four-year mission, and through the studies of Titan by the *Huygens* probe, scientists expect to reveal new discoveries and enrich our understanding of phenomena in fields including biology, atmospheric chemistry and physics, climatology, volcanism, tectonics, the physics of disc systems such as galaxies and solar systems in formation, and more.

Launch

Cassini was launched October 15, 1997, at 4:43 a.m. Eastern Daylight Time (0843 Universal Time) from Cape Canaveral Air Force Station in Florida aboard a Titan IVB / Centaur launch system – the most powerful launch vehicle in the U.S. fleet. After *Cassini* was placed in a parking orbit around Earth, a Centaur upper stage fired to send *Cassini* on the interplanetary trajectory that will deliver the spacecraft to Saturn.

The Mission

In maneuvers called gravity assist swingbys, *Cassini* flew twice past Venus, then once each past Earth and Jupiter. The spacecraft's speed relative to the Sun increased as it approached and swung around each planet, giving *Cassini* the cumulative boost it needs to reach Saturn. The spacecraft will arrive at Saturn on July 1, 2004, when it will enter orbit and begin its detailed scientific observations of the Saturnian system. *Cassini*'s four-year long prime mission ends July 1, 2008.

Cassini executed its first Venus flyby April 26, 1998, at an altitude of 287.2 kilometers (178.5 miles). The second Venus flyby took it within 600 kilometers (373 miles) of the planet on June 24, 1999. Two months later, on August 18, 1999, *Cassini* swung past Earth at an altitude of 1,171 kilometers (727 miles). It flew by Jupiter at an altitude of 9.7 million kilometers (6 million miles) on December 30, 2000. During the six months *Cassini* was closest to Jupiter, it collaborated with NASA's *Galileo* spacecraft, which had been orbiting Jupiter since 1995, for the rare opportunity to study the Jovian system from two different nearby perspectives at the same time. The two craft monitored how fluctuations in the solar wind approaching Jupiter affect the planet's magnetosphere, a vast region of ionized gas under the influence of Jupiter's magnetic field. They also examined Jupiter's moons, rings and storm clouds.

Upon reaching Saturn on July 1, 2004, *Cassini* will fire its main engine for 96 minutes to brake the spacecraft's speed and allow it to be captured as a satellite of Saturn. Passing through the dusty, outermost E-ring, *Cassini* will swing in close to the planet – to an altitude only one-sixth the diameter of Saturn itself – to begin the first of 75 orbits during the rest of its four-year mission.

On December 25, 2004, *Cassini* will release the European-built *Huygens* probe toward Titan. On January 14, 2005, the 2.7-meter (8.9-foot)-diameter *Huygens* will enter Titan's atmosphere, deploy its parachutes and begin its scientific observations during a descent of up to two and a half hours through that moon's dense atmosphere. Instruments on board will measure the temperature, pressure, density and energy balance in the atmosphere. As the probe breaks through the cloud deck, a camera will capture pictures of the Titan panorama. Titan's surface properties will be observed, and about 1,000 images of the clouds and surface will be returned. In the final moments of descent, a spotlight will illuminate the surface for the imaging instrument on board.

If the probe survives landing at a fairly low speed of about 25 km/hr (15 mph), it can possibly return data from Titan's surface, where the atmospheric pressure is 1.6 times that on Earth. The probe could touch down on solid ground, ice or even splash down in a lake of ethane or methane. One instrument on board will discern whether *Huygens* is bobbing in liquid, and other instruments on board would tell the chemical composition of that liquid. Throughout its mission, *Huygens* will radio data collected by its instruments to the *Cassini* orbiter to be stored and then relayed to Earth.

If it lands in liquid ethane, *Huygens* will not be able to return data for long because the extremely low temperature of this liquid, about -180 degrees Celsius (-290 degrees Fahrenheit), would prevent the batteries from operating. In addition, if liquid ethane permeated the probe science instrument packages, the radio would be badly tuned and probably not operable. If the battery-powered probe continues to send data to *Cassini* from Titan's surface, it will be able to do so for only about 30 minutes, when the probe's battery power is expected to run out. With the end of the *Huygens* probe portion of the mission, *Cassini*'s focus will shift to taking measurements with the orbiter's 12 instruments and returning the information to Earth.

During the course of the *Cassini* orbiter's mission, it will execute close flybys of particular bodies of interest – including more than 44 encounters of Titan and seven of selected icy moons of greatest interest. In addition, the orbiter will make at least two dozen more-distant flybys of the Saturnian moons. *Cassini*'s orbits will also allow it to study Saturn's polar regions in addition to the planet's equatorial zone.

Saturn

Saturn is the outermost planet in the solar system readily visible to the naked eye, and was recognized as a slow-moving point of bright yellow light in the sky by ancient cultures around the world. It was not until the invention of the telescope, however, that Saturn's characteristic rings began to come into focus. Italian astronomer Galileo was the first to look at Saturn through a telescope in 1609 and 1610. Saturn, viewed through Galileo's crude instrument was a puzzling sight. Unable to make out the rings, he thought what he saw were two sizable companions close to the planet. Having recently discovered the major moons of Jupiter, it seemed to follow, he supposed, that Saturn could have large moons, too. ". . . [T]o my very great amazement, Saturn was seen to me to be not a single star, but three together, which almost touch each other," he wrote at the time. He was even more astonished when he again looked at Saturn through his telescope two years later to find that the companion bodies had apparently disappeared. "I do not know what to say in a case so surprising, so unlooked-for and so novel," Galileo wrote in 1612. The rings were simply invisible because he was now viewing them edge-on. Two years later, they again reappeared, larger than ever. He concluded that what he saw were some sort of "arms" that grew and disappeared for unknown reasons. He died never knowing that he'd been the first to observe Saturn's rings.

Nearly half a century later, the Dutch scientist Christiaan Huygens solved the puzzle that had vexed Galileo. Thanks to better optics, Huygens was able to pronounce in 1659 that the companions or arms decorating Saturn were in fact a set of rings. The rings were tilted so that, as Saturn orbited the Sun every 29 years, the sheet of rings would occasionally seem to vanish as viewed on-edge from Earth.

While observing Saturn, Huygens also discovered the moon Titan. A few years later, the Italian-French astronomer Jean-Dominique Cassini (also known as Gian Domenico Cassini in his native Italy) added several other key Saturn discoveries. Using new telescopes, Cassini discovered Saturn's four other major moons – Iapetus, Rhea, Tethys and Dione. In 1675, he discovered that Saturn's rings are split largely into two parts by a narrow gap, known since as the "Cassini Division." In the 19th century, James E. Keeler, pursuing theoretical studies by James Clerk Maxwell, showed that the ring system was not a uniform sheet but made up of small particles that orbit Saturn.

The first detection of Saturn's magnetic field came with the flyby of Saturn by NASA's *Pioneer 11* spacecraft in 1979. Then in 1980 and 1981, the NASA *Voyager 1* and *Voyager 2* spacecraft flew through the Saturnian system to reveal storms and subtle latitudinal banding in the planet's atmosphere, several more small moons, and a breathtaking collection of thousands of ringlets. The *Voyagers* found ring particles ranging in size from nearly invisible dust to icebergs the size of a house. The spacing and width of the ringlets were discovered to be orchestrated in part by gravitational tugs from a retinue of orbiting moons and moonlets, some near ring edges but most far beyond the outermost main rings. *Voyager*'s instruments showed that the rings contain water ice, which may cover rocky particles.

Saturn has been a frequent target of the Hubble Space Telescope, which has produced stunning views of long-lived hurricane-like storms in Saturn's atmosphere. The world's major telescopes, including Hubble, were recently trained on Saturn to observe the phenomenon known to astronomers as a Saturn ring plane crossing. The rings were seen edge-on from the Earth's perspective on May 22, 1995, August 10, 1995 and February 11, 1996. Ring plane crossings provide astronomers with unique views of the Saturnian system. With the rings temporarily invisible as viewed from Earth, faint objects near the planet are easier to see. Two new moons observed during these recent opportunities are now believed to be particle swarms within the narrow F-ring. Many of Saturn's 30 known moons were discovered during ring plane crossings.

Saturn is one of four giant gaseous (and ringed) planets in the solar system, the others being Jupiter, Uranus and Neptune. Second in size only to Jupiter, Saturn is shaped like a ball squeezed from opposite sides; its diameter pole-to-pole is only 108,728 kilometers (67,560 miles), compared to about 120,536 kilometers (about 74,898 miles) for the equatorial diameter. This oblateness is caused in part by Saturn's very rapid rotation rate. Combined with the planet's low density, the fast rotation tends to promote a bulge of material near the equator. Saturn's volume would enclose more than 750 Earths. Even so, its mass is only 95 times that of Earth; with a density less than that of water, it would float in an ocean if there were one big enough to hold it.

Unlike rocky inner planets such as Earth, Saturn and the other gas giants have no surface on which to land. A spacecraft pilot foolhardy enough to descend into its atmosphere would simply find the surrounding gases becoming denser and denser, the temperature progressively hotter; eventually the craft would be crushed and melted. Detailed analysis of Saturn's gravitational field leads astronomers to believe that the deepest interior of Saturn must consist of a molten rock core about the same size as Earth, but much more massive than Earth. This rock core may also be surrounded by a layer of melted ices.

Spectroscopic studies by the *Voyager* spacecraft found Saturn to be made up of about 94 percent hydrogen and 6 percent helium. Hydrogen and helium are the primary constituents of all the giant gas planets, the Sun and the stars.

A day on Saturn is about 10.6 hours, and a Saturnian year is about 29.5 Earth years.

A fast equatorial flow like a giant jet stream has been clocked on Saturn. This high-velocity wind of 1,800 km/hr (1,100 mile per hour) remains fairly constant over decades. Saturn also has storms like those seen in the atmosphere of Jupiter, but they are much less visible and perhaps less frequent. They may represent Saturnian weather systems that resemble hurricanes on Earth. On Saturn, however, these storms last much longer, perhaps because they do not encounter continents, which reduce their energy sources. On Earth, hurricane-like storms tend to lose their energy as they come into contact with land.

Saturn is colder than Jupiter, but the colors of Saturn's cloud layers are due to the same basic cloud chemistry as on Jupiter. Near the top of the atmosphere, the ammonia becomes cold enough to crystallize into ice particle clouds, much like high cirrus clouds in Earth's skies. These ammonia clouds are the visible part of Saturn. The primary reason why Saturn is more uniformly colored than Jupiter is because this outermost cloud layer is much deeper in the atmosphere than on Jupiter. The haze layers above the clouds therefore hide or mute the colors of the clouds. Latitudinal banding can be seen on Saturn, but it lacks the distinct color contrasts seen between atmospheric bands on Jupiter.

Gravity at the top of Saturn's clouds is similar to the gravity near the surface of the Earth. The temperature near the cloud tops is about -139 degrees Celsius (-218 degrees Fahrenheit). The temperature increases with depth due to increased atmospheric pressure. At the core, Saturn's temperature is predicted to be about 10,000 degrees Celsius (18,000 degrees Fahrenheit).

Saturn is 9.5 times farther from the Sun than Earth is, so it receives only about 1 percent as much sunlight per square meter as does Earth. Saturn reflects a somewhat smaller fraction of sunlight received than does Earth (34.4 percent compared to 40 percent for Earth, computed across all wavelengths of light). Thus, Saturn absorbs 65.4 percent of the energy it receives from the Sun. Mysteriously, Saturn emits 87 percent more energy than it absorbs from sunlight. Unlike the rocky Earth and the more massive Jupiter, Saturn should not have any heat left over from its original formation. Therefore there must be a source of heat inside Saturn to produce the excess energy. One theory is that the energy comes from the friction of liquid helium raining through lighter liquid hydrogen in the interior of the planet. Cassini scientists will examine Saturn's energy balance for answers to this puzzle.

The Rings

Although the best telescopes on Earth show three nested main rings about Saturn, we now know that the ring system is collection of thousands of ringlets. They are not solid but rather are made up of countless unconnected particles, ranging in size from nearly invisible dust to icebergs the size of a house. The spacing and width of the ringlets are orchestrated by gravitational tugs from a retinue of orbiting moons and moonlets, some near ring edges but most far beyond the outermost main rings. The rings contain water ice, which may cover rocky particles. There are ghostly dark "spokes" in the rings that flicker on and off. Scientists believe they may be electrically charged particles, but we do not really know. Where do the subtle colors in Saturn's rings come from? The Cassini mission may well provide the answers.

And what is the origin of the rings themselves? One theory is that they are the shattered debris of moons broken apart by repeated meteorite impacts. Scientists believe that Saturn's ring system may even serve as a partial model for the disk of gas and dust from which all the planets formed about the early Sun. The Cassini mission will undoubtedly provide important clues to help determine the answers.

Mysterious Moons

Saturn has the most extensive system of known moons of any planet in the solar system – ranging in diameter from less than 20 kilometers (12 miles) to 5,150 kilometers (3,200 miles), larger than the planet Mercury. Most are icy worlds heavily studded with craters caused by impacts very long ago.

Thirty moons have been confirmed in orbit at Saturn, including 12 discovered from ground-based telescopes in 2000. *Cassini* may discover others.

The moon Enceladus poses a mystery. Although covered with water ice like Saturn's other moons, it displays an abnormally smooth surface; there are very few impact craters on the portions seen by *Voyager*. Has much of the surface of Enceladus recently melted to erase craters? Scientists now believe that Enceladus is the likely source for particles, spewed by ice volcanoes, that create Saturn's outermost E-ring.

Saturn's moon Iapetus is equally enigmatic. On one side – the trailing side in its orbit – Iapetus is one of the brightest objects in the solar system, while its leading side is one of the darkest. Scientists surmise that the bright side is water ice and the dark side is an organic material of some kind. But how the dark material got there is a mystery. Did it rise up from the inside of the moon, or was it deposited from the outside? The puzzle is compounded by the fact that the dividing line between the two sides is inexplicably sharp.

Titan

Titan is by far the most intriguing natural satellite of Saturn, and one of the most fascinating in the solar system. Titan lies hidden beneath an opaque atmosphere more than 50 percent denser than Earth's. Titan has two major components of Earth's atmosphere – nitrogen and oxygen – but the oxygen is likely frozen as water ice within the body of the moon. If Titan received more sunlight, its atmosphere might more nearly resemble that of a primitive Earth.

What fascinates scientists about Titan's atmosphere is that it is filled with a brownish orange haze made of complex organic molecules, falling from the sky to the surface. Thus in many ways it may be a chemical factory like the primordial Earth. Most scientists agree that conditions on Titan are too cold for life to have evolved – although there are theories concerning the possibility of life forms in covered lakes of liquid hydrocarbons warmed by the planet's internal heat. Yet even if Titan proves to be lifeless, as expected, understanding chemical interactions on the distant moon may help us understand better the chemistry of the early Earth – and how we came to be.

Saturn has a magnetic field and extensive magnetosphere that shields much of the Saturnian system from the wind of charged particles that flows outward from the Sun. The behavior of charged and other particles trapped in this magnetic bubble around Saturn is of great interest to physicists. *Cassini* will help determine the similarities and differences between the magnetospheres of the planets and possibly moons in the solar system that generate such fields.

The Cassini Spacecraft

The *Cassini* spacecraft, including the orbiter and the *Huygens* probe, is one of the largest, heaviest and most complex interplanetary spacecraft ever built. The orbiter alone weighs 2,125 kilograms (4,685 pounds). When the 320-kilogram (705-pound) *Huygens* probe and a launch vehicle adapter were attached and 3,132 (6,905 pounds) of propellants were loaded, the spacecraft at launch weighed 5,712 kilograms (12,593 pounds). Of all interplanetary spacecraft, only the two *Phobos* spacecraft sent to Mars by the former Soviet Union were heavier.

The *Cassini* spacecraft stands more than 6.7 meters (22 feet) high and is more than 4 meters (13.1 feet) wide. The magnetometer instrument is mounted on an 11-meter (36-foot) boom that extends outward from the spacecraft. Three other 10-meter (32-foot) rod-like booms that act as the antennas for the radio plasma wave subsystem extend outward from the spacecraft in a Y shape. The complexity of the spacecraft is necessitated both by its flight path to Saturn and by the ambitious program of scientific observations to be undertaken once the spacecraft reaches its destination. The spacecraft includes 22,000 wire connections and more than 14 kilometers (8.7 miles) of cabling. Because of the very dim sunlight at Saturn's orbit, solar arrays are not feasible and electrical power is supplied by a set of radioisotope thermoelectric generators, which use heat from the natural decay of plutonium-238 to generate electricity to run *Cassini*'s systems. These power generators are of the same design as those used on the *Galileo* and *Ulysses* missions.

Equipment for a total of 12 science experiments is carried onboard the *Cassini* orbiter. Another six fly on the *Huygens* probe, which will detach from the orbiter to parachute through Titan's atmosphere to its surface.

The *Cassini* orbiter has advanced and extended the United States' technology base with several innovations in engineering and information systems. Whereas previous planetary spacecraft used onboard tape recorders, *Cassini* has pioneered a new solid state data recorder with no moving parts. The solid state recorder eventually will likely replace tape recorders used on space missions. Similarly, the main onboard computer that directs operations of the orbiter uses a novel design drawing on new families of electronic chips. Among them are very high-speed integrated circuit chips developed under a U.S. government-industry research and development initiative. Also part of the computer are powerful new application-specific integrated circuit parts; each component replaces a hundred or more traditional chips.

Elsewhere on the *Cassini* orbiter, the power system benefits from an innovative solid state power switch developed for the mission. This switch eliminates rapid fluctuations called transients that usually occur with conventional power switches, with a significantly improved component lifetime. The solid state power switch holds great promise for use in numerous Earth-based applications as well.

Science Experiments

Orbiter:

- Imaging science subsystem: Takes pictures in visible, near-ultraviolet and near-infrared light.
- *Cassini* radar: Maps surface of Titan using radar imager to pierce veil of haze. Also used to measure heights of surface features.
- Radio science subsystem: Searches for gravitational waves in the universe; studies the atmosphere, rings and gravity fields of Saturn and its moons by measuring telltale changes in radio waves sent from the spacecraft.
- Ion and neutral mass spectrometer: Examines neutral and charged particles near Titan, Saturn and the icy satellites to learn more about their extended atmospheres and ionospheres.
- Visible and infrared mapping spectrometer: Identifies the chemical composition of the surfaces, atmospheres and rings of Saturn and its moons by measuring colors of visible light and infrared energy given off by them.
- Composite infrared spectrometer: Measures infrared energy from the surfaces, atmospheres and rings of Saturn and its moons to study their temperature and composition.
- Cosmic dust analyzer: Studies ice and dust grains in and near the Saturn system.
- Radio and plasma wave spectrometer: Investigates plasma waves (generated by ionized gases flowing out from the Sun or orbiting Saturn), natural emissions of radio energy and dust.
- *Cassini* plasma spectrometer: Explores plasma (highly ionized gas) within and near Saturn's magnetic field.
- Ultraviolet imaging spectrograph: Measures ultraviolet energy from atmospheres and rings to study their structure, chemistry and composition.
- Magnetospheric imaging instrument: Images Saturn's magnetosphere and measures interactions between the magnetosphere and the solar wind, a flow of ionized gases streaming out from the Sun.
- Dual technique magnetometer: Studies Saturn's magnetic field and its interactions with the solar wind, the rings and the moons of Saturn.

Huygens Probe:

- Descent imager and spectral radiometer: Makes images and measures temperatures of particles in Titan's atmosphere and on Titan's surface.
- *Huygens* atmospheric structure instrument: Explores the structure and physical properties of Titan's atmosphere.
- Gas chromatograph and mass spectrometer: Measures the chemical composition of gases and suspended particles in Titan's atmosphere.
- Aerosol collector pyrolyzer: Examines clouds and suspended particles in Titan's atmosphere.
- Surface science package: Investigates the physical properties of Titan's surface.
- Doppler wind experiment: Studies Titan's winds from their effect on the probe during its descent.

The International Team

Hundreds of scientists and engineers from 16 European countries and 33 states of the United States make up the team responsible for designing, building, flying and collecting data from the *Cassini* orbiter and *Huygens* probe.

The Cassini mission is managed by NASA's Jet Propulsion Laboratory in Pasadena, California, where the orbiter was designed and assembled. JPL is a division of the California Institute of Technology. Development of the *Huygens* Titan probe was managed by the European Space Technology and Research Center, whose prime contractor for the probe is Alcatel in France. Equipment and instruments for the probe were supplied from many countries, including the United States.

The Cassini program is an 18-year endeavor; the program received a new start from the U.S. Congress in 1990, and the prime mission extends through July 2008. The *Cassini* orbiter and its instruments represent a $1.422 billion investment by NASA. The agency has budgeted a total of $704 million to support the cruise and orbital operations phase of the mission. Other contributions include $54 million in NASA tracking costs and about $144 million from the U.S. Department of Energy in support of the radioisotope thermoelectric generators and radioisotope heater units for the mission. The launch vehicle, provided to NASA by the U.S. Air Force, cost $422 million.

The European Space Agency's contribution to the Cassini program totals approximately $500 million for the *Huygens* probe, its instruments and probe science and engineering operations. The Italian space agency, Agenzia Spaziale Italiana, has contributed the *Cassini* orbiter's dish-shaped high-gain antenna as well as significant portions of three science instruments; its contribution is $160 million.

The Centaur upper stage and launch vehicle performance analysis were managed by NASA's Lewis Research Center, Cleveland, Ohio.

Communications with *Cassini* during the mission are carried out through stations of NASA's Deep Space Network in California, Spain and Australia. Data from the *Huygens* probe will be received by the network and sent to a European Space Agency operations complex in Darmstadt, Germany.

At NASA Headquarters, the Cassini program executive is Mark Dahl and the program scientist is Dr. Jay Berstralh. At JPL, Robert T. Mitchell is program manager, Dr. Dennis Matson is Cassini project scientist and Dr. Linda Spilker is deputy project scientist.

Cassini Spacecraft and Huygens Probe

CONTENTS

The *Cassini* spacecraft is designed for a daunting – task to leave Earth on a seven year interplanetary voyage to the giant gas planet Saturn. Once there, *Cassini* will spend four years exploring the planet, its rings, and its moons. It will also deploy a probe to the moon Titan – a world with an atmosphere sufficient to allow a parachute to land the probe on the surface. These are the challenges facing the Cassini mission, an international cooperative effort of the National Aeronautics and Space Administration (NASA), the European Space Agency (ESA), and the Italian Space Agency (Agenzia Spaziale Italiana, ASI). The mission is managed for NASA by the Jet Propulsion Laboratory (JPL) of the California Institute of Technology.

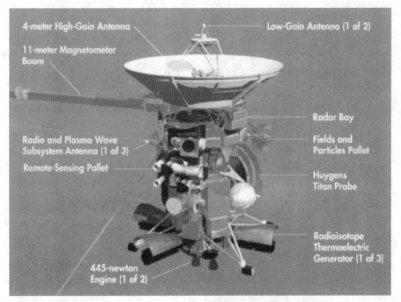

This illustration shows the main design features of the two-story-tall *Cassini* spacecraft.

The *Cassini* spacecraft is composed of the orbiter and the *Huygens* probe. It is the second largest interplanetary spacecraft ever to be launched, owing to its cargo of multiple science instruments and an enormous quantity of propellant. (The former Soviet Union launched the largest interplanetary vehicles: the *Phobos 1* and 2 Mars craft, each weighing 6,220 kilograms, or nearly 7 tons.) The *Cassini* probe and orbiter together weigh about 5,574 kilograms (6.1 tons), more than 50 percent of which is liquid propellant. The propellant mass alone is more than the mass of the *Galileo* and *Voyager* spacecraft combined. *Cassini* stands more than 6.7 meters (22 feet) high and is more than 4 meters (13.1 feet) wide (not including a deployed boom and antennas).

The cone-shaped probe remains dormant during the cruise to Saturn. After arrival at Saturn, the probe will be activated and deployed into the atmosphere of Titan, sending data to the orbiter. The *Cassini* orbiter will relay the probe data to Earth, and then conduct its own scientific investigation of the Saturn system for nearly four years.

The Cassini Orbiter

The design of the orbiter is derived from the special demands of the Cassini mission. Communication, for example, is essential to allow commanding of the spacecraft from Earth and for the return of the scientific data from the orbiter and the probe. This is done by the use of radio systems operating on several different microwave frequencies. The orbiter communicates with controllers on Earth via three separate antennas − a high-gain antenna that is 4 meters (13.1 feet) in diameter, and two low-gain antennas. The rigid high-gain antenna was developed by ASI. Depending on mission phase, data transmission can vary from a low of 5 bits per second up to 249 kilobits per second. The data travel at the speed of light. When the spacecraft is at Saturn, *Cassini* will be 8.6 to 10.6 astronomical units (AU) from Earth. (1 AU is the mean distance from Earth to the Sun: about 150 million kilometers or 93 million miles.) At that distance, it will take from 68 to 84 minutes for radio waves to travel from Earth to the spacecraft or from the spacecraft to Earth.

The orbiter receives electrical power from three radioisotope thermoelectric generators, or RTGs. These produce power by converting heat into electrical energy. Heat is provided by the natural radioactive decay of plutonium dioxide. RTGs have no moving parts and are a very reliable source of energy. Upon the spacecraft's arrival at Saturn, *Cassini*'s three RTGs will provide more than 700 watts of power for engineering and scientific devices.

Propulsion for large changes to the orbiter's trajectory is provided by either of two identical main engines. These powerful engines use nitrogen tetroxide and monomethyl-hydrazine as the oxidizer and fuel, respectively. Sixteen smaller engines, called thrusters, use hydrazine to control orientation and to make small changes to the spacecraft's flight path.

Guidance and control for the spacecraft are governed by sensors that recognize reference stars and the Sun, and by onboard computers that determine the spacecraft's position. Using a new type of gyroscope that vibrates rather than spins, the spacecraft can perform turns, twists, and propulsion engine firings while retaining continuous knowledge of its own orientation. The orbiter is stabilized along all three axes and thus does not normally rotate during its long cruise to Saturn.

Cassini was launched October 15, 1997 on a Titan IVB / Centaur rocket.

The *Cassini* trajectory takes the spacecraft past Venus twice and past Earth once on its trip to Saturn. This poses a challenge of temperature control, because the orbiter is close to the Sun for the first several years of the mission. The high-gain antenna is used as a sunshade to shield the rest of the orbiter and probe. Special paints are also used on the antenna to reflect and radiate much of the Sun's energy. Later in the mission, as the spacecraft approaches Saturn, the sunlight is approximately 1 percent of that received at Earth, and extreme cold becomes a concern. Heat is retained by using a highly efficient insulating blanket that wraps most of the orbiter. The blanket's outer layer is a three-ply membrane composed of a Kapton core with an aluminized inner surface and a metallic outer surface. The translucent Kapton has a yellow color and, when backed by a shiny aluminum layer, appears amber. Additional electrical heaters and 1 watt radioisotope heaters are applied in selected areas. The thermal blankets also provide protection from micrometeoroids that will impact the spacecraft at speeds of 5 to 40 kilometers per second (11,000 to 90,000 mph).

The orbiter requires extensive onboard computing capability because most of the Cassini mission is performed while the orbiter is not in direct communication with ground controllers. Sequences − programs stored on the spacecraft that are detailed computer instructions − direct the activities of the spacecraft. A typical sequence may operate the spacecraft for four weeks without the need for ground controller intervention, but controllers do send at least one command each week to reset the command timer.

The computers must also be designed to withstand the radiation environment of deep space, particularly when the Sun is at peak activity. During solar flares, the effects of which can last for several days, the levels of radiation emitted by the Sun can be 1,000 times higher than normal.

Sophisticated fault-protection software resides in the spacecraft computers to continuously sample and sense the health of the onboard systems. The fault protection system will automatically take corrective action if it determines that *Cassini* is at risk due to an onboard failure.

Orbiter Science Experiments

The orbiter carries 12 science instruments. The Radio Science Investigation will search for gravitational waves in the universe and study the atmosphere, rings, and gravity fields of Saturn and its moons by measuring telltale variations in radio waves sent from the spacecraft. The Imaging Science Subsystem will take pictures in visible, near-ultraviolet, and near-infrared light. The *Cassini* Radar will map Titan's surface using radar and passive microwave imagery to pierce the veil of haze, and will measure heights of surface features. The Ion and Neutral Mass Spectrometer will examine neutral and charged particles near Titan, Saturn's rings, and the icy satellites to learn more about the extended atmospheres and ionospheres of these bodies.

The Visible and Infrared Mapping Spectrometer will identify the chemical composition of the surfaces, the atmospheres, and rings of Saturn and its moons by measuring the colors of the visible light and the infrared energy they reflect. The Composite Infrared Spectrometer will measure infrared energy from the surfaces, atmospheres, and rings of Saturn and its moons to study their temperatures and compositions. The Cosmic Dust Analyzer will study ice and dust grains in and near the Saturn system. Radio and Plasma Wave Science will investigate plasma waves generated by ionized gases flowing out from the Sun or around Saturn, natural emissions of radio energy, and dust.

The *Cassini* Plasma Spectrometer will explore plasma – electrically charged (ionized) gas – within and near Saturn's magnetic field. The Ultraviolet Imaging Spectrograph will measure ultraviolet energy from atmospheres, satellite surfaces, and rings to study structure, chemistry, and composition. The Magnetospheric Imaging Instrument will image Saturn's magnetic environment and measure interactions between the magnetosphere and the solar wind, a flow of ionized gases from the Sun. The Dual Technique Magnetometer will study Saturn's magnetic field and its interactions with the solar wind, the rings, and the moons of Saturn.

The Huygens Titan Probe

The probe is provided by ESA. Except for semiannual health checks, the probe will remain dormant throughout the 6.7 year interplanetary cruise. Prior to the probe's separation from the orbiter, a final health check will be performed, and the orbiter will position the probe on a trajectory to intercept Titan. *Cassini* will set *Huygens'* onboard clock to the precise time necessary to "wake up" the probe systems 15 minutes prior to encountering Titan's atmosphere. For 21 days, *Huygens* will simply coast to Titan with no systems active except for its wake-up clock.

Huygens' main mission phase will occur during its parachute descent through Titan's atmosphere. The batteries and all other resources are sized for a *Huygens* mission duration of three hours, which includes the possibility of up to half an hour or more on Titan's surface. The probe's radio link will be activated early in the descent phase and the data will be relayed to the orbiter for onboard storage and subsequent transmission to Earth. At the end of this three-hour-long

The 2.7-meter-diameter *Huygens* probe undergoing tests in Germany.

communication window, the *Cassini* orbiter will fly out of radio contact with *Huygens* – and shortly thereafter its high-gain antenna will be turned away from Titan and toward Earth.

Huygens Probe Design Features

The probe is ESA's first planetary atmospheric entry mission, and some of the technologies required are very different from those needed for more traditional missions. Special systems such as the thermal protection system and high-speed parachutes have been developed specifically for entry into Titan's atmosphere.

The thermal protection system (TPS) is designed to protect the probe from the extreme heat generated by its rush into Titan's atmosphere at about 6 kilometers per second (13,400 mph). At such a high speed, surface temperatures as hot as 1,700 degrees Celsius (3,000 degrees Fahrenheit) could be reached in less than a minute. The front of the heat shield is covered by tiles similar to those used to protect the space shuttle, made from a material known as AQ60 – a low-density "mat" of silica fibers. The tile thickness on the front shield is calculated to ensure that the structure will not exceed 150 degrees Celsius (302 degrees Fahrenheit), which is below the melting temperature of lead. The rear side of the probe will reach much lower temperatures during atmospheric entry; thus, a spray on layer of "Prosial" (a silicon-based foam) was used. The total mass of the thermal protection system is more than 100 kilograms (220 pounds) – almost one third of the entire probe mass.

The parachute system on board *Huygens* is designed to work in concert with the TPS to ensure that the sensitive instrumentation on the central module will survive atmospheric entry. The main parachute is almost 9 meters (28 feet) in diameter, and will deploy at an altitude of 170 kilometers (105 miles). Thirty seconds after parachute deployment, the probe's heat shield system (including the supporting structure) will be jettisoned, significantly lightening the payload. Finally, a smaller drogue chute will take over for the remainder of *Huygens*' descent.

During the probe's descent, data acquired will be transmitted to the orbiter in real time via an S-band radio link. The data rate required will be 8 kilobits per second for each of two channels. On the orbiter, the data will be received via the high-gain antenna pointed toward the probe's position.

Huygens Scientific Payload

Six scientific instruments comprise the *Huygens* payload. During the descent, these instruments will have both timing and radar attitude data available for their use in sequencing. The Gas Chromatograph and Mass Spectrometer is a versatile gas chemical analyzer designed to identify and quantify various atmospheric constituents. It is equipped with gas samplers that will be filled at high altitude for analysis later in the descent. The Aerosol Collector and Pyrolyzer will collect aerosols for chemical composition analysis. After extension of the sampling device, a pump will draw the atmosphere through filters that capture aerosols. Each sampling device can collect about 30 micrograms of material, which will be vaporized and then passed along to the Gas Chromatograph and Mass Spectrometer for chemical analysis.

The Descent Imager and Spectral Radiometer can take images and make spectral measurements using sensors covering a wide spectral range. A few hundred meters before impact, the instrument will switch on its lamp in order to acquire spectra of surface material. The *Huygens* Atmospheric Structure Instrument consists of sensors for measuring the physical and electrical properties of the atmosphere.

The Doppler Wind Experiment uses radio signals to deduce wind speeds. Winds in Titan's atmosphere will cause the probe to drift, inducing a measurable doppler shift in the carrier signal. The swinging motion of the probe beneath its parachute and other radio signal-perturbing effects, such as atmospheric attenuation, may also be detectable from the signal.

The *Cassini* spacecraft in the test chamber at JPL.

The Surface Science Package is a suite of sensors to determine the physical properties of the surface at the landing site and to provide unique information about its composition. The package includes an accelerometer to measure the impact deceleration and other sensors to measure the index of refraction, temperature, thermal conductivity, heat capacity, speed of sound, and dielectric constant of the (possibly liquid) material at the landing site.

For More Information
Cassini Program Outreach Jet Propulsion
Laboratory, MS 264-441
4800 Oak Grove Drive
Pasadena, CA 91109 8099
Telephone: (818) 354-5011

Visit The Cassini Web Site
http://www.jpl.nasa.gov/cassini/
National Aeronautics and
Space Administration

Jet Propulsion Laboratory
California Institute of Technology
Pasadena, California

JPL 400-777 5/99

GPHS-RTG

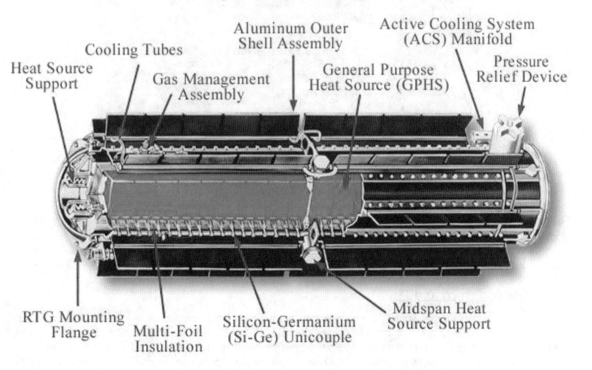

Cassini's Radioisotope Thermoelectric Generator (RTG)

JET PROPULSION LABORATORY

Cassini–Huygens Mission to Saturn

CONTENTS

Saturn and its mysterious moon Titan are the primary destinations of the Cassini–Huygens mission, a project developed by the National Aeronautics and Space Administration (NASA), the European Space Agency (ESA), the Italian Space Agency (Agenzia Spaziale Italiana, ASI), and several separate European academic and industrial contributors. The Cassini–Huygens mission is managed for NASA by the Jet Propulsion Laboratory (JPL) of the California Institute of Technology. The *Cassini* spacecraft was launched on a Titan IVB / Centaur rocket on October 15, 1997, from Cape Canaveral, Florida.

The *Cassini* spacecraft fires one of its two redundant main engines for
94 minutes to slow down enough to be captured into orbit around Saturn.

The Cassini–Huygens mission honors two 17th century astronomers who pioneered observations of Saturn: Jean-Dominique Cassini, who discovered several satellites as well as ring features such as the Cassini division, and Christiaan Huygens, who discovered the planet's largest satellite, Titan. The Cassini mission's objective is a four-year, close-up study of the Saturnian system, including Saturn's atmosphere and magnetic environment, its rings, several moons, and Titan. The mission represents a rare opportunity to gain significant insights into major scientific questions about the creation of the solar system and pre-life conditions on Earth, in addition to a host of questions specific to Saturn.

Mysteries of Saturn and Titan

Saturn, its rings, and its moons hold clues to understanding the origins of our solar system. A detailed study of the rings and icy satellites may reveal information about the material from which Saturn formed and evolved. Saturn's atmosphere shows fewer features than that of Jupiter, perhaps due to more vigorous mixing of various constituents and the fact that temperatures necessary for cloud condensation occur deeper in the atmosphere. Saturn's atmospheric phenomena also include jet streams that are among the fastest in the solar system.

Titan, Saturn's largest moon, also holds its own unique mysteries. What goes on beneath Titan's organically rich, primordial Earth-like atmosphere? *Voyager* spacecraft data and Earth-based radar and infrared observations suggest that there may be continents as well as oceans or lakes of liquid ethane on Titan. What planetary processes might be occurring in such a cold environment, where water ice is so cold that it is as hard as rock? These are among the questions *Cassini* and its *Huygens* Titan probe will address during the mission.

The Path to Saturn

To accomplish *Cassini*'s objectives, the NASA-provided orbiter and ESA-provided probe carry sophisticated instrumentation, for a total of 27 scientific investigations. Funding is provided by NASA and its international partners from Europe. The mission will produce the most complete information about a planetary system ever obtained.

Cassini executes two gravity assist flybys of Venus – one in April 1998 and one in June 1999 – then a flyby of Earth in August 1999 and a flyby of Jupiter in December 2000. *Cassini*'s Sun-relative velocity will increase as it swings around each planet, allowing the spacecraft to reach distant Saturn by July 2004. When *Cassini* arrives at Saturn, an onboard rocket engine will fire to brake the spacecraft into the first of some six dozen orbits around the planet.

Saturn, the sixth planet from the Sun, holds clues to the origin of our solar system.

In late 2004, the orbiter will adjust its trajectory and release the *Huygens* probe for its three-week coast to Titan. After atmospheric entry and parachute deployment, *Huygens* will slowly descend to the surface of Titan. The instrument-laden probe will beam its measurements to the *Cassini* orbiter, where the information will be stored and then relayed to Earth.

During the mission, the orbiter will execute more than four dozen close flybys of particular bodies of interest – including more than three dozen encounters of Titan as well as several close flybys of selected icy satellites. In addition, the orbiter will make at least two dozen more distant flybys of the other Saturnian moons. The changing inclination of the orbits will also allow study of Saturn's polar regions and equatorial zone.

Science at Saturn and Titan

Chief among *Cassini*'s scientific goals is a thorough characterization of Titan. Like Earth, Titan has a nitrogen-rich atmosphere. Complex organic molecules make up the haze that clouds its surface from view. These molecules constantly form and slowly fall to Titan's surface. Determining the chemistry of Titan's atmosphere may be crucial to understanding the evolution of early life on Earth.

A large portion of the Cassini mission's Titan studies will be accomplished by the *Huygens* probe, which will be released from the orbiter and parachute through Titan's opaque atmosphere. During the descent, which may last up to 2½ hours, *Huygens*' camera will capture more than 1,100 images; other instruments will directly sample Titan's atmosphere and determine its composition. There is a possibility that the probe will continue to return science data, including images, directly from Titan's surface.

Titan exploration will also be carried out by imaging radar, which passes signals through clouds or atmospheric haze and showers the surface of its target with a swath of radar pulses. Characteristics of the returned radar signals are processed to create detailed images of the terrain. Imaging radar has been used to great advantage in mapping cloud-covered regions of Earth where most other mapping instruments cannot "see" the surface, and was used on NASA's *Magellan* spacecraft to produce a global terrain map of cloud-enshrouded Venus. Near-infrared mapping may also reveal Titan's surface features.

One *Cassini* instrument will produce an image of reactions within Saturn's magnetosphere – the magnetic "bubble" that surrounds the planet. The instrument will obtain images of the active regions of the plasma envelope surrounding Saturn and its moons, including Titan. This pioneering investigation will open a new observational window into the study of the Saturn system. An array of sensors will study the magnetic field structure, the neutral and charged particles, and the radio waves generated by interactions of the magnetosphere with the solar wind and bodies within the Saturn system.

The *Huygens* probe carries six science instruments to sample Titan's atmosphere and surface properties.

Rings Upon Rings

Saturn's magnificent rings are, of course, a leading target for study. Explorations by the *Voyager 1* and 2 spacecraft showed that the rings are made up of thousands of individual ringlets – which themselves were found to be largely composed of ice particles ranging in size from those of sugar grains to large boulders. Slight color variations indicate that the rings include some rocky material.

Even these short-term *Voyager* studies showed a wide range of unexplained phenomena in the rings, including various wave patterns, small and large gaps, clumping of material, dark "spokes," and small "moonlets" embedded in the rings. Did moons break apart to provide the source of ring material as a result of comet or meteoroid strikes, or from tidal effects within Saturn's strong gravity field? Long-term, close-up investigation of the rings by *Cassini*'s instruments will help scientists answer these questions and better understand the processes that led to the formation and evolution of this relatively young ring system.

As Saturn orbits the Sun, the changing angle of sunlight illuminating the rings dramatically alters their visibility. *Cassini*'s arrival at Saturn is timed for optimum viewing of the rings, during a period when they will be well illuminated by sunlight. Orbiting Saturn, *Cassini* will be able to detect small moonlets inside the rings, determine the compositions of the particles, study the interactions of the rings with the magnetosphere, and conduct intensive observations of ring dynamics over a four-year period.

Saturn has 18 known satellites. This composite image includes six. Seen clockwise from upper right are Titan, Mimas, Tethys, Dione, Enceladus and Rhea.

Applied to larger-scale disk-type systems, the detailed studies of Saturn's rings by *Cassini* will provide important contributions to theories about the origin and evolution of the dust and gas from which the planets formed. Additional consideration of the Saturn system as a microcosm may be applicable to examinations of even larger disk systems so common in the universe, including our home galaxy, the Milky Way.

The Icy Satellites

Among the icy satellites to be explored is Enceladus, which is made almost entirely of water ice and has great regional differences in the number of impact craters on its surface. *Cassini* will determine if Enceladus has some internal heat source that melts the ice enough to erase impact craters. Instruments will search Enceladus for small, geyser-like volcanoes. Some scientists suspect that such volcanoes may shoot ice particles into space, where they are captured by Saturn's gravity and become part of the planet's E-ring.

The moon Iapetus will be studied because of its unique surface. Half the moon is snow-bright, while the other half is as dark as asphalt and thought to contain complex organic compounds. *Cassini* will help determine the surface composition of Iapetus, discover the nature of the dark matter, and determine whether the material came from within the moon or was deposited from another source.

For More Information

Cassini Program Outreach
Jet Propulsion Laboratory, MS 264-441
4800 Oak Grove Drive
Pasadena, CA 91109-8099
Telephone: (818) 354-5011

Visit the Cassini Web Site
http://www.jpl.nasa.gov/cassini/

National Aeronautics and
Space Administration

Jet Propulsion Laboratory
California Institute of Technology
Pasadena, California

JPL 400-776 5/99

Huygens Landed with a Splat

January 18, 2005 (Source: ESA)

Although *Huygens* landed on Titan's surface on 14 January, activity at ESA's European Space Operations Centre (ESOC) in Darmstadt, Germany, continues at a furious pace. Scientists are still working to refine the exact location of the probe's landing site.

While *Huygens* rests frozen at -180 degrees Celsius on Titan's landscape, a symbolic finale to the engineering and flight phase of this historic mission, scientists have taken little time off to eat or sleep.

They have been processing, examining and analysing data, and sometimes even dreaming about it when they sleep. There's enough data to keep *Huygens* scientists busy for months and even years to come.

One of the most interesting early results is the descent profile. Some 30 scientists in the Descent Trajectory Working Group are working to recreate the trajectory of the probe as it parachuted down to Titan's surface.

The descent profile provides the important link between measurements made by instruments on the *Huygens* probe and the *Cassini* orbiter. It is also needed to understand where the probe landed on Titan. Having a profile of a probe entering an atmosphere on a solar system body is important for future space missions.

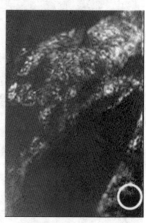

First "Best-Guess" View of Huygens Landing Site
Credit: ESA / NASA / JPL / University of Arizona

After *Huygens*' main parachute unfurled in the upper atmosphere, the probe slowed to a little over 50 metres per second, or about the speed you might drive on a motorway.

In the lower atmosphere, the probe decelerated to approximately 5.4 metres per second, and drifted sideways at about 1.5 metres per second, a leisurely walking pace.

"The ride was bumpier than we thought it would be," said Martin Tomasko, Principal Investigator for the Descent Imager / Spectral Radiometer (DISR), the instrument that provided *Huygens*' stunning images among other data.

The probe rocked more than expected in the upper atmosphere. During its descent through high-altitude haze, it tilted at least 10 to 20 degrees. Below the haze layer, the probe was more stable, tilting less than three degrees.

Tomasko and others are still investigating the reason for the bumpy ride and are focusing on a suspected change in wind profile at about 25 kilometres altitude.

The bumpy ride was not the only surprise during the descent.

Scientists had theorised that the probe would drop out of the haze at between 70 and 50 kilometers. In fact, *Huygens* began to emerge from the haze only at 30 kilometers above the surface.

When the probe landed, it was not with a thud, or a splash, but a "splat." It landed in Titanian "mud."

"I think the biggest surprise is that we survived landing and that we lasted so long," said DISR team member Charles See. "There wasn't even a glitch at impact. That landing was a lot friendlier than we anticipated."

DISR's downward-looking High Resolution Imager camera lens apparently accumulated some material, which suggests the probe may have settled into the surface. "Either that, or we steamed hydrocarbons off the surface and they collected onto the lens," said See.

"The probe's parachute disappeared from sight on landing, so the probe probably isn't pointing east, or we would have seen the parachute," said DISR team member Mike Bushroe.

When the mission was designed, it was decided that the DISR's 20-watt landing lamp should turn on 700 meters above the surface and illuminate the landing site for as long as 15 minutes after touchdown.

"In fact, not only did the landing lamp turn on at exactly 700 meters, but also it was still shining more than an hour later, when *Cassini* moved beyond Titan's horizon for its ongoing exploratory tour of the giant moon and the Saturnian system," said Tomasko.

Cassini-Huygens
MISSION TO SATURN & TITAN

Mosaic of River Channel and Ridge Area on Titan

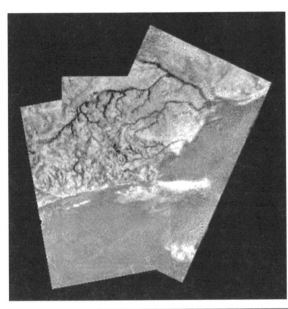

This mosaic of three frames from the Huygens Descent Imager / Spectral Radiometer (DISR) instrument provides unprecedented detail of the high ridge area including the flow down into a major river channel from different sources.

The Descent Imager / Spectral Radiometer is one of two NASA instruments on the probe.

The Cassini-Huygens mission is a cooperative project of NASA, the European Space Agency and the Italian Space Agency. The Jet Propulsion Laboratory, a division of the California Institute of Technology in Pasadena, manages the Cassini-Huygens mission for NASA's Science Mission Directorate, Washington, D.C. The *Cassini* orbiter and its two onboard cameras were designed, developed and assembled at JPL. The Descent Imager / Spectral team is based at the University of Arizona, Tucson, Arizona.

Image Credit: ESA / NASA / JPL / University of Arizona.

Composite of Titan's Surface Seen During Descent

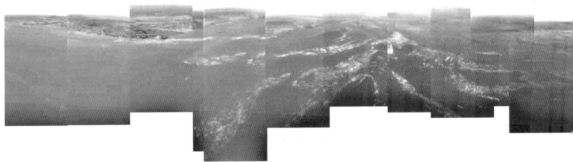

Image Credit: ESA / NASA / University of Arizona.

This composite was produced from images returned January 14, 2005, by the European Space Agency's *Huygens* probe during its successful descent to land on Titan. It shows a full 360-degree view around *Huygens*. The left-hand side, behind *Huygens*, shows a boundary between light and dark areas. The white streaks seen near this boundary could be ground "fog," as they were not immediately visible from higher altitudes.

As the probe descended, it drifted over a plateau (center of image) and was heading towards its landing site in a dark area (right). From the drift of the probe, the wind speed has been estimated at around 6-7 kilometers (about 4 miles) per hour.

These images were taken from an altitude of about 8 kilometers (about 5 miles) with a resolution of about 20 meters (about 65 feet) per pixel. The images were taken by the Descent Imager / Spectral Radiometer, one of two NASA instruments on the probe.

The *Cassini* orbiter and its two onboard cameras were designed, developed and assembled at the Jet Propulsion Laboratory, a division of the California Institute of Technology in Pasadena. The Descent Imager / Spectral team is based at the University of Arizona, Tucson, Arizona.

DEEP SPACE I
MISSION

NATIONAL AERONAUTICS AND SPACE ADMINISTRATION

Deep Space 1 Asteroid Flyby
Press Kit
July 1999

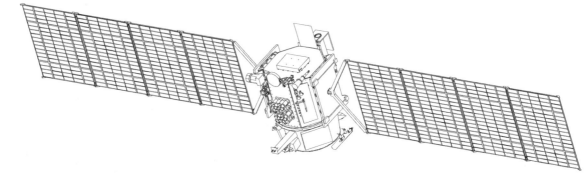

Contacts

Douglas Isbell, Headquarters, Washington, DC	Policy / Program Management	(202) 358-1753
Franklin O'Donnell, Jet Propulsion Laboratory, Pasadena, California	Deep Space 1 Mission	(818) 354-5011
John G. Watson, Jet Propulsion Laboratory, Pasadena, California	Deep Space 1 Mission	(818) 354-0474

CONTENTS

RELEASE:

Deep Space I Set to Fly By Asteroid 9969 Braille

With its technology testing objectives almost fully accomplished, NASA's Deep Space I mission is about to undergo its most comprehensive challenge: the exotic spacecraft is set to fly within 15 kilometers (10 miles) of the newly named asteroid 9969 Braille on July 29 (July 28 Pacific Daylight Time), the closest encounter with an asteroid ever attempted.

Deep Space I will rely on its experimental autonomous navigation system, or AutoNav, to guide the spacecraft past the mysterious, little-known space rock at 04:46 a.m. Universal Time July 29 (9:46 p.m. July 28 Pacific Daylight Time) at a relative speed of nearly 56,000 km/hr (35,000 mph).

"Testing advanced technologies for the benefit of future missions is the purpose of Deep Space I, so we view the flyby and its science return as a bonus," said Dr. Marc Rayman, Deep Space I's chief mission engineer and deputy mission manager. "This ambitious encounter is a high-risk endeavor whose success is by no means guaranteed. But should there be significant data return, the findings will be of great interest to the science community."

Asteroid Braille was previously known by its temporary designation of 1992 KD. The new name was announced today by the Planetary Society, Pasadena, California, as the result of a naming contest focused on inventor themes which drew more than 500 entries from around the world.

The winning entry was submitted by Kerry Babcock of Port Orange, Florida. Eleanor Helin, who co-discovered the asteroid with fellow astronomer Kenneth Lawrence, made the final decision on the name. Helin and Lawrence are astronomers at NASA's Jet Propulsion Laboratory, Pasadena, California, which also manages Deep Space I.

During the encounter, Deep Space I will be in the ecliptic plane (the plane in which Earth and most other planets orbit the Sun), moving more slowly than the asteroid, which will be progressing up through the ecliptic plane from below. Thus it may well be more apt to say that the asteroid will zoom by Deep Space I than the reverse.

The flyby will allow final testing of AutoNav, which enables the spacecraft to use images of distant stars and asteroids within our solar system to keep track of its location in space and to guide trajectory changes it needs to remain on course. Deep Space I has successfully completed testing of its 11 other new technologies.

The asteroid and the space environment surrounding it make scientifically interesting targets for two advanced, highly integrated science instruments aboard Deep Space I. During the flyby, an integrated spectrometer and imaging instrument is scheduled to send back black-and-white photographs as well as images taken in infrared light, while a second instrument that studies the three-dimensional distribution of ions and electrons, or plasma, will conduct several investigations.

In addition to their value for engineering future space missions, images and other data returned from this encounter will greatly assist scientists in their understanding of the fundamental properties of asteroids. Although scientists believe its diameter is approximately 1 to 5 kilometers (0.6 to 3 miles), they know little else about the object. With this flyby, they can learn more about its shape, size, surface composition, mineralogy and terrain.

Launched on October 24, 1998, from Cape Canaveral Air Station, Florida, Deep Space I is the first mission under NASA's New Millennium Program, which tests new technologies for future space and Earth-observing missions. The technologies that have been tested on Deep Space I will help make future science spacecraft smaller, less expensive, more autonomous and capable of more independent decision making so that they rely less on tracking and intervention by ground controllers.

The mission has exceeded almost all of its technology testing requirements by conducting more extensive tests than had been planned. As one dramatic example, the spacecraft's experimental xenon ion engine, which was required to thrust for a minimum of 200 hours, has been operated for more than 1,800 hours to date.

Deep Space I is budgeted at $152 million, including design, development, launch and operations. The mission is managed for NASA's Office of Space Science by the Jet Propulsion Laboratory, Pasadena, California. JPL is a division of the California Institute of Technology.

(End of General Release)

Media Services Information

NASA Television Transmission

NASA Television is broadcast on the satellite GE-2, transponder 9C, C band, 85 degrees west longitude, frequency 3880.0 MHz, vertical polarization, audio monaural at 6.8 MHz. A schedule of programming is available on the Internet at http://www.nasa.gov/ntv.

Status Reports

Status reports on mission activities for Deep Space 1 are issued by the Jet Propulsion Laboratory's Media Relations Office as events dictate. They may be accessed on line as noted below. Audio status reports from the Deep Space 1 project are available by calling (800) 391-6654 or (818) 354-2410.

Briefing

An overview of results from the asteroid flyby will be presented in a news briefing broadcast on NASA Television originating from NASA's Jet Propulsion Laboratory, Pasadena, California, on August 3, 1999, at 10 a.m. PDT.

Internet Information

Extensive information on Deep Space 1, including an electronic copy of this Press Kit, press releases, fact sheets, status reports and images, is available at http://www.jpl.nasa.gov/ds1news. The Deep Space 1 mission also maintains a home page at http://nmp.jpl.nasa.gov/ds1/.

Quick Facts

Spacecraft

- Dimensions: Core bus 1.1 meters deep by 1.1 meters wide by 1.5 meters high (3.6 by 3.6 by 4.9 feet); with all instruments and blankets attached, 2.1 by 1.7 by 2.5 meters (6.9 by 5.6 by 8.2 feet); with solar panels deployed, overall width 11.8 meters (38.6 feet)
- Weight: 486 kilograms (1,071 pounds) total, composed of a 373-kilogram (822-pound) dry spacecraft plus 31 kilograms (68 pounds) hydrazine fuel and 82 kilograms (181 pounds) xenon
- Power: 2,500 watts from two solar array wings

Advanced Technologies

- 1. Ion Propulsion System; 2. Solar Concentrator Arrays

Autonomy:
- 3. Autonomous Navigation; 4. Remote Agent; 5. Beacon Monitor

Science instruments:
- 6. Miniature Integrated Camera Spectrometer; 7. Plasma Experiment for Planetary Exploration

Telecommunications:
- 8. Small Deep-Space Transponder; 9. Ka-Band Solid State Power Amplifier

Microelectronics:
- 10. Low-Power Electronics; 11. Multifunctional Structure; 12. Power Actuation and Switching Module

Mission

- Launch: October 24, 1998, at 8:08 a.m. Eastern Daylight Time from Cape Canaveral Air Station, Florida
- Launch vehicle: Delta II, Model 7326
- Primary technology testing: October-December 1998
- End of prime mission: September 18, 1999

Asteroid

- Object: Asteroid 9969 Braille (formerly temporarily designated as 1992 KD)
- Discovered: May 27, 1992 by Eleanor Helin and Kenneth Lawrence
- Estimated diameter: 1 to 5 kilometers (0.6 to 3 miles)
- Time for asteroid to orbit once around Sun: 1,308.3 days (3.58 years)
- Time for asteroid to rotate once: 9.4 days
- Asteroid's closest approach to Sun: 198 million kilometers (123 million miles)
- Asteroid's farthest distance from Sun: 502 million kilometers (312 million miles)

Asteroid Flyby

- Closest approach: July 29, 1999 at 04:46 Universal Time (July 28 at 9:46 p.m. Pacific Daylight Time)
- Flyby distance: 15 kilometers (9.3 miles)

- Flyby speed: 56,000 km/hr (35,000 mph)
- Distance from Earth at time of flyby: 188 million kilometers (117 million miles)
- One-way speed of light time from spacecraft to Earth during flyby: 10 minutes, 28 seconds
- Distance from Sun at time of flyby: 199 million kilometers (124 million miles)

Program

- Cost of mission: $94.8M prelaunch development; $10.3M mission operations; $3.7M science; total $108.8 million (not including launch service)

The New Millennium Program

NASA has an ambitious plan for space exploration in the next century. The agency foresees launching frequent, affordable missions with spacecraft boasting revolutionary new capabilities compared to those of today. Spacecraft are envisioned as flying in formation, or featuring artificial intelligence to provide the kind of capability that can answer the more detailed level of questions that scientists have about the universe.

The goal of the New Millennium Program is to identify and test advanced technologies that will provide spacecraft with the capabilities they need in order to achieve NASA's vision. Technologies such as ion propulsion and artificial intelligence promise a great leap forward in terms of future spacecraft capability, but they also present a risk to missions that use them for the first time.

Through a series of deep space and Earth-observing flights, the New Millennium Program will demonstrate these promising but risky technologies in space in order to "validate" them – that is, to prove that they work, or to determine what problems may crop up. Once validated, the technologies pose less of a risk to mission teams that would like to use them to achieve their scientific objectives.

The testing of advanced technologies is the basic requirement for these missions. As a bonus, the missions can also collect science data as the advanced technologies are put through their paces. Science, however, is secondary to the technology testing on New Millennium's missions.

Created in 1994, the New Millennium Program forms partnerships among organizations in government, private industry, academia and the nonprofit sector so that the expertise and know-how of scientists, engineers, and managers can be pooled and used as a resource to meet the program's goals.

New Millennium's solicitation of advanced technologies for its missions will also stimulate the development of technologies around the nation and will strengthen the nation's technological infrastructure, making it more competitive in the global market. Many technologies will also have commercial spin-offs that will benefit the public in their daily lives.

Integrated Product Development Teams

The concept of integrated product development teams was developed in the commercial sector by the aircraft and automotive industries. Such teams bring together members of different departments within an organization, such as sales, manufacturing and design, to work together to develop a product. This kind of concurrent decision-making team has made it possible for industries to manufacture products of better quality and competitive costs for their customers.

The New Millennium Program has taken this intra-organizational team concept to a higher level and used it to bring together diverse organizations. For the development of *Deep Space 1* and *Deep Space 2*, it created six integrated product development teams that include technologists from government, private industry, academic and nonprofit sectors across the nation. In effect, they represent the U.S. technology development community.

The teams were formed to develop technologies and concepts for six key areas of space flight:

- Autonomy. If spacecraft are capable of making more decisions on their own, they require less frequent tracking and intervention by ground controllers.
- Telecommunications. These technologies improve the communications link between the spacecraft and Earth.
- Microelectronics. New chips and circuits allow engineers to shrink down science instruments and other spacecraft subsystems, saving size and mass.
- "In Situ" Instruments and Microelectromechanical Systems. "In situ" instruments study a celestial body directly rather than at a distance.
- Instrument Technologies and Architectures. This team develops new technologies for science instruments such as cameras and radiometers, as well as seeking entirely different ways of making the same science observations or measurements.
- Modular and Multifunctional Systems. This team is continuing and accelerating an existing trend toward combining spacecraft's electronics more closely with their mechanical system of trusses, supports, etc.

The technologists were encouraged to search the nation's development programs to find advanced technologies that will provide the capabilities needed to achieve NASA's vision of space exploration in the 21st century.

The membership of each team represents a considerable range of organizations. Technologists come from aerospace companies, small businesses, non-NASA government laboratories, NASA field centers and nonprofit organizations. The diversity of organizations and the resulting interorganizational partnerships capitalize on and effectively take advantage of the nation's overall investment in advanced technology.

Mission Overview

Deep Space 1's mission was most intense during the weeks immediately following its launch in October 1998, when most of the 12 technologies it carries were actively tested. The spacecraft will fly by asteroid 9969 Braille on July 29, 1999 (July 28 Pacific time).

Highlights to Date

Launch. *Deep Space 1* was launched October 24, 1998, at 8:08 a.m. Eastern Daylight Time from Cape Canaveral Air Station, Florida on a variant of the Delta II launch vehicle known as a Delta 7326, the first use of this new and lower-cost model. The launch took place from Space Launch Complex 17A at Cape Canaveral Air Station, Florida.

The Delta 7326 is a liquid fuel rocket augmented by three solid rocket motors. The second stage was restartable, and performed two separate burns during *Deep Space 1*'s launch to place the spacecraft in Earth orbit. A third burn was executed to finalize the orbit for the Delta's secondary payload, a student-built satellite called SEDSat-1 developed by Students for the Exploration and Development of Space. At 49 minutes after launch, the third and final stage of the Delta – a Thiokol Star 37FM solid-fuel booster – fired to send *Deep Space 1* into orbit around the Sun.

Early Cruise. On its way out of Earth's orbit of the Sun into a wider orbit between Earth and Mars, *Deep Space 1* tested all 12 of its new technologies. Final testing of the autonomous navigation experiment, or AutoNav, will take place during the asteroid flyby.

Highlights have included thrusting the ion engine for more than 1,800 hours; progressively increasing AutoNav's autonomy; and observing the Remote Agent software play doctor to self-diagnose a minor glitch that could have otherwise cut short its experimental lifetime.

Asteroid Flyby

Deep Space 1 will fly past asteroid 9969 Braille on July 29, 1999 at 04:46 Universal Time (July 28 at 9:46 p.m. Pacific Daylight Time). The flyby will allow the spacecraft to test its autonomous navigation system, or AutoNav, and will also provide its two science instruments with an interesting target. The encounter is not required for the testing of *Deep Space 1*'s technologies, but will allow engineers an extra chance to observe the performance of the technologies under conditions similar to what would be experienced on future science missions.

The Asteroid. Mission planners considered more than 100 possible asteroids or comets to target before settling on asteroid 9969 Braille, previously known by its temporary designation of 1992 KD. The asteroid was discovered May 27, 1992, by astronomers Eleanor Helin and Kenneth Lawrence of NASA's Jet Propulsion Laboratory using the 46-centimeter (18-inch) Schmidt telescope at Palomar Observatory while scanning the skies as part of the Palomar Planet-Crossing Asteroid Survey. At the time it was discovered, asteroid Braille appeared as a streak traveling north-northeast in the constellation Libra with a magnitude of 15.5, much too dim to see with the unaided eye. At the time, the asteroid was about 38.4 million kilometers (23.9 million miles) from Earth.

In 1999, the Planetary Society sponsored a contest inviting the public to propose names for the asteroid, based on a theme of inventors. The society forwarded 10 finalists to the asteroid's co-discoverer, Eleanor Helin, who selected the name Braille submitted by Kerry Babcock of Port Orange, Florida, a software engineer who works as a contractor at NASA's Kennedy Space Center. The name honors Louis Braille (1809-1852), the blind French educator who developed the system of printing and writing named for him that is extensively used by the blind. The number 9969 in the asteroid's formal name indicates that it is the 9,969th asteroid to be numbered since the first asteroid, Ceres, was discovered in 1801. The name 9969 Braille was accepted by the International Astronomical Union's Small Bodies Naming Committee just a few days before *Deep Space 1*'s flyby.

Asteroid Braille orbits the Sun outside of the main asteroid belt between Mars and Jupiter. Astronomers calculated that its orbit is highly elliptical, or shaped like a big loop. The closest it gets to the Sun (called its "periapsis") is a point midway between Earth and Mars, while at its most distant point (its "apoapsis") it is more than three times further from the Sun than Earth, or more than halfway out to the giant planet Jupiter. Much of the time, the asteroid is a considerable distance above or below the ecliptic plane, the plane in which Earth and most other planets orbit the Sun.

The asteroid's speed changes as it moves through its orbit. When it is close to the Sun, it moves at 31 kilometers per second (69,480 mph) relative to the Sun, whereas at its most distant it slows down to 12.2 kilometers per second (27,360 mph) relative to the Sun. One loop around the Sun takes 1,308.3 days, or 3.58 years. At the time of *Deep Space 1*'s flyby, when asteroid Braille will be 6.3 days past its closest approach to the Sun, the asteroid's velocity will be 30.9 kilometers per second (68,400 mph) relative to the Sun. Its diameter is estimated as approximately 1 to 5 kilometers (0.6 to 3 miles). Scientists know that the asteroid is elongated, but they don't know if it's one large chunk or a rubble pile.

Astronomers know little else with certainty about asteroid Braille; it is the least-studied asteroid that has ever been the target of a spacecraft flyby. It is very dim, making it difficult to observe. In late 1998, a team of astronomers led by Drs. Michael D. Hicks and Bonnie Buratti of the Jet Propulsion Laboratory used a number of telescopes equipped with special filters and detectors to study the asteroid. Scientists scrutinize the colors of light given off by celestial objects to look for clues called absorption lines that reveal what the object is made out of. In the 1998 observations, Braille exhibited colors that suggest it is may contain the greenish mineral olivine and/or pyroxene, a group of minerals that vary from white to dark green or black. This would suggest that Braille is more similar in makeup to the stony meteorites that fall to Earth than most asteroids in the main asteroid belt are.

Some astronomers theorize that asteroid Braille may have been knocked off of Vesta. Discovered in 1807, Vesta is the brightest asteroid (though not the largest), sometimes visible to the unaided eye; it is about 390 kilometers (240 miles) in diameter, circling the Sun in the main asteroid belt. One theory holds that a hole was knocked into Vesta when it collided with another body, and that the debris from this collision provides the raw materials for most of the meteorites that fall to Earth.

Primary Mission Trajectory

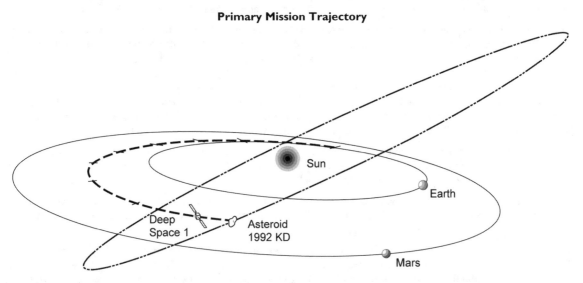

The fact that Braille's orbit is tilted so much relative to the solar system's ecliptic plane shows that it must have been disrupted at some point. If it was in fact knocked off an asteroid in the main belt like Vesta, it would have joined other debris or particles from that larger body that spewed off in all directions. Astronomers determined in June 1999 – only a month and a half before *Deep Space 1*'s flyby – that Braille rotates at a relatively slow rate of once every 9.4 days. Some scientists think that the fact that it rotates this slowly suggests it is a young object, lending more evidence to the theory that it was knocked off a larger body relatively recently in terms of the age of the solar system. Most asteroids rotate once every 6 to 9 hours; if one is found to spin much faster or slower, astronomers assume that it must have been sped up or slowed down by a relatively recent collision.

With the *Deep Space 1* flyby, space scientists will be able to learn more about asteroid Braille's shape, size, surface composition, terrain and, perhaps, how it interacts with the solar wind, a flow of charged particles constantly streaming away from the Sun. The spacecraft's two science instruments will be used to gather images and other data on the asteroid and its environment.

Approach Sequence. In the final 30 days of the spacecraft's approach to the asteroid, *Deep Space 1*'s AutoNav experiment has been taking more frequent pictures of so-called "beacon" asteroids to help it navigate. Most of these are objects in the main asteroid belt that are bright enough to be seen in the camera and whose positions are sufficiently well known to be useful for navigation. Asteroid Braille itself is too dim to be seen by the spacecraft until perhaps a day before the closest approach. During this month, the spacecraft has also been firing its thrusters in a series of increasingly frequent trajectory correction maneuvers to control the targeting of the final encounter.

Deep Space 1 will send its final session of data to Earth about seven hours before closest approach. During this transmission, the dish-shaped high-gain antenna must be pointed toward Earth. Following this session, the spacecraft will turn to point its camera toward the asteroid. Six minor trajectory correction maneuvers are scheduled as needed in the 48 hours preceding the flyby, when the spacecraft is likely to see the asteroid for the first time.

The last opportunity for a trajectory correction maneuver will be three hours before closest approach. During the final approach, the spacecraft's camera will both take navigation pictures for AutoNav and collect images and spectra for scientific purposes. The late navigation images will contain information that AutoNav needs to provide rapid updates to its estimates of the distance to the asteroid, critical for keeping the asteroid in the camera's field of view. The camera may be able to obtain pictures of the asteroid's surface with resolutions as high as about 30 to 50 meters (100 to 150 feet) per pixel.

Although the spacecraft will not linger as it flashes past the asteroid, there will be opportunities for it to take scientifically interesting pictures during the last five minutes as it approaches asteroid Braille. As the asteroid and spacecraft close in on each other, *Deep Space 1*'s autonomous navigation system will change the spacecraft's orientation in the last five minutes before the encounter by approximately 2 to 3.5 degrees to keep its camera oriented toward the quickly approaching body.

In planning the encounter, scientists had to select tradeoffs between observations by the spacecraft's camera and plasma instrument. In order to take pictures, the spacecraft must keep its camera pointed at the asteroid by frequently firing its onboard thrusters. The plasma instrument, however, cannot collect high-quality data when the thrusters are fired often. Scientists decided to take pictures as the spacecraft approaches the asteroid, and then switch to collecting plasma data at the closest moments of the flyby. At this point, pictures would very probably be blurry, while this is the only time that the plasma instrument would be able to detect any faint magnetic field at the asteroid, if any exists.

With that in mind, the camera will stop taking photos and the spacecraft will reduce the frequency of its thruster firings about 20 to 25 seconds before closest approach, when the spacecraft is about 350 kilometers (220 miles) from the asteroid. The plasma instrument will then collect data as *Deep Space 1* sails past the asteroid.

Flyby Geometry. At the time of the flyby, the spacecraft will be in the ecliptic plane, the plane in which Earth and most of the other planets orbit the Sun. The spacecraft will be moving more slowly than the asteroid, which, in broad terms, will be progressing up through the ecliptic plane from below. Thus in some ways it may be more apt to say that the asteroid will zoom by *Deep Space 1* than the other way around.

Once the asteroid has passed by, the spacecraft will take about 15 minutes to execute a complex 180-degree turn to watch and record the asteroid as it moves away, sailing above the ecliptic plane. This maneuver gives the spacecraft the opportunity to study the opposite side of the asteroid.

The asteroid has far too little gravity to cause the spacecraft's flight path to turn or bend in any measurable way, as other spacecraft do when they execute "gravity assist" flybys of planets. Although scientists are unsure of Braille's surface gravity, it is estimated to be in the range of $1/10,000$ that of Earth.

During the encounter, *Deep Space 1*'s autonomous navigation experiment will attempt to guide the spacecraft to within 15 kilometers (about 10 miles) of the asteroid's center, making it the closest flyby of a solar system body ever attempted. During the flyby, the spacecraft will be moving at a velocity of about 24.2 kilometers per second (54,000 mph) relative to the Sun and 15.5 kilometers per second (34,560 mph) relative to the asteroid.

Deep Space 1 will fly past the dark side of the asteroid facing away from the Sun, passing through its shadow or umbra. Much as the asteroid has an optical shadow, it also casts a "shadow" marked by an absence of the charged particles that make up the solar wind constantly streaming outward from the Sun. The spacecraft will sample this area as it flies by.

Post-Encounter Period. The spacecraft is expected to take two to three days to transmit to Earth all of its technology testing and science data from the flyby. Starting a day and a half after the flyby, the spacecraft will start thrusting with its ion engine while keeping its dish-shaped high-gain antenna pointed at Earth. Twelve days after the flyby, the spacecraft will turn to a new orientation so that thrusting from the ion engine will shape its flight path for a possible comet flyby in 2001 if NASA elects to extend the spacecraft's mission.

End of Prime Mission

Deep Space 1's prime mission will conclude September 18, 1999. By this time, all of the technology testing will have been accomplished.

If the spacecraft is healthy and NASA chooses to continue the mission, an extended mission may be conducted. At the end of the primary mission, *Deep Space 1* will be on a trajectory that could result in two scientifically interesting flybys within the following two years. One would be of an object known as comet Wilson-Harrington; this body is believed to be either a dormant comet or a "transition object" that is in the process of changing from a comet to an asteroid.

Wilson-Harrington, which *Deep Space 1* could fly by in January 2001, has not been observed to behave like a comet – spewing gas with a coma and tail – since 1949; it is very unusual for a comet to exhibit this type of change in behavior.

The second possible flyby target of an extended mission, comet Borrelly, is one of the most active comets that regularly visit the inner solar system. *Deep Space 1* could fly by Borrelly in September 2001.

Deep Space Network and Ground Support

Communication with *Deep Space 1* is enabled by two major systems on the ground: the Deep Space Network and the Advanced Multimission Operations System.

The Deep Space Network provides the vital two-way communications link that both controls the spacecraft and receives telemetry. The network consists of three complexes located approximately 120 degrees of longitude apart around the world: at Goldstone in California's Mojave Desert; near Madrid, Spain; and near Canberra, Australia. This spacing insures that one antenna is always within view of a spacecraft as Earth rotates. The stations feature precision-pointed, high-gain, parabolic reflector antennas; some of these giant dishes are 34 meters (112 feet) in diameter, while the most sensitive are 70 meters (230 feet) in diameter. They work together with high-power transmitters and ultra-low-noise amplifiers to optimize the communication link with spacecraft millions of kilometers (miles) away. All data gathered by antennas at the three complexes are communicated directly to the control center at the Jet Propulsion Laboratory, the operations hub for the network. Voice and data traffic between various locations is sent via land lines, submarine cable, microwave links or communications satellites.

The Advanced Multimission Operations System is an integrated ground system at JPL which provides a common set of mission operations services and tools to space projects. These allow engineers to carry out mission planning and analysis, develop pre-planned sets of commands to be sent to the spacecraft, perform trajectory calculations to check the autonomous navigation system on the spacecraft, and process data transmitted to Earth from the spacecraft. The system also provides capabilities to display and analyze key measurements from the spacecraft, such as readings of temperature, pressure and power. Other mission operations services include simulation of telemetry and command data, data management and retrieval of all data types. *Deep Space 1* is the first mission to rely heavily on the operations system's multimission capability for mission operations.

Compared to some other solar system spacecraft, *Deep Space 1* has made many decisions on its own and has required relatively infrequent intervention from ground controllers. For long periods of the mission, *Deep Space 1* will be tracked only once per week for normal telemetry dumps and command loads during non-prime hours. This ability to use tracking passes as available will help reduce competition between spacecraft projects for antenna time on the Deep Space Network system.

Deep Space Network engineers are also very interested in *Deep Space 1*'s successful tests of its small deep space transponder and Ka-band solid state power amplifier, technologies that will improve telecommunications for future spacecraft missions.

Student Involvement

Boys & Girls Clubs. In an effort to reach children who traditionally have not had much access to the marvels of space exploration, the Deep Space 1 mission has sponsored a "Picture Yourself in the New Millennium" activity with the Boys & Girls Clubs of America. The 1,800 clubs nationwide have a membership of 2.6 million children from inner city, underprivileged backgrounds.

Based on a discussion of what a millennium means, the progress of technology in the past century, and how dramatically they expect life to change because of technology in the next century, youths in the program have created drawings and written poems about life in the new millennium. Approximately 1,000 entries and the names of all the children participating were scanned onto a CD-ROM that is flying on *Deep Space 1*. Spectrum Astro Inc. designed certificates to acknowledge the children's contributions. Deep Space 1 and the New Millennium Program will be sponsoring more activities with the clubs in the future.

ITEA. The International Technology Education Association (ITEA), the nation's largest professional education association devoted to technology education in kindergarten through 12th grade, has formed an educational partnership with JPL. Among the joint projects resulting from this alliance is a web site, "The Space Place," designed to introduce students and their teachers to some of the latest, most advanced technologies being tested on New Millennium Program missions for use on space missions of the future. The Space Place is accessible at http://spaceplace.jpl.nasa.gov. The Deep Space 1 project team has developed several curriculum supplements which have been published in ITEA's publication Technology Teacher. These include "Ion-Drive Your Way Through Space" and "We've Got Algo-Rhythm."

In the summer of 1998, both the Boys and Girls Clubs of America and ITEA participated in a Deep Space 1 "National Countdown" activity, through which thousands of school children took part in a three-tiered exercise: setting goals, identifying ways of achieving them and, finally, pledging to take action to turn these goals into reality. Youths from four Boys and Girls Clubs from the area near Cape Canaveral, Florida, participated in a pre-launch event at Kennedy Space Center. The kids shared their goals with several NASA scientists and then packed a time capsule to be opened in the year 2048.

Science Objectives

Unlike most solar system missions, Deep Space 1's main purpose is to test new technologies rather than visit or make observations of celestial objects. Even so, the spacecraft's science instruments will collect valuable data, particularly during the asteroid flyby. This encounter will allow engineers an extra chance to observe the performance of the technologies under conditions similar to what would be experienced on future science missions.

Among the 12 new technologies that Deep Space 1 has been testing are two advanced science instruments that will collect information during the flyby. One, the Miniature Integrated Camera Spectrometer (MICAS), is a package that combines two cameras with spectrometers that analyze ultraviolet and infrared energy thrown off by celestial objects. The other experiment, called the Plasma Experiment for Planetary Exploration (PEPE), combines multiple instruments that study space plasma, or ions and electrons that flow outward from the Sun and elsewhere in space. Detailed information on the hardware of these experiment packages is found in the section on technologies in this Press Kit.

The Camera-Spectrometer Team

Pictures taken by Deep Space 1's advanced camera will allow scientists to study the asteroid's general physical characteristics, such as its dimensions, shape, surface texture, brightness and diversity, and to make estimates of its mass, volume and density. High-resolution images will be particularly useful for studying the size, number and distribution of craters.

The teams using the camera-spectrometer have the following goals:

- Understand the asteroid's composition. Scientists will use the instrument's infrared spectrometer to look for the spectral signatures of the minerals that make up the asteroid. The instrument will allow them to do this at a much higher resolution than possible from ground-based telescopes. They hope that this analysis will allow them to pinpoint similarities between the asteroid and meteorites that fall to Earth. By studying how light is reflected from the asteroid, scientists will also try to form conclusions about its surface texture.
- Understand how Braille's spectral signatures correspond to those of asteroids in the main asteroid belt, based on observations with the infrared spectrometer.
- Determine whether Braille contains any "pre-biotic" materials – meaning any precursors for life such as carbon, oxygen, hydrogen or nitrogen – again using the infrared channel.
- Understand the asteroid's "morphology" – which is to say, its dimensions, form and surface relief – using both of the experiment's optical cameras, the active pixel sensor and charge coupled device.

These investigations are designed to answer questions that fall into two key categories:

- What is the overall structure, or "morphology," of this asteroid? Is it a solid chunk, or a rubble pile? Perhaps Braille and similar near-Earth asteroids are chips off of large asteroids in the main belt that have somehow drifted closer to Earth. Once the mystery of the asteroid's structure has been solved, scientists may be able to answer questions about how asteroids are formed and what kinds of events, such as collisions, lead to their transport from the asteroid belt to Earth.
- What is the asteroid's composition? There is evidence that the asteroid is unusually bright, indicating that the surface is fresher and younger than other asteroids. Its color suggests it may be closer in composition to terrestrial meteorites than to main-belt asteroids.

The Plasma Team

Electrically charged ions and electrons constantly flow through the solar system. Many of them stream outward from the Sun in a flow called the solar wind. Some gather in the powerful magnetic fields of giant planets such as Jupiter and Saturn. Still others are believed to enter the solar system from regions beyond in the galaxy at large. Deep Space 1's plasma experiment will study whether any charged particles are thrown off from the asteroid's surface as it interacts with sunlight or the solar wind. The team will also investigate the composition of any such particles that may be expelled by the asteroid. Knowing the composition of such materials would reveal information about the composition of the asteroid itself. If successful, this will be the first time that material of any kind has been directly measured coming off an asteroid. This information can be directly compared with other information about the composition of the asteroid's surface obtained from the infrared channel on Deep Space 1's camera-spectrometer.

Scientists will also attempt to measure any effects that the asteroid might have on the solar wind. This might be caused, for example, if the asteroid has a magnetic field – considered unlikely – or if it gives off a sufficient amount of charged particles. Scientists would detect such an effect if the instrument's count of electrons in the solar wind varies as the spacecraft passes the asteroid.

Deep Space I Science Team

Dr. Robert Nelson,   Jet Propulsion Laboratory, Pasadena, California, Project Scientist
Dr. Laurence Soderblom,   U.S. Geological Survey, Flagstaff, Arizona, Camera-Spectrometer, Group Leader
Dr. David T. Young,   Southwest Research Institute, San Antonio, Texas, Plasma Experiment Group Leader

Team Members:

Dr. Frances Bagenal,   University of Colorado, Boulder, Colorado
Dr. Daniel Boice,   Southwest Research Institute, San Antonio, Texas
Dr. Daniel Britt,   University of Arizona, Tucson, Arizona
Dr. Robert H. Brown,   Lunar and Planetary Laboratory, University of Arizona, Tucson, Arizona
Dr. Bonnie Buratti,   Jet Propulsion Laboratory, Pasadena, California
Dr. Wing Ip,   Max-Planck-Institut für Aeronomie, Katlenburg-Lindau, Germany
Dr. Jurgen Oberst,   DLR Institute of Planetary Exploration, Berlin, Germany
Dr. Tobias Owen,   University of Hawaii, Honolulu, Hawaii
Dr. Bill Sandel,   Lunar and Planetary Laboratory, University of Arizona, Tucson, Arizona
Dr. Alan Stern,   Southwest Research Institute, Boulder, Colorado
Dr. Nicholas Thomas,   Max-Planck-Institut für Aeronomie, Katlenburg-Lindau, Germany
Dr. Joseph J. Wang,   Jet Propulsion Laboratory, Pasadena, California
Dr. Roger Yelle,   Boston University, Boston, Massachusetts

The 12 Technologies

The technologies that have been tested on *Deep Space I* will contribute to spacecraft of the future in several ways. A number of the technologies are designed to help make future spacecraft smaller and less expensive. Others make spacecraft less dependent on tracking by NASA's Deep Space Network and intervention from ground controllers.

Ion Engine

Ion propulsion has been a technology favored by science fiction writers for decades. As imagined in television's *Star Trek* or the *Star Wars* movie series, ion drives are highly advanced devices delivering an extremely powerful thrust allowing spaceships to outrun routine vessels of the future.

The reality of ion propulsion, at least today, is very different. The ion drive on *Deep Space I* combines a gas found in photo flash units with some of the technologies that make television picture tubes work to deliver a thrust only as powerful as the pressure of a sheet of paper resting on the palm of a hand. Despite the almost imperceptible level of thrust, however, over the long haul *Deep Space I*'s ion engine can deliver up to 10 times more thrust than a conventional liquid or solid fuel rocket for a given amount of fuel.

This year, *Deep Space I*'s ion propulsion system and the team that developed it have been honored with awards from both *Popular Science* and *Discover* magazines.

Ion Propulsion Basics. The fuel used in *Deep Space I*'s ion engine is xenon, a colorless, odorless and tasteless gas more than 4½ times heavier than air. Xenon was discovered in 1898 by British chemists Sir William Ramsay and Morris W. Travers when they were distilling krypton, another noble gas that had been discovered only six weeks earlier. Xenon occurs naturally in air, but only accounts for 1 part in 10 million of Earth's atmosphere by volume. The gas is used commercially in products such as photo flash units and some lasers. The *Deep Space I* system carried a total of 82 kilograms (181 pounds) of xenon at launch.

When the ion engine is running, electrons are emitted from a hollow bar called a cathode into a chamber ringed by magnets, much like the cathode in a TV picture tube or computer monitor. The electrons strike atoms of xenon, knocking away one of the 54 electrons orbiting each atom's nucleus. This leaves each atom one electron short, giving it a net positive charge – making the atom what is known as an ion.

At the rear of the chamber are a pair of metal grids, one of which is charged positively and the other charged negatively, with up to 1,280 volts of electric potential. The force of this electric charge exerts a strong "electrostatic" pull on the xenon ions – much like the way that bits of lint are pulled to a pocket comb that has been given a static electricity charge by rubbing it on wool on a dry day. The electrostatic force in the ion engine's chamber, however, is much more powerful – causing the xenon ions to shoot past at a speed of more than 100,000 km/hr (62,000 mph), continuing right on out the back of the engine and into space. In order to keep the xenon ions from being attracted back into the engine's chamber, an electrode at the very rear of the engine emits electrons which rejoin with many of the xenon atoms speeding past to neutralize their electrical charge.

At full throttle, the ion engine would consume about 2,500 watts of electrical power and put out 90 millinewtons ($1/50$th of a pound) of thrust. This is comparable to the force exerted by a single sheet of paper resting on the palm of a hand. At the minimum possible throttling level, the engine uses 500 watts of power and puts out about 20 millinewtons ($1/200$th of a pound) of thrust. *Deep Space 1*'s solar arrays can supply a maximum of 2,500 watts of power, not enough to cover both the engine at full throttle in addition to the nearly 400 watts that the spacecraft consumes, so the ion engine will not be run at full thrust during the mission.

The engine's magic, however, lies in its staying power. Such a minute force could never be used to launch a spacecraft from Earth's surface, but it is ideal for the cruise segment of interplanetary journeys lasting months or years. With constant operation over time, an ion engine can provide substantial thrust to a spacecraft. Over the life of the primary mission, the ion engine on *Deep Space 1* will change the spacecraft's speed by a total of nearly 13,000 km/hr (more than 8,000 mph). Even more significantly, the ion engine can deliver more than 10 times more thrust than a conventional liquid or solid fuel motor for a given amount of fuel. The ion engine on *Deep Space 1* is 30 centimeters (12 inches) in diameter.

Development History. Ion propulsion – also known as solar electric propulsion because of its dependence on electricity from solar panels – has been under development since the 1950s. Dr. Harold Kaufman, a now-retired engineer at NASA's Glenn Research Center (formerly the Lewis Research Center), Cleveland, Ohio, built the first ion engine in 1959.

In 1964, a pair of NASA Lewis Research Center ion engines were launched on a Scout rocket from Wallops Island, Virginia, under the name Space Electric Rocket Test 1 (SERT 1); one of the two thrusters onboard did not work, but the other operated for 31 minutes. A follow-up mission, SERT 2, was launched in 1970 on a Thor Agena rocket from Vandenberg Air Force Base, California. SERT 2 carried two ion thrusters, one operating for more than five months and the other for nearly three months.

Many early ion engines used mercury or cesium instead of xenon. SERT 1 carried one mercury and one cesium engine, while SERT 2 had two mercury engines. Apart from the fuel, these ion drives were similar to *Deep Space 1*'s; the mercury or cesium would be turned into a gas, bombarded with electrons to ionize it, then electrostatically accelerated out the rear of the engine. But mercury and cesium proved to be difficult to work with. At room temperature, mercury is a liquid and cesium is a solid; both must be heated to turn them into gases. After exiting the ion engine, many mercury or cesium atoms would cool and condense on the exterior of the spacecraft. Eventually researchers turned to xenon as a cleaner and simpler fuel for ion engines.

Ion Propulsion System

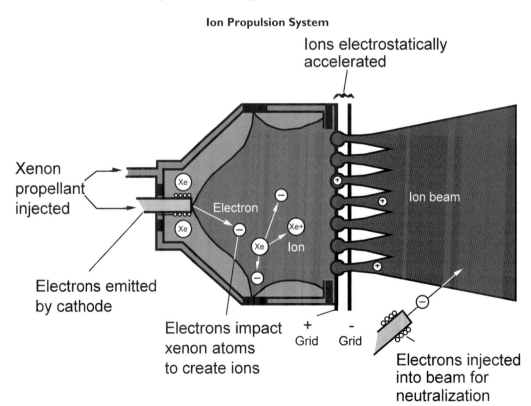

Beginning in the 1960s, the Hughes Research Laboratories, Malibu, California, conducted development work on ion engines. The first xenon ion drive ever flown was a Hughes engine launched in 1979 on the Air Force Geophysics Laboratory's Spacecraft Charging at High Altitude (SCATHA) satellite. In August 1997, Hughes launched the first commercial use of a xenon ion engine on PanAmSat 5 (PAS-5), a communications satellite launched on a Russian Proton rocket from the Baikonur Cosmodrome in Kazakhstan. This ion engine is used to maintain the position of the communications satellite in its proper orbit and orientation. Ion engines for such purposes are smaller than systems like *Deep Space 1*'s, which is designed for long-term interplanetary thrusting.

In the early 1990s, JPL and NASA Lewis partnered on an effort called the NASA Solar Electric Propulsion Technology Application Readiness (NSTAR) project. The purpose of NSTAR was to develop xenon ion engines for deep space missions. In June 1996, a prototype engine built by NASA Lewis began a long-duration test in a vacuum chamber at JPL simulating the conditions of outer space. The test concluded in September 1997 after the engine successfully logged more than 8,000 hours of operation. Results of the NSTAR tests were used to define the design of flight hardware.

Testing on Deep Space 1. After an initial problem attempting to start the ion engine on November 10, 1998, the system was successfully fired up two weeks later on November 24, and continued thrusting for two weeks uninterrupted. With this, *Deep Space 1* had already operated its thruster for a much longer uninterrupted time than any deep space probe using any other propulsion system.

From November 24 through November 30, throttle levels were commanded up from level 6 to 90, the maximum being 111, in order to gauge power consumption at various levels. At throttle level 90, it appeared that the spacecraft was near its limit for providing power to the ion propulsion system, drawing approximately 2.4 kilowatts of power.

The engine was turned off on December 8. By running the ion engine for more than the required 200 hours and successfully testing the spacecraft's solar array and transponder, the team achieved the minimum criteria that NASA established for overall mission success.

The engine was turned back on on December 11, when the team determined that the highest throttle level that could be supported was approximately 83 at the spacecraft's distance from the Sun at that time (the solar array's capabilities change as the spacecraft recedes from the Sun); at level 85, it became necessary for the spacecraft's batteries to supplement solar array power.

On January 5, 1999, the autonomous navigation system turned off the ion engine, completing the first thrust segment of the *Deep Space 1* mission. During that period, the engine accumulated more than 850 operating hours and experienced 59 recycles, which are momentary automatic interruptions, or shutoffs, of the system, primarily for the system to protect itself from damage due to drifting particulates. By contrast, in the first 850 hours of ground testing of the flight-spare ion engine, approximately 240 recycles were experienced. The lower number of recycles in flight suggests that in-space operation of an ion thruster is more benign than operation in a vacuum chamber.

On April 27, *Deep Space 1* completed a six-week period of thrusting with its ion propulsion system. It took less than 5 kilograms (under 11 pounds) of xenon to provide the steady push for the six weeks of thrusting that ended on April 27, during which the spacecraft's speed was increased by nearly 300 meters per second (about 650 mph). If the spacecraft had expended the same amount of standard rocket propellant instead of using ion propulsion, the speed would have changed by a mere 80 km/hr (50 mph). *Deep Space 1* has now completed a total of more than 75 days of thrusting.

Partners. The xenon ion flight engine was built for *Deep Space 1* by Hughes Electron Dynamics Division, Torrance, California, and Spectrum Astro Inc., Gilbert, Arizona. Other partners in the development of the flight engine included Moog Inc., East Aurora, New York, and Physical Science Inc., Andover, Massachusetts. Development of the xenon ion propulsion system was supported by NASA's Office of Space Science and Office of Aeronautics and Space Transportation Technology, Washington, DC. A portion of the NSTAR program was supported by the Advanced Space Transportation Program, managed by NASA's Marshall Space Flight Center, Huntsville, Alabama.

Solar Concentrator Arrays

Because of the ion engine's power requirements, *Deep Space 1* requires a high-power solar array. Designers met this need by combining high-performance solar cells with lenses designed to focus sunlight on them.

The spacecraft is equipped with two solar wings, each of which is composed of four panels measuring about 113 by 160 centimeters (44 by 63 inches). At launch, the wings were folded up so that the spacecraft fit into the launch vehicle's fairing; when fully extended, the wings measure 11.8 meters (38.6 feet) from tip to tip. A total of 720 cylindrical Fresnel lenses made of silicone concentrate sunlight onto 3,600 solar cells made of a combination of gallium indium phosphide, gallium arsenide and germanium.

The arrays produce 15 to 20 percent more power than most modern solar arrays of the same size – about 2,500 watts at the beginning of the mission (declining over the life of the mission as the array ages and the spacecraft recedes from the Sun) with a voltage of 100 volts.

An earlier version the solar array was included as a test on a satellite on the unsuccessful launch of the Conestoga launch vehicle in October 1995, so it was never tested in space.

Because power from the arrays was needed almost immediately, the arrays were tested within just two hours of launch. Telemetry was received from the spacecraft through NASA's Deep Space Network at 1 hour, 37 minutes after launch, and 3 minutes later it was determined that the spacecraft's arrays had been deployed.

On October 31, 1998, one day ahead of schedule, the operations team measured the arrays' electrical characteristics, particularly eight modules comprised of five solar cells each that are specially instrumented for measuring current and voltage. The following day, a more complex test was executed successfully in a 10-hour activity. The intricate choreography included rotations of the pair of solar arrays and turns of the spacecraft to vary the angle of the sunlight incident upon the arrays, all designed to determine the best angle to maximize collection of sunlight.

When the ion engine was powered on, beginning November 24, the team gathered additional test data on the arrays. This marked the first opportunity that the arrays had to provide high power, since the propulsion system has the greatest power requirements of any unit on board.

To test exactly how much power the array could generate, the operations team needed to draw as much power from the array as it could deliver. On January 22, 1999, the ion engine was powered on and raised to a high throttle level. The data that were generated contributed to understanding this advanced technology's detailed performance.

The solar concentrator array was developed by AEC-Able Engineering Inc., Goleta, California; Entech, Keller, Texas; NASA's Glenn Research Center, Cleveland, Ohio; and JPL. Technology development was sponsored by the Ballistic Missile Defense Organization, Washington, DC.

Autonomy Technologies

As more planetary spacecraft are launched into the solar system more frequently in NASA's era of smaller and more rapidly developed missions, competition increases for tracking time on the giant dish antennas of the agency's Deep Space Network. Several new technologies on *Deep Space 1* are designed to make spacecraft more self-reliant, depending less on tracking and the intervention of ground controllers. Autonomy also helps when spacecraft are too far away from Earth for rapid assistance from ground controllers.

• *Autonomous Navigation* (AutoNav). In a traditional solar system mission, ground controllers track the radio signal from a spacecraft to determine its position in space. They may also periodically command the spacecraft's camera to take pictures of the target planet, asteroid or comet to check the position of the craft in space. Based on these measurements, engineers command the spacecraft to execute thruster firings to fine tune its flight path. *Deep Space 1* has dramatically improved this process by allowing the spacecraft to take over the parts of the navigation job formerly carried out by ground controllers.

Deep Space 1 has been finding its location in the solar system by taking images of known asteroids and comparing their positions to background stars. The orbits of 42 asteroids and the positions of 250,000 stars were stored in computer memory at launch. Using the positions of the asteroids and stars, the actual spacecraft location can be determined.

In most cases, the trajectory changes for which AutoNav called were implemented through changes in the ion engine's thrust profile. In some cases, small maneuvers were achieved with dedicated firings of the ion engine or by firings of the spacecraft's separate hydrazine thrusters.

One of many challenges in the use of AutoNav was its reliance on another new technology, ion propulsion, to achieve the thrusting necessary to fulfill its navigation decisions. Because of this uncertainty, the onboard navigator was designed to be able to cope with a wide range of propulsion performance.

Engineers consider AutoNav to be 95 percent validated, with the final test taking place with the flyby, when this new technology will be called upon to locate asteroid Braille and to navigate the spacecraft to within 15 kilometers (10 miles) of the asteroid's center.

Test activities began in mid-November, when AutoNav first turned the spacecraft to point the spacecraft's camera to take pictures of various asteroids. The flight team watched as AutoNav took the spacecraft to several different orientations to collect its data.

AutoNav transitioned into further spacecraft control on December 18, 1998, when it directed the ion propulsion system to pressurize its xenon tanks for thrusting, commanding the spacecraft's attitude control system to turn the spacecraft to thrust in the direction that AutoNav desired and, finally, starting the thruster.

When the ion propulsion system is thrusting, AutoNav updates both the direction and the throttle level for the thrusting every 12 hours in order to follow the flight profile stored on board. So far, the AutoNav system has operated flawlessly.

On December 21, thrusting was suspended for a few hours, during which AutoNav commanded the spacecraft to turn to point its camera at asteroids and stars and take images of them. The images taken allowed AutoNav's designers to improve onboard computer routines for processing such pictures. Previously, all they had was prelaunch predictions of the camera's performance; now, with actual images, the routines could be updated.

During the week of February 1, 1999, AutoNav's weekly optical navigation imaging session was different than usual; for the first time, the operations team allowed the session to proceed without monitoring it. In addition, much of the normal, careful ground testing that precedes most spacecraft activities was avoided.

Until April 12, AutoNav controlled the ion engine without using its full autonomy potential, as it was merely following a plan generated by the mission team. Starting on April 12, however, AutoNav was allowed to use its own calculations to change that plan for the first time ever. It began autonomously commanding the ion engine to thrust as needed – an important further step toward thorough testing of this autonomy technology.

On May 4, the mission team successfully conducted further AutoNav tests and has conducted intermittent experiments ever since.

AutoNav was developed by JPL.

• **Remote Agent**. This experiment takes an even bigger step toward spacecraft autonomy with onboard computer software designed to make a wider variety of decisions. It is capable of planning and executing many onboard activities with only general direction from the ground.

The software is an autonomous "remote agent" of ground controllers in the sense that they rely on the agent to achieve particular goals. Ground controllers do not tell the agent exactly what to do at each instant; rather, they assign it more generalized tasks.

The software package includes a "planner / scheduler" that generates a set of time-based and event-based activities, known as tokens, that are delivered to an "executive" that is also a part of the software system. The executive makes decisions by taking into account knowledge of the spacecraft state, constraints on spacecraft operations and the high-level goals provided by the ground. The executive expands the tokens to a sequence of commands that are issued directly to the appropriate subsystems on the spacecraft. The executive monitors responses to these commands, and reissues or modifies them if the response is not what was planned.

Remote Agent's design is flexible enough to handle a variety of unexpected situations on board. Because of its access to a much more complete description of the spacecraft state than would be available to ground controllers in a traditional operations concept, it can make better use of onboard resources.

Remote Agent software was not designed to control *Deep Space 1* throughout the mission; software was transmitted to the spacecraft after launch to control the ion engine and selected other systems during a 48-hour test period starting on the morning of May 17, 1999. On the morning of May 18, however, the experiment team detected an anomaly that interrupted execution of the experiment. The first indication occurred when Remote Agent did not command the spacecraft's ion propulsion system to shut down as expected. While some portions of the Remote Agent software continued to operate, the component that issues commands had suspended operation.

The spacecraft was determined to be safe and healthy. The experiment software appeared active and was allowed to continue processing while the experiment team and *Deep Space 1* flight team analyzed the problem. After retrieving diagnostic data from the spacecraft, a ground command was issued that afternoon that halted the experiment.

By the time it was halted, the experiment had already achieved about 70 percent of its test objectives. A small bug in the very complex software was identified as the probable cause of the suspension, with the significant assistance of Remote Agent's own self-diagnostic software.

A successful follow-up experiment on May 21 completed the remaining objectives for the Remote Agent test. Presented with three simulated failures on the spacecraft, the experimental software correctly handled each event. The simulations included: a failed electronics unit, which Remote Agent fixed by reactivating the unit; a failed sensor providing false information, which Remote Agent recognized as unreliable and therefore correctly ignored; and a thruster stuck in the "off" position, which Remote Agent detected and for which it compensated by switching to a mode that did not depend on that thruster.

Remote Agent was developed by NASA's Ames Research Center, Moffett Field, California; JPL; and Carnegie Mellon University, Pittsburgh, Pennsylvania.

• **Beacon Monitor Operations Experiment**. This experiment simplifies the way that the spacecraft communicates information about its condition to ground controllers. In a traditional planetary mission, spacecraft send information to Earth as part of telemetry transmitted as digital information in radio signals. Such digital signals are relatively demanding for the antennas of NASA's Deep Space Network to receive and process.

The Beacon Monitor experiment, by contrast, translates overall spacecraft health and status into one of four general states. The monitor then radios one of four tones to Earth to notify ground controllers of the spacecraft's state. A so-called "green" tone indicates that the spacecraft is operating within acceptable conditions. An "orange" tone indicates that an anomaly was resolved by the spacecraft but conditions are acceptable. A "yellow" tone indicates a desire to send data to the ground or to request help with a problem that may escalate to jeopardize the mission. Finally, a "red" tone indicates that the spacecraft has a critical anomaly it cannot resolve and requires urgent assistance from the ground. A substantial portion of the system is the onboard artificial intelligence software that allows it to summarize the spacecraft's condition succinctly.

The beacon monitor makes communication with spacecraft easier in two ways. First, the beacon's tones are much simpler to receive and understand than traditional complex, encoded digital telemetry. Instead of requiring one of the mammoth 70-meter (230-foot) antennas of the Deep Space Network to track a spacecraft, a mission might get by with an antenna only 3 to 10 meters (10 to 30 feet) in diameter. Second, a ground station can receive the beacon monitor's tone, understand it and move on to another spacecraft much more quickly than it could receive digital telemetry. A spacecraft might go for weeks or months without sending digital telemetry, instead broadcasting only pre-arranged simple beacon status checks.

During *Deep Space 1*'s primary mission, mission managers haven't relied on the beacon monitor continuously, instead using it during selected test periods. In February 1999, for instance, it successfully transmitted four different beacon signals, allowing engineers to verify predictions of how well experimental instruments could detect them.

The Beacon Monitor Operations Experiment was developed by JPL.

Science Instruments

- *Miniature Integrated Camera Spectrometer* (MICAS). This package is one of two next-generation science instruments being flown on *Deep Space 1*. MICAS includes several imaging systems within a single 12-kilogram (26-pound) package.

The instrument has a total of four sensors that share a single 10-centimeter (4-inch)-diameter telescope. Two are black-and-white cameras – one with a conventional charge coupled device (CCD) detector, and the other with a more exotic device called an active pixel sensor. The active pixel sensor is a next-generation imaging device that combines support electronics that are usually separate in conventional cameras onto the same chip as the detector itself.

The other two systems combined into MICAS are imaging spectrometers; scientists use these to analyze the light reflected by a celestial object in order to determine what it is made of. One of the spectrometers operates in the ultraviolet spectrum, while the other works in the infrared spectrum. Both of the spectrometers work in "push-broom" mode, meaning that the instrument must sweep across the target body to collect data.

In-flight data collection about MICAS's capabilities began on December 11, 1998. In early March 1999, MICAS observed a variety of targets, including Mars and selected stars, and sent back a large volume of data which have helped to evaluate the instrument's performance. In early May, the mission team successfully completed a calibration of the instrument.

Through these activities, the charge coupled device (CCD) and active pixel sensor cameras and the infrared spectrometer have been well exercised. Testing has revealed that the instrument's ultraviolet detector is not functioning properly and is thus unable to return meaningful data. The optical cameras are less sensitive than expected, and pictures exhibit some geometric distortion, but scientists have been able to compensate partly for these problems by developing new software. In addition, stray light has been found in images taken by MICAS, caused partly by how the instrument is mounted on the spacecraft and partly because of its internal design. According to engineers, all of these issues will be easy to correct in future versions of the camera; they consider the debugging process to be a successful example of the flight testing that *Deep Space 1* was designed for.

MICAS was developed by the U.S. Geological Survey, Flagstaff, Arizona; SSG Inc., Waltham, Massachusetts; the University of Arizona Lunar & Planetary Laboratory, Tucson, Arizona; Boston University Center of Space Physics, Boston, Massachusetts; Rockwell International Science Center, Thousand Oaks, California; and JPL.

- *Plasma Experiment for Planetary Exploration* (PEPE). The second of *Deep Space 1*'s two advanced science experiments, PEPE combines several instruments that study space plasma – charged particles, most of which flow outward from the Sun – in one compact, 6-kilogram (13-pound) package.

PEPE was powered up for the first time on December 8, 1998. Within two days, the instrument demonstrated the ability to measure both electrons and ions in the solar wind, and, within a few weeks, engineers began using data from PEPE to assess the effect of ion propulsion on the instrument itself.

On January 6, 1999, new software for PEPE was tested. On January 8, PEPE was turned to its highest data rate so that it and a plasma instrument on the Saturn-bound *Cassini* spacecraft could make simultaneous observations of the solar wind.

On January 22, the operations team tested new software designed to allow the plasma experiment to operate in the presence of the xenon ions produced by the advanced propulsion system. Because this plasma experiment is designed to observe the more tenuous solar wind, it could be overwhelmed by the xenon, but results showed that this versatile instrument can be adjusted to accommodate the ion propulsion system.

The instrument was developed by the Southwest Research Institute, San Antonio, Texas, and the Los Alamos National Laboratory, Los Alamos, New Mexico.

Telecommunications Technologies

• **Small Deep Space Transponder**. This is one of two technologies designed to improve spacecraft telecommunications hardware. *Deep Space 1*'s transponder, or radio, combines a number of different functions – receiver, command detector, telemetry modulation, exciters, beacon tone generation and control functions – into one small, 3-kilogram (6.6-pound) package. The unit can receive and transmit in the microwave X band, and transmit in the higher-frequency Ka band. The small size and low mass is enabled by the use of advanced gallium arsenide monolithic microwave integrated circuit chips, high-density packaging techniques and silicon application-specific integrated circuit chips.

Because this transponder is used for communications to and from the spacecraft, it was validated within two hours of launch; the first telemetry was received from the spacecraft through NASA's Deep Space Network at 1 hour, 37 minutes after launch.

Key testing took place in December 1998, with some further tests in February 1999; this instrument passed all of its tests perfectly. During these periods, the transponder was used in a successful preliminary test of the beacon monitor experiment. The small deep space transponder successfully transmitted four different beacon signals, allowing engineers to verify predictions of how well experimental instruments could detect them.

The transponder was developed by the Motorola Government Space Systems Division's Space and Systems Technology Group, Scottsdale, Arizona.

• **Ka-Band Solid State Power Amplifier**. This is the second of two technologies concerned with telecommunications hardware. This amplifier allows the spacecraft's radio to transmit in the microwave Ka band.

Engineers are interested in the seldom-used Ka band because it allows the same amount of data to be sent over smaller antennas with less power as compared with missions using lower-frequency transmitters in the X band. The Ka band, however, is more vulnerable to interference from weather on Earth. During the Deep Space 1 mission, engineers have not only tested performance of the amplifier but also have conducted general experiments in Ka-band communications.

The mission's first use of Ka band took place on December 8, 1998, when *Deep Space 1* sent data to Earth in the Ka band at a frequency four times higher than that used for most communications with solar system exploration spacecraft today. During these preliminary tests, which have been followed up with further testing throughout the mission, the spacecraft's transponder sent telemetry at 14 different data rates.

On January 10, 1999, *Deep Space 1* once again transmitted to the Deep Space Network using Ka band, with regularly scheduled tests having taken place successfully since then.

The Deep Space Network's complex at Goldstone in California's Mojave Desert is the only station equipped to receive Ka-band signals, so all of *Deep Space 1*'s tests have been conducted through Goldstone.

The power amplifier was developed by Lockheed-Martin, Valley Forge, Pennsylvania.

Microelectronics Technologies

These three new technologies were first tested in flight on February 25, 1999, and have been successfully tested intermittently since then.

• **Low-Power Electronics**. This is one of three experiments concerned with microelectronics. The experiment involves low-voltage technologies, low-energy architectures and micro-power management. Devices being tested include a ring oscillator and transistors, and are designed to consume very little electrical power.

The low-power electronics experiment was developed by the Massachusetts Institute of Technology's Lincoln Laboratory, Cambridge, Massachusetts, and JPL.

• **Multifunctional Structure**. The structural, thermal and electronic functions of a spacecraft have traditionally been designed and fabricated into separate elements. These single-function elements are bolted together during the final assembly of a spacecraft. Power distribution and signal transmission between the elements are accomplished by the use of bulky connectors and cable bundles.

On *Deep Space 1*, however, the multifunctional structure combines thermal management and electronics in one load-bearing structural element. It consists of a composite panel that has copper polyimide patches bonded to one side and embedded heat-transferring devices. The panel's outer surface acts as a thermal radiator. Electrical circuitry is designed in the copper polyimide layer; flex jumpers serve as electrical interconnects for power distribution and data transmission.

The second of three microelectronics experiments on *Deep Space 1*, the multifunctional structure was developed by the U.S. Air Force's Phillips Laboratory, Kirtland Air Force Base, New Mexico, and Lockheed-Martin Astronautics, Denver, Colorado.

• **Power Actuation and Switching Module**. The third of the mission's three microelectronics experiments, this technology is a smart power switch. The module actually consists of a total of eight power switches capable of monitoring a total of four electrical loads. The switches sense voltage and current, and also limit current if necessary.

The module was developed by Lockheed-Martin Missiles and Space Inc., Sunnyvale, California; the Boeing Co., Seattle, Washington; and JPL.

Spacecraft

There are not enough advanced technologies on *Deep Space 1* to compose an entire spacecraft. Because the focus of the New Millennium Program is on the advanced technologies and not on overall spacecraft design, the remainder of the hardware uses off-the-shelf, low-cost components.

Traditional spacecraft have used redundant systems to lower risk; if an onboard computer or star sensor gives out, the spacecraft can switch to a backup. NASA's philosophy in launching *Deep Space 1*, however, is to mount a technologically challenging, low-cost mission. Most of the spacecraft therefore is "single-string," with no backup systems or redundancy. The design does include limited internal redundancy in some devices and some functional redundancy at the subsystem level.

The central spacecraft structure, or "bus," is aluminum. Most components are mounted on the exterior of the bus, making them easy to access and replace during pre-launch integration and testing. A boom is attached to help technicians reach the battery plug and hydrazine, helium and xenon lines when the spacecraft is in the Delta launch vehicle's payload fairing. Thermal control is accomplished with standard multilayer insulation or thermal blanketing, as well as with electrical heaters and radiators.

Sensors used for attitude control – which is to say, control of the spacecraft's orientation – include a star sensor, an inertial measurement unit or gyro, and a Sun sensor. Most of the electronics are enclosed in an integrated electronics module.

Power is provided from the time of launch until the solar arrays are deployed by a 24-amp-hour nickel hydrogen battery provided by the U.S. Air Force's Phillips Laboratory. The battery also supplements the solar array power during ion engine thrusting to cover transients in the spacecraft's power consumption. It will also be used if the geometry of the spacecraft's asteroid flyby requires the solar arrays to be pointed too far away from the Sun for them to collect sufficient energy.

Spectrum Astro Inc., Gilbert, Arizona, was JPL's primary industrial partner in spacecraft development.

What's Next

Deep Space 1 is the first flight project to be launched under NASA's New Millennium Program. Several more missions to test new technologies are planned in the years ahead.

Deep Space 2

Only two and a half months after *Deep Space 1*'s launch, *Deep Space 2* departed on a mission to Mars in January 1999. In order to test a variety of new technologies, *Deep Space 2* is sending two small probes weighing 2 kilograms (4.5 pounds) each aboard the Mars '98 project's Mars Polar Lander. Shortly before the lander reaches Mars on December 3, 1999, it will release the *Deep Space 2* microprobes; each is designed to impact and penetrate the Martian surface up to a depth of about 1 meter (3 feet). The microprobes' technologies include temperature sensors for measuring the thermal properties of the Martian soil, and a sub-surface soil collection and analysis instrument. The microprobes will join instruments on Mars Polar Lander in searching for subsurface water on Mars. Deep Space 2 is managed by JPL.

Earth Observing 1

Just as the "Deep Space" series of New Millennium missions tests technologies on spacecraft headed out into the solar system, the program also includes an "Earth Observing" series that will evaluate technologies in local orbit. An advanced, light-weight scientific instrument designed to produce visible and short-wave infrared images of Earth's land surfaces was selected as Earth Observing 1. The mission is scheduled for launch in December 1999.

Deep Space I Spacecraft

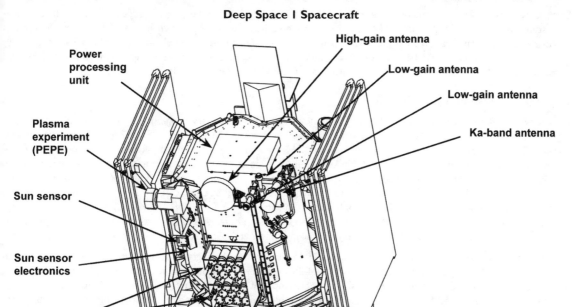

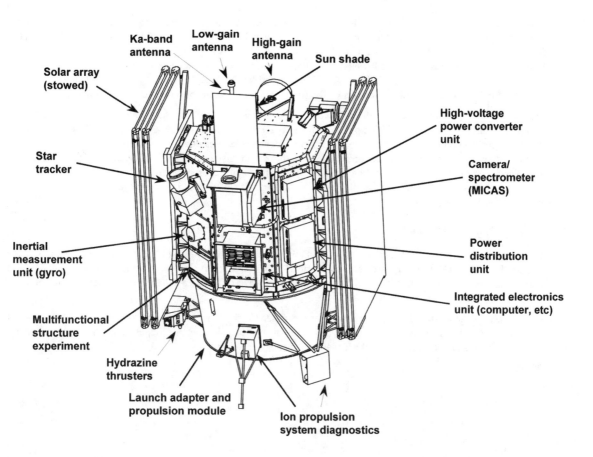

Earth Observing 1 will serve multiple purposes, including providing remote-sensing measurements of Earth that are consistent with data collected since 1972 by the Landsat series of satellites, which is used by farmers, foresters, geologists and city planners. In addition, it will acquire data with finer spectral resolution, a capability long sought by many scientists, and it will lay the technological groundwork for inexpensive, more compact imagers in the future. Earth Observing 1 is managed by NASA's Goddard Space Flight Center, Greenbelt, Maryland.

Other NASA Missions to Small Bodies

- *Near-Earth Asteroid Rendezvous* (NEAR). The first in NASA's Discovery program of lower-cost, highly-focused planetary science missions, NEAR was launched in February 1996. On June 27, 1997, NEAR became the second spacecraft to fly by an asteroid when it encountered 253 Mathilde (the first asteroid flybys were conducted by the Jupiter-bound *Galileo* in 1991 and 1993). NEAR found Mathilde to be composed of extremely dark material, with numerous large impact craters, including one nearly 10 kilometers (6 miles) deep. A subsequent thruster firing in July 1997 brought NEAR back around Earth for a slingshot gravity assist that put the spacecraft on a trajectory for its main mission: a rendezvous with the Manhattan-sized asteroid 433 Eros. NEAR will arrive at Eros in February 2000 and become the first spacecraft ever to orbit an asteroid, studying Eros from as close as 15 kilometers (9.3 miles). NEAR was built and is managed by Johns Hopkins University's Applied Physics Laboratory, Laurel, Maryland.

- *Stardust*. This technically daring mission under NASA's Discovery program will fly a spacecraft to within 160 kilometers (100 miles) of the nucleus of comet Wild-2 to capture comet dust particles in a material called "aerogel" and return the sample to Earth for analysis. A direct sample of a known comet has been long sought by planetary scientists because comets are thought to be nearly pristine examples of the original material from which the Sun and planets were formed 4.6 billion years ago. The spacecraft launched in February 1999, and the sample will be returned to Earth in 2006. The mission is led by principal investigator Dr. Donald Brownlee of the University of Washington and is managed by JPL.

- *Mu Space Engineering Spacecraft C* (MUSES-C). This innovative mission, led by the Japanese space agency ISAS, will use ion propulsion to send a spacecraft to asteroid 4660 Nereus. Although the ion drive is built entirely in Japan, it is functionally the same as – though less powerful than – its *Deep Space 1* counterpart. MUSES-C will deliver a NASA nanorover to the asteroid's surface, collect samples of the asteroid and return them to Earth for laboratory analysis. The nanorover, which is being built by JPL, is so small it can be held in the palm of the hand. The mission is scheduled for launch in 2002.

- *Comet Nucleus Tour* (CONTOUR). This Discovery program mission will take images and comparative spectral maps of at least three comet nuclei and analyze the dust flowing from them. CONTOUR is scheduled for launch in July 2002, with its first comet flyby to occur in November 2003. This flyby of Comet Encke at a distance of about 60 miles (100 kilometers) will be followed by similar encounters with comet Schwassmann-Wachmann-3 in June 2006 and comet d'Arrest in August 2008. CONTOUR is led by Dr. Joseph Veverka of Cornell University, Ithaca, New York, and is managed by Johns Hopkins University's Applied Physics Laboratory, Laurel, Maryland.

- *Deep Impact*. This mission under NASA's Discovery program will launch in January 2004 toward an explosive July 4, 2005 encounter with Comet P/Tempel 1. The spacecraft will send a 500-kilogram (1,100-pound) copper projectile into the comet, creating a crater as big as a football field and as deep as a seven-story building. A camera and infrared spectrometer on the spacecraft, along with ground-based observatories, will study the resulting icy debris blasted off the comet, as well as the pristine interior material exposed by the impact. Deep Impact is led by Dr. Michael A'Hearn of the University of Maryland, College Park, Maryland; managed by JPL; and built by Ball Aerospace, Boulder, Colorado.

Program / Project Management

The Deep Space 1 mission is managed by the Jet Propulsion Laboratory for NASA's Office of Space Science, Washington, DC. At NASA Headquarters, Dr. Edward Weiler is associate administrator for space science. Ken Ledbetter is director of the Mission and Payload Development Division. Lia LaPiana is program executive for Deep Space 1. Dr. Tom Morgan is program scientist.

At the Jet Propulsion Laboratory, Dr. Fuk Li is program manager for the New Millennium Program. For Deep Space 1, David Lehman is project manager, Dr. Marc Rayman is chief mission engineer and deputy mission manager and Dr. Philip Varghese is mission manager; Dr. Robert Nelson is project scientist.

7-26-99 JPL

NATIONAL AERONAUTICS AND SPACE ADMINISTRATION

Stardust Comet Flyby
Press Kit

January 2004

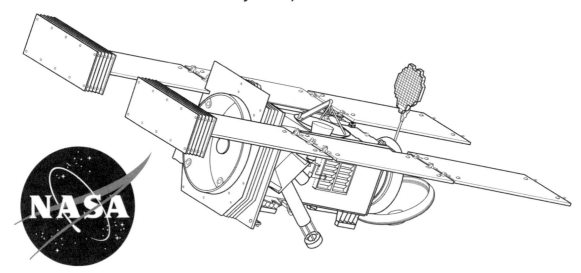

Contacts

Don Savage, NASA Headquarters, Washington

D.C. Agle, Jet Propulsion Laboratory, Pasadena, California

Vince Stricherz, University of Washington, Seattle, Washington

Policy / Program Management	(202) 358-1727
Stardust Mission	(818) 393-9011
Science Investigation	(206) 543-2580

CONTENTS

GENERAL RELEASE:

NASA Comet Hunter Closing on Quarry

Having trekked 3.2 billion kilometers (2 billion miles) across cold, radiation-charged and interstellar dust-swept space in just under five years, NASA's *Stardust* spacecraft is closing in on the main target of its mission – a comet flyby.

"As the saying goes, 'We are good to go,'" said project manager Tom Duxbury at NASA's Jet Propulsion Laboratory, Pasadena, California. "There are significant milestones ahead that we need to achieve before we reach the comet on January 2, but we have a great team of engineers and scientists that have trained hard for this moment, and we have a spacecraft that is in great shape."

All this intense Earthly preparation is directed at Wild-2 (pronounced Vilt-2), a ball of dirty ice and rock, about as big as 20 HMS Titanics laid end to end. Discovered in 1978, Wild-2 orbits the Sun once every 6.39 years on a trajectory that carries it nearly as close to the Sun as Mars is, and as far away from the Sun as Jupiter.

Though comet Wild-2 is the target, it is not the goal of Team Stardust to actually hit their target. Instead, *Stardust* is tasked to have a close encounter of the cometary kind. On January 2 at 11:40:35 am PST, the 5.4-kilometer (3.3-mile)-wide comet will sail past the 5-meter (16-foot)-long *Stardust* spacecraft at a distance of about 300 kilometers (188 miles) and at a relative speed of 21,960 km/hr (13,650 mph). The plan is thus because *Stardust* is a sample return mission.

"In recent decades, spacecraft have passed fairly close to comets and provided us with excellent data," said Dr. Don Brownlee of the University of Washington, principal investigator for the Stardust mission. "*Stardust*, however, marks the first time that we have ever collected samples from a comet and brought them back to Earth for study."

Clad for battle behind specially designed armored shielding, *Stardust* will document its passage through the hailstorm of comet debris with two scientific instruments that will scrutinize the size, number and composition of dust particles in the coma – the region of dust and gas surrounding the comet's nucleus. Along with these instruments, the spacecraft's optical navigation camera will be active during the flyby and should provide images of the dark mass of the comet's nucleus. Data from all three will be recorded on board *Stardust* and beamed back to Earth soon after the encounter.

The chain of events begins nine days out from the comet when *Stardust* deploys its "cometary catcher's mitt," a tennis-racket-shaped particle catcher of more than 1,000 square centimeters (160 square inches) of collection area filled with a material called aerogel. Made of pure silicon dioxide, like sand and glass, aerogel is a thousand times less dense than glass because it is 99.8 percent air. The high-tech material has enough "give" in it to slow and stop particles without altering them radically.

"The samples we will collect are extremely small, 10 to 300 microns in diameter, and can only be adequately studied in laboratories with sophisticated analytical instruments," said Brownlee. "Even if a ton of sample were returned, the main information in the solids would still be recorded at the micron level, and the analyses would still be done a single grain at a time."

After the sample has been collected, the collector will fold down into a return capsule, which will close like a clamshell to secure the sample for a soft landing at the U.S. Air Force's Utah Test and Training Range in January 2006. The capsule, holding microscopic particles of comet and interstellar dust, will be taken to the planetary material curatorial facility at NASA's Johnson Space Center, Houston, Texas, where the samples will be carefully stored and examined.

Scientists believe in-depth terrestrial analysis of cometary samples will reveal a great deal not only about comets but also related to the earliest history of the solar system. Locked within the cometary particles is unique chemical and physical information that could provide a record of the formation of the planets and the materials from which they were made.

Stardust, a project under NASA's Discovery Program of low-cost, highly focused science missions, was built by Lockheed-Martin Space Systems, Denver, Colorado, and is managed by the Jet Propulsion Laboratory, Pasadena, California, for NASA's Office of Space Science, Washington, D.C. JPL is a division of the California Institute of Technology in Pasadena. More information on the Stardust mission is available at http://stardust.jpl.nasa.gov.

Media Services Information

NASA Television Transmission

NASA Television is carried on the satellite AMC-9, transponder 9C, at 85 degrees west longitude, vertical polarization. Frequency is on 3880.0 megahertz with audio on 6.8 megahertz. The schedule for Stardust Encounter television transmissions will be available from the Jet Propulsion Laboratory, Pasadena, California, and from NASA Headquarters, Washington.

Media Credentialing

News media representatives who wish to cover the *Stardust* encounter with comet Wild-2 at NASA's Jet Propulsion Laboratory must be accredited through the JPL Media Relations Office. Journalists may phone the news room at (818) 354-5011 for more information.

Briefings

News briefings will be held at JPL before and after the encounter. Information about upcoming briefings will be available on the Internet as noted below.

Internet Information

Extensive information on the Stardust project including an electronic copy of this Press Kit, press releases, fact sheets, status reports, briefing schedule and images, is available from the Jet Propulsion Laboratory's Stardust news web site, http://www.jpl.nasa.gov/stardust. The Stardust project web site is at http://stardust.jpl.nasa.gov.

Quick Facts

Spacecraft

- Dimensions: Main bus 1.7 meters (5.6 feet) high, 0.66 meter (2.16 feet) wide, 0.66 meter (2.16 feet) deep; length of solar arrays 4.8 meters (15.9 feet) tip to tip; sample return capsule 0.8 meter (32 inches) diameter and 0.5 meter (21 inches) high
- Weight: 385 kilograms (848 pounds) total at launch, consisting of 254-kilogram (560-pound) spacecraft and 46-kilogram (101-pound) return capsule, and 85 kilograms (187 pounds) fuel
- Power: Solar panels providing from 170 to 800 watts, depending on distance from Sun. At time of comet flyby, solar panels will generate about 330 watts

Mission Milestones

- Launch: February 7, 1999 from Cape Canaveral Air Force Station, Florida
- Launch Vehicle: Delta II (model 7426) with Star 37 upper stage
- Earth-comet distance at time of launch: 820 million kilometers (508 million miles)
- Interstellar dust collection: February 22 - May 1, 2000; August 5 - December 9, 2002
- Earth gravity assist flyby: January 15, 2001
- Altitude at Earth gravity assist: 6,008 kilometers (3,734 miles)
- Asteroid Annefrank flyby: November 2, 2002
- Comet Wild-2 encounter: January 2, 2004
- Earth-comet distance at time of encounter: 389 million kilometers (242 million miles)
- Total distance traveled Earth to comet: 3.22 billion kilometers (2 billion miles)
- Spacecraft speed relative to comet at closest approach: 22,023 km/hr (13,684 mph)
- Earth return: January 15, 2006
- Landing site: Utah Test and Training Range
- Earth-comet distance at Earth landing: 860 million kilometers (534 million miles)
- Total distance traveled comet to Earth: 1.14 billion kilometers (708 million miles)
- Total distance traveled entire mission (Earth to comet to Earth): 4.37 billion kilometers (2.7 billion miles)

Program

- Cost: $168.4 million total (not including launch vehicle), consisting of $128.4 million spacecraft development and $40 million mission operations

Why Stardust?

Far beyond the orbits of the planets on the outer fringes of the solar system, a vast swarm of perhaps a trillion dormant comets circles the Sun. Frozen balls of ice, rocks and dust, they are the undercooked leftovers that remained after a sprawling cloud of gas and dust condensed to form the Sun and planets about 4.6 billion years ago. From time to time, the gravitational pull of a passing star will nudge some of them out of their orbits, plunging them into the inner solar system, where they erupt with glowing tails as they loop around the Sun.

Closer to home, a stream of interstellar dust flows continuously through the solar system. Each perhaps 1/50th the width of a human hair, these tiny particles are the pulverized flotsam of the galaxy, bits of ancient stars that exploded as they died. This "stardust" is literally the stuff of which we are all made, being the source of nearly all of the elements on Earth heavier than oxygen.

These two niches bearing clues of the dawn of the solar system are the target for NASA's Stardust mission. The spacecraft will use a collector mechanism that employs a unique substance called aerogel to snag comet particles as well as interstellar dust flowing through the solar system, returning them to Earth for detailed study in laboratories.

Data returned from the *Stardust* spacecraft and the precious samples it returns to Earth will provide opportunities for significant breakthroughs in areas of key interest to astrophysics, planetary science and astrobiology. The samples will provide scientists with direct information on the solid particles that permeate our galaxy.

Stardust's cometary dust and interstellar dust samples will help provide answers to fundamental questions about the origin of solar systems, planets and life – How and when did the elements that led to life enter the solar system? How were these materials transformed within the solar system by forces such as heating and exposure to ultraviolet light? How were they distributed among planetary bodies, and in what molecular and mineral-based forms? These questions are of major importance for astrobiology and the search for life-generating processes and environments elsewhere in the universe.

Comets

Though frequently beautiful, comets traditionally have stricken terror as often as they have generated excitement as they wheel across the sky during their passage around the Sun. Astrologers interpreted the sudden appearances of the glowing visitors as ill omens presaging famine, flood or the death of kings. Even as recently as the 1910 appearance of Halley's Comet, entrepreneurs did a brisk business selling gas masks to people who feared Earth's passage through the comet's tail.

In the 4th century B.C., the Greek philosopher Aristotle concluded that comets were some kind of emission from Earth that rose into the sky. The heavens, he maintained, were perfect and orderly; a phenomenon as unexpected and erratic as a comet surely could not be part of the celestial vault. In 1577, Danish astronomer Tycho Brahe carefully examined the positions of a comet and the Moon against the stars during the evening and predawn morning. Due to parallax, a close object will appear to change its position against the stars more than a distant object will, similar to holding up a finger and looking at it while closing one eye and then the other. The Moon appeared to move more against the stars from evening to morning than the comet did, leading Tycho to conclude that the comet was at least four times farther away.

A hundred years later, the English physicist Isaac Newton established that a comet appearing in 1680 followed a nearly parabolic orbit. The English astronomer Edmund Halley used Newton's method to study the orbits of two dozen documented cometary visits. Three comet passages in 1531, 1607 and 1682 were so similar that he concluded they in fact were appearances of a single comet wheeling around the Sun in a closed ellipse every 75 years. He successfully predicted another visit in 1758-9, and the comet thereafter bore his name.

Since then, astronomers have concluded that some comets return relatively frequently, in intervals ranging from 3 to 200 years; these are the so-called "short period" comets. Others have enormous orbits that bring them back only once in many centuries.

In the mid-1800s, scientists also began to turn their attention to the question of comets' composition. Astronomers noted that several major meteor showers took place when Earth passed through the known orbits of comets, leading them to conclude that the objects are clumps of dust or sand. By the early 20th century, astronomers studied comets using the technique of spectroscopy, breaking down the color spectrum of light given off by an object to reveal the chemical makeup of the object. They concluded that comets also emitted gases as well as molecular ions.

In 1950, the American astronomer Fred L. Whipple authored a major paper proposing the "dirty snowball" model of the cometary nucleus. This model, which has since been widely adopted, pictures the nucleus as a mixture of dark organic material, rocky grains and water ice. ("Organic" means that the compound is carbon-based, but not necessarily biological in origin.) Most comets range in size from about 2 to 7 kilometers (1 to 5 miles) in diameter. Shields that protect the *Stardust* spacecraft from dust impacts were named for Whipple in honor of his role in cometary science.

If comets contain icy material, they must originate somewhere much colder than the relatively warm inner solar system. In 1950, the Dutch astronomer Jan Hendrick Oort (1900-1992) used indirect reasoning from observations to establish the existence of a vast cloud of comets orbiting many billions of miles from the Sun – perhaps 50,000 astronomical units (AU) away (one AU is the distance from Earth to the Sun), or nearly halfway to the nearest star. This region has since become known as the Oort Cloud.

A year later, the Dutch-born American astronomer Gerard Kuiper (1905-1973) made the point that the Oort Cloud is too distant to act as the nursery for short period comets. He suggested the existence of a belt of dormant comets lying just outside the orbits of the planets at perhaps 30 to 100 AU from the Sun; this has become known as the Kuiper Belt. Jupiter's gravity periodically influences one of these bodies to take up a new orbit around the Sun. The Oort Cloud, by contrast, would be the home of long period comets. They are periodically nudged from their orbits by any one of several influences – perhaps the gravitational pull of a passing star or giant molecular cloud, or tidal forces of the Milky Way Galaxy.

In addition to the length of time between their visits, another feature distinguishes short and long period comets. The orbits of short period comets are all fairly close to the ecliptic plane, the plane in which Earth and most other planets orbit the Sun. Long period comets, by contrast, dive inwards toward the Sun from virtually any part of the sky. This suggests that the Kuiper Belt is a relatively flat belt, whereas the Oort Cloud is a three-dimensional sphere surrounding the solar system.

Where did the Oort Cloud and Kuiper Belt come from? Most astronomers now believe that the material that became comets condensed in the outer solar system around the orbits of Uranus and Neptune. Gravitational effects from those giant planets flung the comets outward.

Residing at the farthest reaches of the Sun's influence, comets did not undergo the same heating as the rest of the objects in the solar system, so they retain, largely unchanged, the original composition of solar system materials. As the preserved building blocks of the outer solar system, comets offer clues to the chemical mixture from which the planets formed some 4.6 billion years ago.

The geologic record of the planets shows that about 3.9 billion years ago a period of heavy comet and asteroid bombardment tapered off. The earliest evidence of life on Earth dates from just after the end of this heavy bombardment. The constant barrage of debris had vaporized Earth's oceans, leaving the planet too hot for the survival of fragile carbon-based molecules upon which life is based. Scientists therefore wonder: How could life form so quickly when there was so little liquid water or carbon-based molecules on Earth's surface? The answer may be that comets, which are abundant in both water and carbon-based molecules, delivered essential ingredients for life to begin.

Comets are also at least partially responsible for the replenishment of Earth's ocean after the vaporization of an early ocean during the late heavy bombardment. While Earth has long been regarded as the "water planet," it and the other terrestrial planets (Mercury, Venus and Mars) are actually poor in the percentage of water and in carbon-based molecules they contain when compared to objects that reside in the outer solar system at Jupiter's orbit or beyond. Comets are about 50 percent water by weight and about 10 to 20 percent carbon by weight. It has long been suspected that what little carbon and water there is on Earth was delivered here by objects such as comets that came from a more water-rich part of the solar system.

While comets are a likely source for life's building blocks, they have also played a devastating role in altering life on our planet. A comet or asteroid is credited as the likely source of the impact that changed Earth's climate, wiped out the dinosaurs and gave rise to the age of mammals 65 million years ago. A catastrophic collision between a comet or asteroid and Earth is estimated to happen at intervals of several tens of millions of years.

Other Comet Missions

Comets have been studied by several other spacecraft, not all of which were originally designed for that purpose. Several new missions to comets are being developed for launch in coming years.

Past comet missions include:

- In 1985, NASA modified the orbit of the International Sun-Earth *Explorer* spacecraft to execute a flyby of Comet Giacobini-Zinner. At that point, the spacecraft was renamed International Comet Explorer. It successfully executed a flyby of comet Giacobini-Zinner in 1985 and comet Halley in 1986.
- An international armada of robotic spacecraft flew out to greet Halley's Comet during its return in 1986. The fleet included the European Space Agency's *Giotto*, the Soviet Union's *Vega 1* and *Vega 2*, and Japan's *Sakigake* and *Suisei* spacecraft.
- Comet Shoemaker-Levy 9's spectacular collision with Jupiter in 1994 was observed by NASA's Hubble Space Telescope, the Jupiter-bound *Galileo* spacecraft and the Sun-orbiting *Ulysses* spacecraft.
- *Deep Space 1* launched from Cape Canaveral on October 24, 1998. During a highly successful primary mission, it tested 12 advanced, high-risk technologies in space. In an extremely successful extended mission, it encountered comet Borrelly and returned the best images and other science data taken from a comet to date.
- The Comet Nucleus Tour, or Contour, mission launched from Cape Canaveral on July 3, 2002. Six weeks later, on August 15, contact with the spacecraft was lost after a planned maneuver that was intended to propel it out of Earth orbit and into its comet-chasing solar orbit.

Future comet missions are:

- *Deep Impact* will purposely slam a 370-kilogram (approximately 820-pound) cylinder into a comet's nucleus so experts can study the interior of the comet. The launch of the *Deep Impact* spacecraft is planned for December 2004 and it is scheduled to reach out and touch its target, comet Tempel 1, on July 4, 2005.
- A European Space Agency mission, *Rosetta* will be launched in February 2004 to orbit comet 67P/Churyumov-Gerasimenko and deliver a science package to its surface via a lander. NASA is providing science instruments for the comet orbiter.

Right Place, Right Time, Right Snowball

Comet 81P/Wild-2 is a fresh periodic comet – meaning that it moves about the Sun in an elliptic orbit. In Wild-2's case that is once every 6.39 years. Its nucleus is thought to be of low density, with a diameter of about 5.4 kilometers (3.3 miles).

Until September 10, 1974, comet Wild-2's orbit lay between Jupiter and a point near Uranus. But on that date nearly 30 years ago, the comet passed within 897,500 kilometers (557,735 miles) of the solar system's biggest planet, Jupiter. That encounter with Jupiter forever altered the comet's orbit, carrying it for the first time into the inner solar system. The new flight path carried it as close to the Sun as just beyond the distance of Mars and far from the Sun as about Jupiter. On January 6, 1978, astronomer Paul Wild (pronounced "Vilt") discovered the comet during its first passage relatively near to the Earth – passing within 181,014,000 kilometers (112,476,679 miles).

When a comet comes close enough to the Sun to get heated up, it loses some of its material through a process called sublimation. This happens when a solid becomes a vapor without first melting into a liquid. After about 1,000 trips past the Sun, a comet loses most of its volatile materials and no longer generates a coma, which is made up of the gases that escape off its surface. Since it is the escaping gases that drive the dust particles from the nucleus – the solid part of the comet – the comet no longer creates the long beautiful dust tail that we can sometimes see in the night sky.

An important aspect of *Stardust*'s exploration of comet Wild-2 is that by the time *Stardust* encounters it, the comet will have made only five trips around the Sun in its new orbit. By contrast, Comet Halley has passed close to the Sun more than 100 times, coming close enough to have been greatly altered from its original condition.

Another important aspect is the comet's orbit. *Stardust* navigators were able to plot a flight path that allowed the spacecraft to encounter the comet at a relatively sedate closing speed. Because of this low-velocity meeting (passing each other at 6.1 kilometers per second, or about 13,600 mph), the spacecraft can capture comet dust, rather than having it blow right through the collectors. The dust samples can then be brought back to the Earth to be analyzed.

NASA's Discovery Program

Stardust is a mission under NASA's Discovery Program, which sponsors low-cost solar system exploration projects with highly focused science goals. Created in 1992, the Discovery Program competitively selects proposals submitted by teams led by scientists called principal investigators and supported by organizations which provide project management and build and fly the spacecraft. In recent years, NASA has identified several finalists from dozens of mission proposals submitted. These finalists receive funding to conduct feasibility studies for an additional period of time before a final selection is made.

Other missions in the Discovery Program are:

- The *Near Earth Asteroid Rendezvous* spacecraft was launched in February 1996 and became the first spacecraft to orbit an asteroid when it reached Eros in February 2000. A year later, it became the first spacecraft to land on an asteroid when it put down on Eros, providing the highest resolution images ever obtained of an asteroid, showing features as small as one centimeter across. The mission was managed by Johns Hopkins University's Applied Physics Laboratory.
- *Mars Pathfinder* was launched in December 1996 and reached Mars on July 4, 1997, demonstrating a unique way of landing with airbags to deliver a small robotic rover. Mars Pathfinder was managed by the Jet Propulsion Laboratory.
- Launched in January 1998, *Lunar Prospector* entered orbit around Earth's Moon five days later, circling at an altitude of about 100 kilometers (60 miles). Principal investigator was Dr. Alan Binder of the Lunar Research Institute, Gilroy, California, with project management by NASA's Ames Research Center.
- Launched in August of 2001, *Genesis* is a mission that will return samples of solar wind particles. Principal investigator is Dr. Donald Burnett of the California Institute of Technology, with project management by the Jet Propulsion Laboratory. Genesis is NASA's first sample return mission since the Apollo Moon landings ended in December of 1972.
- The Comet Nucleus Tour, or Contour, mission launched from Cape Canaveral on July 3, 2002. Six weeks later, on August 15, contact with the spacecraft was lost after a planned maneuver that was intended to propel it out of Earth orbit and into its comet-chasing solar orbit.
- *Messenger* will be launched in May 2004 and will enter Mercury orbit in July 2009, after three flybys of Venus along the way. Multiple flybys of the planet will provide opportunities for global mapping and detailed characterization of the surface, interior, atmosphere, and magnetosphere.
- Scheduled for launch in December of 2004, *Deep Impact* is a spacecraft that will travel to comet Tempel 1 and release a small impactor, creating a crater in the side of the comet. The main spacecraft will measure and observe the gas released from the crater, to discover what makes up fresh comet material, and to understand what's inside a comet. The scheduled date for these celestial fireworks is July 4, 2005.
- Planned for launch in May 2006, *Dawn* will reach the asteroid Vesta in 2010 and the asteroid Ceres in 2014. These minor planets have remained intact since the earliest time of solar system formation. Their surfaces are believed to contain a snapshot of the conditions present in the solar system's first 10 million years, allowing *Dawn* to investigate both the origin of the solar system and its present state.

- Planned for launch in the fall of 2007, *Kepler* will monitor 100,000 stars similar to our Sun for four years. *Kepler* could detect many habitable, Earth-size planets and provide a stepping stone to the next extensive search for habitable planets and life.
- The Analyzer of Space Plasma and Energetic Atoms, or Aspera 3, is one of seven scientific instruments flying aboard the *Mars Express* spacecraft. NASA is contributing hardware and software to the instrument, led by Sweden's Institute of Space Physics. *Mars Express* is a European Space Agency mission scheduled to enter Martian orbit in late December 2003.

Interstellar Dust

In 1990, NASA launched the *Ulysses* spacecraft on a flight path that would take it close to Jupiter, in turn flinging it into an orbit around the Sun far above and below the ecliptic, the plane in which most planets orbit the Sun. While en route from Earth to Jupiter, the spacecraft's dust detector measured a constant flow of particles – each about a micron in size, or 1/50th the diameter of a human hair – entering the solar system from interstellar space. This observation was corroborated by a similar dust detector on the *Galileo* spacecraft, which reached Jupiter in 1995.

Scientists believe that interstellar dust is ubiquitous in the space between the stars of the Milky Way Galaxy. The dust curtains huge areas of the sky; the broad, dark line across the length of the Milky Way that can be seen with the naked eye is a blanket of interstellar dust.

As the Sun orbits the galactic center, it cuts through the dust like a ship passing through waves. From our perspective within the solar system, the dust seems to be flowing from approximately the direction that the Sun is moving toward – a point called the "solar apex" in the constellation Hercules. Outward pressure from the solar wind sweeps the inner solar system near Earth clean, but interstellar dust is more easily detected beyond the orbit of Mars.

Interstellar dust provided the building blocks for solid materials on Earth and other planets. In the life cycle of stars, light elements such as hydrogen coalesce to form the star, which in time develops enough mass to burst into an ongoing nuclear reaction. Many stars die in a spectacular explosion, converting light elements into heavier elements. The resulting interstellar particles contain a record of the processes at work in their parent stars as well as the environments they have passed through in the galaxy. This information is retained in particles at a scale smaller than a micron.

Interstellar dust forms by condensation in circumstellar regions around evolved stars of many different types, including red giants, carbon stars, novas and supernovas. The process gives rise to silicate grains when there is more oxygen than carbon in the star, and carbon-based grains when the carbon content exceeds that of oxygen. Pristine grains will retain the radioactive signatures of the environment they formed in.

In the past decade, scientists have gained new understanding about the formation and early evolution of the solar system and the role of interstellar dust and comets in that process. Studies of interstellar dust have been conducted with Earth- and space-based telescopes; in addition, scientists have collected and studied dust in the stratosphere and in Earth orbit. Since the late 1960s, collections of ocean sediment have brought up microscopic glass and metallic spherules, space particles that melted during atmospheric entry. The sample of interstellar dust returned by *Stardust* will be compared with these and others to help define how dust evolves from its interstellar state to help create stars, planets and life in the universe.

Infrared observations have also provided new knowledge of star formation and the role that dust plays in that process. Scientists have found many similarities between interstellar dust and cometary composition. The same gases, ice particles and silicates believed to be in comets also are found in interstellar clouds.

Even though the interstellar dust samples will be small and partly eroded, they will open a significant new window of information on galactic and nebular processes, materials and environments. Having actual samples in hand provides many unique advantages. Just as the return of lunar samples by the Apollo missions of the 1960s and 1970s revolutionized our understanding of the Moon, scientists expect that the *Stardust* mission's sample return will also have a profound impact on our knowledge of comets and stars.

Earth Assist

Assisting the Stardust team in their celestial pursuit of comet Wild-2 are several teams of Earth-based astronomers. These observatories will make observations and measurements of Wild-2 from mid- to late-December 2003. These observations will assist the Stardust team in validating the size of coma and amount of activity taking place on comet Wild-2, as well as assist in further reducing any errors in the plotting of the comet's orbit.

Mission Overview

Launched in 1999, the *Stardust* spacecraft will loop a total of three times around the Sun over seven years before it returns to Earth in 2006. On the way to its comet encounter, it collected interstellar dust on two different solar orbits. On January 2, 2004, *Stardust* will fly through the coma, or cloud of dust and debris, surrounding comet Wild-2.

Finally the spacecraft will approach Earth, ejecting an entry capsule that will descend into the U.S. Air Force's Utah Test and Training Range carrying cometary and interstellar dust samples.

Stardust's trajectory was calculated to allow the spacecraft to fly past Wild-2 at a relatively low speed at a time when the comet is active – but not too active. The trajectory also minimized the energy needed to launch the spacecraft, allowing for a smaller, less expensive launch vehicle; maximizes the time for favorable collection of interstellar dust; and makes the spacecraft approach Earth at a relatively low speed when it returns.

Launch

Stardust began its voyage on February 7, 1999 from Space Launch Complex 17A at Cape Canaveral Air Station, Florida, on a variant of the Delta II launch vehicle known as a Delta 7426, one of a new series of rockets procured under NASA's Med-Lite program.

Launch events occurred in three phases. First, the Delta lifted off and entered a 185-kilometer (115-mile)-high parking orbit; then it coasted for about a half hour until its position was properly aligned; and finally an upper-stage engine fired to send *Stardust* out of Earth orbit.

Cruise

Stardust's first two years of flight carried it on the first of its three orbital loops around the Sun. In January 2000, when *Stardust* was between the orbits of Mars and Jupiter – the most distant point from the Sun that it reached during that orbit – the spacecraft's thrusters fired to place it on course for a later gravity assist swingby of Earth.

As *Stardust* traveled back inward toward the Sun on the latter part of that first orbit, it passed through a region where interstellar particles flow through the solar system. From February through May 2000, the spacecraft deployed its collector to capture these interstellar particles. One part of collector mechanism called its "B side" was employed to collect the particles, reserving the other half, called the "A side," for the spacecraft's later dust collection mission at comet Wild-2.

The spacecraft completed its first solar orbit when it flew by the Earth on January 15, 2001. The flyby altitude at its closest approach distance was approximately 6,008 kilometers (3,773 miles). The effect of Earth's gravity increased the size of *Stardust's* orbit so that it circled the Sun once each 2½ years, and placed it on a flight path leading to an intercept of its quarry, comet Wild-2.

Beginning in August 2002, as the spacecraft traveled back inward toward the Sun on the latter part of its second orbit, *Stardust* again exposed the "B side" of its collector to interstellar particles flowing through the solar system. Then, in December 2002, *Stardust* concluded its second and final collection period of interstellar particles. The total time it spent collecting interstellar particles over the entire mission was 195 days.

During this second period collecting interstellar particles, *Stardust* flew within 3,100 kilometers (1927 miles) of asteroid Annefrank. This encounter took place on November 2, 2002, and was used as an engineering test of the ground and spacecraft operations that will be implemented at the primary scientific target, comet Wild-2.

Comet Flyby Overview

On January 2, 2004, at about 11:41 a.m. PST (2:40 p.m. EST), after traveling the solar system for almost four years – and 3.22 billion kilometers (2 billion miles) – the *Stardust* spacecraft will make its closest approach to comet Wild-2 as the comet sails past *Stardust* at a relative speed of 21,960 km/hr (13,650 mph). In terrestrial terms such velocity is guaranteed to smoke any police officer's radar gun, but in cosmic terms such relative speed between spacecraft and comet is relatively leisurely, allowing *Stardust* to "soft-catch" samples of comet dust without changing them greatly.

The passage through the most intensive rain of cometary particles within the coma is expected to last about 8 minutes. Within minutes after the closest approach, *Stardust* will begin to transmit images and other scientific data collected and stored in its computer memory during the flyby.

Flyby Miss Distance

Planning for the near-miss of a 5.4-kilometer (3.4-mile)-wide comet and a 4.8-meter (15.9-foot)-long spacecraft closing on each other at a velocity much faster than any rifle bullet – all while 389 million kilometers (242 million miles) away from home – is a full-time job. The flight team was tasked to come up with a flight plan that balances the maximum opportunity for collecting the freshest cometary particles off the comet's nucleus while providing the greatest likelihood of keeping the spacecraft out of harm's way.

Defining "harm's way" in an environment as astronomically foreign as the coma of a comet is a challenge, to say the least. To help understand the hazards posed by the flyby, scientists and engineers developed a dust model for Wild-2.

This model spells out the team's best estimates for the location, quantity and size of the dust surrounding the comet's nucleus. *Stardust*'s managers have been carefully weighing this information as they fine tune the "miss distance" for the spacecraft's pass by the comet.

Based on the that analysis, mission managers established a miss distance of 300 kilometers (about 186 miles). That distance, however, can be tweaked as late as December 31, 2003 – just two days before the flyby – when the spacecraft fires its thrusters to fine tune its trajectory for the encounter. The team will gather images taken by *Stardust*'s navigation camera, as well as data from ground-based observatories, to help make a final decision on the comet flyby distance.

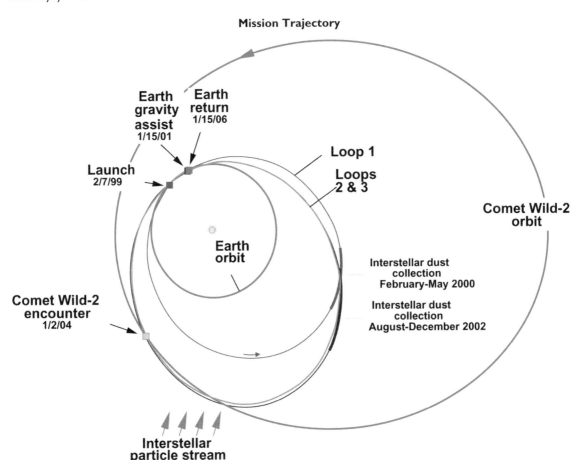

Mission Trajectory

Encounter Sequences

For planning purposes, the Stardust mission team has defined the spacecraft's encounter of comet Wild-2 as a 119-day period spanning from 88 days prior to comet flyby to 31 days after. This encounter is further divided into five time segments during which various mission activities are planned:

- Far Encounter: Minus 88 days to minus 1 day
- Near Encounter: Minus 1 day to minus 5 hours
- Close Encounter: Minus 5 hours to plus 5 hours
- Closest Encounter: Minus 360 seconds to plus 360 seconds
- Post-Encounter: Encounter day to plus 31 days

Far Encounter: Minus 88 Days to Minus 1 Day

During this phase, the Stardust team has been using the spacecraft's navigation camera to scour the skies in the area where they expected to find Wild-2. Acquiring the comet visually is considered of paramount importance for a successfully executed flyby, as it will further refine knowledge of the exact location and orbit of the comet.

On November 13, 2003, some 50 days before the encounter, an important milestone was passed when Wild-2 was detected by the spacecraft's navigation camera, several weeks earlier than anticipated. These initial images and later images leading up to encounter are used by navigators to help plan thruster firings to fine tune *Stardust*'s flight path.

During the final month of the spacecraft's approach to the comet, thruster firings have been scheduled to take place December 3, December 23 and December 31, 2003, as well as on January 2, 2004 (January 1 PST). The December 31 maneuver will set *Stardust*'s final targeted miss distance, and also change the spacecraft's orientation, or "attitude," so that its protective shields will safeguard *Stardust* from the onslaught of cometary particles. The January 2 thruster firing will clean up any execution errors from the previous maneuver and improve targeting precision.

On December 24, when it is 9 days out, *Stardust*'s schedule calls for it to deploy its aerogel-laden collector grid with the "A side" facing the direction of incoming comet particles. After a heat shield is removed from the sample return capsule's opening, the collector emerges from the capsule and is extended fully sticking above the spacecraft's shielding exposed to the stream of comet particles. The sequence will take about 30 minutes to complete.

Near Encounter: Minus 1 Day to Minus 5 Hours

As the spacecraft nears the comet, it will take pictures of the comet more frequently, and spend more time transmitting these images to Earth. To support the increased data traffic, ground stations of the Deep Space Network will be in touch with *Stardust* continuously during these hours.

Around 5:18 p.m. PST on January 1, the spacecraft will fire its thrusters for a final time before encountering the comet. After this maneuver, the spacecraft will roll back to its protective "encounter attitude" or orientation.

Close Encounter Milestones
January 2-3, 2004

January 2:
- 5:49 a.m.: "Encounter sequence" of onboard computer commands begins
- 6:19 a.m.: Cometary and interstellar dust analyzer instrument configured
- 9:19 a.m.: Navigation camera takes approach image
- 10:19 a.m.: Navigation camera takes approach image
- 11:04 a.m.: Dust flux monitor instrument turned on
- 11:12 a.m.: Stops sending data, transmits carrier signal only
- 11:13 a.m.: Final roll maneuver to adjust encounter orientation or "attitude"
- 11:19 a.m.: Closest approach to comet
- 11:25 a.m.: Navigation camera ends period of highest frequency imaging
- 11:25 a.m.: Roll maneuver to take spacecraft out of encounter orientation
- 11:26 a.m.: Resumes sending data instead of carrier signal
- 11:27 a.m.: Navigation camera takes final picture
- 11:29 a.m.: Navigation camera turned off
- 11:36 a.m.: Begins transmitting images, dust flux monitor data

January 3:
- 5:19 a.m.: Cometary and interstellar dust analyzer returned to cruise mode
- 5:19 a.m.: "Encounter sequence" of computer commands ends

All times listed as actual event time at the spacecraft, in Pacific Standard Time. Signals confirming each event will take 20 minutes to reach Earth.

Close Encounter: Minus 5 Hours to Plus 5 Hours

In all of *Stardust*'s journey spanning almost seven years, the majority of scientific data will be collected in a period of just 10 hours – what mission managers call the "close encounter." As this time window opens, when the spacecraft is about 100,000 kilometers (60,000 miles) away from the nucleus, *Stardust* will begin to enter the comet's coma. Because of its flight path, it will enter the coma in front of the nucleus as it flies through the halo of gases and dust at the head of comet Wild-2.

During this phase the comet's nucleus should begin to emerge in the navigation camera's field-of-view as an extended dark body. At 30 minutes out, the navigation camera will begin taking images at the rate of nearly one every 30 seconds. Over the next 38 minutes, 72 images will be stored in the spacecraft's computer memory. Fifteen minutes before closest approach, the spacecraft will turn on its dust flux monitor.

Closest Encounter: Minus 360 Seconds to Plus 360 Seconds

Stardust will be flying through the most concentrated region of cometary particles at this point. At 5 minutes out, the navigation camera will increase its rate of picture taking to one every 10 seconds. This rate will continue for the next 10 minutes.

Comet Wild-2 is flying through space faster than the spacecraft is. To pull off the encounter, navigators had to calculate a location in time and space where the spacecraft needed to be so that the comet could essentially "run over" it.

Navigators were further tasked to have this brief meeting of spacecraft and comet occur on the sunlit side of the comet, so that the Sun could act as a natural flashbulb for picture taking and because that is where the majority of comet particles are expected to be found.

The comet will approach *Stardust* from outside and below the spacecraft's orbital plane. At this point the spacecraft will be flying "backwards" with its nose pointed down so its defensive shielding can protect it from the expected hailstorm of particles. The comet will overtake the spacecraft and continue on its orbital path, which will carry it above and inside the trajectory of *Stardust*.

One major challenge for *Stardust*'s mission designers was to figure out a way to keep the comet nucleus within the navigation camera's field-of-view during encounter. To accomplish this, they programmed the navigation camera's scanning mirror to gradually pan as the comet passes by. During closest encounter, the spacecraft will perform a roll to keep the camera pointed at the comet.

This roll, lasting about 30 seconds, could result in loss of the signal from the spacecraft's high-gain antenna. To minimize the chance of loss of signal during this maneuver, mission planners will command the spacecraft's medium-gain antenna to take over from the high-gain dish during the 12 minutes surrounding closest encounter. The spacecraft will stop sending scientific or engineering data and instead will transmit a simple carrier signal during this closest passage. After the 12 minutes of closest encounter are over, the spacecraft will roll to point its high-gain antenna at Earth once again, and will resume transmitting encoded data.

As *Stardust* flies through the most concentrated region of cometary particles, its dust flux monitor instrument will be logging into the spacecraft's computer memory measures of the size and frequency of dust particle hits. At the same time, the comet and interstellar dust analyzer will be performing instantaneous compositional analysis of material and also storing its results in the spacecraft's onboard memory.

Post-Encounter: Flyby Day to Plus 31 Days

After pointing its high-gain antenna at Earth once again, *Stardust* will use its navigation camera will take one parting snapshot of Wild-2. Selected images and other science and engineering data recorded on board during the coma fly-through will begin to be transmitted to Earth.

About five hours after closest approach, the sample collector with its cometary pickings will begin a 30-minute process of stowing itself. After the sample return capsule's lid closes, it will remain sealed until Earth return in January 2006.

Earth Return

Stardust is scheduled to use its thrusters to fine tune its flight path three times as it makes its final return to Earth on January 15, 2006. These maneuvers are scheduled 13 days, one day and 12 hours before Earth entry, respectively.

Soon after the final trajectory maneuver at an altitude of 110,728 kilometers (68,805 miles), *Stardust* will release its sample return capsule. A spring mechanism will impart a spin to the capsule as it is pushed away from the spacecraft in order to stabilize it. After the capsule has been released, the main spacecraft will perform a maneuver to divert itself to avoid entering Earth's atmosphere. The spacecraft will remain in orbit around the Sun.

The capsule will enter Earth's atmosphere at a velocity of approximately 12.8 kilometers per second (28,600 mph). The capsule's aerodynamic shape and center of gravity are designed like a badminton shuttlecock so that the capsule will automatically orient itself with its nose down as it enters the atmosphere.

As the capsule descends, its speed will be reduced by friction on its heat shield, a 60-degree half-angle blunt cone made of a graphite-epoxy composite covered with a new, lightweight thermal protection system.

The capsule will slow to a speed about 1.4 times the speed of sound at an altitude of about 30 kilometers (100,000 feet), at which time a small pyrotechnic charge will be fired, releasing a drogue parachute. After descending to about 3 kilometers (10,000 feet), a line holding the drogue chute will be cut, allowing the drogue to pull out a larger parachute that will carry the capsule to its soft landing. At touchdown, the capsule will be traveling at approximately 4.5 meters per second (14.8 feet per second), or about 16 km/hr (10 mph). In all, about 10 minutes will elapse between the beginning of the entry into Earth's atmosphere until the parachute is deployed.

The landing site at the Utah Test and Training Range near Salt Lake City was chosen because the area is a vast, desolate and unoccupied salt flat controlled by the U.S. Air Force in conjunction with the U.S. Army. The landing footprint for the sample return capsule will be about 30 by 84 kilometers (18 by 52 miles), an ample space to allow for aerodynamic uncertainties and winds that might affect the direction the capsule travels in the atmosphere. To land within the footprint, the capsule's trajectory must achieve an entry accuracy of 0.08 degree. The sample return capsule will approach the landing zone on a heading of approximately 122 degrees on a northwest to southeast trajectory. Landing time will take place at about 3 a.m. Mountain Standard Time on January 15, 2006.

The actual landing footprint will be predicted by tracking the spacecraft just before the capsule's release. Roughly six hours before entry, an updated footprint will be provided to the capsule recovery team.

Ground Recovery

A UHF radio beacon on the capsule will transmit a signal as the capsule descends to Earth, while the parachute and capsule will be tracked by radar. A helicopter will be used to fly the retrieval crew to the landing site. Given the small size and mass of the capsule, mission planners do not expect that its recovery and transportation will require extraordinary handling measures or hardware other than a specialized handling fixture to cradle the capsule during transport.

Sample Curation

Once the sample return capsule is recovered, its contents will be immediately transported to its final destination, the planetary material curatorial facility at NASA's Johnson Space Center in Houston. The Johnson Space Center's curation laboratory is a special facility designed for payload cleaning and curation of samples returned from space missions. It includes facilities for the Apollo lunar samples, Antarctic meteorites, cosmic dust, samples from NASA's Genesis mission collecting solar particles, and hardware exposed to the environment of space.

The laboratory consists of numerous clean rooms maintained at varying degrees of cleanliness. The laboratory currently being developed for *Stardust* samples will be class 100. A class 100 clean room maintains less than 100 particles larger than 0.5 microns in each cubic foot of air space (or about 3,530 particles per cubic meter of air).

Once safely at the curation laboratory, technicians will open the *Stardust* sample return capsule. The aerogel and its collection of comet and interstellar dust will be inspected, extracted, characterized, and made available to the scientific community for analysis.

The Stardust curation team has developed exacting techniques for the removal and analysis of captured grains from the silica aerogel in which it is embedded. They will continue to improve and practice these techniques before the comet samples are in their hands in 2006.

Most particles from a comet are smaller than the diameter of a human hair. The expected total mass of the sample returned by *Stardust* will probably be about 1 milligram – less than a thimbleful. Though this sample quantity could seem small, to cometary scientists this celestial acquisition is nearly an embarrassment of riches. Abundant evidence indicates that solid samples from both cometary and interstellar sources are very fine-grained, most of them on the scale of a micron (1/50th the diameter of a human hair) or smaller. Because the Stardust science team is focused on these grains, they do not require a large sample mass. A single 100-micron cometary particle could be an aggregate composed of millions of individual interstellar grains.

The key information in these samples is retained at the micron level, and even aggregates of 10 microns in size are considered giant samples.

Planetary Protection

The U.S. is a signatory to the United Nations' 1966 Treaty of Principles Governing the Activities of States in the Exploration and Use of Outer Space, Including the Moon and Other Celestial Bodies. Known as the "Outer Space Treaty," this document states in part that exploration of the Moon and other celestial bodies shall be conducted "so as to avoid their harmful contamination and also adverse changes in the environment of the Earth resulting from the introduction of extraterrestrial matter."

Comets are believed to be primordial bodies made up of material that is virtually unchanged since their creation when the solar system formed 4.6 billion years ago. This means that any evolutionary processes leading to the emergence of life have not occurred. There is no scientific reason to believe that bacteria or viruses or any other life exist on comets. One of the objectives of the Stardust mission is to investigate whether the chemical building blocks of life exist on comets. But even if such building blocks do reside there, comets have not provided the hospitable environment required over millions of years to accommodate the complex processes that could result in the emergence of even single-celled organic life.

On *Stardust*, all comet particles that are collected will be heated to extremely high temperature due to their impact speed on the aerogel collector. The temperature caused by the compression interaction between the aerogel and any given particle is calculated to be at least 1,000° C (roughly 1,800° F). In fact, the collector material literally melts to encapsulate the captured particles. Such high temperatures are naturally sterilizing. As a particle hits the aerogel sample collector, it will come to a dead stop within a microsecond, having traveled about 3 centimeters (1.2 inches) into the aerogel. By that point, the aerogel, which is silica-based, will have melted around the particle, trapping it in glass.

It should be noted that particles from space, including material from comets, fall onto Earth's surface at a rate of approximately 40,000 tons per year, and some of this material is believed to survive atmospheric entry without severe heating.

Names Microchips

The Stardust project sponsored a "Send Your Name to a Comet" campaign that invited people from around the world to submit their names via the Internet to fly onboard the Stardust spacecraft. Two microchips bearing the names of more than 1.1 million people are on board the Stardust spacecraft.

The names were electronically etched onto fingernail-sized silicon chips at a microdevices laboratory at JPL. The lettering on the microchips is so small that about 80 letters would equal the width of a human hair. The names can be read only with the aid of an electron microscope.

The first Stardust microchip contains 136,000 names collected in 1997 from persons all over the world. That microchip has been placed inside the sample return capsule. The second microchip contains more than a million names from members of the public, and was placed on the back of the arm that holds the dust collector. In addition to holding names from the public at large, the second microchip contains all 58,214 names inscribed on the Vietnam Veterans Memorial in Washington, D.C., as a tribute to those who died in that war.

Spacecraft

The Stardust spacecraft incorporates innovative, state-of-the-art technologies pioneered by other recent missions with off-the-shelf spacecraft components and, in some cases, spare parts and instrumentation left over from previous missions.

The Stardust spacecraft is derived from a rectangular deep space bus called SpaceProbe developed by Lockheed-Martin Space Systems, Denver, Colorado. Total weight of the spacecraft, including the sample return capsule and propellant carried on board for trajectory adjustments, is 385 kilograms (848 pounds). The main bus is 1.7 meters (5.6 feet) high, 0.66 meter (2.16 feet) wide and 0.66 meter (2.16 feet) deep, about the size of an average office desk. Panels are made of a core of aluminum honeycomb, with outer layers of graphite fibers and polycyanate face sheets. When its two parallel solar panels are deployed in space, the spacecraft takes on the shape of a letter H.

There are three dedicated science packages on Stardust – the two-sided dust collector, the comet and interstellar dust analyzer, and the dust flux monitor. Science data will also be obtained without dedicated hardware. The navigation camera, for example, will provide images of the comet both for targeting accuracy and scientific analysis.

Aerogel Dust Collectors

To collect particles without damaging them, Stardust will use an extraordinary substance called aerogel – a silicon-based solid with a porous, sponge-like structure in which 99 percent of the volume is empty space. Originally invented in 1930 by a researcher at the College of the Pacific in Northern California, aerogel is made from fine silica mixed with a solvent. The mixture is set in molds of the desired shape and thickness, and then pressure-cooked at high temperature.

Over the past several years, aerogel has been made and flight-qualified at the Jet Propulsion Laboratory for space missions. A cube of aerogel looks like solid, pale-blue smoke. It is the lightest-weight, lowest-mass solid known, and has been found to be ideal for capturing tiny particles in space. There is extensive experience, both in laboratory and space flight experiments, in using aerogel to collect hypervelocity particles. Eight Space Shuttle flights have been equipped with aerogel collectors.

The exotic material has many unusual properties, such as uniquely low thermal and sound conductivity, in addition to its exceptional ability to capture hypervelocity dust. Aerogel was also used as a lightweight thermal insulator on Mars Pathfinder's Sojourner rover. When Stardust flies through the comet's coma, the impact velocity of particles as they are captured will be up to six times the speed of a bullet fired from a high-powered rifle. The Whipple shields can protect the spacecraft from impacts of particles the size of a pea, but larger particles present a more severe hazard.

Although the particles captured in aerogel will each be smaller than a grain of sand, high-speed capture in most substances would alter their shape and chemical composition – or vaporize them entirely. With aerogel, however, particles are softly caught in the material and slowed to a stop. When a particle hits the aerogel, it will bury itself, creating a carrot-shaped track in the aerogel up to 200 times its own length as it slows down and comes to a stop. The aerogel made for the Stardust mission has extraordinary, water-like clarity that will allow scientists to locate a particle at the end of each track etched in the substance. Each narrow, hollow cone leading to a particle will easily be seen in the aerogel with a stereo microscope.

The sizes of the particles collected in the aerogel are expected to range mostly from about a micron (a millionth of a meter, or $1/25,000^{th}$ of an inch, or about $1/50^{th}$ of the width of a human hair) to 100 microns (a tenth of a millimeter, or $1/250^{th}$ of an inch, or about twice the width of a human hair). Stardust scientists anticipate that the aerogel will collect a few particles at the upper end of this size range, and many more particles in the submicron range. Most of the scientific analysis will be devoted to particles that are 15 microns (about $1/1,700^{th}$ of an inch, or about one-third the width of a human hair) in size. The Stardust science team expects that the samples returned will be profoundly complex, and each particle will be probed for years in research labs.

One side of the dust collection module, called the "A side," will be used for the comet encounter, while the opposite side ("B side") will be used for interstellar collection. More than 1,000 square centimeters (160 square inches) of collection area is provided on each side. Each of *Stardust*'s two collectors has 130 rectangular blocks of aerogel measuring 2 by 4 centimeters (0.8 by 1.6 inches), plus two slightly smaller rhomboidal blocks.

The thickness of the aerogel on the cometary particle collection side is 3 centimeters (1.2 inches), while the thickness of the aerogel on the interstellar dust particle collection side is 1 centimeter (0.4 inch). The density of the aerogel is graded – less dense at the point of particle entry, and progressively denser deeper in the material. Each block of aerogel is held in a frame with thin aluminum sheeting.

Overall, the collection unit resembles a metal ice tray set in an oversize tennis racket. It is similar to previous systems used to collect particles in Earth orbit on *SpaceHab* and other Space Shuttle-borne experiments. The sample return capsule is a little less than a meter (or yard) in diameter, and opens like a clamshell to extend the dust collector into the dust stream. After collecting samples, the cell assembly will fold down for stowage into the sample return capsule.

Stardust Spacecraft

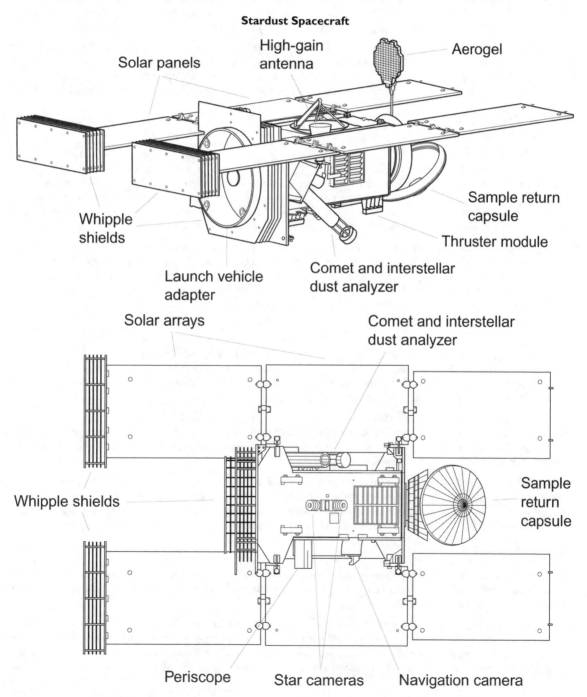

Comet and Interstellar Dust Analyzer

The comet and interstellar dust analyzer is derived from the design of an instrument that flew on the European Space Agency's *Giotto* spacecraft and the Soviet Union's *Vega* spacecraft when they encountered Comet Halley in 1986. The instrument obtained unique data on the chemical composition of individual particulates in Halley's coma. *Stardust*'s version of the instrument will study the chemical composition of particulates in the coma of comet Wild-2.

The purpose of the analyzer instrument is to intercept and perform instantaneous compositional analysis of dust as it is encountered by the spacecraft. Data will be transmitted to Earth as soon as a communication link is available.

The instrument is what scientists call a "time-of-flight" mass spectrometer, which separates the masses of ions by comparing differences in their flight times. When a dust particle hits the instrument's target, the impact creates ions which are extracted from the particle by an electrostatic grid. Depending on the polarity of the target, positive or negative ions can be extracted. As extracted ions move through the instrument, they are reflected and then detected. Heavier ions take more time to travel through the instrument than lighter ones, so the flight times of the ions are then used to calculate their masses. From this information, the ion's chemical identification can be made. In all, the instrument consists of a particle inlet, a target, an ion extractor, a mass spectrometer and an ion detector.

Co-investigator in charge of the comet and interstellar dust analyzer is Dr. Jochen Kissel of the Max-Planck-Institut für Extraterrestrische Physik, Garching, Germany. The instrument was developed and fabricated by von Hoerner & Sulger GmbH, Schwetzingen, Germany, under contract to the German Space Agency and the Max-Planck-Institut. Software for the instrument was developed by the Finnish Meteorological Institute, Helsinki, Finland, under subcontract to von Hoerner & Sulger.

Dust Flux Monitor

The dust flux monitor measures the size and frequency of dust particles in the comet's coma. The instrument consists of two film sensors and two vibration sensors. The film material responds to particle impacts by generating a small electrical signal when penetrated by dust particles. The mass of the particle is determined by measuring the size of the electrical signals. The number of particles is determined by counting the number of signals. By using two film sensors with different diameters and thicknesses, the instrument will provide data on what particle sizes were encountered and what the size distribution of the particles is.

The two vibration sensors are designed to provide similar data for larger particles, and are installed on the Whipple shield that protect the spacecraft's main bus. These sensors will detect the impact of large comet dust particles that penetrate the outer layers of the shield. This system, essentially a particle impact counter, will give mission engineers information about the potential dust hazard as the spacecraft flies through the coma environment. Co-investigator in charge of the dust flux monitor is Dr. Anthony Tuzzolino of the University of Chicago, where the monitor was developed.

Navigation Camera

Stardust's navigation camera is an amalgam of flight-ready hardware left over from other NASA solar system exploration missions. The main camera is a spare wide-angle unit left over from the two *Voyager* spacecraft missions launched to the outer planets in 1977. The camera uses a single clear filter, thermal housing, and spare optics and mechanisms. For *Stardust*, designers added a thermal radiator.

Also combined with the camera is a modernized sensor head left over from the *Galileo* mission to Jupiter launched in 1989. The sensor head uses the existing *Galileo* design updated with a 1024 by 1024-pixel array charge coupled device (CCD) from the *Cassini* mission to Saturn, but has been modified to use new miniature electronics. Other components originated for NASA's *Deep Space 1* program.

During distant imaging of the comet's coma, the camera will take pictures through a periscope in order to protect the camera's primary optics as the spacecraft enters the coma. In the periscope, light is reflected off mirrors made of highly polished metals designed to minimize image degradation while withstanding particle impacts. During close approach, the nucleus is tracked and several images taken with a rotating mirror that no longer views through the periscope.

Propulsion System

The *Stardust* spacecraft needs only a relatively modest propulsion system because it is on a low-energy trajectory for its flyby of comet Wild-2 and subsequent return to Earth, and because it was aided by a gravity assisted boost maneuver when it flew past Earth in January 2001.

The spacecraft is equipped with two sets of thrusters that use hydrazine as a monopropellant. Eight larger thrusters, each of which puts out 4.4 newtons (1 pound) of thrust, will be used for trajectory correction maneuvers or turning the spacecraft. Eight smaller thrusters producing 0.9 newton (0.2 pound) of thrust each will be used to control the spacecraft's attitude, or orientation. The thrusters are in four clusters located on the opposite side of the spacecraft from the deployed aerogel. At launch the spacecraft carried 85 kilograms (187 pounds) of hydrazine propellant. When *Stardust* flies by the comet, it will be carrying about 31 kilograms (68.3 pounds) of fuel. This is well within the mission's fuel budget.

Attitude Control

The attitude control system manages the spacecraft's orientation in space. Like most solar system exploration spacecraft, *Stardust* is three-axis stabilized, meaning that its orientation is held fixed in relation to space, as opposed to spacecraft that stabilize themselves by spinning.

Stardust determines its orientation at any given time using a star camera or one of two inertial measurement units, each of which consists of three ring-laser gyroscopes and three accelerometers. The spacecraft's orientation is changed by firing thrusters. The inertial measurement units are needed only during trajectory correction maneuvers and during the fly-through of the cometary coma when stars may be difficult to detect. Otherwise, the vehicle can be operated in a mode using only stellar guidance for spacecraft positioning. Two Sun sensors will serve as backup units, coming into play if needed to augment or replace the information provided by the rest of the attitude control system's elements.

Command and Data Handling

The spacecraft's computer is embedded in the spacecraft's command and data handling subsystem, and provides computing capability for all spacecraft subsystems. At its heart is a RAD6000 processor, a radiation-hardened version of the PowerPC chip used on some models of Macintosh computers. It can be switched between clock speeds of 5, 10 or 20 MHz. The computer includes 128 megabytes of random access memory (RAM); unlike many previous spacecraft, *Stardust* does not have an onboard tape recorder, but instead stores data in its RAM for transmission to Earth. The computer also has 3 megabytes of programmable memory that can store data even when the computer is powered off.

The spacecraft uses about 20 percent of the 128 megabytes of data storage for its own internal housekeeping. The rest of the memory is used to store science data and for computer programs that control science observations. Memory allocated to specific instruments includes about 75 megabytes for images taken by the navigation camera, 13 megabytes for data from the comet and interstellar dust analyzer, and 2 megabytes for data from the dust flux monitor.

Power

Two solar array panels affixed to the spacecraft were deployed shortly after launch. Together they provide 6.6 square meters (7.9 square yards) of solar collecting area using high-efficiency silicon solar cells. One 16 amp-hour nickel-hydrogen battery provides power when the solar arrays are pointed away from the Sun and during peak power operations.

Thermal Control

Stardust's thermal control subsystem uses louvers to control the temperature of the inertial measurement units and the telecommunications system's solid state power amplifiers. Thermal coatings and multi-layer insulation blankets and heaters are used to control the temperature of other parts of the spacecraft.

Telecommunications

Stardust is equipped with a transponder (radio transmitter / receiver) originally developed for the *Cassini* mission to Saturn, as well as a 15-watt radio frequency solid state amplifier. Data rates will range from 40 to 22,000 bits per second.

During cruise, communications are mainly conducted through the spacecraft's medium-gain antenna. Three low-gain antennas are used for initial communications near Earth and to receive commands when the spacecraft is in nearly any orientation.

A 0.6-meter (2-foot)-diameter high-gain dish antenna is used primarily for communication immediately following closest approach to the comet. *Stardust* will use it to transmit images of the comet nucleus, as well as data from the comet and interstellar dust analyzer and the dust flux monitor, at a high data rate to minimize the transmission time and the risk of losing data during the extended time that would be required to transmit the data through the medium-gain antenna. Most data from the spacecraft will be received through the Deep Space Network's 34-meter (112-foot)-diameter ground antennas, but 70-meter (230-foot) antennas will be used during some critical telecommunications phases, such as when *Stardust* transmits science data during and after the comet encounter.

Redundancy

Virtually all spacecraft components are redundant, with critical items "cross-strapped" or interconnected so that they can be switched in or out most efficiently. The battery includes an extra pair of cells. Fault protection software is designed so that the spacecraft is protected from reasonable, credible faults without unnecessarily putting the spacecraft into a safe mode due to unanticipated but probably benign glitches.

Whipple Shields

The shields that will protect *Stardust* from the blast of cometary particles is named for American astronomer Dr. Fred L. Whipple, who in 1950 accurately predicted the "dirty snowball" model of the cometary nucleus as a mixture of dark organic material, rocky grains and water ice. Whipple came up with the idea of shielding spacecraft from high-speed collisions of the bits and pieces that are ejected from comets as they circle the Sun.

The system includes two bumpers at the front of the spacecraft – which protect the solar panels – and another shield protecting the main spacecraft body. Each of the shields is built around composite panels designed to disperse particles as they impact, augmented by blankets of a ceramic cloth called Nextel that further dissipate and spread particle debris.

Sample Return Capsule

The sample return capsule is a blunt-nosed cone with a diameter of 81 centimeters (32 inches). It has five major components: a heat shield, back shell, sample canister, parachute system and avionics. The total mass of the capsule, including the parachute system, is 45.7 kilograms (101 pounds).

A hinged clamshell mechanism opens and closes the capsule. The dust collector fits inside, extending on hinges to collect samples and retracting to fold down back inside the capsule. The capsule is encased in ablative materials to protect the samples stowed in its interior from the heat of reentry.

The heat shield is made of a graphite-epoxy composite covered with a thermal protection system. The thermal protection system is made of a phenolic-impregnated carbon ablator developed by NASA's Ames Research Center for use on high-speed reentry vehicles. The capsule's heat shield will remain attached to the capsule throughout descent and serves as a protective cover for the sample canister at touchdown.

The backshell structure is also made of a graphite-epoxy composite covered with a thermal protection system that is made of a cork-based material called SLA 561V. The material was developed by Lockheed-Martin for use on the Viking missions to Mars in the 1970s, and has been used on several space missions including NASA's Mars Pathfinder, Genesis and Mars Exploration Rover missions. The backshell structure provides the attach points for the parachute system.

The sample canister is an aluminum enclosure that holds the aerogel and the mechanism used to deploy and stow the aerogel collector during the mission. The canister is mounted on a composite equipment deck suspended between the backshell and heat shield. The parachute system incorporates a drogue and main parachute inside a single canister.

As the capsule descends toward Earth, a gravity switch sensor and timer will trigger a pyrotechnic gas cartridge that will pressurize a mortar tube and expel the drogue chute. The drogue chute will be deployed to provide stability to the capsule when it is at an altitude of approximately 30 kilometers (100,000 feet) moving at a speed of about mach 1.4. Based on information from timer and backup pressure transducers, a small pyrotechnic device will cut the drogue chute from the capsule at an altitude of approximately 3 kilometers (10,000 feet). As the drogue chute moves away, it will extract the 8.2-meter (27-foot)-diameter main chute from the canister. Upon touchdown, cutters will fire to cut the main chute cables so that winds do not drag the capsule across the terrain.

The capsule carries a UHF radio locator beacon to be used in conjunction with locator equipment on the recovery helicopters. The beacon will be turned on at main parachute deployment and will remain on until turned off by recovery personnel. The beacon is powered by redundant sets of lithium sulfur dioxide batteries, which have long shelf life and tolerance to wide temperature extremes, and are safe to handle. The capsule carries sufficient battery capacity to operate the UHF beacon for at least 40 hours.

Science Objectives

The primary goal of the Stardust mission is to collect samples of a comet's coma and return them to Earth. In addition, interstellar dust samples will be gathered en route to the comet.

In laboratories, the samples will be scrutinized to understand their elemental makeup; presence of isotopes; mineralogical and chemical properties; and possible biogenic properties.

During the encounter itself, comet dust will be studied by a mass spectrometer. This instrument will provide data on organic particle materials that might not survive aerogel capture. Since an identical instrument flew on the European Space Agency's *Giotto* spacecraft during its comet Halley flyby in 1986, data from the two missions can be compared to understand how different comets compare and contrast.

During selected portions of its cruise phase, *Stardust* collected interstellar grains now passing through the solar system. Interstellar gas samples should also be absorbed in the aerogel, allowing direct measurement of isotopes of elements such as helium and neon.

Laboratory investigation of the returned samples using instruments such as electron microscopes, ion microprobes, atomic force microscopes, synchrotron microprobes and laser probe mass spectrometers will allow examination of cometary matter and interstellar grains at the highest possible level of detail. Advances in microanalytical instruments provide unprecedented capabilities for analysis on the micron and submicron level, even to the atomic scale for imaging.

Stardust Science Team

Dr. Donald Brownlee, University of Washington, Seattle, Washington, Principal Investigator
Dr. Peter Tsou, Jet Propulsion Laboratory, Pasadena, California, Deputy Principal Investigator

Co-Investigators:
Dr. John Anderson, Jet Propulsion Laboratory, Pasadena, California
Dr. Benton C. Clark, Lockheed-Martin Astronautics, Denver, Colorado
Dr. Martha S. Hanner, Jet Propulsion Laboratory, Pasadena, California
Dr. Friedrich Horz, NASA Johnson Space Center, Houston, Texas
Dr. Jochen Kissel, Max Planck Institut, Garching, Germany
Dr. J.A.M. McDonnell, University of Kent, Canterbury, United Kingdom
Ray L. Newburn, Jet Propulsion Laboratory, Pasadena, California (retired)
Dr. Scott Sandford, NASA Ames Research Center, Moffett Field, California
Dr. Zdenek Sekanina, Jet Propulsion Laboratory, Pasadena, California
Dr. Anthony Tuzzolino, University of Chicago, Chicago, Illinois
Dr. Michael E. Zolensky, NASA Johnson Space Center, Houston, Texas

These instruments will provide direct information on the nature of the interstellar grains that constitute most of the solid matter in the galaxy, and they will provide a highly intimate view of both pre-solar dust and nebular condensates contained in comets. Such materials will be compared with primitive meteorites and interplanetary dust samples to understand the solids involved in solar system formation, the solids that existed in the outer regions of the nebula where comets formed, and solids in the inner regions of the nebula where asteroids formed. The data will provide fundamental insight into the materials, processes and environments that existed during the origin and early evolution of the solar system.

Interstellar grains are currently studied mainly by astronomical techniques capable only of revealing general physical properties such as size and shape. The recent discovery and study of rare interstellar grains preserved in meteorites has shown that they contain excellent records about the nature of their parent stars, including details of the complex nuclear reactions that occur within the stars. The interstellar grains that have been identified in meteorites are predominantly grains that formed in gas flows from carbon-rich stars such as red giants and what are called ABG stars, while the more typical grains from oxygen-rich stars have not been found. It is expected *Stardust* will collect grains produced by star types that are major sources of interstellar dust.

Comets are now known to contain large quantities of volatiles, including organic compounds and a rich variety of microparticles of various types (pure organic particles, silicates, sulfides and mixed particles) with sizes ranging as low as submicron diameters. Organic particulates actually consist of several sub-populations, which can be described based on the elements that they are made up of. These include particles containing:

- Hydrogen, carbon and nitrogen
- Hydrogen, carbon and oxygen
- Hydrogen and carbon
- Hydrogen, carbon, nitrogen and oxygen, with and without magnesium (termed "CHON" particles)

Since comets are rich in water and other volatiles, it has been postulated that they carried to Earth elements critical to the origin of life. The study of cometary material is essential for understanding the formation of the solar system and the role of organic matter from interstellar sources. Astronomers have identified some 60 compounds in interstellar clouds, three-fourths of which are organic. ("Organic" means that the compound is carbon-based, but not necessarily biological in origin.) There is compelling evidence that four of the first five interstellar molecules detected by astronomers are present in comets, and the fifth might be also.

The volatiles and silicates that appear to be in comets also are found in interstellar clouds. How the elements necessary for life entered the solar system, were transformed by solar system processes, were distributed among planetary bodies, and what molecular and mineral forms they took during this history are questions of major importance for exobiology. Comparing the composition of the volatiles from cometary material with those found in carbonaceous meteorites and interplanetary dust will provide a basis to determine which particles, if any, have common source regions.

Finally, the discovery of iridium in rocks at Earth's Cretaceous-Tertiary geologic layer marking the end of the age of the dinosaurs about 65 million years ago has, along with other evidence, raised the probability that an impact of an asteroid-sized body with Earth was responsible for the demise of the giant creatures. Although the chance of finding a unique elemental signature in captured cometary coma material might be slight, such a discovery would be enormously valuable in distinguishing whether it was an asteroid or a comet that made the impact.

Program / Project Management

Stardust's principal investigator is Dr. Donald Brownlee of the University of Washington, Seattle, Washington. Dr. Peter Tsou of NASA's Jet Propulsion Laboratory, Pasadena, California, is deputy principal investigator.

The Stardust mission is managed by the Jet Propulsion Laboratory for NASA's Office of Space Science, Washington. At NASA Headquarters, Dr. Edward Weiler is associate administrator for space science. Orlando Figueroa is the director of NASA's Solar System Exploration Division, Kenneth Ledbetter is deputy associate administrator for programs, Barry Geldzahler is Stardust program executive, and Dr. Thomas Morgan is Stardust program scientist.

At the Jet Propulsion Laboratory, Tom Duxbury is project manager, and Bob Ryan is mission manager. JPL is a division of the California Institute of Technology, Pasadena, California.

At Lockheed Martin Space Systems, Denver, Colorado, Joseph M. Vellinga is the company's Stardust program manager, and Dr. Benton C. Clark is the company's chief scientist for space exploration systems. Lockheed-Martin Space Systems designed, built and operates the spacecraft, and will recover the sample return capsule.

12-29-03

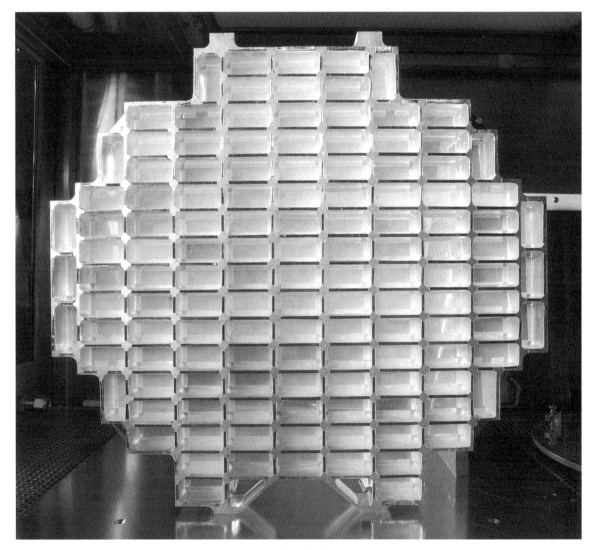

Aerogel Dust Collectors

FUTURE
DEEP SPACE
MISSIONS

DEEP IMPACT
FIRST LOOK INSIDE A COMET!

http://deepimpact.jpl.nasa.gov
http://deepimpact.umd.edu

What's Deep Inside a Comet?

Comets are time capsules that hold clues about the formation and evolution of the solar system. They are composed of ice, gas and dust, primitive debris from the solar system's distant and coldest regions that formed 4.5 billion years ago. Deep Impact, a NASA Discovery Mission, is the first space mission to probe beneath the surface of a comet and reveal the secrets of its interior.

On July 4, 2005, the *Deep Impact* spacecraft arrives at Comet Tempel 1 to impact it with a 370-kilogram (~820-pound) mass. On impact, the crater produced is expected to range in size from that of a house to that of a football stadium, and two to fourteen stories deep. Ice and dust debris is ejected from the crater revealing fresh material beneath. Sunlight reflecting off the ejected material provides a dramatic brightening that fades slowly as the debris dissipates into space or falls back onto the comet. Images from cameras and a spectrometer are sent to Earth covering the approach, the impact and its aftermath. The effects of the collision with the comet will also be observable from certain locations on Earth and in some cases with smaller telescopes. The data is analyzed and combined with that of other NASA and international comet missions. Results from these missions will lead to a better understanding of both the solar system's formation and implications of comets colliding with Earth.

The Mission

The Deep Impact mission lasts six years from start to finish. Planning and design for the mission took place from November 1999 through May 2001. The mission team is proceeding with the building and testing of the two-part spacecraft. The larger "flyby" spacecraft carries a smaller "impactor" spacecraft to Tempel 1 and releases it into the comet's path for a planned collision.

In December 2004, a Delta II rocket launches the combined *Deep Impact* spacecraft, which leaves Earth's orbit and is directed toward the comet. The combined spacecraft approaches Tempel 1 and collects images of the comet before the impact. In early July 2005, 24 hours before impact, the flyby spacecraft points high-precision tracking telescopes at the comet and releases the impactor on a course to hit the comet's sunlit side.

The impactor is a battery-powered spacecraft that operates independently of the flyby spacecraft for just one day. It is called a "smart" impactor because, after its release, it takes over its own navigation and maneuvers into the path of the comet. A camera on the impactor captures and relays images of the comet's nucleus just seconds before collision. The impact is not forceful enough to make an appreciable change in the comet's orbital path around the Sun.

After release of the impactor, the flyby spacecraft maneuvers to a new path that, at closest approach, passes 500 kilometers (300 miles) from the comet. The flyby spacecraft observes and records the impact, the ejected material blasted from the crater, and the structure and composition of the crater's interior. After its shields protect it from the comet's dust tail passing overhead, the flyby spacecraft turns to look at the comet again. The flyby spacecraft takes additional data from the other side of the nucleus and observes changes in the comet's activity. While the flyby spacecraft and impactor do their jobs, professional and amateur astronomers at both large and small telescopes on Earth observe the impact and its aftermath, and results are broadcast over the Internet.

Comet Tempel 1

Comet Tempel 1 was discovered in 1867 by Ernst Tempel. The comet has made many passages through the inner solar system orbiting the Sun every 5.5 years. This makes Tempel 1 a good

NASA

National Aeronautics and
Space Administration

Jet Propulsion Laboratory
California Institute of
Technology
Pasadena, California

JPL 400-936, rev. 1 04/03

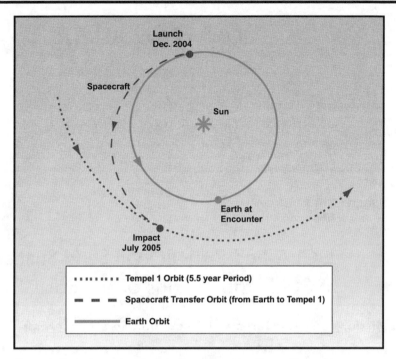

Deep Impact's orbital path to encounter Comet Tempel 1.

target to study evolutionary change in the mantle, or upper crust. Comets are visible for two reasons. First, dust driven from a comet's nucleus reflects sunlight as it travels through space. Second, certain gases in the comet's coma, stimulated by the Sun, give off light like a fluorescent bulb. Over time, a comet may become less active or even dormant. Scientists are eager to learn whether comets exhaust their supply of gas and dust to space or seal it into their interiors. They would also like to learn about the structure of a comet's interior and how it is different from its surface. The controlled cratering experiment of this mission provides answers to these questions.

Technical Implementation

The flyby spacecraft carries a set of instruments and the smart impactor. Two instruments on the flyby spacecraft observe the impact, crater and debris with optical imaging and infrared spectral mapping. The flyby spacecraft uses an X-band radio antenna (transmission at about eight gigahertz) to communicate to Earth as it also listens to the impactor on a different frequency. For most of the mission, the flyby spacecraft communicates through the 34-meter antennae of NASA's Deep Space Network. During the short period of encounter and impact, when there is an increase in volume of data, overlapping antennas around the world are used. Primary data is transmitted immediately and other data is transmitted over the following week. The impactor spacecraft is composed mainly of copper, which is not expected to appear in data from a comet's composition. For its short period of operation, the impactor uses simpler versions of the flyby spacecraft's hardware and software – and fewer backup systems.

The Team

The Deep Impact mission is a partnership among the University of Maryland (UMD), the California Institute of Technology's Jet Propulsion Laboratory (JPL) and Ball Aerospace and Technology Corp. The scientific leadership of the mission is based at UMD. Engineers at Ball Aerospace design and build the spacecraft under JPL's management. Engineers at JPL control the spacecraft after launch and relay data to scientists for analysis. The entire team consists of more than 250 scientists, engineers, managers, and educators. Deep Impact is a NASA Discovery Mission, eighth in a series of low-cost, highly-focused space science investigations. Deep Impact offers an extensive outreach program in partnership with other comet and asteroid missions and institutions to benefit the public, educational and scientific communities.

Dawn Press Release

FOR IMMEDIATE USE
December 2001

UCLA
Harlan Lebo hlebo@college.ucla.edu
Stuart Wolpert stuartw@college.ucla.edu
(310) 206-0510

JPL
Martha Heil martha.heil@jpl.nasa.gov
(818) 354-0850

NASA DISCOVERY PROGRAM
Don Savage dsavage@hq.nasa.gov
(202) 358-1727

UCLA-Led Project Will Send Spacecraft to Study the Origins of the Solar System

Dawn Mission to Study Ceres and Vesta, the Largest Asteroids

The Office of Space Science at NASA has approved the "Dawn Mission," a UCLA-led project that will develop a spacecraft to orbit and study Ceres and Vesta, the two largest asteroids (minor planets) in our solar system.

The Dawn Mission marks the first time that a spacecraft will orbit two planetary bodies on the same mission.

Christopher T. Russell, professor in the Department of Earth and Space Sciences at UCLA, will direct the Dawn mission.

According to current theories, the differing properties of Vesta and Ceres are the result of these minor planets being formed and evolving in different parts of the solar system. By observing both minor planets with the same set of instruments, *Dawn* will provide new answers to questions about the formation and evolution of the early solar system.

The Dawn mission will launch in May 2006. It will study Vesta beginning in July 2010, and Ceres beginning in August 2014.

This press release contains general background material about the Dawn mission. For the web site of the Dawn science team, visit http://www-ssc.igpp.ucla.edu/dawn.

The Dawn Mission is part of NASA's Discovery Program, an initiative for lower-cost, highly-focused, rapid-development scientific spacecraft. The Discovery Program is managed by NASA's Jet Propulsion Laboratory, for the Office of Space Science, Washington, D.C. For more information about the Discovery Program, visit http://discovery.nasa.gov.

The Discovery Program is part of NASA's initiative for lower-cost, highly-focused, rapid-development scientific spacecraft. The Discovery Program is managed by NASA's Jet Propulsion Laboratory, for the Office of Space Science, Washington, D.C.

This press release contains background material about the Dawn mission. The sections include:

- The Dawn Mission
- The Dawn Mission: Why Ceres and Vesta?
- Ceres
- Vesta
- Dawn Mission Overview
- The *Dawn* Spacecraft
- Dawn: Mission Management
- Dawn: Science Team
- Ion Propulsion
- The Discovery Program
- Attempting to Answer the Fundamental Questions
- The Discovery Program Objectives
- Asteroids, Discovery Program and the Dawn Mission: More Information

The Dawn Mission

The goal of the Dawn mission is to understand the conditions and processes during the earliest history of our solar system. To accomplish this mission, *Dawn* will explore the structure and composition of Ceres and Vesta, two minor planets that have many contrasting characteristics and have remained intact since their formation more than 4.6 billion years ago.

The Dawn mission will launch from Cape Canaveral on May 27, 2006. After more than four years of travel, the spacecraft will rendezvous with Vesta on July 30, 2010. *Dawn* will then orbit Vesta for almost a year, studying its basic structure and composition.

On July 3, 2011, *Dawn* will leave orbit around Vesta for a three-year cruise to Ceres. *Dawn* will rendezvous with Ceres and begin orbit on August 20, 2014, and conduct studies and observations until July 26, 2015. After the exploration of Ceres, and more than nine years of space travel, *Dawn* may continue with additional exploration of the asteroid belt.

The Dawn mission is led by UCLA space scientist Christopher T. Russell. Team members include scientists from the German Aerospace Center (DLR), The Institute for Space Astrophysics in Rome, the Jet Propulsion Laboratory, Los Alamos National Laboratory, University of Hawaii, University of Maryland, University of Tennessee (Knoxville), Brown University, NASA Goddard Space Flight Center, University of Arizona and Massachusetts Institute of Technology.

Orbital Sciences Corporation will construct the spacecraft, and the Jet Propulsion Laboratory will provide the ion engines and management of the overall flight system development. The German Aerospace Center will provide the framing camera, and the Institute for Space Astrophysics in Rome will provide the mapping spectrometer.

Educators from New Roads School in Santa Monica, California, and Mid-continent Research for Education and Learning (McREL), Aurora, Colorado, will develop standards-based learning materials for people of all ages, helping to bring the mission into classrooms and homes across the nation.

The Dawn Mission: Why Ceres and Vesta?

The Dawn mission focuses on two of the first bodies formed in our solar system, the minor planets Ceres and Vesta.

Studies of meteorites believed to be from Vesta that were found on Earth suggest that this body formed from dust in the solar nebula within 5-15 million years of the time the solar system evolved about 4.6 billion years ago. Although no meteorites from Ceres have yet been found, this body also formed during the first 10 million years of the solar system's existence. Today Ceres and Vesta represent two of the few large protoplanets that have not been heavily damaged by collisions with other bodies.

Ceres and Vesta feature striking contrasts in composition. Scientists believe many of these differences stem from the conditions under which Ceres and Vesta formed, with Ceres forming wet and Vesta dry. Evidence of water – frost or vapor on the surface, and possibly liquid water under the surface – still exist on Ceres; this water kept Ceres cool throughout its evolution. At the same time, Vesta was hot, melted internally and became volcanic early in its development. As a result of these two different evolutionary paths, Ceres remains in its primordial state, while Vesta evolved and changed over millions of years.

Because these bodies lie near the plane of the Earth's orbit, they can both be studied with a single mission. The Dawn mission will help us understand the evolution of the interior structure and thermal history of Ceres and Vesta – information that provides keys to the secrets of the creation of our solar system.

Dawn Mission: The Minor Planets – Ceres

Ceres, the largest asteroid in our solar system, is a roughly round object about 600 miles in diameter. It orbits the sun in the asteroid belt between Mars and Jupiter approximately 258 million miles from Earth.

The year 2001 marks the 200th anniversary of the discovery of Ceres by Giuseppe Piazzi in 1801 with a small telescope atop the royal palace in Palermo. At first Piazzi believed he had found the missing planet expected to be in the region we now call the asteroid belt. However this minor planet turned out to be very small indeed, only one-quarter of the diameter of the Earth's Moon.

Ceres was the first asteroid to be discovered in our solar system. Additional observations by Piazzi were cut short due to illness. Carl Friedrich Gauss, at the age of 24, was able to solve a system of 17 linear equations to allow Ceres to be rediscovered, a remarkable feat for this time. Within one year of its initial discovery, both H. Olbers and Franz von Zach were also able to re-identify Ceres.

Ceres was named after the Roman goddess of agriculture. It circles the sun in 4.6 terrestrial years.

Dawn Mission: The Minor Planets – Vesta

Vesta is the brightest asteroid in our solar system, and is the only one visible with the unaided eye; its oval, pumpkin-like shape has an average diameter of about 320 miles. Vesta is the second most massive minor planet. Found on March 29, 1807, by Heinrich Wilhelm Matthäus Olbers, it was the fourth minor planet to be discovered.

Named for the ancient Roman goddess of the hearth, Vesta is approximately 220 million miles from Earth. It circles the sun in 3.6 terrestrial years.

Dawn Mission Overview

- May 27, 2006:
  Launch from Cape Canaveral on a Delta 7925H rocket

- July 30, 2010 - July 3, 2011:
  Rendezvous with Vesta (nine month study)
  Orbit at 420 and 80 miles

- July 3, 2011 - August 20, 2014:
  Cruise to Ceres

- August 20, 2014 - July 26, 2015:
  Rendezvous with Ceres (nine-month study)
  Orbit at 530 and 80 miles

- July 26, 2015:
  Observation of Ceres concludes
  Possible continuing exploration in the asteroid belt

2006 Vesta-Ceres Rendezvous

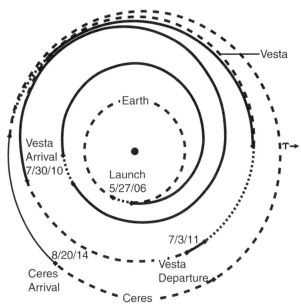

····Non thrusting ▬▬Thrusting ▬ ▪Earth, Vesta, Ceres

The Dawn Spacecraft

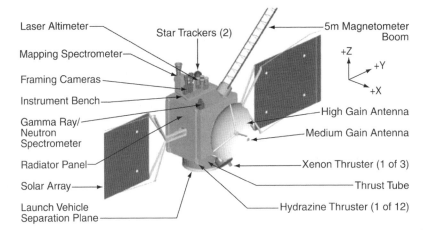

With a two-stage, nine-year journey to study the minor planets Ceres and Vesta, *Dawn* will become the first spacecraft to orbit two planetary bodies during the same mission.

To accomplish its mission to study the physical structure and evolution of asteroids Ceres and Vesta, *Dawn* would carry:

- A framing camera provided by the German Aerospace Center (DLR), Institute of Space Sensor Technology and Planetary Exploration
- A mapping spectrometer provided by the Institute for Space Astrophysics in Rome
- A laser altimeter provided by the NASA Goddard Space Flight Center
- A gamma ray spectrometer from the Department of Energy's Los Alamos National Laboratory
- A magnetometer provided by UCLA

The *Dawn* spacecraft will be built using construction methods and components that have been used successfully in many other satellites, including *Deep Space 1*, *Orbview*, the Topex-Poseidon ocean topography mission and the Far Ultraviolet Spectrum Explorer. The *Dawn* spacecraft will also be the first fully scientific space mission to use ion propulsion to power the spacecraft's journey.

Dawn: Mission Management

UCLA:
• Science lead – Science operations, data products, archiving and analysis

German Aerospace Center (DLR), Institute of Space Sensor Technology and Planetary Exploration:
• Framing camera and mapping spectrometer: Integration, testing, sequencing and analysis

Jet Propulsion Laboratory (JPL):
• Project management, systems engineering, mission assurance, payload, SEP, navigation, mission operations, level zero data

Orbital Science Corporation:
• Spacecraft design and construction

Dawn: Science Team

A. Coradini	Institute for Space Astrophysics (IAS), Rome
W.C. Feldman	Los Alamos National Laboratory (LANL)
R. Jaumann	Institute of Space Sensor Technology and Planetary Exploration, German Aerospace Center (DLR)
A.S. Konopliv	Jet Propulsion Laboratory (JPL)
T.B. McCord	University of Hawaii
L.A. McFadden	University of Maryland
H.Y. McSween	University of Tennessee, Knoxville
S. Mottola	DLR
G. Neukum	DLR
C.M. Pieters	Brown University
C.A. Raymond	JPL
C.T. Russell	UCLA (project director)
D.E. Smith	NASA Goddard Space Flight Center
M.V. Sykes	University of Arizona
B. Williams	JPL
M.T. Zuber	Massachusetts Institute of Technology

Ion Propulsion

The *Dawn* spacecraft would be the first purely scientific mission to be powered by ion propulsion, an advanced technology now being used by NASA's *Deep Space 1* mission.

The principle behind the ion engines is much the same as the phenomenon you experience when you pull hot socks out of the clothes dryer on a cold winter day. The socks push away from each other because they are electrostatically charged, and like charges repel. The challenge in electric space propulsion is to charge a fluid so its atoms can be expelled in one direction, and thus propel the spacecraft in the other direction.

Unlike chemical rocket engines, ion engines accelerate nearly continuously, giving each ion a tremendous burst of speed. The fuel used by an ion engine is xenon, a gas also used in photo flash units, that is more than 4 times heavier than air.

When the ion engine is running, electrons are emitted from a hollow tube called a cathode. These electrons enter a magnet-ringed chamber, where they strike the xenon atoms. The impact of an electron on a xenon atom knocks away one of xenon's 54 electrons. This results in a xenon atom with a positive charge – a xenon ion.

At the rear of the chamber, a pair of metal grids is charged positively and negatively. The force of this electric charge exerts a strong electrostatic pull on the xenon ions. The xenon ions shoot out the back of the engine at a speed of 68,000 mph.

At full throttle, the ion engine consumes 2,500 watts of electrical power, and produces $1/50^{th}$ of a pound of thrust – about the same pressure as a sheet of paper resting on the palm of a hand. That is far less thrust than is produced by even small chemical rockets. But an ion engine can run for months or even years, and despite the almost imperceptible thrust, this engine, for a given amount of fuel, can gradually increase a spacecraft's velocity 10 times more than can a conventional rocket powered by liquid or solid fuel.

The ion propulsion system on NASA's *Deep Space 1* spacecraft won *Discover* Magazine's Award for Technological Innovation.

For more information about NASA developments in ion propulsion, visit NASA's site.

The Discovery Program

NASA's Discovery program offers the scientific community opportunities to accomplish frequent scientific investigations using innovative and efficient management approaches. It seeks to keep performance high and expenses low by using new technologies and strict cost caps. Proposals require careful tradeoffs between science and cost to produce investigations with the highest possible science value for the price.

Discovery solicits mission proposals that are assembled by a team from industry, small businesses, government laboratories and universities. The goal is to launch these smaller missions every 12-24 months at a cost of less than $299 million.

The Discovery Program is managed at NASA's Jet Propulsion Laboratory, a division of the California Institute of Technology, Pasadena, for the Office of Space Science, Washington, D.C. More information on the Discovery Program is available at http://discovery.nasa.gov.

Attempting to Answer the Fundamental Questions

The Space Science Enterprise Strategic Plan published by NASA in November 1997 addresses the concept of cosmic origins, evolution and destiny – how the universe began, how life on Earth originated and what fate awaits our planet and our species. All that we do in space science is part of a quest to understand our cosmic origins and destiny, and how these are linked by cycles of evolution. In response, the space science community has formulated the following fundamental questions that lie at the core of gaining this understanding:

- How did the universe, galaxies, stars and planets evolve? How can our exploration of the universe and our solar system revolutionize our understanding of physics, chemistry and biology?

- Does life in any form exist elsewhere than on planet Earth? Are there Earth-like planets beyond our solar system?

- How can we use knowledge of the Sun, Earth and other planetary bodies to improve the quality of life on Earth? What cutting edge technologies must we develop to conduct our research in the most productive, economical and timely manner?

The Discovery Program Objectives

The Discovery Program's prime objective is to enhance our understanding of the solar system by providing answers to these fundamental questions. Discovery missions explore the planets, their moons and other small bodies within the solar system, either by traveling to them or by remote examination. Discovery also studies planetary systems beyond our solar system.

Discovery also offers a Mission of Opportunity program which allows investigators to participate in a non-NASA mission, typically sponsored by non-U.S. governments, other U.S. government agencies or private sector organizations. This participation could include providing a complete science instrument, hardware components of a science instrument or expertise in critical areas of the mission.

Eight Discovery Missions have been chosen to date, including the Near Earth Asteroid Rendezvous, the Mars Pathfinder Mission, the *Lunar Prospector*, the *Stardust* spacecraft, the *Genesis* spacecraft, the Comet Nucleus Tour, the MESSENGER mission and the Deep Impact Mission.

More information about the Discovery program and its missions is available at http://discovery.nasa.gov.

More Information

Asteroids, Discovery Program and the Dawn Mission:

Dawn Mission – on-line media information:
  http://www.college.ucla.edu/dawn

Dawn Mission – technical background:
  http://www-ssc.igpp.ucla.edu/dawn

Interviews with Dawn team members:
  UCLA: Harlan Lebo,   (310) 206-0510
  UCLA: Stuart Wolpert, (310) 206-0511

JPL contact:
    Martha Heil, (818) 354-0850

Information about the Discovery Program and the NASA Office of Space Science:
    NASA: Don Savage, (202) 358-1727

Asteroids – An Introduction:
    http://www.solarviews.com/eng/asteroid.htm

Ceres:
    See the paper "Bode's Law and the Discovery of Ceres" by Michael Hoskin, at
    http://www.astropa.unipa.it/versione_inglese/Hystory/BODE'S_LAW.htm

Vesta:
    http://www.solarviews.com/eng/vesta.htm

Ion Propulsion – Background:
    http://www.science.nasa.gov/newhome/headlines/prop06apr99_2.htm

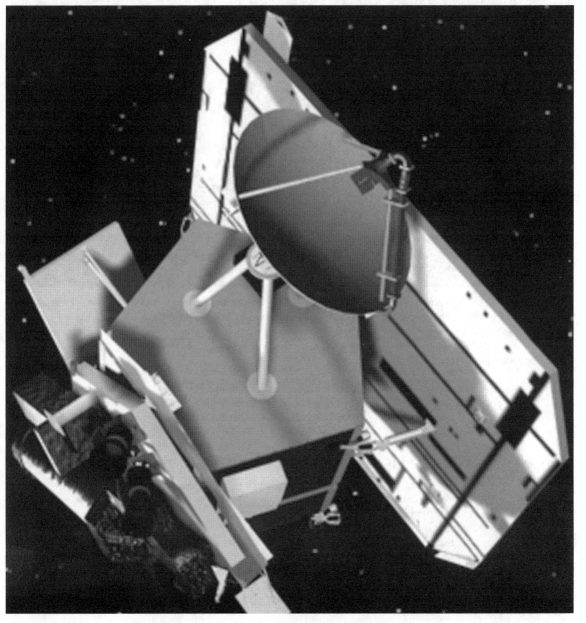

Artist's concept of the Deep Impact Flyby Spacecraft

Jupiter Icy Moons Orbiter
Fact Sheet

What is the Jupiter Icy Moons Orbiter?

NASA is developing plans for an ambitious mission to orbit three planet-sized moons of Jupiter – Callisto, Ganymede and Europa – which may harbor vast oceans beneath their icy surfaces. The mission, called the Jupiter Icy Moons Orbiter, would orbit each of these moons for extensive investigations of their makeup, their history and their potential for sustaining life. NASA's *Galileo* spacecraft found evidence for these subsurface oceans, a finding that ranks among the major scientific discoveries of the Space Age.

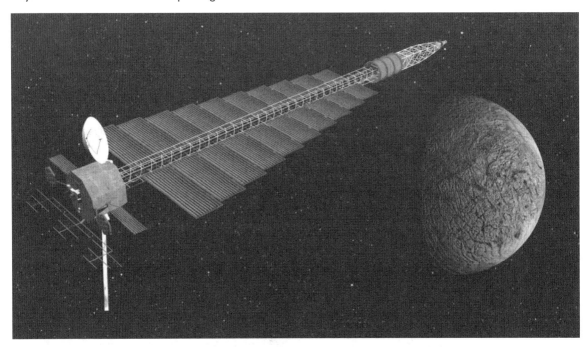

A technology-pioneering spacecraft called Jupiter Icy Moons Orbiter approaches the moon Europa in this artist's version of a proposed NASA mission. The spacecraft would orbit three different moons of Jupiter where earlier spacecraft discovered evidence for vast saltwater oceans hidden beneath icy surface layers: Europa, Ganymede and Callisto. The proposed mission would also raise NASA's capability for space exploration to a revolutionary new level by demonstrating safe and reliable use of electric propulsion powered by a nuclear fission reactor.

The JIMO mission also will raise NASA's capability for space exploration to a revolutionary new level by pioneering the use of electric propulsion powered by a nuclear fission reactor. This technology not only makes it possible to consider a realistic mission for orbiting three of the moons of Jupiter, one after the other, it also would open the rest of the outer Solar System to detailed exploration in later missions.

Why are Jupiter's Icy Moons a Priority for a Major NASA Program?

To explore the universe and search for life is central to the agency's mission. Jupiter's large icy moons appear to have three ingredients considered essential for life: water, energy, and the necessary chemical elements. The evidence from *Galileo* suggests melted water on Europa has been in contact with the surface in geologically recent times and may still lie relatively close to the surface. The National Research Council completed a report last year drawing on input from scores of planetary scientists to prioritize potential projects for exploring the solar system. It ranked a Europa orbiter proposal as top priority for a "flagship" mission, because of the recent discovery of Europa's ocean and the potential that it might harbor life. The *Jupiter Icy Moons Orbiter* would fulfill NASA's science goals for the exploration of Europa and also examine Callisto and Ganymede, providing comparisons key to understanding the evolution of all three.

What are the Science Goals?

The mission has three top-level science goals:

1. Scout the potential for sustaining life on these moons. This would include determining whether the moons do indeed have subsurface oceans; mapping where organic compounds and other chemicals of biological interest lie on the surface; and determining the thicknesses of ice layers, with emphasis on locating potential future landing sites.
2. Investigate the origin and evolution of these moons. This would include determining their interior structures, surface features and surface compositions in order to interpret their evolutionary histories (geology, geochemistry, geophysics) and how this illuminated the understanding of the origin and evolution of the Earth.
3. Determine the radiation environments around these moons and the rates at which the moons are weathered by material hitting their surfaces. Callisto, Ganymede and Europa all orbit within the powerful magnetic environment that surrounds Jupiter. They display varying effects from the natural radiation, charged particles and dust within this environment. Understanding this environment has implications for understanding whether life could have arisen on these distant moons.

What Instruments Would it Carry?

NASA would choose the final suite of instruments through a competitive process open to proposals from scientists worldwide. Two highly probable ones are a radar instrument for mapping the thickness of surface ice and a laser instrument for mapping surface elevations. Others would likely include a camera, an infrared imager, a magnetometer, and instruments to study charged particles, atoms and dust that the spacecraft encounters near each moon. A generous electrical power supply available from the onboard nuclear system could run higher-powered instruments than have flown on other spacecraft and would boost the data transmission rate back to Earth. The expanded scientific capacities would allow mapping the entire surfaces of Callisto and Ganymede, and more than half of Europa, in enough detail to see features as small as a house.

How Would it Get There?

The proposed baseline for the spacecraft incorporated a form of electric propulsion, called ion propulsion. NASA's Deep Space 1 mission, that successfully demonstrated ion propulsion for interplanetary travel, drew electricity for its thrusters from solar panels. The *Jupiter Icy Moons Orbiter*, a more heavily instrumented craft traveling farther from the Sun, would power its ion thrusters with a nuclear fission reactor and a system for converting the reactor's heat to electricity. This could give the craft more than 100 times as much power as a non-fission system of comparable weight.

What is the Mission Timeline?

To allow sufficient development and ground testing time, the mission is not proposed for launch before the year 2011. A heavy lift expendable launch vehicle would lift the spacecraft into high-Earth orbit. The ion propulsion thrusters would spiral the spacecraft away from Earth and then on its trip to Jupiter. After entering orbit around Jupiter, the spacecraft would then orbit Callisto, then Ganymede, and finally Europa. The intensity of the radiation belts at Europa limits how long a spacecraft's electronics are able to operate in orbit around Europa, even with advances in radiation-resistant electronics that would be used on this mission. The instruments on board the spacecraft will take uniform measurements, using the same complement of instruments, of all three moons. Special attention will be paid to identifying high-potential future landing sites.

Several NASA centers around the country, the Department of Energy, national laboratories, industrial contractors, and university researchers, could collaborate on this mission.

What is NASA's Project Prometheus?

NASA's Project Prometheus was established this year to develop technology and conduct advanced studies in the areas of radioisotope power system and nuclear power and propulsion for the peaceful exploration of the Solar System. Project Prometheus, organized within the NASA Office of Space Science, has the goal of developing the first reactor powered spacecraft capability and demonstrating that it can be operated safely and reliably in deep space on long duration missions. The proposed Jupiter Icy Moons Orbiter has been identified as the first space science mission to potentially incorporate this new revolutionary capability.

#

DVD-Video / DVD-ROM

Side One of the attached disc will play on a home DVD-Video player*
To view Side Two requires a machine with DVD-ROM capability.

DVD video includes:

      And Then There Was Voyager
      STS-34 / Galileo Mission Launch Highlights
      Cassini-Titan Mission Launch & Preflight
      Deep Space 1 Mission Launch & Preflight
      Stardust Mission Launch & Preflight

DVD-ROM includes complete NASA publications about deep space exploration missions:

      Pioneer to Jupiter
      Pioneer - First To Jupiter, Saturn and Beyond (SP-446)
      Voyager to Jupiter (SP-439)
      Voyages to Saturn (SP-451)
      Atlas of Six Saturnian Satellites (SP-474)
      Galileo - Arrival at Jupiter
      Cassini-Huygens Mission to Saturn and Titan Reference
      Cassini - Outward to the Beginning
      And much more . . .

PLUS!

      Hundreds of color pictures of the outer solar system
      MPEG** video of the Pioneer 10 5th Anniversary press conference
      MPEG** video of Voyager Encounter Highlights
      MPEG** video Stardust Mission animations
      Complete Cassini-Huygens mission press release archive, and program updates

* NTSC Region 0, ** MPEG-1 standard video format